Harald Stumpf
Thomas Borne

# Composite Particle Dynamics in Quantum Field Theory

Harald Stumpf
Thomas Borne

# Composite Particle Dynamics in Quantum Field Theory

vieweg

Softcover reprint of the hardcover 1st edition 1994

Vieweg is a subsidiary company of Bertelsmann Professional Information.



Printed on acid-free paper

ISBN-13: 978-3-528-06498-3 e-ISBN-13: 978-3-322-83901-5
DOI: 10.1007/ 978-3-322-83901-5

# Preface

Each atomistic theory of matter is based on the idea that agglomerations of their constituents are to be identified with observable objects of physical reality and that in this way the diversity of physical phenomena and reactions can be reduced to the interplay of a few elementary entities. This means theoretically that the formation of observable objects and their reactions have to be derived from the dynamics of their atomistic, i.e. elementary constituents.

At present such atomistic theories of matter are formulated by quantum field theories. In view of the above described aim it is thus one of the most important tasks in quantum field theory to derive composite particles as bound states and to explain their dynamics as an effective dynamics induced by the elementary fields. In the development of quantum field theory many attempts have been made to solve this problem. So far, however, these attempts have been unsatisfactory and in this book some comments will be made on what the reasons are and where the difficulties arise. Roughly speaking the latter are closely connected with the idea to describe the effective dynamics of composite particles by means of the dynamics of field operator products. To avoid these difficulties we alternatively developed the method of weak mapping of quantum fields. The presentation of this method and of some applications to current problems is the object of this monography.

One of the main ingredients of this alternative method is the use of the Hamilton formalism, i.e. a one–time formalism, in combination with general features of algebraic representation theory of quantum fields and Heisenberg's equation of motion. In this way one can nonperturbatively treat the field dynamics and avoid the free–field representations of the interaction picture which is inconsistent for a proper description of systems with an infinite number of degrees of freedom. In conventional gauge theories the use of the Hamilton formalism is connected with noncovariant gauges and is thus not in accordance with the commonly accepted covariant formalism of quantum field theories. But it will be demonstrated in this book that for a successful composite particle theory the use of this Hamilton formalism is imperative. The reason for this is due to the fact that a correct construction of the various representation spaces of quantum fields is only possible in this formalism.

It was Dirac who for a long time advocated to proceed along this line in quantum field theory. In his lectures on quantum field theory, see Dirac [Dir 66], he wrote: "The treatment of quantum field theory given here is based on the Hamilton formalism and the theory is developed as far as possible in a logical manner with results following from equations of motion. This involves a considerable departure from the usual treatment of quantum field theory in which one rather abandons the hope of logical deduction and is content with setting up working rules and showing their consistency, the equation of motion getting lost in the process."

Apart from the latter judgement, in the modern development of gauge theories noncovariant gauges, and in connection with them the Hamilton formalism are of increasing interest, and in attempts to quantize classical gravity the Hamilton formalism plays a decisive role. So Dirac's opinion becomes more and more justified, and the method presented in this book is a further step in this direction.

For simplicity we first examplify our method for nonlinear spinorfields. Such fields have a wide range of applications: Nonrelativistic nonlocal (or even local) spinorfields are fundamental for the electronic motion in atoms, molecules and solids, i.e. for quantum chemistry, solid state physics and atomic physics. Relativistic local or nonlocal nonlinear spinorfields are used in nuclear physics to explain nuclear properties in the quark model, and in high energy physics these fields are used for attempts to explain the structure of matter. Although these fields are nonrenormalizable in the usual form, they are frequently used with cut–offs and other regularization prescriptions. In this book we apply a new nonperturbative Pauli–Villars regularization which induces an indefinite metric state space; we will, however, demonstrate that a probabilistic interpretation is possible.

Having discussed the general framework for the case of spinorfields we proceed to applications and in a first step demonstrate the power of our method by deriving Yang–Mills theories as effective theories for vector bosons being bound states of two subfermions and by putting forth de Broglie's proposals of gravity as an outgrow of subfermion interactions with four subfermions in spin–2 bound states. Later on we treat the bound state problem in quantum electrodynamics and by means of our formalism develop a suitable algebraic framework for quantum chromodynamics. Then a discussion follows of the BCS superconductivity model in a new version, which is suggested by our method, and finally we devote two chapters to critical comments on the ability of path integration to grasp composite particle dynamics and on bosonization in Fock space. In this way we demonstrate the rather universal use which can be made of the weak mapping method and satisfy a further demand on the theory imposed by Dirac, [Dir 66]: "It is necessary that quantum field theory be based on concepts and methods that can be unified with those used in the rest of physics. This necessity forces one to think of quantum field theory in terms of equations of motion."

We want to add that an appropriate modern treatment of the equations of motion which is in accordance with algebraic representation theory consists in a map of the corresponding state space in a functional state space, whereby the Heisenberg equations of motion are mapped into functional equations. From the very beginning we will use this formulation which extends the usual functional calculus of statistical mechanics and quantum field theory by the explicit construction of linear functional state spaces isomorphic to the underlying state space of the fields and which may be called Functional Quantum Theory.

Furthermore we would like to mention that to a certain degree the chapters containing applications are written with redundance so that a reader who is mainly interested in applications should be able to select the corresponding chapter and to understand its content.

Finally we emphasize that we have based the first part of the book on propositions and proofs, as far as this is possible for the very difficult mathematical structure of quantum fields. Nevertheless, by such propositions we want to express our intention to provide a rigorous basis for our composite particle theory, which can claim for greater generality as for instance approximation methods. In particular the central idea of weak mapping is based on a sound mathematical basis by means of weak mapping theorems.

As already mentioned, a central point in our formalism is the use of Heisenberg equations of motion. Because of the fundamental role of these operator equations one could ask about a mathematical proof of their existence; with respect to this problem in this book we take the existence of the Heisenberg equations for granted and refer to Baez, Segal and Zhou [Baez 92] for a discussion in the framework of constructive quantum field theory. We want to add that the use of the Hamilton formalism does not mean loss of covariance, because there exists a biunique map between Hamilton perturbation theory and covariant Feynman rules for the calculation of the $S$–matrix, cf. Halpern [Halp 67].

While preparing this book, further work was finished; in particular we refer to the theses of Grimm [Grim 94] and Kerschner [Ker 94] which could not be incorporated here, but which contain interesting contributions to this topic.

## Acknowledgement:

The original papers on the weak mapping method were mainly published in Journal of Physical Sciences (Z. f. Naturforsch.); we want to thank its chief editor Prof. A. Klemm for the promotion of our research project. Furthermore the president of the University of Tübingen, Prof. A. Theis, supported the preparation of this book by administrative arrangement, for which we want to thank him.

For encouragement and (or) various support and (or) valuable discussions one of the authors (H. St.) is indepted to Profs. T. Ahrens, J. Audretsch, M. Borneas, H. Dehnen, D. Grosser, F.W. Hehl, P. Kramer, G. Möllenstädt, H. Müther, A. Rieckers, H. Saller, E. Schmid, P. Urban, W. Walcher, K. Wildermuth and M. Wolff.

For partial reading of the manuscript and corrections as well as for improvements and numerous discussions both authors want to thank Dipl. Phys. B. Fauser, R. Kerschner, K. Weinrich, Dr. W. Pfister and the other members of our group.

Furthermore the preparation of a book is not only scientific work but also a matter of organization. For this book this organization was almost completely done by Mrs. Regine Adler. She ordered, copied, collected and arranged the great number of original papers and typed several versions of the manuscript with LaTeX in an excellent way. So we want to express our gratitude for this effectful cooperation.

Finally we want to express our thanks to all the members of the Vieweg Verlag who made the publication of this book possible.

# Contents

# Notation

## General conventions:

In the first four chapters, which provide the framework of our composite particle theory, we use *cgs*–units, while the applications discussed in Chapters 5–7 are formulated in natural units, i.e. with $\hbar = c = 1$.

As a rule we use the summation convention and sum over double discrete indices and integrate over two continous variables. Exceptions from this rule will be explicitly quoted.

A contravariant Minkowski space–time vector is given by its coordinates

$$x = (x^\mu) = (x_0, x_1, x_2, x_3) = (ct, \mathbf{r}) .$$

The Minkowski metric is described by the diagonal tensor

$$\eta_{\mu\nu} = \mathrm{diag}(1, -1, -1, -1) ,$$

while the metrical tensor of a Riemann–Cartan space is denoted by $g_{\mu\nu}(x)$.

Partial derivatives with respect to space–time coordinates are written as

$$\partial_\mu \equiv \partial_\mu(x) = \frac{\partial}{\partial x^\mu} .$$

In general small greek indices $\alpha, \beta \ldots$ take the values $0, 1, 2, 3$ and small latin indices $i, j, k \ldots$ the values $1, 2, 3$; exceptions from this usage will be explicitly announced.

For Fourier transformation we use the convention

$$\tilde{f}(p) = \frac{1}{(2\pi)^4} \int d^4x\, f(x) e^{ipx}$$

with

$$f(x) = \int d^4p\, \tilde{f}(p) e^{-ipx} .$$

The complex conjugated of a complex number $z$ is given by $z^*$; for an operator $\mathbf{A}$ the hermitian conjugation is denoted by $\mathbf{A}^+$ and the transponation of a matrix $A$ by ${}^TA$.

Furthermore, for spinors $\psi \equiv (\psi_\alpha)$ we introduce $\bar{\psi} := \psi^+ \gamma^0$ and the charge conjugated spinor $\psi^C := C\bar{\psi}$ with the charge conjugation matrix $C$, see below.

Commutators and anticommutators are given by

$$[A, B]_- = AB - BA, \qquad [A, B]_+ = AB + BA.$$

Finally symmetrization and antisymmetrization of a quantity $M_{kl}$ in its indices is denoted by $M_{[kl]}$ or $M_{\{kl\}}$, respectively. Alternatively we use the symmetrization operator $S$ or the antisymmetrization operator $A$.

## Pauli– and Dirac matrices:

We use the representation

$$\sigma_1 = \begin{pmatrix} 0 & 1 \\ 1 & 0 \end{pmatrix}, \quad \sigma_2 = \begin{pmatrix} 0 & -i \\ i & 0 \end{pmatrix}, \quad \sigma_3 = \begin{pmatrix} 1 & 0 \\ 0 & -1 \end{pmatrix}$$

of the Pauli matrices. The Dirac matrices fulfill the relations

$$[\gamma^\mu, \gamma^\nu]_+ = 2\eta_{\mu\nu} ,$$

and we choose the Dirac representation

$$\gamma^0 = \begin{pmatrix} 1 & 0 \\ 0 & -1 \end{pmatrix}, \quad \gamma^k = \begin{pmatrix} 0 & \sigma_k \\ -\sigma_k & 0 \end{pmatrix} .$$

Furthermore we use the matrices $\gamma^5 = i\gamma^0\gamma^1\gamma^2\gamma^3$ and the charge conjugation matrix $C = i\gamma^2\gamma^0$. In the Dirac representation they read

$$\gamma^5 = \begin{pmatrix} 0 & 1 \\ 1 & 0 \end{pmatrix}, \quad C = -i \begin{pmatrix} 0 & \sigma_2 \\ \sigma_2 & 0 \end{pmatrix} .$$

In the following we want to list further important symbols used in the text. This list is not complete; for the remaining symbols, indices and abbreviations we refer to the text.

## General symbols:

| | |
|---|---|
| N | set of natural numbers |
| $\mathbb{R}$ | set of real numbers |
| $\mathbb{C}$ | set of complex numbers |
| $\otimes$ | direct product |
| $\oplus$ | direct sum |
| $S_n$ | n–dimensional permutation group |
| $p_i$ | permutation of $i$ |
| $T$ | time–ordering operator |

## State space symbols:

| | |
|---|---|
| $\mathbb{H}$ | (indefinite) inner product space |
| $\|a\rangle, \dots$ | vectors from $\mathbb{H}$ |
| $\|0\rangle$ | vacuum or groundstate of $\mathbb{H}$ |
| $\mathbf{S}, \dots$ | linear operators on $\mathbb{H}$ |
| $\mathbb{L}$ | Liouville space |
| $\|a), \dots$ | vectors from $\mathbb{L}$ |
| $\|\Omega_T)$ | thermal groundstate of $\mathbb{L}$ |

## Function symbols:

| | |
|---|---|
| $\tau^{(n)}(I_1 \dots I_n \vert a)$ | $\tau$–functions, time–ordered transition matrix elements |
| $\varphi^{(n)}(I_1 \dots I_n \vert a)$ | $\varphi$–functions, normal–ordered transition matrix elements |
| $\tau_t^{(n)}, \dots$ | corresponding one–time functions |
| $\rho^{(n)}, \dots$ | correlation functions of effective theories |
| $\Delta(x, m), F_{I_1 I_2}$ | propagators |
| $K_i(r)$ | modified Bessel functions |
| $G_{I_1 I_2}$ | Green's functions |
| $C_J^I, C_k^{I_1 I_2}, \dots$ | Cluster functions |
| $R_I^J, R_{I_1 I_2}^k, \dots$ | corresponding duals |

## Functional symbols:

| | |
|---|---|
| $\mathbb{K}_F$ | fermion functional space |
| $\mathbb{K}_B$ | boson functional space |
| $\vert 0\rangle_F, \dots$ | functional groundstates |
| $\vert \mathcal{T}(j; a)\rangle$ | functional state for $\tau^{(n)}$–functions |
| $\vert \mathcal{F}(j; a)\rangle$ | functional state for $\varphi^{(n)}$–functions |
| $\vert \mathcal{A}(j; a)\rangle$ | functional state for one–time $\tau^{(n)}$–functions |
| $\vert \mathcal{B}(b; a)\rangle$ | effective boson functional state |
| $\vert \mathcal{G}(b, f; a)\rangle$ | effective boson–fermion functional state |
| $j_I, \partial_I \equiv \delta/\delta j_I$ | generating functional fermion operators |
| $b_k, \partial_k^b \equiv \delta/\delta b_k$ | generating functional boson operators |
| $f_l, \partial_l^f \equiv \delta/\delta f_l$ | functional dressed fermion operators |
| $\mathcal{H}, \mathcal{P}, \dots$ | linear operators on functional space |

# 1 The Spinorfield Model

## 1.1 Introduction

The essential symmetry group of relativistic quantum fields is the Poincaré group. All representations of this group can be generated by direct products of elementary spin–1/2 representations, cf. Wigner [Wign 39], Fonda and Ghirardi [Fon 70], Bogoliubov, Logunov and Todorov [Bogo 75], Haag [Haag 92]. Therefore, as this group is fundamental one expects that nature reflects this composition principle, i.e. that spin–1/2 fermions are the basic quantities for any attempt to generate the hierarchy of observable and nonobservable physical particles. This fusion principle originates from de Broglie [Brog 34], [Brog 54]. The mathematical description of spin–1/2 fermions is performed by spinorfields, and as nontrivial interactions are required, the corresponding equations have to be nonlinear spinorfield equations. Thus, at first we will extensively discuss nonlinear spinorfield theory and resulting effective composite particle theories. Later on we will pass to gauge theories, i.e. theories which couple elementary fermions with gauge bosons and which at present are one of the main topics of theoretical research.

In nonrelativistic physics nonlinear spinorfield models are widely used in solid state theory, for instance in formulating superconductivity by BCS– or Hubbard models, see, e.g., Fetter and Walecka [Fett 71], Balachandrian, Ercolesi, Morandi and Srivastava [Bal 90], Micmas, Ranninger and Robaszkiewicz [Mic 90]. In most cases such models are treated as thermofield systems and a lot of more or less intuitive approximations are applied. We will also investigate the effective composite particle dynamics for the BCS model in Chapter 9, but first start with the relativistic theory.

Before going into details some general remarks are necessary. In contrast to nonrelativistic models the relativistic theories are generally formulated by means of Dirac operators with first order derivatives, and in addition they have to contain local interaction terms in order to guarantee relativistic invariance and causality. However, first order spinorfield equations with nonlinear local selfinteraction terms are nonrenormalizable in (perturbative) quantum field theory. In spite of this drawback such models are frequently used in high and middle energy physics, see for instance Heisenberg [Heis 66], [Heis 84], Dürr and Saller [Dürr 77a,b,], [Dürr 78], Saller [Sall 75], [Sall 76], Vogl and Weise [Vogl 91], Klevanski [Klev 92], Dickhoff and Müther [Dick 92], Hatsuda and Kunihiro [Hat 94], Alkofer, Reinhardt and Weigel [Alk 94].

To circumvent nonrenormalizability, in the great majority of applications the spinorfields are regularized by various cut–off prescriptions and (or) nonlocal interactions. An exception is the regularization prescription used by Heisenberg and coworkers who suggested a vanishing anticommutator of canonical conjugated spinorfields for equal times, cf. [Heis 66]. While the first kind of regularizations leads to the breakdown either of relativistic invariance, or relativistic causality, or both of them, Heisenberg's prescription destroys the canonical structure of the theory which produces considerable difficulties. In order to avoid these difficulties we apply a new nonperturbative Pauli–Villars regularization which respects relativistic invariance and does not lead to nonlocal interactions. It can be introduced in a systematic way by a decomposition of a higher order nonlinear spinorfield equation which preserves the canonical structure of the theory.

The prize of this kind of regularization which is independent of any calculation scheme, is the introduction of indefinite metric into the theory. Indefinite metric also appears in the relativistic invariant formulation of gauge theories where it is removed by constraints. Constraints may be interpreted as measuring limitations formulated theoretically and imposed on the state space. In the spinorfield model such constraints cannot be analogously derived at present. Nevertheless, the interpretation that indefinite metric imposes limitations of measurement remains possible, cf. also Feynman [Feyn 87]. Assuming the indefinite metric state space to be a Krein space, in Chapter 3 we will show the way for a statistical interpretation of spinorfield models regularized in this manner.

However, in the following we shall make use of this exact method only for the normalization of dressed fermion states in Chapter 5. For the rest we take a pragmatic point of view and investigate effective composite particle theories resulting from the spinorfield model without further discussion of the indefinite metric. If these effective theories are identical with phenomenological quantum field theories which allow a statistical interpretation, then in an indirect way in this sector of the spinorfield state space the compatibility of the spinorfield metric with a statistical interpretation is secured.

Finally it should be mentioned that we equip our spinorfield model with a somewhat richer algebraic structure by adding isospin degrees of freedom. Figuratively speaking, this corresponds to the introduction of subfermion "protons" and "neutrons", or equivalently "preon protons" and "preon neutrons". These isospin properties are needed for a successful derivation of phenomenological effective theories.

## 1.2 Spinorfield Regularization

The basic quantities of the model are two four–component Dirac spi"-nor–iso"-spi"-nor fields $\Psi_{A\alpha}(x), \alpha = 1, \ldots, 4, A = 1, 2$ where $\alpha$ is the spinor index and $A$ the isospinor index. For these fields we postulate the following nonlinear field equation:

$$(i\hbar\gamma^\mu\partial_\mu - cm)^{\text{reg}}_{\alpha\beta}\delta_{AB}\Psi_{B\beta}(x) = gV_{\substack{ABCD\\ \alpha\beta\gamma\delta}}\Psi_{B\beta}(x)\bar{\Psi}_{C\gamma}(x)\Psi_{D\delta}(x) \qquad (1.1)$$

with the adjoint spinor–isospinor field

$$\bar{\Psi}_{A\alpha}(x) := \Psi^+_{B\beta}(x)\gamma^0_{\beta\alpha}\delta_{AB} \quad , \qquad (1.2)$$

which satisfies the adjoint equation

$$(i\hbar\, {}^T\gamma^\mu\partial_\mu - cm)^{\text{reg}}_{\alpha\beta}\delta_{AB}\bar{\Psi}_{B\beta}(x) = gV_{\substack{ABCD\\ \alpha\beta\gamma\delta}}\bar{\Psi}_{D\delta}(x)\Psi_{C\gamma}(x)\bar{\Psi}_{B\beta}(x) \quad . \qquad (1.3)$$

The vertex is defined by

$$\begin{aligned} V_{\substack{ABCD\\ \alpha\beta\gamma\delta}} &:= \frac{1}{2}\sum_{h=1}^{2}\left(\delta_{AB}\delta_{CD}v^h_{\alpha\beta}v^h_{\gamma\delta} - \delta_{AD}\delta_{CB}v^h_{\alpha\delta}v^h_{\gamma\beta}\right) \quad , \qquad (1.4)\\ v^1_{\alpha\beta} &:= \delta_{\alpha\beta} \quad , \\ v^2_{\alpha\beta} &:= i\gamma^5_{\alpha\beta} \quad . \end{aligned}$$

The coupling constant $g$ is real and will be fixed later. With respect to the choice of the vertex we refer to Grimm [Gri 94].

The superindex "reg" of the Dirac operator in (1.1) and (1.3) indicates that the two equations have to be regularized if applied in their quantized version. We define the regularization by

$$(i\hbar\gamma^\mu\partial_\mu - cm)^{\text{reg}}_{\alpha\beta} := \Big[(i\hbar\gamma^\mu\partial_\mu - cm_0)(i\hbar\gamma^\nu\partial_\nu - cm_1)(i\hbar\gamma^\rho\partial_\rho - cm_2)\Big]_{\alpha\beta} \tag{1.5}$$

with

$$(m_i)_{\alpha\beta} := m_i\delta_{\alpha\beta}\,, \quad i = 0,1,2, \tag{1.6}$$

where $m_i$, $i = 0,1,2$, are real and $m_0 \neq m_1 \neq m_2$, i.e. we regularize (1.1) by a third order differential operator. For later applications we set $m_0 > m_1 > m_2$. The units of the regularized fields will be discussed in Chapter 6.

The field equations (1.1) and (1.3) are forminvariant under Poincaré–, $C$–, $P$–, and global $SU(2)$ isospin–transformations, but the mass terms break chiral symmetry.

Higher order differential equations were first introduced in quantum electrodynamics by Bopp [Bopp 40] and Podolski [Pod 40]. The canonical formalism for such higher order theories was developed by various authors, e.g. de Wet [Wet 48], Chang [Cha 48], Borneas [Bor 69], Riewe and Green [Riew 72], Bigi, Dürr and Winter [Bigi 74], Rodrigues [Rod 77], Aldaya and Azcarraga [Ald 80]. The conventional way to embed higher order differential equations into the canonical formalism is based on the introduction of derivatives of the original fields as independent fields which are "canonically" quantized. This seems to be a somewhat arbitrary procedure which leads to commutation relations between the original fields and their derivatives and to rather complicated expressions for the conserved quantities like energy–momentum tensor etc. Thus we have developed a new method for canonical quantization which is based on a decomposition theorem.

A theorem which decomposes a linear Dirac equation with higher order derivatives, i.e. (1.1) with $V \equiv 0$, into a set of linear Dirac equations with first order derivatives was first proved by Wildermuth [Wild 50]. The extension to the decomposition of nonlinear field equations is nontrivial. As a preliminary step we prove the following proposition:

- **Proposition 1.1:** Let $\{m_i, i = 1,\dots,N,\ N \geq 2\}$ be a set of real parameters with $m_i \neq m_k$, $\forall\, i \neq k$ and $D$ a complex variable. For the decomposition of the following fraction into partial fractions

$$\prod_{i=1}^{N}(D - m_i)^{-1} = \sum_{i=1}^{N}\lambda_i(D - m_i)^{-1} \tag{1.7}$$

the relation

$$\sum_{i=1}^{N}\lambda_i\prod_{\substack{k=1\\k\neq i}}^{N}(D - m_k) = 1 \tag{1.8}$$

holds with

$$\lambda_i = \prod_{\substack{k=1\\k\neq i}}^{N}(m_i - m_k)^{-1} \quad . \tag{1.9}$$

**Proof:** Let $F(D)$ be defined by $F(D) := \prod_{i=1}^{N}(D - m_i)$. If (1.7) is multiplied with $F(D)(D - m_k)$, one obtains

$$D - m_k = \lambda_k F(D) + \sum_{\substack{i=1\\ i\neq k}}^{N} \frac{\lambda_i}{D - m_i}(D - m_k)F(D) \quad . \tag{1.10}$$

Differentiation with respect to $D$ yields for $D = m_k$

$$1 = \lambda_k F'(m_k) \tag{1.11}$$

with

$$F'(m_k) = \prod_{\substack{i=1\\ i\neq k}}^{N}(m_k - m_i) \quad . \tag{1.12}$$

Thus $\lambda_i$ is given by (1.9). Multiplication of (1.7) with $F(D)$ gives the relation (1.8). ◇

We mention that Proposition 1.1 holds for operator–valued $D$, too.

The relation between a higher order differential (field) equation and first order equations can now be established by the following "decomposition theorem" proven for $N = 3$ by Stumpf [Stu 82] and for general $N$ by Grosser [Gro 83].

- **Proposition 1.2:** Under the conditions of Proposition 1.1 the following statements hold:
  i) Let $\Psi \equiv \Psi(x)$ be a solution of an $N$th order equation

  $$\prod_{i=1}^{N}(D - m_i)\Psi = V[\Psi] \quad , \tag{1.13}$$

  where $V[\Psi]$ is a (nonlinear) interaction term, and let $\psi_i \equiv \psi_i(x)$, $i = 1, \ldots N$, be auxiliary fields which are defined by

  $$\psi_i := \lambda_i \prod_{\substack{k=1\\ k\neq i}}^{N}(D - m_k)\Psi \ , \quad i = 1, \ldots N, \tag{1.14}$$

  with

  $$\lambda_i := \prod_{\substack{k=1\\ k\neq i}}^{N}(m_i - m_k)^{-1} \ , \quad i = 1, \ldots N. \tag{1.15}$$

  Then it holds

  $$\Psi = \sum_{i=1}^{N} \psi_i \tag{1.16}$$

  and the auxiliary fields are solutions of the first order equations

$$(D - m_i)\psi_i = \lambda_i V\left[\sum_{j=1}^{N}\psi_j\right] \quad , \quad i = 1, \dots N, \tag{1.17}$$

with no summation over $i$ on the left–hand side.

ii) If the fields $\psi_i \equiv \psi_i(x)$, $i = 1, \dots N$, are solutions of equations (1.17) with $\lambda_i$ of (1.15), and if $\Psi \equiv \Psi(x)$ is defined by (1.16), then the relations (1.14) hold and $\Psi$ is a solution of equation (1.13).

**Proof:** i) With (1.8) and (1.14) we have

$$\Psi = \sum_{i=1}^{N}\lambda_i \prod_{\substack{k=1\\k\neq i}}^{N}(D - m_k)\Psi = \sum_{i=1}^{N}\psi_i \quad , \tag{1.18}$$

and due to (1.13) and (1.14) it is

$$(D - m_i)\psi_i = \lambda_i \prod_{k=1}^{N}(D - m_k)\Psi = \lambda_i V\left[\sum_{j=1}^{N}\psi_j\right] \tag{1.19}$$

with no summation over $i$ on the left–hand side. Thus (1.16) and (1.17) hold.

ii) With (1.8), (1.16) and (1.17) we have

$$\begin{aligned}\prod_{i=1}^{N}(D - m_i)\Psi &= \prod_{i=1}^{N}(D - m_i)\sum_{j=1}^{N}\psi_j \\ &= \sum_{j=1}^{N}\prod_{\substack{i=1\\i\neq j}}^{N}(D - m_i)\lambda_j V\left[\sum_{k=1}^{N}\psi_k\right] \\ &= V\left[\sum_{k=1}^{N}\psi_k\right] \\ &= V\left[\Psi\right] ,\end{aligned} \tag{1.20}$$

i.e. equation (1.13) holds.

For $i = 1$ it follows from (1.17)

$$\lambda_1^{-1}(D - m_1)\psi_1 = \lambda_j^{-1}(D - m_j)\psi_j \; , \; j = 2, \dots N, \tag{1.21}$$

with no summation over $j$ on the right–hand side. With (1.21) and (1.8) it is

$$\begin{aligned}\lambda_1 \prod_{k=2}^{N}(D - m_k)\Psi &= \lambda_1 \prod_{k=2}^{N}(D - m_k)\sum_{j=1}^{N}\psi_j \\ &= \lambda_1 \prod_{k=2}^{N}(D - m_k)\psi_1 + \sum_{j=2}^{N}\lambda_1 \prod_{\substack{k=2\\k\neq j}}^{N}(D - m_k)\frac{\lambda_j}{\lambda_1}(D - m_1)\psi_1 \\ &= \sum_{j=1}^{N}\lambda_j \prod_{\substack{k=1\\k\neq j}}^{N}(D - m_k)\psi_1 \\ &= \psi_1 \quad .\end{aligned} \tag{1.22}$$

The same conclusions can be drawn for $i = 2, \dots N$. Thus (1.14) holds. $\diamond$

Proposition 1.2 can be reformulated as follows: Proposition 1.2 states exactly that equation (1.14) defines a bijective map between the sets of solutions of equations (1.13) and (1.17). The inverse map of (1.14) is given by (1.16).

For the proof of this statement let $M_1$ be the set of solutions of (1.13), $M_2^i$ be the set of solutions of (1.17) for $i = 1, \dots N$, and $f_1^i$ and $f_2$ the maps which are defined by (1.14) and (1.16). Then according to Proposition 1.2 we have for $M_2 := \oplus_{i=1}^N M_2^i$ and $f_1 = (f_1^1, \dots f_1^N)$

$$f_1 : M_1 \to M_2 \,, \quad f_2 : M_2 \to M_1 \quad , \tag{1.23}$$

and due to (1.14) and (1.16) we have

$$f_2 \circ f_1 = \mathbf{1}_{M_1} \,, \quad f_1 \circ f_2 = \mathbf{1}_{M_2} \quad . \tag{1.24}$$

Thus $f_1$ defines a bijective map from $M_1$ onto $M_2$ with $f_1^{-1} = f_2$.

Conversely we can regain the statements of Proposition 1.2 if the relations (1.23) and (1.24) are valid.

The decomposition theorem is also valid for operator–valued $D$ and can be easily extended to multi–component equations. Thus we can apply it to the field equations (1.1) and (1.3) by defining the auxiliary fields

$$\psi_{A\alpha i}(x) \;:=\; \lambda_i \prod_{\substack{k=0\\k\neq i}}^{2} (i\hbar\gamma^\mu\partial_\mu - cm_k)_{\alpha\beta}\delta_{AB}\Psi_{B\beta}(x) \quad , \tag{1.25}$$

$$\bar\psi_{A\alpha i}(x) \;:=\; \lambda_i \prod_{\substack{k=0\\k\neq i}}^{2} (-i\hbar\,{}^T\gamma^\mu\partial_\mu - cm_k)_{\alpha\beta}\delta_{AB}\bar\Psi_{B\beta}(x) \tag{1.26}$$

with

$$\lambda_i := \prod_{\substack{k=0\\k\neq i}}^{2} (cm_i - cm_k)^{-1} \quad . \tag{1.27}$$

Then according to Proposition 1.1 we have the inverse relations

$$\Psi_{A\alpha}(x) = \sum_{i=0}^{2} \psi_{A\alpha i}(x) \,, \quad \bar\Psi_{A\alpha}(x) = \sum_{i=0}^{2} \bar\psi_{A\alpha i}(x) \tag{1.28}$$

and the auxiliary fields satisfy the first order differential equations

$$\begin{aligned}(i\hbar\gamma^\mu\partial_\mu - cm_i)_{\alpha\beta}\delta_{AB}\psi_{B\beta i}(x) \\ = \; g\lambda_i V_{\substack{ABCD\\ \alpha\beta\gamma\delta}} \sum_{j,k,l=0}^{2} \psi_{B\beta j}(x)\bar\psi_{C\gamma k}(x)\psi_{D\delta l}(x)\end{aligned} \tag{1.29}$$

and

$$
\begin{aligned}
&(-i\hbar\, {}^T\gamma^\mu \partial_\mu - cm_i)_{\alpha\beta}\delta_{AB}\bar{\psi}_{B\beta i}(x) \\
&= g\lambda_i V_{\substack{ABCD\\ \alpha\beta\gamma\delta}} \sum_{j,k,l=0}^{2} \bar{\psi}_{D\delta l}(x)\psi_{C\gamma k}(x)\bar{\psi}_{B\beta j}(x)
\end{aligned} \tag{1.30}
$$

with no summation over $i$ on the left–hand sides. Due to (1.2) from (1.28) the relation

$$
\bar{\psi}_{A\alpha i}(x) = \psi^+_{A\beta i}(x)\gamma^0_{\beta\alpha} \quad , \tag{1.31}
$$

follows, i.e. $\bar{\psi}$ has to be the adjoint auxiliary field spinor.

Finally we mention that the $\lambda_i$ of (1.27) satisfy the Pauli–Villars regularization conditions, see Pauli and Villars [Pau 49], Rayski [Ray 48], Bogoliubov and Shirkov [Bogo 59]:

$$
\sum_{i=0}^{2} \lambda_i = 0\,, \quad \sum_{i=0}^{2} \lambda_i m_i = 0 \quad . \tag{1.32}
$$

Thus in combination with the Lagrange formulation for the auxiliary fields our regularization (1.5) turns out to be a nonperturbative Pauli–Villars regularization.

## 1.3 Lagrange Formalism

In order to derive conservation laws and to carry out the canonical quantization procedure, we need the Lagrangian of the fields. This program can be successfully performed only if the field dynamics is based on first order differential equations. Thus the decomposition theorem serves as the means to simultaneously apply spinorfield regularization and the canonical Lagrangian formalism. So in the following all derivations and conclusions will be based on the auxiliary spinorfield dynamics given by equations (1.29) and (1.30). Performing the regularization then means to sum over the auxiliary fields (and to specify the masses $m_0$, $m_1$ and $m_2$).

One can easily verify that (1.29) and (1.30) follow by variation with respect to $\psi_{A\alpha i}(x)$ or $\bar{\psi}_{A\alpha i}(x)$ from an action with the Lagrangian density

$$
\begin{aligned}
\mathcal{L}[\psi,\bar{\psi}] &= \mathcal{L}_{\text{kin}}[\psi,\bar{\psi}] + \mathcal{L}_{\text{int}}[\psi,\bar{\psi}] \\
&= \sum_{i=0}^{2} \lambda_i^{-1}\bar{\psi}_{A\alpha i}(x)(i\hbar\gamma^\mu\partial_\mu - cm_i)_{\alpha\beta}\delta_{AB}\psi_{B\beta i}(x) \\
&\quad -\frac{g}{2} V_{\substack{ABCD\\ \alpha\beta\gamma\delta}} \sum_{i,j,k,l=0}^{2} \bar{\psi}_{A\alpha i}(x)\psi_{B\beta j}(x)\bar{\psi}_{C\gamma k}(x)\psi_{D\delta l}(x) \quad ,
\end{aligned} \tag{1.33}
$$

where $\mathcal{L}_{\text{kin}}$ has the structure of a Pauli–Villars regularized Lagrangian. But with respect to the regularization of the interaction term in conventional quantum field theory $\mathcal{L}_{\text{int}}$ is regularized by hand and not automatically by the formalism itself, see, e.g., Itzykson and Zuber [Itz 80]. In contrast to this common usage the Lagrangian (1.33) automatically provides also a regularization of the interaction term, and this regularization is independent of the kind of calculation, in particular it does not depend on the application of perturbation theory, as will be demonstrated in the following. Thus by the decomposition theorem we have achieved a nonperturbative Pauli–Villars regularization.

In order to examine the conserved quantities we define

$$\pi^{\mu}_{A\alpha i}(x) := \frac{\delta \mathcal{L}}{\delta \partial_\mu \psi_{A\alpha i}(x)} \quad . \tag{1.34}$$

Due to the translational invariance of the Lagrangian and its action, according to Noether's theorem it follows that the energy–momentum vector

$$P^{\mu} := \int \mathcal{T}^{0\mu}(\mathbf{r}, t) d^3 r \tag{1.35}$$

with the canonical energy–momentum tensor density

$$\mathcal{T}^{\mu\nu}(x) := \pi^{\mu}_{A\alpha i}(x)\partial^{\nu}\psi_{A\alpha i}(x) - \eta^{\mu\nu}\mathcal{L}(x) \tag{1.36}$$

is a conserved quantity.

The canonically conjugated momentum $\pi_{A\alpha i}(x) := \pi^{0}_{A\alpha i}(x)$ to $\psi_{A\alpha i}(x)$ can be calculated by means of (1.33) and (1.34) and reads

$$\pi_{A\alpha i}(x) = \frac{i\hbar}{\lambda_i}\bar{\psi}_{A\beta i}(x)\gamma^{0}_{\beta\alpha} = \frac{i\hbar}{\lambda_i}\psi^{+}_{A\alpha i}(x) \tag{1.37}$$

with no summation over $i$ on the right–hand side. The hamiltonian density follows from $\mathcal{H}(x) := c\,\mathcal{T}^{00}(x)$ and is given by

$$\begin{aligned}
\mathcal{H}(x) &= c\sum_{i=0}^{2}\pi_{A\alpha i}(x)\partial_0\psi_{A\alpha i}(x) - c\mathcal{L}(x) \qquad (1.38)\\
&= c\sum_{i=0}^{2}\lambda_i^{-1}\bar{\psi}_{A\alpha i}(x)(-i\hbar\gamma^k\partial_k + cm_i)_{\alpha\beta}\delta_{AB}\psi_{B\beta i}(x) - c\mathcal{L}_{\text{int}}(x)\\
&= \frac{c}{\hbar}\sum_{i=0}^{2}\pi_{C\gamma i}(x)\gamma^{0}_{\gamma\alpha}\delta_{CA}(-\hbar\gamma^k\partial_k - icm_i)_{\alpha\beta}\delta_{AB}\psi_{B\beta i}(x) - c\mathcal{L}_{\text{int}}(x) \,.
\end{aligned}$$

If by means of Noether's theorem one evaluates the consequence of the invariance of the action under homogeneous Lorentz transformations one obtains a conservation law for the angular momentum tensor

$$M_{\mu\nu} := \int \mathcal{M}^{0}_{\mu\nu}(\mathbf{r}, t) d^3 r \tag{1.39}$$

with the canonical angular momentum tensor density

$$\begin{aligned}
\mathcal{M}^{\rho}_{\mu\nu}(x) &:= \pi^{\rho}_{A\alpha i}(x)\Sigma^{\alpha\beta}_{\mu\nu}\psi_{A\beta i}(x) + \pi^{\rho}_{A\alpha i}(x)(x_\mu\partial_\nu - x_\nu\partial_\mu)\psi_{A\alpha i}(x) \qquad (1.40)\\
&\quad +(\delta^{\rho}_{\mu}x_\nu - \delta^{\rho}_{\nu}x_\mu)\mathcal{L}(x) \quad ,
\end{aligned}$$

where $\Sigma_{\mu\nu} := -\frac{1}{4}[\gamma_\mu, \gamma_\nu]_-$ is the representation of generators of the homogeneous Lorentz group in the space of Dirac spinors.

So far we have derived some important properties and consequences of the quasiclassical spinorfield theory given by equation (1.1).

## 1.4 Canonical Spinorfield Quantization

With the decomposition theorem we are prepared to canonically quantize the regularized spinorfield which is dynamically characterized by the Lagrangian density (1.33). Canonical quantization is performed by considering the spinorfields $\psi_{A\alpha i}(x)$ and their conjugate momenta $\pi_{A\alpha i}(x)$ as the constituting elements of a CAR–algebra with corresponding anticommutation relations.

Because according to (1.37) we have $\pi_{A\alpha i}(x) = i\hbar\lambda_i^{-1}\psi^+_{A\alpha i}(x)$ (with no summation over $i$), the adjoint spinorfields $\bar{\psi}_{A\alpha i}(x) = \psi^+_{A\beta i}(x)\gamma^0_{\beta\alpha}$ depend linearly on the conjugate momenta $\pi_{A\alpha i}(x)$ and thus need not be additionally included in the algebra of fields. From now on we assume the $\psi_{A\alpha i}$ etc. to be field operators on a state space and no longer to be classical quantities.

The canonical anticommutation relations read

$$\begin{aligned}[\pi_{A\alpha i}(\mathbf{r},t), \psi_{B\beta j}(\mathbf{r}',t)]_+ &= i\hbar\delta_{ij}\delta_{AB}\delta_{\alpha\beta}\delta(\mathbf{r}-\mathbf{r}')\,, \\ [\psi_{A\alpha i}(\mathbf{r},t), \psi_{B\beta j}(\mathbf{r}',t)]_+ &= 0\,, \\ [\pi_{A\alpha i}(\mathbf{r},t), \pi_{B\beta j}(\mathbf{r}',t)]_+ &= 0\end{aligned} \tag{1.41}$$

for equal times. These equal time relations indicate that the complete CAR–algebra is already defined at an arbitrary, but fixed moment of time. The dynamical evolution of a quantum system is then described by a continuous group of automorphism of this algebra. This property of canonical quantization, in connection with a corresponding time translation generator (Hamiltonian **H**), will turn out to be a central point of our treatment of composite particle dynamics.

If one substitutes (1.37) into (1.41) these relations go over into the spinorfield relations

$$\begin{aligned}\left[\psi^+_{A\alpha i}(\mathbf{r},t), \psi_{B\beta j}(\mathbf{r}',t)\right]_+ &= \lambda_i\delta_{ij}\delta_{AB}\delta_{\alpha\beta}\delta(\mathbf{r}-\mathbf{r}')\,, \\ [\psi_{A\alpha i}(\mathbf{r},t), \psi_{B\beta j}(\mathbf{r}',t)]_+ &= 0\,, \\ \left[\psi^+_{A\alpha i}(\mathbf{r},t), \psi^+_{B\beta j}(\mathbf{r}',t)\right]_+ &= 0\,.\end{aligned} \tag{1.42}$$

Due to (1.32) not all $\lambda_i$ can be positive quantities. Thus (1.42) shows that regularization enforces indefinite metric by producing wrong signs in the anticommutation relations of the fields. This feature of spinorfield theory will be discussed in detail in Chapter 3.

The regularization effect can be drastically seen if one derives the anticommutation relations for the original field $\Psi_{A\alpha}(x)$ from (1.42). With (1.16) we obtain from (1.42) by means of (1.32)

$$\begin{aligned}\left[\Psi^+_{A\alpha}(\mathbf{r},t), \Psi_{B\beta}(\mathbf{r}',t)\right]_+ &= \sum_{i,j=0}^{2} \left[\psi^+_{A\alpha i}(\mathbf{r},t), \psi_{B\beta j}(\mathbf{r}',t)\right]_+ \\ &= \sum_{i,j=0}^{2} \lambda_i\delta_{ij}\delta_{AB}\delta(\mathbf{r}-\mathbf{r}') \\ &= 0\quad,\end{aligned} \tag{1.43}$$

$$\left[\Psi^+_{A\alpha}(\mathbf{r},t), \Psi^+_{B\beta}(\mathbf{r}',t)\right]_+ = [\Psi_{A\alpha}(\mathbf{r},t), \Psi_{B\beta}(\mathbf{r}',t)]_+ = 0\quad. \tag{1.44}$$

Thus the canonical quantization of the auxiliary fields is equivalent to a noncanonical quantization of the original field. The latter relations (1.43) or equivalently (1.32) were already assumed by Heisenberg [Heis 66], but without the background of a decomposition theorem and thus without relation to a strictly canonically quantized system. It is just the combination of our regularization with CAR–algebra and Hamilton formalism which allows a successful treatment of spinorfield dynamics.

Finally we observe that the canonical quantization has an effect on the vertex of equations (1.1) and (1.3), or (1.29) and (1.30), respectively. From (1.2) and (1.43) it follows that

$$\left[\bar{\Psi}_{A\alpha}(\mathbf{r},t), \Psi_{B\beta}(\mathbf{r}',t)\right]_+ = 0 \tag{1.45}$$

holds, too. Together with (1.44) this means that the local operator product of the regularized spinorfields $\Psi_{B\beta}(x)\bar{\Psi}_{C\gamma}(x)\Psi_{D\delta}(x)$ in (1.1) and (1.3), or (1.29) and (1.30), respectively, is antisymmetric under commutation of the pairs $(B\beta)$ and $(D\delta)$. Therefore, if this product is contracted with the vertex $V_{\substack{ABCD\\ \alpha\beta\gamma\delta}}$ an antisymmetric part is projected out. In order to secure this antisymmetry in any step of calculation we assumed $V_{\substack{ABCD\\ \alpha\beta\gamma\delta}}$ to be antisymmetrized in $(B\beta)$ and $(D\delta)$ from the beginning.

## 1.5 Superindexing

Formally, superindexing is a very compact notation for the spinorfields and their equations. Later on, however, it will turn out that superindexing is indispensable for the derivation of composite particle dynamics with an appropriate phenomenological interpretation, i.e. for the physical understanding of effective theories.

We introduce the charge conjugated spinorfields

$$\psi^c_{A\alpha i}(x) := C_{\alpha\beta}\bar{\psi}_{A\beta i}(x) \tag{1.46}$$

with the charge conjugation matrix $C$, and rewrite equations (1.29) and (1.30) in terms of $\psi$ and $\psi^c$. This gives

$$\begin{aligned}
&(i\hbar\gamma^\mu\partial_\mu - cm_i)_{\alpha\beta}\delta_{AB}\psi_{B\beta i}(x) \\
&\quad = g\lambda_i V_{\substack{ABCD\\ \alpha\beta\gamma\delta}} C^{-1}_{\delta\epsilon} \sum_{j,k,l=0}^{2} \psi_{B\beta j}(x)\psi^c_{C\epsilon k}(x)\psi_{D\delta l}(x)
\end{aligned} \tag{1.47}$$

and

$$\begin{aligned}
&(i\hbar\gamma^\mu\partial_\mu - cm_i)_{\alpha\beta}\delta_{AB}\psi^c_{B\beta i}(x) \\
&\quad = -g\lambda_i V_{\substack{ABCD\\ \alpha'\beta'\gamma\delta'}} C_{\alpha\alpha'} C^{-1}_{\beta'\beta} C^{-1}_{\delta'\delta} \sum_{j,k,l=0}^{2} \psi^c_{B\beta j}(x)\psi_{C\gamma k}(x)\psi^c_{D\delta l}(x)
\end{aligned} \tag{1.48}$$

with no summation over $i$ on the left–hand sides. From (1.43) it follows that also $\Psi^c \equiv \sum_i \psi^c_i$ and $\Psi$ anticommute for equal times. Hence the vertices in (1.47) and (1.48) have the same antisymmetry properties as the old ones.

Equations (1.47) and (1.48) can now be combined into one superspinor equation if we introduce superspinors by the definition

$$\psi_{A\alpha i\Lambda}(x) := \begin{cases} \psi_{A\alpha i}(x)\,, & \Lambda = 1 \\ \psi^c_{A\alpha i}(x)\,, & \Lambda = 2 \end{cases} \quad . \tag{1.49}$$

Together with the superindex $Z := (A, \alpha, i, \Lambda)$, where

$$\begin{aligned} A &= 1,2 && \text{(isospinor index)} \\ \alpha &= 1,2,3,4 && \text{(spinor index)} \\ i &= 0,1,2 && \text{(auxiliary field index)} \\ \Lambda &= 1,2 && \text{(superspinor index)} \quad , \end{aligned}$$

equations (1.47) and (1.48) can be rewritten as

$$\left(D^{\mu}_{Z_1 Z_2}\partial_\mu - m_{Z_1 Z_2}\right)\psi_{Z_2}(x) = U_{Z_1 Z_2 Z_3 Z_4}\psi_{Z_2}(x)\psi_{Z_3}(x)\psi_{Z_4}(x) \tag{1.50}$$

with

$$\begin{aligned} D^{\mu}_{Z_1 Z_2} &:= i\hbar\gamma^{\mu}_{\alpha_1\alpha_2}\delta_{A_1A_2}\delta_{i_1i_2}\delta_{\Lambda_1\Lambda_2}\,, \\ m_{Z_1 Z_2} &:= cm_{i_1}\delta_{A_1A_2}\delta_{\alpha_1\alpha_2}\delta_{i_1i_2}\delta_{\Lambda_1\Lambda_2}\,, \\ U_{Z_1Z_2Z_3Z_4} &:= \sum_{h=1}^{2} g\lambda_{i_1}B_{i_2i_3i_4}v^h_{\alpha_1\alpha_2}\delta_{A_1A_2}\delta_{\Lambda_1\Lambda_2}(v^hC)_{\alpha_3\alpha_4}\delta_{A_3A_4}\delta_{\Lambda_31}\delta_{\Lambda_42} \\ B_{i_2i_3i_4} &= 1 \ \text{ for } i_2, i_3, i_4 = 0,1,2, \end{aligned} \tag{1.51}$$

where $B_{i_2i_3i_4}$ is a purely formal quantity expressing the auxiliary field summation in the vertex, and with

$$U_{Z_1Z_2Z_3Z_4} \equiv U_{Z_1\{Z_2Z_3Z_4\}} \tag{1.52}$$

due to the total anticommutativity of the sum of all auxiliary fields in the vertex.

In contrast to the sum of all auxiliary fields the single auxiliary fields fulfill nontrivial anticommutation relations (1.42) which read in superindex notation

$$[\psi_Z(\mathbf{r},t), \psi_{Z'}(\mathbf{r}',t)]_+ = A_{ZZ'}\delta(\mathbf{r}-\mathbf{r}') \tag{1.53}$$

with

$$A_{ZZ'} := \lambda_i\delta_{ii'}\delta_{AA'}\sigma^1_{\Lambda\Lambda'}(C\gamma^0)_{\alpha\alpha'} \quad , \tag{1.54}$$

no summation over $i$ on the left–hand side and $\sigma^1 = \begin{pmatrix} 0 & 1 \\ 1 & 0 \end{pmatrix}$.

Moreover as often as possible we still use a more compact notation by combining $(Z, x)$ into only one general index $I := (Z, x)$ and denote the underlying index set with $\mathcal{I}$. In this case (1.53) can be rewritten as

$$[\psi_I, \psi_{I'}]_+^{t=t'} = A_{II'} \tag{1.55}$$

with

$$A_{II'} := A_{ZZ'}\delta(\mathbf{r}-\mathbf{r}') \quad . \tag{1.56}$$

In a similar way equation (1.50) can be compactified and reads

$$K_{I_1 I_2}\psi_{I_2} = W_{I_1 I_2 I_3 I_4}\psi_{I_2}\psi_{I_3}\psi_{I_4} \tag{1.57}$$

where the following abbreviations are used:

$$K_{I_1 I_2} := (D^\mu \partial_\mu - m)_{I_1 I_2} \tag{1.58}$$

with

$$D^\mu_{I_1 I_2} \quad := \quad D^\mu_{Z_1 Z_2}\delta(x_1 - x_2) \tag{1.59}$$

$$m_{I_1 I_2} \quad := \quad m_{Z_1 Z_2}\delta(x_1 - x_2) \tag{1.60}$$

and

$$W_{I_1 I_2 I_3 I_4} := U_{Z_1 Z_2 Z_3 Z_4}\delta(x_1 - x_2)\delta(x_1 - x_3)\delta(x_1 - x_4) \tag{1.61}$$

and where the summation convention is assumed to range over the whole set of indices in $I = (Z, x)$.

In the course of our investigations we will also need a version of equation (1.57), where (1.57) is resolved with respect to $\dot{\psi}_I$, and $t$ is considered as an evolution parameter. In this case we transform (1.57) into

$$i\frac{\hbar^2}{c}\frac{\partial}{\partial t}\psi_{I_1} = \hat{K}_{I_1 I_2}\psi_{I_2} + \hat{W}_{I_1 I_2 I_3 I_4}\psi_{I_2}\psi_{I_3}\psi_{I_4} \tag{1.62}$$

where all field operators have to be taken at the same time $t$, and exclude $t$ from the summation convention. The corresponding definitions are

$$\hat{K}_{I_1 I_2} \quad := \quad iD^0_{Z_1 Z_3}\left(D^k_{Z_3 Z_2}\partial_k - m_{Z_3 Z_2}\right)\delta(\mathbf{r}_1 - \mathbf{r}_2) \quad , \tag{1.63}$$

$$\hat{W}_{I_1 I_2 I_3 I_4} \quad := \quad -iD^0_{Z_1 Z_5}U_{Z_5 Z_2 Z_3 Z_4}\delta(\mathbf{r}_1 - \mathbf{r}_2)\delta(\mathbf{r}_1 - \mathbf{r}_3)\delta(\mathbf{r}_1 - \mathbf{r}_4) \quad , \tag{1.64}$$

and the summation convention is assumed to range over the indices $(Z, \mathbf{r})$ in $I = (Z, \mathbf{r}, t)$. The transition to equal times $t$ on the quantum level will be discussed in Chapter 3.

## 1.6 Symmetry Conditions

The state space of a quantum theory is usually assumed to be spanned by a complete set of eigenstates for a maximal set of commuting observables. (For a discussion of the difficulties arising in connection with indefinite state spaces we refer to Chapter 3.) In general such a maximal set of commuting observables is constituted by a commuting set of generators of the corresponding symmetry group transformations. In the case of a relativistic quantum field theory the most important symmetry group is the Poincaré group, and in the following we will discuss the resulting symmetry conditions connected with this group. If not otherwise stated we will restrict ourselves to the proper orthochronous part of the Lorentz group. Moreover we will not discuss the discrete symmetries but refer to Grimm [Gri 94].

We assume that with respect to the spinorfield operators $\psi_{A\alpha i\Lambda}(x)$ there exists a state space $\mathbb{H}$ which is an (indefinite) inner product space. In order to realize symmetries for quantum systems we additionally have to assume that the state space is a representation space of the corresponding symmetry group, i.e. in our case of the Poincaré group. Then for any Poincaré transformation in coordinate space

$$x'^{\mu} = \Lambda^{\mu}_{\nu} x^{\nu} + b^{\mu} \tag{1.65}$$

there exists an operator $\mathbf{U}(\Lambda, b)$ which is unitary on a subspace of $\mathbb{H}$ (see Chapter 3) and induces a corresponding transformation in $\mathbb{H}$, i.e.

$$|a'\rangle = \mathbf{U}(\Lambda, b)|a\rangle \ , \quad |a\rangle, |a'\rangle \in \mathbb{H} \quad . \tag{1.66}$$

Furthermore, we assume that there exists a unique state $|0\rangle \in \mathbb{H}$, the "vacuum state", with positive norm $\langle 0|0\rangle = 1$ being invariant under Poincaré transformations, i.e.

$$\mathbf{U}(\Lambda, b)|0\rangle = |0\rangle \quad . \tag{1.67}$$

According to the correspondence principle one postulates that matrix elements of spinorfields transform as classical spinorial functions. Therefore, if for the transformed system the matrix elements are given by $\langle a'|\psi_{A\alpha i}(x')|b'\rangle$, while in the original system they are defined to have the form $\langle a|\psi_{A\alpha i}(x)|b\rangle$, the transformation law can be expressed by

$$\langle a'|\psi_{A\alpha i}(x')|b'\rangle = S_{\alpha\beta}\langle a|\psi_{A\beta i}(x)|b\rangle \quad , \tag{1.68}$$

where $S_{\alpha\beta}$ is the classical transformation matrix for Dirac spinors. If (1.66) is substituted into (1.68) one obtains the transformation law for the spinorfield operators themselves:

$$\psi'_{A\alpha i}(x') := \mathbf{U}(\Lambda, b)\psi_{A\alpha i}(x)\mathbf{U}^{-1}(\Lambda, b) = S^{-1}_{\alpha\beta}(\Lambda)\psi_{A\beta i}(x') \,. \tag{1.69}$$

The Lagrangian (1.33) is invariant under these transformations.

In superindex formulation equation (1.69) reads

$$\mathbf{U}(\Lambda, b)\psi_Z(x)\mathbf{U}^{-1}(\Lambda, b) = L_{ZZ'}(\Lambda^{-1})\psi_{Z'}(x') \tag{1.70}$$

with

$$L_{ZZ'}(\Lambda) := \delta_{AA'}\delta_{ii'}\delta_{\Lambda\Lambda'}S_{\alpha\alpha'}(\Lambda) \,. \tag{1.71}$$

Generators are obtained from infinitesimal transformations of continuous groups. For infinitesimal Poincaré transformations

$$x'^{\mu} = x^{\mu} + \omega^{\mu}_{\nu} x^{\nu} + \epsilon^{\mu} \ , \quad \epsilon^{\mu}, \omega^{\mu}_{\nu} \in \mathbb{R} \ , \quad \omega^{\mu}_{\nu} = -\omega^{\nu}_{\mu} \quad , \tag{1.72}$$

we write for the group element of the corresponding representation in $\mathbb{H}$

$$\mathbf{U}(\omega, \epsilon) = 1 + \frac{i}{\hbar}\mathbf{P}^{\mu}\epsilon_{\mu} + \frac{i}{2\hbar}\mathbf{M}^{\mu\nu}\omega_{\mu\nu} \quad , \tag{1.73}$$

where $\mathbf{P}^{\mu}$ and $\mathbf{M}^{\mu\nu}$ are the generators of the Poincaré group in this representation. In this case the general transformation law (1.69) reads

i) for infinitesimal translations, i.e. $\omega^{\mu}_{\nu} = 0, \ \epsilon^{\mu} \neq 0$ :

$$\left[\mathbf{P}^\mu, \psi_{A\alpha i}(x)\right]_- = P^\mu(x)\psi_{A\alpha i}(x) \tag{1.74}$$

with

$$P^\mu(x)\psi_{A\alpha i}(x) := -i\hbar\partial^\mu\psi_{A\alpha i}(x) \quad ; \tag{1.75}$$

ii) for infinitesimal rotations, i.e. $\omega^\mu_\nu \neq 0$; $\epsilon^\mu = 0$ :

$$\left[\mathbf{M}^{\mu\nu}, \psi_{A\alpha i}(x)\right]_- = M^{\mu\nu}(x)\psi_{A\alpha i}(x) \tag{1.76}$$

with

$$M^{\mu\nu}(x)\psi_{A\alpha i}(x) := -i\hbar(x^\mu\partial^\nu - x^\nu\partial^\mu)\psi_{A\alpha i}(x) - i\hbar\Sigma^{\mu\nu}_{\alpha\beta}\psi_{A\beta i}(x) \; . \tag{1.77}$$

Equations (1.74) and (1.76) are conditions which have to be satisfied by the generators $\mathbf{P}^\mu$ and $\mathbf{M}^{\mu\nu}$. We show that these conditions are just fulfilled when the classical expressions for four–momentum and angular momentum of the spinorfields (1.35), (1.36) and (1.39), (1.40) are transferred into quantum expressions by the correspondence principle and identified with $\mathbf{P}^\mu$ and $\mathbf{M}^{\mu\nu}$, see Lauxmann [Lau 86], Borne [Bor 91].

- **Proposition 1.3:** Let $\psi_{A'\alpha'i'}(x')$ be a field operator satisfying equations (1.29) and (1.30). Then the Heisenberg equation

$$\left[\mathbf{P}^\mu, \psi_{A'\alpha'i'}(x')\right]_- = -i\hbar\partial^\mu\psi_{A'\alpha'i'}(x') \tag{1.78}$$

holds, with

$$\mathbf{P}^\mu := \int d^3r \left[\pi_{A\alpha i}(\mathbf{r},t)\partial^\mu\psi_{A\alpha i}(\mathbf{r},t) - \eta^{0\mu}\mathcal{L}(\mathbf{r},t)\right] \quad . \tag{1.79}$$

**Proof:** With (1.35), (1.36) and (1.37) we have

$$\begin{aligned}\left[\mathbf{P}^\mu, \psi_{A'\alpha'i'}(x')\right]_- &= \sum_{i=0}^{2}\frac{i\hbar}{\lambda_i}\int d^3r\left[\psi^+_{\alpha i}(x)\partial^\mu\psi_{A\alpha i}(x), \psi_{A'\alpha'i'}(x')\right]_- \\ &\quad -\eta^{0\mu}\int d^3r\left[\mathcal{L}(x), \psi_{A'\alpha'i'}(x')\right]_- \quad .\end{aligned} \tag{1.80}$$

Due to the conservation law for $\mathbf{P}^\mu$ we can choose an arbitrary moment $t$ in $x = (\mathbf{r}, t)$ without loss of generality. We thus choose $t = t'$ in (1.80) and can directly evaluate the commutators in (1.80). One obtains

$$\begin{aligned}\left[\mathbf{P}^\mu, \psi_{A'\alpha'i'}(x')\right]_- &= \sum_{i=0}^{2}\frac{i\hbar}{\lambda_i}\int d^3r\Big\{\psi^+_{\alpha i}(x)\left[\partial^\mu\psi_{A\alpha i}(x), \psi_{A'\alpha'i'}(x')\right]_+^{t=t'} \\ &\quad -\left[\psi^+_{A\alpha i}(x), \psi_{A'\alpha'i'}(x')\right]_+^{t=t'}\partial^\mu\psi_{A\alpha i}(x)\Big\} \\ &\quad -\eta^{0\mu}\int d^3r\left[\mathcal{L}(x), \psi_{A'\alpha'i'}(x')\right]_-^{t=t'} \quad .\end{aligned} \tag{1.81}$$

To evaluate the anticommutators we observe that with (1.42) we have for spatial derivatives

$$\partial^k \left[\psi_{A\alpha i}(x), \psi_{A'\alpha' i'}(x')\right]_+^{t=t'} = \left[\partial^k \psi_{A\alpha i}(x), \psi_{A'\alpha' i'}(x')\right]_+^{t=t'} = 0 \quad . \tag{1.82}$$

Together with the $\psi^+$–$\psi$–anticommutator this yields

$$\begin{aligned}\left[\mathbf{P}^\mu, \psi_{A'\alpha' i'}(x')\right]_- &= -i\hbar\partial^\mu \psi_{A\alpha' i'}(x') \\ &\quad +\eta^{0\mu} \sum_{i=0}^{2} \frac{i\hbar}{\lambda_i} \int d^3 r \psi^+_{A\alpha i}(x) \left[\partial^0 \psi_{A\alpha i}(x), \psi_{A'\alpha' i'}(x')\right]_+^{t=t'} \\ &\quad -\eta^{0\mu} \int d^3 r \left[\mathcal{L}(x), \psi_{A'\alpha' i'}(x')\right]_-^{t=t'} \quad .\end{aligned} \tag{1.83}$$

Hence we have to show that the last two terms cancel each other. We decompose $\mathcal{L}$ into $\mathcal{L}_{\text{kin}} + \mathcal{L}_{\text{int}}$. With (1.33) it is

$$\begin{aligned}&\left[\mathcal{L}_{\text{kin}}(x), \psi_{A'\alpha' i'}(x')\right]_-^{t=t'} \\ &= \sum_{i=0}^{2} \lambda_i^{-1} \left[\bar{\psi}_{A\alpha i}(x)(i\hbar\gamma^\mu\partial_\mu - cm_i)_{\alpha\beta}\psi_{A\beta i}(x), \psi_{A'\alpha' i'}(x')\right]_-^{t=t'} \\ &= \sum_{i=0}^{2} \lambda_i^{-1} \bar{\psi}_{A\alpha i}(x) \left[(i\hbar\gamma^\mu\partial_\mu - cm_i)_{\alpha\beta}\psi_{A\beta i}(x), \psi_{A'\alpha' i'}(x')\right]_+^{t=t'} \\ &\quad - \sum_{i=0}^{2} \lambda_i^{-1} \left[\bar{\psi}_{A\alpha i}(x), \psi_{A'\alpha' i'}(x')\right]_+^{t=t'} (i\hbar\gamma^\mu\partial_\mu - cm_i)_{\alpha\beta}\psi_{A\beta i}(x) \quad .\end{aligned} \tag{1.84}$$

Using the field equations (1.29) and the properties of the anticommutators we obtain

$$\begin{aligned}&\left[\mathcal{L}_{\text{kin}}(x), \psi_{A'\alpha' i'}(x')\right]_-^{t=t'} \\ &= \sum_{i=0}^{2} \frac{i\hbar}{\lambda i} \psi^+_{A\beta i}(x) \left[\partial_0 \psi_{A\beta i}(x), \psi_{A'\alpha' i'}(x')\right]_+^{t=t'} \\ &\quad -g \sum_{i=0}^{2} \left[\bar{\psi}_{A\alpha i}(x), \psi_{A'\alpha' i'}(x')\right]_+^{t=t'} V_{\substack{ABCD \\ \alpha\beta\gamma\delta}} \sum_{j,k,l=0}^{2} \psi_{B\beta j}(x)\bar{\psi}_{C\gamma k}(x)\psi_{D\delta l}(x) \quad .\end{aligned} \tag{1.85}$$

Finally we evaluate the commutator for $\mathcal{L}_{\text{int}}$. According to (1.33) it is

$$\begin{aligned}&\left[\mathcal{L}_{\text{int}}(x), \psi_{A'\alpha' i'}(x')\right]_-^{t=t'} \\ &= -\frac{g}{2} V_{\substack{ABCD \\ \alpha\beta\gamma\delta}} \sum_{i,j,k,l=0}^{2} \left[\bar{\psi}_{A\alpha\imath}(x)\psi_{B\beta j}(x)\bar{\psi}_{C\gamma k}(x)\psi_{D\delta l}(x), \psi_{A'\alpha' i'}(x')\right]_-^{t=t'} \\ &= \frac{g}{2} V_{\substack{ABCD \\ \alpha\beta\gamma\delta}} \sum_{i,j,k,l=0}^{2} \Big\{ \bar{\psi}_{A\alpha i}(x)\psi_{B\beta j}(x) \left[\bar{\psi}_{C\gamma k}(x), \psi_{A'\alpha' i'}(x')\right]_+^{t=t'} \psi_{D\delta l}(x) \\ &\quad + \left[\bar{\psi}_{A\alpha i}(x), \psi_{A'\alpha' i'}(x')\right]_+^{t=t'} \psi_{B\beta j}(x)\bar{\psi}_{C\gamma k}(x)\psi_{D\delta l}(x) \Big\} \quad .\end{aligned} \tag{1.86}$$

If one substitutes (1.85) and (1.86) into (1.83), the first term on the right–hand side of (1.85) and the second term on the right–hand side of (1.83) cancel each other. Hence the resulting expression reads

$$\begin{aligned}
\left[\mathbf{P}^\mu, \psi_{A'\alpha' i'}(x')\right]_- &= -i\hbar\partial^\mu \psi_{A\alpha' i'}(x') \qquad (1.87)\\
&+\eta^{0\mu} g \int d^3r V_{\substack{ABCD\\ \alpha\beta\gamma\delta}} \sum_{i,j,k,l=0}^{2} \left[\bar\psi_{A\alpha i}(x), \psi_{A'\alpha' i'}(x')\right]_+^{t=t'} \psi_{B\beta j}(x)\bar\psi_{C\gamma k}(x)\psi_{D\delta l}(x)\\
&-\eta^{0\mu}\frac{g}{2} \int d^3r V_{\substack{ABCD\\ \alpha\beta\gamma\delta}} \sum_{i,j,k,l=0}^{2} \Big\{\bar\psi_{A\alpha i}(x)\psi_{B\beta j}(x)\left[\bar\psi_{C\gamma k}(x), \psi_{A'\alpha' i'}(x')\right]_+^{t=t'} \psi_{D\delta l}(x)\\
&+\left[\bar\psi_{A\alpha i}(x), \psi_{A'\alpha' i'}(x')\right]_+^{t=t'} \psi_{B\beta j}(x)\bar\psi_{C\gamma k}(x)\psi_{D\delta l}(x)\Big\} \quad .
\end{aligned}$$

According to (1.43) the anticommutators for the sums of auxiliary fields, i.e. for the $\Psi$–fields, vanish. We use this property for the rearrangement

$$\begin{aligned}
& V_{\substack{ABCD\\ \alpha\beta\gamma\delta}} \sum_{i,j,k,l=0}^{2} \left[\bar\psi_{C\gamma k}(x), \psi_{A'\alpha' i'}(x')\right]_+^{t=t'} \bar\psi_{A\alpha i}(x)\psi_{B\beta j}(x)\psi_{D\delta l}(x) \qquad (1.88)\\
&= V_{\substack{ABCD\\ \alpha\beta\gamma\delta}} \sum_{i,j,k,l=0}^{2} \left[\bar\psi_{C\gamma k}(x), \psi_{A'\alpha' i'}(x')\right]_+^{t=t'} \psi_{D\delta l}(x)\bar\psi_{A\alpha i}(x)\psi_{B\beta j}(x)\\
&= V_{\substack{ABCD\\ \alpha\beta\gamma\delta}} \sum_{i,j,k,l=0}^{2} \left[\bar\psi_{A\alpha i}(x), \psi_{A'\alpha' i'}(x')\right]_+^{t=t'} \psi_{B\beta j}(x)\bar\psi_{C\gamma k}(x)\psi_{D\delta l}(x) \quad ,
\end{aligned}$$

where in the last step we performed a reindexing $(A\alpha i), (B\beta j) \rightleftharpoons (C\gamma k), (D\delta l)$ and used $V_{\substack{ABCD\\ \alpha\beta\gamma\delta}} = V_{\substack{CDAB\\ \gamma\delta\alpha\beta}}$. By substitution of (1.88) into (1.87), apart from the first term on the right–hand side all other terms cancel each other. ◊

In an analogous way one can show that equations (1.76) can be satisfied with

$$\begin{aligned}
\mathbf{M}^{\mu\nu} &= \int d^3r \Big[\pi_{A\alpha i}(x)\Sigma^{\mu\nu}_{\alpha\beta}\psi_{A\beta i}(x) + \pi_{A\alpha i}(x)\left(x^\mu\partial^\nu - x^\nu\partial^\mu\right)\psi_{A\alpha i}(x)\\
&\quad +\left(x^\nu\eta^{0\mu} - x^\mu\eta^{0\nu}\right)\mathcal{L}(x)\Big] \quad . \qquad (1.89)
\end{aligned}$$

In addition the operators $\mathbf{P}^\mu$ and $\mathbf{M}^{\mu\nu}$ satisfy the Lie algebra relations

$$\begin{aligned}
\left[\mathbf{P}^\mu, \mathbf{P}^\nu\right]_- &= 0 \qquad (1.90)\\
\left[\mathbf{P}^\rho, \mathbf{M}^{\mu\nu}\right]_- &= i\hbar\left(\eta^{\rho\nu}\mathbf{P}^\mu - \eta^{\rho\mu}\mathbf{P}^\nu\right)\\
\left[\mathbf{M}^{\mu\nu}, \mathbf{M}^{\rho\sigma}\right]_- &= i\hbar\left(\eta^{\mu\rho}\mathbf{M}^{\nu\sigma} + \eta^{\nu\sigma}\mathbf{M}^{\mu\rho} - \eta^{\mu\sigma}\mathbf{M}^{\nu\rho} - \eta^{\nu\rho}\mathbf{M}^{\mu\sigma}\right) \;,
\end{aligned}$$

i.e. $\mathbf{P}^\mu$ and $\mathbf{M}^{\mu\nu}$ given by (1.79) and (1.81) are the generators of a representation of the Poincaré group in the state space $\mathbb{H}$. Furthermore $\mathbf{P}^\mu$ and $\mathbf{M}^{\mu\nu}$ can be shown to be selfadjoint operators, so they are global observables in our theory.

For the Poincaré group the maximal set of commuting observables is given by the set of generators $\{\mathbf{P}^2, \mathbf{P}_\mu, \mathbf{W}^2, \mathbf{W}_3\}$, where the Casimir operators of the Poincaré group are defined by

$$\mathbf{P}^2 = \mathbf{P}_\mu \mathbf{P}^\mu \ , \ \ \mathbf{W}^2 = \mathbf{W}_\mu \mathbf{W}^\mu \tag{1.91}$$

with

$$\mathbf{W}_\mu := \frac{1}{2} \left(\mathbf{P}^2\right)^{-\frac{1}{2}} \epsilon_{\mu\nu\rho\sigma} \mathbf{M}^{\nu\rho} \mathbf{P}^\sigma \quad . \tag{1.92}$$

As commuting observables can have common eigenstates, we consider those states $|p\rangle$ which simultaneously diagonalize $\mathbf{P}^2$, $\mathbf{P}_\mu$, $\mathbf{W}^2$ and $\mathbf{W}_3$, i.e.

$$\begin{aligned} \mathbf{P}^2 |p\rangle &= (mc)^2 |p\rangle \ , \\ \mathbf{P}_\mu |p\rangle &= p_\mu |p\rangle \ , \\ \mathbf{W}^2 |p\rangle &= \hbar^2 s(s+1) |p\rangle \ , \\ \mathbf{W}_3 |p\rangle &= \hbar s_3 |p\rangle \ , \end{aligned} \tag{1.93}$$

where the $\mathbf{W}_3$–equation is referred to the rest system, i.e. for $\mathbf{p} = 0$.

Due to (1.78) and (1.79) all operators of the set of equations (1.93) are referred to a fixed, but arbitrary moment of time. Hence if we consider the common eigenstates of these operators, they depend on this moment of time, i.e. $|p\rangle \equiv |p, t\rangle$. The time–translation generator $\mathbf{P}^0$ is a conserved quantity, so we are allowed to set $t = 0$. As the eigenstates $|p\rangle$ of (1.93) are assumed to form a basis of the full state space $\mathbb{H}$ (for a detailed discussion concerning the indefinite metric see Chapter 3) this means that we work with time independent states, in other words with Heisenberg states, while the field operators carry the full time dependence. For our further investigations it is sufficient to work in the Heisenberg picture. On the other hand we will give a description of quantum fields which replaces the Schrödinger equation of quantum mechanics in Chapter 3. So our formulation may be regarded as a kind of fusion of Heisenberg and Schrödinger picture.

Finally, in order to fulfill (1.66) the eigenvalues of $\mathbf{P}_\mu$ have to be renormalized with respect to the vacuum, i.e. $\mathbf{P}_\mu|0\rangle = 0$ must hold. We will explicitly perform this renormalization in Chapter 3.

# 2 Covariant Quantum Field Dynamics

## 2.1 Introduction

Following the rules of quantum mechanics the theoretical description of quantum systems is based on the explicit knowledge of the corresponding state space. For quantum field theories these state spaces are assumed to be spanned by the sets of eigenstates of the commuting Poincaré generators with the corresponding eigenvalue equations (1.93). However, in contrast to quantum mechanics, these eigenvalue equations cannot be straightforwardly integrated in quantum field theory; rather the infinite number of degrees of freedom of quantum fields enforces a more sophisticated kind of treatment. This is due to the fact that, in contrast to quantum mechanics, systems with an infinite number of degrees of freedom possess infinitely many inequivalent representations which means infinitely many different state representations. In particular, the only explicitly known state representation, the Fock representation, is a definitely wrong representation for quantum fields with mutual interaction. This fact is known as Haag's theorem, cf. Haag [Haag 92].

In conventional quantum field theory this fundamental difficulty has promoted several attempts to extract information without explicit state construction and therefore without determination of the representation. Apart from many very specialized considerations there are two mainstreams of dealing with this problem, which are denoted as Axiomatic or General Quantum Field Theory.

The first one originates from perturbation theory which was covariantly reformulated by Dyson [Dys 49a,b], [Dys 51a–d]. Its basic elements are vacuum expectation values (VEV) of time–ordered products of field operators for which corresponding equations can be derived. Referring to Dyson [Dys 49a,b], Schwinger [Schwi 51] and Freese [Free 51] the latter may be called Schwinger–Dyson–Freese equations. The explicit state space construction for the interacting quantum fields is circumvented by considering only asymptotic states of scattering processes which are assumed to describe free ingoing or outgoing particles where the state representations are referred to the corresponding free fields and thus are Fock representations. The connection of the VEV with these asymptotic states, i.e. the $S$–matrix, was formulated in an axiomatic way by Lehmann, Symanzik and Zimmermann [Leh 55] and is quoted as LSZ–formalism. In the subsequent development of the theory the Schwinger–Dyson–Freese equations were replaced by path integrals which, however, did not basically change this line of approach, cf., e.g., Faddeev and Slavnov [Fadd 80].

The second mainstream is the Wightman formalism. In this case the basic entities are the Wightman functions which are defined as vacuum expectation values of ordinary products of field operators, see Wightman [Wigh 56]. But so far the axiomatic approach to relativistic quantum field theory "has not enabled us yet to construct or even characterize a specific theory", Haag [Haag 92], apart from models in lower dimensions, see, e.g., Glimm and Jaffe [Gli 87]. The formalism yields an abstract reconstruction theorem, but no explicit state space construction method.

The common feature of these two approaches is the emphasis laid on a covariant formulation and the fact that they use vacuum expectation values, i.e. classical generalized functions for the characterization of a physical system. This appearance of VEV as the basic quantities of a quantum field indicates a change of the notion of a "state" in the theory which is consequently elaborated in the formalism of Algebraic Quantum Theory. In this formalism a "state" is a positive linear functional on a $C^*$–algebra, cf. Bratteli and Robinson, [Brat 79]. The link between algebraic theory and conventional quantum theory is the GNS–construction, see Gelfand, Naimark [Gel 43] and Segal [Seg 47], and also Bogoliubov, Logunov and Todorov [Bogo 75], Bratteli and Robinson [Brat 79], Haag [Haag 92], which depends on an algebraic state and allows the representation of the algebra of observables on a suitable cyclic Hilbert space and of an algebraic "state" as a set of vacuum expectation values in this Hilbert space. Thus for a definite algebraic state the existence of a suitable corresponding state space is assured. With respect to our problem of inequivalent representations this fact makes algebraic theory an attractive candidate for a framework to formulate our theory.

However, the problem of an explicit construction of the GNS–Hilbert space is not solved in a general and constructive way, and if one looks somewhat closer there are some additional handicaps of the algebraic theory due to the presuppositions which have to be fulfilled in order to apply it.

As a consequence algebraic theory has so far been applied only to simple models of statistical mechanics. For a review of this line of research see Emch [Emch 72], Bratteli and Robinson [Brat 79], Sewell [Sew 86], Thirring [Thir 80], Primas [Prim 83], Rieckers [Rie 91], Haag [Haag 92], while there were no convincing applications in relativistic quantum field theory. Thus, in order to treat composite particles theoretically and practically in relativistic quantum field theory, we are forced to develop a new formalism which takes an intermediate position between algebraic and conventional theory and which may be called Functional Quantum Theory. Once this theory is established it can of course be applied to nonrelativistic models, too.

Functional Quantum Theory is based on the language of conventional quantum field theory. To see in which respect it is referred to the algebraic theory we shortly review the statement of the GNS–construction.

The GNS–theorem points out that for each algebraic state, i.e. for each positive, linear and normalized functional $\omega$ on the algebra $\mathcal{A}$ of observables, there exists a cyclic Hilbert space $\mathbb{H}_\omega$ with a cyclic vector $|\Omega\rangle$, so that $\omega(A) = \langle\Omega|\pi_\omega(A)|\Omega\rangle$, where $\pi_\omega(A)$ is the representation of the algebra element $A \in \mathcal{A}$ in $\mathbb{H}_\omega$. Associated with this $|\Omega\rangle$ there is a set of (algebraic) states $\omega_a$, with $\omega_a(A) = \langle a|\pi_\omega(A)|a\rangle$ for each $|a\rangle = B|\Omega\rangle$, $B \in \Pi_\omega(\mathcal{A})$. So, by means of these relations we may consider the algebraic states $\omega$ and $\omega_a$ as being represented by vectors $|\Omega\rangle \in \mathbb{H}_\omega$ or $|a\rangle \in \mathbb{H}_\omega$, respectively, and if we put $|\Omega\rangle = |0\rangle$ we obtain the usual vacuum expectation values as states $\omega(A)$.

We now deviate from this concept and assume that each state $|a\rangle$ within a generalized state space $\mathbb{H}$ is represented by the set of projections onto a cyclic basis of $\mathbb{H}$. In the algebraic language these projections are given by the functionals $\tilde{\omega}(A) = \langle 0|\pi_\omega(A)|a\rangle$, where we assumed the cyclic vector $|\Omega\rangle$ to be identical with the physical vacuum $|0\rangle$. These functionals may be negative, so they are no longer proper algebraic states, and with respect to the formulation of a covariant theory it is necessary to extend such functionals to time–ordered expressions.

For these transition matrix elements we will succeed in deriving corresponding Dyson–

Schwinger–Freese equations and we will assume the representation of the theory to be fixed by these equations, including additional boundary conditions.

Within the covariant theory Nishijima [Nish 53] tried to apply covariant state representations based on a formal GNS–construction for cyclic basis vectors with time–ordered monomials of field operators. But due to the difficulties of the covariant formalism, Nishijima was not able to define and to evaluate the corresponding state representations unambigously and he therefore returned to Green functions techniques. A more successful use of covariant transition matrix elements was made by the calculation of bound state Bethe–Salpeter amplitudes. As these amplitudes are very special matrix elements we will discuss this approach in more detail in the introduction of Chapter 5. Furthermore, as already mentioned above, Schwinger–Dyson–Freese equations can be applied to the calculation of transition matrix elements, too. To solve these equations suitable truncation procedures were proposed which were frequently referred to as New (covariant) Tamm–Dancoff method, cf. Zimmermann [Zim 54], Heisenberg [Heis 66]. But this approach did not lead to widespread applications.

In analogy to statistical mechanics the theory of quantum fields is frequently formulated by means of generating functionals, see for instance Schwinger [Schwi 54a,b], Symanzik [Sym 54], Lurie [Luri 68], Rzewuski [Rze 69], Fried [Frie 72]. We follow this line and discuss our problem in terms of such functionals. Due to our new conception we are, however, forced to deviate from the usual practice for functionals which considers functionals only as collections of an infinite number of matrix elements. In contrast to common use we shall enlarge the concept of the generating functionals by embedding the latter into corresponding functional spaces. In this way we define functional states and use these states to establish a map between the original state space $\mathbb{H}$ and its functional image $\mathbb{K}_F$. As we want each "state" from this functional space $\mathbb{K}_F$ to represent one state in $\mathbb{H}$, this map has to conserve the quantum numbers of a state in $\mathbb{H}$ and it has to be an isometric map.

This program is performed in this and the next chapter and leads to the Functional Quantum Theory. In this formalism we can synonymously speak of a state in $\mathbb{H}$ or of the corresponding state in the functional space $\mathbb{K}_F$ or of a suitable set of transition matrix elements. As we will derive equations for the functional images of the eigenstates of the Poincaré generators in $\mathbb{H}$, in principle the state space can be explicitly constructed.

In this chapter we will try to perform the above–mentioned program under the additional condition of covariance.

## 2.2 Construction of Functional States

One of the aims of modern quantum field theory is to provide a manifestly relativistic covariant calculation scheme. For such a formalism matrix elements of Heisenberg operators are an appropriate starting point. In the usual treatment time–ordered vacuum expectation values are considered, see for instance Lurie [Luri 68], Itzykson and Zuber [Itz 80]. However, with regard to our conception the suitable quantities are transition matrix elements of time–ordered products of field operators.

We define these matrix elements by

$$\tau^{(n)}\left(I_1 \ldots I_n|a\right) := \langle 0|T\left(\psi_{I_1} \ldots \psi_{I_n}\right)|a\rangle\ , \quad \forall\, |a\rangle \in \mathbb{H}\ , \tag{2.1}$$

where $|0\rangle$ is a cyclic state of $\mathbb{H}$, the "vacuum state", and the time–ordering operator $T$ is given by

$$T\left(\psi_{I_1}\dots\psi_{I_n}\right) := \sum_{p\in S_n} \mathrm{sgn}(p)\Theta\left(t_{p_1},\dots,t_{p_n}\right)\psi_{I_{p_1}}\dots\psi_{I_{p_n}} \tag{2.2}$$

with

$$\Theta(t_1,\dots,t_n) := \Theta(t_1-t_2)\Theta(t_2-t_3)\dots\Theta(t_{n-1}-t_n) \quad . \tag{2.3}$$

For later applications we introduce the alternative notation

$$\tau^{(n)}(I_1\dots I_n|a) \equiv \tau^{(n)}_{Z_1\dots Z_n}(x_1\dots x_n|a) \quad . \tag{2.4}$$

We add that the time–ordered products (2.2) are well–defined only for unequal times $t_i \neq t_j$ for $i \neq j$.

The matrix elements (2.1) have remarkable properties. For their transformation properties under Poincaré transformations the following proposition holds:

- **Proposition 2.1:** The transformed matrix elements

$$\tau^{(n)}\left(I'_1\dots I'_n|a'\right) := \langle 0|T\left[\psi_{Z'_1}(x'_1)\dots\psi_{Z'_n}(x'_n)\right]|a'\rangle \tag{2.5}$$

with

$$x' := \Lambda x + b\,, \quad |a'\rangle := \mathbf{U}(\Lambda,b)|a\rangle\,,$$

where $\mathbf{U}$ is given in (1.66) and $\Lambda$ orthochronous, obey the relation

$$\tau^{(n)}\left(I'_1\dots I'_n|a'\right) = L_{Z'_1Z_1}\dots L_{Z'_nZ_n}\tau^{(n)}(I_1\dots I_n|a) \quad , \tag{2.6}$$

with $I=(Z,x)$ and $L_{Z'Z}\equiv L_{Z'Z}(\Lambda)$ given by (1.71).

**Proof:** Equation (1.70) can be resolved to give

$$\psi_{Z'}(x') = L_{Z'Z}(\Lambda)\mathbf{U}(\Lambda,b)\psi_Z(x)\mathbf{U}^{-1}(\Lambda,b) \quad . \tag{2.7}$$

Substitution of (2.7) into (2.5) yields with $I'=(Z',x')$

$$\begin{aligned}\tau^{(n)}\left(I'_1\dots I'_n|a'\right) &= \sum_{p\in S_n}\mathrm{sgn}(p)\Theta\left(t'_{p_1}\dots t'_{p_n}\right)L_{Z'_1Z_1}\dots L_{Z'_nZ_n}\times \\ &\quad\langle 0|\mathbf{U}(\Lambda,b)\psi_{Z_{p_1}}(x_{p_1})\mathbf{U}^{-1}(\Lambda,b)\dots\mathbf{U}(\Lambda,b)\psi_{Z_{p_n}}(x_{p_n})\mathbf{U}^{-1}(\Lambda,b)|a'\rangle \\ &= \sum_{p\in S_n}\mathrm{sgn}(p)\Theta\left(t'_{p_1}\dots t'_{p_n}\right)L_{Z'_1Z_1}\dots L_{Z'_nZ_n}\langle 0|\psi_{Z_{p_1}}(x_{p_1})\dots\psi_{Z_{p_n}}(x_{p_n})|a\rangle\,.\end{aligned} \tag{2.8}$$

For orthochronous transformations we have $\Theta(t'_{p_1}\dots t'_{p_n})\equiv\Theta(t_{p_1}\dots t_{p_n})$, and substitution of this identity into (2.8) leads to (2.6). ◊

Furthermore we observe that due to definition (2.2) for $t_1\neq t_2\dots\neq t_n$ the matrix elements (2.1) are antisymmetric in their arguments:

$$\tau^{(n)}(I_1\dots I_i\dots I_j\dots I_n|a) = -\tau^{(n)}(I_1\dots I_j\dots I_i\dots I_n|a) \tag{2.9}$$

for all $i, j = 1, \dots n$.
Thus these functions not only transform correctly under Poincaré transformations, i.e. as direct products of spinors, but also provide a representation of the permutation group. The only flaw of this definition is the difficulty to continue (2.1) to equal times. This can only be achieved by a definite limiting process.

As one sees from definition (2.1) the number $n$ of field operators can take an arbitrary value. Roughly speaking, by (2.1) a state $|a\rangle \in \mathbb{H}$ is characterized by its projections onto the set of basis vectors $\langle 0|T(\psi_{I_1} \dots \psi_{I_n})$, $n = 1 \dots \infty$, where as in a Hilbert space we assume that the state space $\mathbb{H}$ is isomorphic to its dual space $\mathbb{H}^d$. This concept will be discussed more precisely in Chapter 3. But before entering into the discussion of these projections it is convenient to introduce the generating functionals as a means to formulate the dynamics of quantum fields and of composite particles later on.

In accordance with our intention we equip the commonly used functionals with appropriate transformation properties and take them as states in a functional space with an inner product structure. The direct construction of representations of such spaces were given by Stumpf and Scheerer [Stu 75]. Here we only quote the results of this investigation.

Let $j_Z(x)$ and $\partial_Z(x)$, $x \in \mathbb{M}^4$, be the generators of a CAR–algebra with the anticommutation relations

$$[j_Z(x), j_{Z'}(x')]_+ = [\partial_Z(x), \partial_{Z'}(x')]_+ = 0 \quad , \tag{2.10}$$

$$[j_Z(x), \partial_{Z'}(x')]_+ = \delta_{ZZ'}\delta(x - x') \quad . \tag{2.11}$$

In addition we assume that these generators fulfill the following transformation rules under Poincaré transformations:

$$\mathcal{V}^{-1}(\Lambda, b) j_Z(x') \mathcal{V}(\Lambda, b) = j_{Z'}(x) L^{-1}_{Z'Z}(\Lambda) \quad , \tag{2.12}$$

$$\mathcal{V}^{-1}(\Lambda, b) \partial_Z(x') \mathcal{V}(\Lambda, b) = \partial_{Z'}(x) L_{Z'Z}(\Lambda) \quad , \tag{2.13}$$

where $\mathcal{V}(\Lambda, b)$ is a unitary representation of the Poincaré group in functional space and $L_{ZZ'}$ is the transformation matrix (1.71). The anticommutators (2.10) and (2.11) remain invariant under the transformations (2.12) and (2.13).

According to the above quoted paper the compatibility of the postulates (2.10)–(2.13) is guaranteed and the explicit construction of a suitable representation shows that this Poincaré CAR–algebra can be represented in a cyclic inner product space space $\mathbb{K}_F$, i.e., there exists a cyclic "functional groundstate" $|0\rangle_F$ with ${}_F\langle 0|0\rangle_F = 1$ and

$${}_F\langle 0| j_Z(x) = \partial_Z(x)|0\rangle_F = 0 \tag{2.14}$$

and with

$$\mathcal{V}(\Lambda, b)|0\rangle_F = |0\rangle_F \quad . \tag{2.15}$$

For brevity we again denote the elements of the abstract algebra and of their representations by the same symbols; this will not give rise to misunderstandings.

By means of the cyclic vector $|0\rangle_F$ we obtain a set of basis vectors of the functional space $\mathbb{K}_F$ by

$$\mathbb{B}_F := \left\{ \frac{i^n}{n!} j_{I_1} \dots j_{I_n} |0\rangle_F \, , \ I_1 \dots I_n \in \mathcal{I}, \ n \in \mathrm{N} \right\} \tag{2.16}$$

with $j_I := j_Z(x)$ and where the factors $\frac{i^n}{n!}$ are introduced for convenience.

Using this construction we can now introduce generating functional states by the definition

$$|\mathcal{T}(j;a)\rangle := \sum_{n=0}^{\infty} \frac{i^n}{n!} \tau^{(n)}(I_1 \dots I_n|a) j_{I_1} \dots j_{I_n}|0\rangle_F \tag{2.17}$$

for all $|a\rangle \in \mathbb{H}$. With this definition we have collected all possible projections of a state $|a\rangle$ onto the dual states $\langle 0|T(\psi_{I_1} \dots \psi_{I_n})$, $n = 1 \dots \infty$, into one functional state which from now on we consider as a representation of $|a\rangle \in \mathbb{H}$ by its map into functional space $\mathbb{K}_F$.

In order to work with such functional states and to elaborate their properties we need some calculation rules in functional space:

- **Proposition 2.2:** For functional states

$$|\mathcal{T}(j;a)\rangle := \sum_{n=0}^{\infty} \frac{i^n}{n!} \tau^{(n)}(I_1 \dots I_n|a) j_{I_1} \dots j_{I_n}|0\rangle_F$$

the following relations hold :

i) $$i^{-1}\partial_K|\mathcal{T}(j;a)\rangle = \sum_{n=0}^{\infty} \frac{i^n}{n!} \tau^{(n+1)}(K, I_1 \dots I_n|a) j_{I_1} \dots j_{I_n}|0\rangle_F \tag{2.18}$$

ii) $$\begin{aligned} &ij_K|\mathcal{T}(j;a)\rangle \\ &= \sum_{n=0}^{\infty} \frac{i^n}{n!} \sum_{l=1}^{n} (-1)^{l-1} \delta_{KI_l} \tau^{(n-1)}(I_1 \dots \overset{o}{I_l} \dots I_n|a) j_{I_1} \dots j_{I_n}|0\rangle_F \end{aligned} \tag{2.19}$$

iii) $$i^{-m}\,{}_F\langle 0|\partial_{K_m} \dots \partial_{K_1}|\mathcal{T}(j;a)\rangle = \tau^{(m)}(K_1 \dots K_m|a) \tag{2.20}$$

with

$$\tau^{(n-1)}(I_1 \dots \overset{o}{I_l} \dots I_n|a) := \tau^{(n-1)}(I_1 \dots I_{l-1}, I_{l+1} \dots I_n|a)$$

and $\delta_{I_1 I_2} := \delta_{Z_1 Z_2}\delta(x_1 - x_2)$.

**Proof:** For illustration we give a rather detailed treatment of these functional calculation rules.
i) With (2.11) and (2.14) it is

$$\partial_K j_{I_1} \dots j_{I_n}|0\rangle_F = \sum_{l=1}^{n} (-1)^{l-1} \delta_{KI_l} j_{I_1} \dots \overset{o}{j}_{I_l} \dots j_{I_n}|0\rangle_F \quad . \tag{2.21}$$

With that and by use of the antisymmetry (2.9) and reindexing it follows

$$\begin{aligned} &i^{-1}\partial_K|\mathcal{T}(j;a)\rangle \\ &= \sum_{n=1}^{\infty} \frac{i^{n-1}}{n!} \sum_{l=1}^{n} (-1)^{l-1} \tau^{(n)}(I_1 \dots I_{l-1}, K, I_{l+1} \dots I_n|a) \times \\ &\quad j_{I_1} \dots \overset{o}{j}_{I_l} \dots j_{I_n}|0\rangle_F \end{aligned} \tag{2.22}$$

$$
\begin{aligned}
&= \sum_{n=1}^{\infty} \frac{i^{n-1}}{n!} \sum_{l=1}^{n} \tau^{(n)}(K, I_1 \dots \overset{o}{I_l} \dots I_n | a) j_{I_1} \dots \overset{o}{j}_{I_l} \dots j_{I_n} |0\rangle_F \\
&= \sum_{n=1}^{\infty} \frac{i^{n-1}}{(n-1)!} \tau^{(n)}(K, I_1 \dots I_{n-1} | a) j_{I_1} \dots j_{I_{n-1}} |0\rangle_F \\
&= \sum_{n=0}^{\infty} \frac{i^{n}}{n!} \tau^{(n+1)}(K, I_1 \dots I_n | a) j_{I_1} \dots j_{I_n} |0\rangle_F \quad .
\end{aligned}
$$

ii) Due to (2.10) it is

$$
\begin{aligned}
& i j_K | \mathcal{T}(j; a)\rangle \qquad (2.23)\\
&= \sum_{n=0}^{\infty} \frac{i^{n+1}}{n!} \tau^{(n)}(I_1 \dots I_n | a) j_K j_{I_1} \dots j_{I_n} |0\rangle_F \\
&= \sum_{n=0}^{\infty} \frac{i^{n+1}}{n!} \tau^{(n)}(I_1 \dots I_n | a) \delta_{KJ} (-1)^{l-1} j_{I_1} \dots j_{I_{l-1}} j_J j_{I_l} \dots j_{I_n} |0\rangle_F \\
&= \sum_{n=0}^{\infty} \frac{i^{n+1}}{n!} \tau^{(n)}(I_1 \dots \overset{o}{I_l} \dots I_{n+1} | a) \delta_{KI_l} (-1)^{l-1} j_{I_1} \dots j_{I_{n+1}} |0\rangle_F \\
&= \sum_{n=1}^{\infty} \frac{i^{n}}{(n-1)!} \tau^{(n-1)}(I_1 \dots \overset{o}{I_l} \dots I_n | a) \delta_{KI_l} (-1)^{l-1} j_{I_1} \dots j_{I_n} |0\rangle_F \\
&= \sum_{n=1}^{\infty} \frac{i^{n}}{n!} \sum_{l=1}^{n} \tau^{(n-1)}(I_1 \dots \overset{o}{I_l} \dots I_n | a) \delta_{KI_l} (-1)^{l-1} j_{I_1} \dots j_{I_n} |0\rangle_F \quad .
\end{aligned}
$$

iii) We define the dual set of functional basis vectors by

$$
\mathbb{B}_F^d := \left\{ {}_F\langle 0 | (\partial_{K_1} \dots \partial_{K_m}),\ K_1 \dots K_m \in \mathcal{I},\ m \in \mathbb{N} \right\} \quad . \qquad (2.24)
$$

From (2.10), (2.11) and (2.14) and from ${}_F\langle 0|0\rangle_F = 1$ it follows that

$$
{}_F\langle 0 | \partial_{K_m} \dots \partial_{K_1} j_{I_1} \dots j_{I_n} |0\rangle_F = \delta_{mn} \sum_{p \in S_n} \operatorname{sgn}(p) \delta_{I_1 K_{p_1}} \dots \delta_{I_n K_{p_n}} \quad . \qquad (2.25)
$$

Thus we obtain for ${}_F\langle 0|0\rangle_F = 1$ the customary Fock space scalar product between the basis states from (2.16) and their duals (2.24).

With (2.25) and the antisymmetry (2.9) we obtain

$$
\begin{aligned}
& i^{-m} {}_F\langle 0 | \partial_{K_m} \dots \partial_{K_1} | \mathcal{T}(j; a)\rangle \\
&= \sum_{n=0}^{\infty} i^{-m} \frac{i^n}{n!} \tau^{(n)}(I_1 \dots I_n | a) {}_F\langle 0 | \partial_{K_m} \dots \partial_{K_1} j_{I_1} \dots j_{I_n} |0\rangle_F \\
&= \frac{1}{m!} \tau^{(m)}(I_1 \dots I_m | a) \sum_{p \in S_m} \operatorname{sgn}(p) \delta_{I_1 K_{p_1}} \dots \delta_{I_m K_{p_m}} \\
&= \tau^{(m)}(K_1 \dots K_m | a) \quad .
\end{aligned}
$$

◊

By means of (2.20) the individual $\tau^{(n)}$–functions can be regained from the functional state $|\mathcal{T}(j;a)\rangle$; we call this procedure the projection of functional states into coordinate space.

Finally it should be mentioned: As in ordinary Minkowski space the scalar products between the original basis states (2.16) and the dual basis states (2.24) give no hint about the metrical structure of $\mathbb{K}_F$. For this purpose rather we have to consider the scalar products between the original basis states themselves, i.e. we have to calculate the scalar products ${}_F\langle 0|(j_{K_1}\dots j_{K_m})^+(j_{I_1}\dots j_{I_n})|0\rangle_F$. It was shown by Stumpf and Scheerer [Stu 75] that $\partial_Z(x) \neq j_Z(x)^+$ and the metrical tensor for these basis states was derived. Thus with $\langle\mathcal{T}(j,a)| := |\mathcal{T}(j,a)\rangle^+$ one can evaluate scalar products $\langle\mathcal{T}(j,a)|\mathcal{T}(j,a)\rangle$.

However, these scalar products in functional space are completely independent from those of the state space $\mathbb{H}$ and cannot be used for the construction of an isometric map between $\mathbb{H}$ and $\mathbb{K}_F$. The functional space is an artificial construction and one cannot expect to guess the physical scalar products by a priori constructions in functional space, as these physical scalar products depend on the system dynamics itself. The theory of isometric maps will be developed in Chapter 3.

## 2.3 Symmetries in Functional Space

By means of definition (2.17) a map between the physical state space $\mathbb{H}$ and the functional state space $\mathbb{K}_F$ is established. According to our assumption that the basis states of $\mathbb{H}$ are eigenstates of equations (1.93), $\mathbb{H}$ is a representation space of the Poincaré group. Because we want to describe a quantum system in the functional space $\mathbb{K}_F$ the image state $|\mathcal{T}(j;p)\rangle \in \mathbb{K}_F$ of a basis state $|p\rangle \in \mathbb{H}$ should be characterized by the same quantum numbers as $|p\rangle$. This implies that $\mathbb{K}_F$ also has to be a representation space of the Poincaré group.

By the postulates (2.12), (2.13), (2.15) $\mathbb{K}_F$ is equipped with transformation properties. In this section we explore the consequences of these transformation properties in more detail based on theorems which were given by Stumpf, Scheerer and Märtl in [Stu 70b] and [Stu 72]. It will turn out that $\mathbb{K}_F$ indeed carries a representation of the Poincaré group.

- **Proposition 2.3:** The Poincaré transformed functional state

$$|\mathcal{T}(j;a')\rangle := \sum_{n=0}^{\infty}\frac{i^n}{n!}\tau^{(n)}(I_1'\dots I_n'|a')j_{I_1'}\dots j_{I_n'}|0\rangle_F \tag{2.26}$$

with $I' = (Z',x'),\ x' = \Lambda x + b$ and $\Lambda$ orthochronous is related to the original state $|\mathcal{T}(j;a)\rangle$ by the transformation rule

$$|\mathcal{T}(j;a')\rangle = \mathcal{V}(\Lambda,b)|\mathcal{T}(j,a)\rangle\ . \tag{2.27}$$

**Proof:** With (2.6), (2.12) and (2.15) we have

$$|\mathcal{T}(j;a')\rangle = \tag{2.28}$$

$$= \sum_{n=0}^{\infty}\frac{i^n}{n!}\tau^{(n)}_{Z_1\dots Z_n}(x_1\dots x_n|a)L_{Z_1'Z_1}(\Lambda)\dots L_{Z_n'Z_n}(\Lambda)\times$$

$$j_{Z_1'}(x_1')\dots j_{Z_n'}(x_n')|0\rangle_F$$

$$= \sum_{n=0}^{\infty} \frac{i^n}{n!} \tau^{(n)}_{Z_1 \ldots Z_n}(x_1 \ldots x_n|a) L_{Z_1' Z_1}(\Lambda) \ldots L_{Z_n' Z_n}(\Lambda) \times$$
$$L^{-1}_{Z_1'' Z_1'} \mathcal{V}(\Lambda, b) j_{Z_1''}(x_1) \mathcal{V}^{-1}(\Lambda, b) \ldots L^{-1}_{Z_n'' Z_n'} \mathcal{V}(\Lambda, b) j_{Z_n''}(x_n) \mathcal{V}^{-1}(\Lambda, b)|0\rangle_F \quad .$$

The products $\mathcal{V}^{-1}\mathcal{V}$ and $LL^{-1}$ cancel out and we obtain

$$|\mathcal{T}(j; a')\rangle = \sum_{n=0}^{\infty} \frac{i^n}{n!} \tau^{(n)}(I_1 \ldots I_n|a) \mathcal{V}(\Lambda, b) j_{I_1} \ldots j_{I_n}|0\rangle_F \tag{2.29}$$
$$= \mathcal{V}(\Lambda, b)|\mathcal{T}(j; a)\rangle \quad .$$

◊

In order to derive the functional counterpart of equations (1.93) we first prove the following proposition:

- **Proposition 2.4:** The effect of the Poincaré generators $P_\mu(x)$ and $M_{\mu\nu}(x)$ in function space on matrix elements of time–ordered products of field operators (2.1) is given by the relations

$$\text{i)} \qquad \sum_{i=1}^{n} P_\mu(x_i) \langle 0| T\left(\psi_{I_1} \ldots \psi_{I_n}\right)|a\rangle \tag{2.30}$$
$$= \langle 0| T \sum_{i=1}^{n} P_\mu(x_i) \left(\psi_{I_1} \ldots \psi_{I_n}\right)|a\rangle \quad ,$$

$$\text{ii)} \qquad \sum_{i=1}^{n} M_{\mu\nu}(x_i) \langle 0| T\left(\psi_{I_1} \ldots \psi_{I_n}\right)|a\rangle \tag{2.31}$$
$$= \langle 0| T \sum_{i=1}^{n} M_{\mu\nu}(x_i) \left(\psi_{I_1} \ldots \psi_{I_n}\right)|a\rangle \quad .$$

**Proof:** i) We have to show that $\sum_{i=1}^{n} P_\mu(x_i)$ commutes with time–ordering. This is nontrivial only for $\mu = 0$. With $P_0(x_i) := -i\hbar\partial_0(x_i) \equiv -i\frac{\hbar}{c}\partial/\partial t_i$ and (2.2) we obtain

$$\sum_{i=1}^{n} \partial_0(x_i) \sum_{p \in S_n} \text{sgn}(p) \Theta(t_{p_1}, \ldots, t_{p_n}) \langle 0|\psi_{I_{p_1}} \ldots \psi_{I_{p_n}}|a\rangle \tag{2.32}$$
$$= \sum_{p \in S_n} \text{sgn}(p) \langle 0|\psi_{I_{p_1}} \ldots \psi_{I_{p_n}}|a\rangle \sum_{i=1}^{n} \partial_0(x_{p_i}) \Theta(t_{p_1}, \ldots, t_{p_n})$$
$$+ \langle 0| T \left[\sum_{i=1}^{n} \partial_0(x_i) \psi_{I_1} \ldots \psi_{I_n}\right] |a\rangle \quad .$$

Furthermore with (2.3) we have

$$\left[\sum_{i=1}^{n} \partial_0(x_{p_i})\right] \Theta(t_{p_1}, \ldots t_{p_n}) \tag{2.33}$$
$$= \sum_{\alpha=1}^{n-1} \left\{ \Theta(t_{p_1} - t_{p_2}) \ldots \left[\sum_{i=1}^{n} \partial_0(x_{p_i}) \Theta(t_{p_\alpha} - t_{p_{\alpha+1}})\right] \ldots \Theta(t_{p_{n-1}} - t_{p_n}) \right\}$$

where the differentiation is applied to any factor in the product. Because of $\frac{d}{dt}\Theta(t) = \delta(t)$ it follows

$$\left[\sum_{i=1}^{n} \partial_0(x_{p_i})\right] \Theta(t_{p_k} - t_{p_{k+1}}) = 0 \qquad \text{for} \quad k = 1, \ldots n-1\,, \tag{2.34}$$

so relation (2.30) is proven.

ii) We have to show that $\sum_{i=1}^{n} M_{\mu\nu}(x_i)$ commutes with time–ordering. With (1.77) the only nontrivial part of the commutator comes from the term $\sum_{i=1}^{n} x_\mu^i \partial_0(x_i)$. The proof runs along the same lines as for i). We replace in (2.32) $\sum_{i=1}^{n} \partial_0(x_{p_i})$ by $\sum_{i=1}^{n} x_\mu^{p_i} \partial_0(x_{p_i})$. Then the analogue to (2.34) is

$$\left[\sum_{i=1}^{n} x_\mu^{p_i} \partial_0(x_{p_i})\right] \Theta(t_{p_k} - t_{p_{k+1}}) = 0 \qquad \text{for} \quad k = 1, \ldots n-1, \tag{2.35}$$

which is valid because we have restricted the definition of time–ordered products to unequal times $t_i \neq t_j,\ \forall\, i \neq j$. ◇

By means of Proposition 2.4 we can now derive the symmetry conditions for $\tau^{(n)}$–functions (2.1).

- **Proposition 2.5:** For all states $|p\rangle \in \mathbb{H}$ being eigenstates of equations (1.93) the corresponding $\tau^{(n)}$–functions satisfy the symmetry conditions

i)
$$\left[\sum_{i=1}^{n} P_\mu(x_i) + p_\mu\right] \tau^{(n)}(I_1 \ldots I_n|p) = 0\,, \tag{2.36}$$

ii)
$$\left[\sum_{i=1}^{n} P_\mu(x_i) \sum_{j=1}^{n} P^\mu(x_j) - (mc)^2\right] \tau^{(n)}(I_1 \ldots I_n|p) = 0\,, \tag{2.37}$$

iii)
$$\left[\frac{1}{4m^2}\epsilon_{\mu\nu\rho\sigma}\epsilon^{\mu\nu'\rho'\sigma'} \sum_{i,j,k,l=1}^{n} P^\nu(x_i) M^{\rho\sigma}(x_j) P_{\nu'}(x_k) M_{\rho'\sigma'}(x_l) - \hbar^2 s(s+1)\right] \tau^{(n)}(I_1 \ldots I_n|p) = 0\,, \tag{2.38}$$

iv)
$$\left[\frac{1}{2m}\epsilon_{3\nu\rho\sigma} \sum_{i,j=1}^{n} P^\nu(x_i) M^{\rho\sigma}(x_j) - \hbar s_3\right] \tau^{(n)}(I_1 \ldots I_n|p) = 0 \tag{2.39}$$

**Proof:** According to (1.93) it is $\mathbf{P}_\mu|p\rangle = p_\mu|p\rangle$ and therefore

$$\langle 0|T(\psi_{I_1} \ldots \psi_{I_n})\mathbf{P}_\mu|p\rangle = p_\mu \langle 0|T(\psi_{I_1} \ldots \psi_{I_n})|p\rangle \quad . \tag{2.40}$$

Due to the commutator of $\mathbf{P}_\mu$ and $\psi_I$ given by (1.74) we can derive the commutator of $\mathbf{P}_\mu$ with $T(\psi_{I_1} \ldots \psi_{I_n})$ and obtain

$$T(\psi_{I_1}\dots\psi_{I_n})\mathbf{P}_\mu = \mathbf{P}_\mu T(\psi_{I_1}\dots\psi_{I_n}) - \sum_{i=1}^{n} P_\mu(x_i)T(\psi_{I_1}\dots\psi_{I_n}) \quad . \tag{2.41}$$

Substitution of (2.41) into (2.40) and observing $\langle 0|\mathbf{P}_\mu = 0$ we get equation (2.36).

Taking into account that the commutators of the rotation operators $\mathbf{M}_{\mu\nu}$ with $\psi_I$ are given by (1.76), we can derive equations (2.37)–(2.39) in an analogous way to (2.36). ◊

In the next step we formulate the symmetry conditions for the generating functional states (2.17).

- **Proposition 2.6:** For all states $|p\rangle \in \mathbb{H}$ being eigenstates of equations (1.93) the corresponding generating functional states $|\mathcal{T}(j;p)\rangle$ satisfy the symmetry conditions

$$\mathcal{P}_\mu|\mathcal{T}(j;p)\rangle = p_\mu|\mathcal{T}(j;p)\rangle \quad , \tag{2.42}$$

$$\mathcal{P}_\mu\mathcal{P}^\mu|\mathcal{T}(j;p)\rangle = (mc)^2|\mathcal{T}(j;p)\rangle \quad , \tag{2.43}$$

$$\mathcal{W}_\mu\mathcal{W}^\mu|\mathcal{T}(j;p)\rangle = \hbar^2 s(s+1)|\mathcal{T}(j;p)\rangle \quad , \tag{2.44}$$

$$\mathcal{W}_3|\mathcal{T}(j;p)\rangle = \hbar s_3|\mathcal{T}(j;p)\rangle \tag{2.45}$$

with

$$\mathcal{P}_\mu := -\int j_Z(x)P_\mu(x)\partial_Z(x)d^4x \quad , \tag{2.46}$$

$$\mathcal{W}_\mu := \frac{1}{2mc}\epsilon_{\mu\nu\rho\sigma}\mathcal{P}^\nu\mathcal{M}^{\rho\sigma} \quad , \tag{2.47}$$

$$\mathcal{M}^{\rho\sigma} := -\int j_Z(x)M^{\rho\sigma}(x)\partial_Z(x)d^4x \quad . \tag{2.48}$$

**Proof:** We apply (2.21) to evaluate $\mathcal{P}_\mu|\mathcal{T}(j;p)\rangle$. This gives

$$\int d^4x j_Z(x)P_\mu(x)\partial_Z(x)|\mathcal{T}(j;p)\rangle \tag{2.49}$$

$$= -i\hbar\int d^4x j_Z(x)\partial_\mu(x)\sum_{n=0}^{\infty}\frac{i^n}{n!}\sum_{l=1}^{n}(-1)^{l-1}\delta_{ZZ_l}\delta(x-x_l)\times$$
$$\langle 0|T(\psi_{I_1}\dots\psi_{I_n})|p\rangle j_{I_1}\dots\overset{o}{j}_{I_l}\dots j_{I_n}|0\rangle_F$$
$$= -i\hbar\sum_{n=0}^{\infty}\frac{i^n}{n!}\sum_{l=1}^{n}(-1)^l j_Z(x_l)\delta_{ZZ_l}\partial_\mu(x_l)\langle 0|T(\psi_{I_1}\dots\psi_{I_n})|p\rangle\times$$
$$j_{I_1}\dots\overset{o}{j}_{I_l}\dots j_{I_n}|0\rangle_F$$
$$= \sum_{n=0}^{\infty}\frac{i^n}{n!}\sum_{l=1}^{n}\Big[P_\mu(x_l)\langle 0|T(\psi_{I_1}\dots\psi_{I_n})|p\rangle\Big]j_{I_1}\dots j_{I_n}|0\rangle_F \quad .$$

Combination with equation (2.36) yields (2.42).

In the same way one can proceed to prove the remaining equations (2.43), (2.44) and (2.45). ◊

- **Proposition 2.7:** The operators $\mathcal{P}_\mu$ and $\mathcal{M}_{\mu\nu}$ defined by (2.46) and (2.48) satisfy the Lie algebra relations

$$
\begin{aligned}
[\mathcal{P}_\mu, \mathcal{P}_\nu]_- &= 0 \quad , & (2.50)\\
[\mathcal{P}_\lambda, \mathcal{M}_{\mu\nu}]_- &= -i\hbar\,(\eta_{\lambda\nu}\mathcal{P}_\mu - \eta_{\lambda\mu}\mathcal{P}_\nu) \quad , & (2.51)\\
[\mathcal{M}_{\mu\nu}, \mathcal{M}_{\rho\sigma}]_- &= -i\hbar\,(\eta_{\mu\rho}\mathcal{M}_{\nu\sigma} + \eta_{\nu\sigma}\mathcal{M}_{\mu\rho} - \eta_{\mu\sigma}\mathcal{M}_{\nu\rho} - \eta_{\nu\rho}\mathcal{M}_{\mu\sigma}) & (2.52)
\end{aligned}
$$

of the Poincaré group in functional space.

**Proof:** Let $\{G_a(x)\}$ be a set of generators satisfying the Lie algebraic relations

$$[G_a(x), G_b(x)]_- = f^c{}_{ab}G_c(x) \tag{2.53}$$

in function space. Then the corresponding representation in functional space is given by $\mathcal{G}_a := -\int \jmath_Z(x)G_a(x)\partial_Z(x)d^4x$. Application of the anticommutation relations (2.10) and (2.11) yields for $\mathcal{G}_a\mathcal{G}_b$ the expression

$$
\begin{aligned}
\mathcal{G}_a\mathcal{G}_b &= \int \jmath_Z(x)G_a(x)\delta_{ZZ'}\delta(x-x')G_b(x')\partial_{Z'}(x')d^4x d^4x' \qquad (2.54)\\
&\quad - \int \jmath_Z(x)\jmath_{Z'}(x')G_a(x)G_b(x')\partial_Z(x)\partial_{Z'}(x')d^4x'd^4x \quad .
\end{aligned}
$$

In the last term of equation (2.54) the operators $G_a(x)$ and $G_b(x')$ are referred to different coordinates and thus commute. Hence we obtain

$$
\begin{aligned}
[\mathcal{G}_a, \mathcal{G}_b]_- &= \int \jmath_Z(x)G_a(x)G_b(x)\partial_Z(x)d^4x \qquad (2.55)\\
&\quad - \int \jmath_Z(x)G_b(x)G_a(x)\partial_Z(x)d^4x\\
&= \int \jmath_Z(x)\,[G_a(x), G_b(x)]_-\,\partial_Z(x)d^4x \quad ,
\end{aligned}
$$

and with (2.53) this yields

$$[\mathcal{G}_a, \mathcal{G}_b]_- = -f^c{}_{ab}\mathcal{G}_c \, .$$

◇

Therefore a representation of the Poincaré group in functional space is generated by the infinitesimal transformation

$$\mathcal{V}(\omega, \epsilon) := 1 + \frac{i}{\hbar}\epsilon_\mu\mathcal{P}^\mu + \frac{i}{2\hbar}\omega_{\mu\nu}\mathcal{M}^{\mu\nu} \quad . \tag{2.56}$$

- **Proposition 2.8:** In the restsystem with $\mathbf{p} = 0$ the symmetry conditions (2.42)–(2.45) go over into the equations

$$
\begin{aligned}
\mathcal{P}_0|\mathcal{T}(\jmath;p)\rangle &= mc|\mathcal{T}(\jmath;p)\rangle \quad , & (2.57)\\
\mathcal{P}_k|\mathcal{T}(\jmath;p)\rangle &= 0 \quad , & (2.58)\\
\mathcal{P}_\mu\mathcal{P}^\mu|\mathcal{T}(\jmath;p)\rangle &= (mc)^2|\mathcal{T}(\jmath;p)\rangle \quad , & (2.59)\\
\left(\mathcal{M}_{23}^2 + \mathcal{M}_{31}^2 + \mathcal{M}_{12}^2\right)|\mathcal{T}(\jmath;p)\rangle &= \hbar^2 s(s+1)|\mathcal{T}(\jmath;p)\rangle \quad , & (2.60)\\
\mathcal{M}_{12}|\mathcal{T}(\jmath;p)\rangle &= \hbar s_3|\mathcal{T}(\jmath;p)\rangle \quad . & (2.61)
\end{aligned}
$$

**Proof:** The definition of the restsystem is given by (2.57) and (2.58) which immediately yield (2.59). By means of this definition we obtain from (2.47) for $\mu = 3$

$$\begin{aligned} \mathcal{W}_3|\mathcal{T}(j;p)\rangle &= \frac{1}{2cm}\mathcal{P}_0\left(\mathcal{M}^{12} - \mathcal{M}^{21}\right)|\mathcal{T}(j;p)\rangle \\ &= \mathcal{M}^{12}|\mathcal{T}(j;p)\rangle \\ &= \hbar s_3|\mathcal{T}(j;p)\rangle \quad . \end{aligned} \tag{2.62}$$

In the same way (2.60) can be derived from (2.44). ◊

## 2.4 Functional Field Equations

So far we have discussed the functional formulation of symmetry constraints. To complete the map from $\mathbb{H}$ to $\mathbb{K}_F$ we have to represent the field equations in functional space. As a first step we prove a proposition, see Stumpf [Stu 70a], concerning time–ordering and time derivation.

- **Proposition 2.9:** The time derivative of time–ordered products is given by

$$\begin{aligned} &\partial_t\left[T(\psi_J\psi_{I_1}\dots\psi_{I_n})\right] \\ &= T\left[(\partial_t\psi_J)\psi_{I_1}\dots\psi_{I_n}\right] - \sum_{l=1}^{n}(-1)^l A_{JI_l}\delta(t-t_l)T(\psi_{I_1}\dots\overset{o}{\psi}_{I_l}\dots\psi_{I_n}) \end{aligned} \tag{2.63}$$

with

$$A_{I_1I_2} = [\psi_{I_1},\psi_{I_2}]_+^{t=t'}$$

and $J = (Z, \mathbf{r}, t)$, $I_i = (Z_i, \mathbf{r}_i, t_i)$.

**Proof:** For brevity we replace $\psi_{I_i}$ by $I_i$ for $i = 1\dots n$ and $\psi_J$ by $J$. Then for $n \geq 2$ it is

$$\begin{aligned} T(JI_1\dots I_n) &= \sum_{p\in S_n}\operatorname{sgn}(p)JI_{p_1}\dots I_{p_n}\Theta(t,t_{p_1}\dots,t_{p_n}) \\ &+ \sum_{p\in S_n}\sum_{k=1}^{n-1}\operatorname{sgn}(p)(-1)^k I_{p_1}\dots I_{p_k}JI_{p_{k+1}}\dots I_{p_n}\Theta(t_{p_1}\dots,t_{p_k},t,t_{p_{k+1}}\dots,t_{p_n}) \\ &+ \sum_{p\in S_n}\operatorname{sgn}(p)(-1)^n I_{p_1}\dots I_{p_n}J\Theta(t_{p_1}\dots,t_{p_n},t) \quad . \end{aligned} \tag{2.64}$$

With

$$\Theta(t_{p_k}-t)\Theta(t-t_{p_{k+1}}) = \Theta(t_{p_k}-t_{p_{k+1}})\left[\theta(t-t_{p_{k+1}}) - \theta(t-t_{p_k})\right] \tag{2.65}$$

and

$$\theta(t-t_i) := \frac{1}{2}\left[\Theta(t-t_i) - \Theta(t_i-t)\right] \tag{2.66}$$

one obtains

$$
\begin{aligned}
&T(JI_1 \dots I_n) \qquad (2.67)\\
&= \sum_{p\in S_n} \mathrm{sgn}(p)\Theta(t_{p_1}\dots,t_{p_n})\Big[JI_{p_1}\dots I_{p_n}\Theta(t-t_{p_1})\\
&\quad +\sum_{k=1}^{n-1}(-1)^k I_{p_1}\dots I_{p_k}JI_{p_{k+1}}\dots I_{p_n}\theta(t-t_{p_{k+1}})\\
&\quad -\sum_{k=1}^{n-1}(-1)^k I_{p_1}\dots I_{p_k}JI_{p_{k+1}}\dots I_{p_n}\theta(t-t_{p_k})\\
&\quad +(-1)^n I_{p_1}\dots I_{p_n}J\Theta(t_{p_n}-t)\Big]\\
&= \sum_{p\in S_n} \mathrm{sgn}(p)\Theta(t_{p_1}\dots,t_{p_n})\Big\{\left[JI_{p_1}\Theta(t-t_{p_1})+I_{p_1}J\theta(t-t_{p_1})\right]I_{p_2}\dots,I_{p_n}\\
&\quad +(-1)^n I_{p_1}\dots I_{p_{n-1}}\left[I_{p_n}J\Theta(t_{p_n}-t)-JI_{p_n}\theta(t-t_{p_n})\right]\\
&\quad +\sum_{k=2}^{n-1}(-1)^{k+1}I_{p_1}\dots I_{p_{k-1}}\left[I_{p_k},J\right]_+ I_{p_{k+1}}\dots I_{p_n}\theta(t-t_{p_k})\Big\} \quad ,
\end{aligned}
$$

and with

$$\partial_t\theta(t-t_i)=\delta(t-t_i) \qquad (2.68)$$

it follows

$$
\begin{aligned}
&\partial_t T(JI_1\dots I_n) = T(\dot{J}I_1\dots I_n) \qquad (2.69)\\
&\quad -\sum_{p\in S_n}\mathrm{sgn}(p)\Theta(t_{p_1}\dots,t_{p_n})\sum_{k=1}^{n}(-1)^k I_{p_1}\dots I_{p_{k-1}}[I_{p_k},J]_+ I_{p_{k+1}}\dots I_{p_n}\delta(t-t_{p_k})\\
&= T(\dot{J}I_1\dots I_n)\\
&\quad -\sum_{p\in S_n}\mathrm{sgn}(p)\Theta(t_{p_1}\dots,t_{p_n})\sum_{k=1}^{n}(-1)^k I_{p_1}\dots \overset{o}{I}_{p_k}\dots I_{p_n}A_{I_{p_k}J}\delta(t-t_{p_k}) \quad .
\end{aligned}
$$

If one transforms the sequence of indices $(1,2,\dots k-1,k,k+1,\dots n)$ in the last term of (2.69) into the sequence $(2,3,\dots k,1,k+1,\dots n)$, then (2.69) goes over into

$$
\begin{aligned}
&\partial_t T(JI_1\dots I_n) = T(\dot{J}I_1\dots I_n) \qquad (2.70)\\
&\quad +\sum_{p\in S_n}\mathrm{sgn}(p)\sum_{k=1}^{n}A_{I_{p_1}J}\delta(t-t_{p_1})I_{p_2}\dots I_{p_n}\Theta(t_{p_2},t_{p_3},\dots t_{p_k},t_{p_1},t_{p_{k+1}},\dots t_{p_n})\, .
\end{aligned}
$$

Renewed permutation of $(1,2,\dots l,l+1,\dots n)$ into $(l,1,2,\dots l-1,l+1,\dots n)$ for $l=1,\dots,n$ results in

$$
\begin{aligned}
&\partial_t T(JI_1\dots I_n) = T(\dot{J}I_1\dots I_n) \qquad (2.71)\\
&\quad +\sum_{p\in S_n}\mathrm{sgn}(p)\sum_{l=1}^{n}(-1)^{l+1}A_{JI_l}\delta(t-t_l)(I_{p_1}\dots \overset{o}{I}_l\dots I_{p_n}) \quad\times\\
&\left[\sum_{k=1}^{l}\Theta(t_{p_1}\dots t_{p_{k-1}},t_l,t_{p_k}\dots \overset{o}{t}_l\dots t_{p_n})+\sum_{k=l+1}^{n}\Theta(t_{p_1}\dots \overset{o}{t}_l\dots t_{p_k},t_l,t_{p_{k+1}}\dots t_{p_n})\right] .
\end{aligned}
$$

For any $l, 1 \leq l \leq n$ and for any $p \in S_{n-1}$ there exists just one $k, 1 \leq k \leq n$, for which the sum over $k$ does not vanish. Thus it is

$$\partial_t T(JI_1 \dots I_n) = T(\dot{J}I_1 \dots I_n) \tag{2.72}$$
$$+ \sum_{p \in S_{n-1}} \mathrm{sgn}(p) \sum_{l=1}^{n} (-1)^{l+1} A_{JI_l} \delta(t - t_l) I_{p_1} \dots \overset{o}{I_l} \dots I_{p_n} \Theta(t_{p_1}, \dots \overset{o}{t_l}, \dots t_{p_n})$$
$$= T(\dot{J}I_1 \dots I_n) - \sum_{l=1}^{n} (-1)^l A_{JI_l} \delta(t - t_l) T(I_{p_1} \dots \overset{o}{I_l} \dots I_{p_n}) \quad .$$

The case $n = 1$ can easily be verified by direct calculation. ◇

Proposition 2.9 enables us to connect the field equations (1.50) with time–ordered products (2.1). Multiplying (1.50) from the right by $\psi_{I_1'} \dots \psi_{I_n'}$ and applying the time–ordering operator $T$ from the left, we obtain by means of (2.63) and with $I' = (Z', x')$

$$T\left[\left(D^{\mu}_{Z_1 Z_2} \partial_\mu - m_{Z_1 Z_2}\right) \psi_{Z_2}(x) \psi_{I_1'} \dots \psi_{I_n'}\right] \tag{2.73}$$
$$= \left(D^{\mu}_{Z_1 Z_2} \partial_\mu - m_{Z_1 Z_2}\right) T \left[\psi_{Z_2}(x) \psi_{I_1'} \dots \psi_{I_n'}\right]$$
$$+ D^0_{Z_1 Z_2} \sum_{l=1}^{n} (-1)^l A_{Z_2 Z_l'} \delta(x - x_l') T(\psi_{I_1'} \dots \overset{o}{\psi}_{I_l'} \dots \psi_{I_n'})$$
$$= U_{Z_1 Z_2 Z_3 Z_4} T \left[\psi_{Z_2}(x) \psi_{Z_3}(x) \psi_{Z_4}(x) \psi_{I_1'} \dots \psi_{I_n'}\right] \quad .$$

The intention of this procedure is to transform equations (1.50) into a system of equations for $\tau^{(n)}$–functions which in turn shall be written as a functional equation. But this intention conflicts with the interaction term of (2.73), where the operator product $\psi_{Z_2}(x)\psi_{Z_3}(x)\psi_{Z_4}(x)$ cannot be expressed in functional space. We circumvent this difficulty by a formal extension of the interaction term to a multi–time expression, cf. Pfister, [Pfi 87].

- **Proposition 2.10:** With the definition of the limit

$$\lim_U := \lim_{\substack{t_1, t_2, t_3 \to t \\ t_1 > t_2 > t_3}} \tag{2.74}$$

and for $t \neq t_i'$, $i = 1, \dots n$ the interaction term of (2.73) can be expressed by

$$\lim_U U_{Z_1 Z_2 Z_3 Z_4} T \left[\psi_{Z_2}(\mathbf{r}, t_1) \psi_{Z_3}(\mathbf{r}, t_2) \psi_{Z_4}(\mathbf{r}, t_3) \psi_{I_1'} \dots \psi_{I_n'}\right] \tag{2.75}$$
$$= U_{Z_1 Z_2 Z_3 Z_4} T \left[\psi_{Z_2}(\mathbf{r}, t) \psi_{Z_3}(\mathbf{r}, t) \psi_{Z_4}(\mathbf{r}, t) \psi_{I_1'} \dots \psi_{I_n'}\right]$$

where $I' = (Z', \mathbf{r}', t')$.

**Proof:** We assume that the time–ordered product $T\left[\psi_I\psi_{I'_1}\dots\psi_{I'_n}\right]$ with $I=(Z,x)$ is well defined. This includes that for any admitted $(n+1)$–tupel $(II'_1\dots I'_n)$ the corresponding time coordinates can be arranged in a sequence with increasing times $t'_{p_1}<\dots t<\dots t'_{p_n}$. Now consider the same set of arguments $t'_{p_1}\dots t'_{p_n}$, but in time ordering with $t_1,t_2,t_3$. For finite differences between $t_1,t_2,t_3$ we obtain an arrangement $t'_{p_1}<\dots t_1<\dots t_2<\dots t_3<\dots t'_{p_n}$ where in the intervals $[t_1,t_2]$ and $[t_2,t_3]$ other times $t'_{p_i}$ may be included. But in the limit $t_1,t_2,t_3\to t$ these intervals are null sequences and this enforces the limit $t'_{p_i}\to t$, which contradicts the assumption $t\neq t'_i$. So no $t'_{p_i}$ can occur in the intervals $[t_1,t_2]$, etc.

Thus in the $U$–limit all terms of the time–ordered product $T\left[\psi_{I_1}\psi_{I_2}\psi_{I_3}\psi_{I'_1}\dots\psi_{I'_n}\right]$ except for the terms containing the expression $\Theta\left(t'_{p_1}\dots t_1,t_2,t_3\dots t'_{p_n}\right)$ vanish. With the definition of the $\Theta$–functions this limit yields (2.75). ◇

By means of this proposition we can derive from equations (2.73) a system of equations for $\tau^{(n)}$–functions:

$$\begin{aligned}&\left(D^{\mu}_{Z_1Z_2}\partial_\mu-m_{Z_1Z_2}\right)\langle 0|T\left[\psi_{Z_2}(x)\psi_{I'_1}\dots\psi_{I'_n}\right]|a\rangle \\ &\quad+D^0_{Z_1Z_2}\sum_{l=1}^{n}(-1)^l A_{Z_2Z'_l}\delta(x-x'_l)\langle 0|T(\psi_{I'_1}\dots\overset{o}{\psi}_{I'_l}\dots\psi_{I'_n})|a\rangle \\ &=\lim_U U_{Z_1Z_2Z_3Z_4}\langle 0|T\left[\psi_{Z_2}(\mathbf{r},t_2)\psi_{Z_3}(\mathbf{r},t_3)\psi_{Z_4}(\mathbf{r},t_4)\psi_{I'_1}\dots\psi_{I'_n}\right]|a\rangle\quad.\end{aligned}\tag{2.76}$$

In a last step we express equations (2.76) in functional space as an equation for functional states (2.17) by means of

- **Proposition 2.11:** The map of the field equations (1.50) into functional space yields the equation

$$\begin{aligned}&\left[-\left(D^{\mu}_{Z_1Z_2}\partial_\mu-m_{Z_1Z_2}\right)\partial_{Z_2}(x)-D^0_{Z_1Z_2}A_{Z_2Z_3}j_{Z_3}(x)\right]|\mathcal{T}(j;a)\rangle \\ &=\lim_U U_{Z_1Z_2Z_3Z_4}\partial_{Z_4}(\mathbf{r},t_4)\partial_{Z_3}(\mathbf{r},t_3)\partial_{Z_2}(\mathbf{r},t_2)|\mathcal{T}(j;a)\rangle\end{aligned}\tag{2.77}$$

  for the determination of the time–ordered functional states (2.17).

**Proof:** We multiply equation (2.76) with $i^n(n!)^{-1}j_{I'_1}\dots j_{I'_n}|0\rangle_F$ and sum over $I'_1\dots I'_n$ and $n$. The first and the second term on the left–hand side of (2.76) can be mapped into functional space by means of formulas (2.18) and (2.19). The interaction term of (2.76) needs a repeated application of (2.18). We define the functional state

$$|\hat{\mathcal{T}}_{K_1}(j;a)\rangle:=i^{-1}\partial_{K_1}|\mathcal{T}(j;a)\rangle\quad.\tag{2.78}$$

For

$$|\hat{\mathcal{T}}_{K_1}(j;a)\rangle=\sum_{n=0}^{\infty}\frac{i^n}{n!}\hat{\tau}^{(n)}_{K_1}(I_1\dots I_n|a)j_{I_1}\dots j_{I_n}|0\rangle_F$$

and $|\mathcal{T}(j;a)\rangle$ given by (2.17) it holds

$$\hat{\tau}^{(n)}_{K_1}(I_1\dots I_n|a)=\tau^{(n+1)}(K_1,I_1\dots I_n|a)\quad.\tag{2.79}$$

By iteration of this procedure one can derive

$$\begin{aligned} & i^{-3}\partial_{K_3}\partial_{K_2}\partial_{K_1}|\mathcal{T}(j;a)\rangle \\ &= \sum_{n=0}^{\infty}\frac{i^n}{n!}\tau^{(n+3)}(K_3,K_2,K_1,I_1\dots I_n|a)j_{I_1}\dots j_{I_n}|0\rangle_F \quad . \end{aligned} \tag{2.80}$$

With (2.80) the interaction term of (2.76) can be transformed into functional space to yield the right–hand side of (2.77). ◊

If (2.20) is applied to (2.77) we can regain equations (2.76).

Equation (2.77) corresponds to the functional version of the Schwinger–Dyson–Freese equations of the spinorfield if $|\mathcal{T}(j;a)\rangle$ is replaced by the state functional $\mathcal{T}(j)$ for $|a\rangle \equiv |0\rangle$. Originally such equations were derived by perturbation theory, see e.g. Lurié [Luri 68], or by a direct calculation of the time derivative of the generating functional, cf. Symanzik [Sym 54], which is a dubious operation. In contrast the derivation by means of Proposition 2.9 and the embedding of these equations into functional space are completely transparent mathematical operations.

It should, however, be emphasized that this nonperturbative derivation is restricted either to a pure fermion or to a pure boson field theory. The definition of time ordering for combined boson–fermion fields which appear in coupling theories is possible, but the derivation of corresponding functional equations beyond perturbation theory encounters considerable difficulties. A nonperturbative treatment of coupling theories will thus be performed in the Hamilton formalism of Chapter 3, cf. also Chapter 7.

Finally we derive a functional energy eigenvalue equation from equation (2.77). Energy equations in the covariant formalism were formally derived by Zimmermann [Zim 54] and Symanzik [Sym 54]. But as we shall see in Chapter 3 the real problem consists in performing the limit to equal times, as it will turn out that in order to give the energy equation a definite meaning it has to be considered in a one–time formalism; see also the model discussion for the covariantly formulated anharmonic oscillator by Maison and Stumpf [Mai 66].

- **Proposition 2.12:** The functional energy eigenvalue equation of the spinor theory reads

$$\begin{aligned} E_{p0}|\mathcal{T}(j;p)\rangle &= \frac{ic}{\hbar}\int d^4x j_{Z_0}(x)D^0_{Z_0Z_1}\Big[(D^k\partial_k - m)_{Z_1Z_2}\partial_{Z_2}(x) \\ &\quad + \lim_{U} U_{Z_1Z_2Z_3Z_4}\partial_{Z_4}(\mathbf{r},t_4)\partial_{Z_3}(\mathbf{r},t_3)\partial_{Z_2}(\mathbf{r},t_2)\Big]|\mathcal{T}(j;p)\rangle \quad , \end{aligned} \tag{2.81}$$

where $E_{p0} := E_p - E_0$ is the renormalized energy eigenvalue with $E_p$ energy value of $|p\rangle$ and $E_0$ energy value of $|0\rangle$.

**Proof:** We combine the constraint equation (2.42) for $\mu = 0$ with equation (2.77). The symmetry condition (2.42) reads for $\mu = 0$

$$i\hbar c\int d^4x j_Z(x)\partial_0(x)\partial_Z(x)|\mathcal{T}(j;p)\rangle = E_{p0}|\mathcal{T}(j;p)\rangle \tag{2.82}$$

if we assume $\mathbf{P}_0|0\rangle = E_0|0\rangle$. Then by construction equation (2.82) explicitly shows the renormalization shift.

We multiply (2.77) from the left by $j_{Z_0}(x)D^0_{Z_0 Z_1}$ and integrate over $x$. Observing $D^0_{Z_0 Z_1} D^0_{Z_1 Z_2} = -\hbar^2 \delta_{Z_0 Z_2}$ we can replace the $\partial_0$–term of (2.77) by (2.82) with the result

$$\begin{aligned} & E_{p0}|T(j;p)\rangle \\ = & \frac{ic}{\hbar}\int d^4x j_{Z_0}(x)\Big[D^0_{Z_0 Z_1}(D^k\partial_k - m)_{Z_1 Z_2}\partial_{Z_2}(x) - \hbar^2 A_{Z_0 Z_1} j_{Z_1}(x) \\ & + \lim_U D^0_{Z_0 Z_1} U_{Z_1 Z_2 Z_3 Z_4}\partial_{Z_4}(\mathbf{r},t_4)\partial_{Z_3}(\mathbf{r},t_3)\partial_{Z_2}(\mathbf{r},t_2)\Big]|T(j;p)\rangle \quad . \end{aligned} \tag{2.83}$$

Furthermore it is

$$j_{Z_0}(x)A_{Z_0 Z_1} j_{Z_1}(x) = -j_{Z_1}(x)A_{Z_1 Z_0} j_{Z_0}(x) = 0 \tag{2.84}$$

due to the properties of the anticommuting sources and the symmetry $A_{ZZ'} = A_{Z'Z}$. Thus the $A$–term in (2.83) vanishes and (2.81) holds. ◇

## 2.5 Nonperturbative Normalordering

In conventional quantum field theory normalordering is an operator rearrangement which enables one to evaluate the $S$–matrix in the interaction picture, i.e. in perturbation theory with free–field representation. In our presentation of quantum field theory we strictly work beyond perturbation theory; in particular we do not fix the representation of the field operator algebra a priori. Therefore, if we introduce "normalordering" in this context, this rearrangement needs a new physical and mathematical interpretation.

It is a peculiarity of quantum field theory and statistical mechanics that the field or system dynamics expressed by the functional states (2.17), etc., contain so–called "uncorrelated" parts. These uncorrelated parts are vacuum expectation values appearing in the decomposition of matrix elements (2.1) by intermediate states. If we try to establish a composite particle theory, such vacuum expectation values prevent a consistent composite particle interpretation and thus partially have to be removed. This elimination serves as the definition of nonperturbative "normalordering".

If, for instance, we consider transition matrix elements $\langle 0|\psi_{I_1}\psi_{I_2}|a\rangle$ as representatives of two–fermion composite particle states, then in the composite particle interpretation these states are mapped onto one–particle states of a phenomenological boson field $\chi(x)$ expressed by the set $\{\langle 0|\chi(x)|b\rangle\}$. But the vacuum expectation values $\langle 0|\psi_{I_1}\psi_{I_2}|0\rangle$ have no counterparts in the set of one–boson states $\langle 0|\chi(x)|b\rangle$ if symmetry breaking is excluded. Thus, in order to provide a bijective map between the original $\psi$–field dynamics and the phenomenological $\chi$–field dynamics, the quantities $\langle 0|\psi_{I_1}\psi_{I_2}|0\rangle$ have to be removed from the original $\psi$–field theory.

On the other hand, higher vacuum expectation values like $\langle 0|\psi_{I_1}\psi_{I_2}\psi_{I_3}\psi_{I_4}|0\rangle$ find their counterparts in terms like $\langle 0|\chi_{k_1}\chi_{k_2}|0\rangle$ and therefore need not be eliminated from the original $\psi$–field theory. This example shows that nonperturbative normalordering depends on the kind of composite particle dynamics under consideration.

In the following we discuss the elimination of two–point functions $\langle 0|T(\psi_{I_1}\psi_{I_2})|0\rangle$ from the original $\psi$–field theory. The corresponding elimination procedure by normalordering is in accordance with the general structure of matrix element decomposition by intermediate states and holds independently of perturbation theory. However, it is not restricted to our spinorfield theory, and in general it allows the derivation of an effective dynamics for two– and three–particle bound states. In high energy physics these are the configurations which are of the utmost physical interest.

We introduce normalordering for two–point functions by the transformation

$$|\mathcal{F}(j;a)\rangle := \exp\,(\frac{1}{2}j_{I_1}F_{I_1I_2}j_{I_2})|\mathcal{T}(j;a)\rangle \tag{2.85}$$

with the exponential function defined by its series and $F_{I_1I_2}$ given by

$$F_{I_1I_2} \equiv F_{Z_1Z_2}(x_1,x_2) := \langle 0|T\left[\psi_{Z_1}(x_1)\psi_{Z_2}(x_2)\right]|0\rangle \quad . \tag{2.86}$$

With (2.17) and the power series expansion of the exponential it follows that $|\mathcal{F}(j;a)\rangle$ can be represented in the form

$$|\mathcal{F}(j;a)\rangle = \sum_n \frac{i^n}{n!}\varphi^{(n)}(I_1\dots I_n|a)j_{I_1}\dots j_{I_n}|0\rangle_F \quad , \tag{2.87}$$

where the $\varphi^{(n)}$–functions are completely antisymmetric in their arguments $I_1\dots I_n$.

For the special case $\psi_I \equiv \psi_I^{\text{int}}$, i.e. in the interaction picture with free–field representation, $F_{I_1I_2}$ is the free–fermion propagator and (2.85) goes over into the functional formulation of the Wick rule, see Rampacher, Stumpf and Wagner [Ram 65]. But, independently of any representation, (2.87) is a functional state from which all propagator parts (2.86) are extracted. This can be explicitly seen by projecting (2.85) in coordinate space according to (2.20). Thus the remaining four–point functions etc. do not contain the propagator parts. We introduce the notation

$$\varphi^{(n)}(I_1\dots I_n|a) \equiv \langle 0|N(\psi_{I_1}\dots\psi_{I_n})|a\rangle \tag{2.88}$$

for all states $|a\rangle \in \mathbb{H}$ with inclusion of the vacuum state $|0\rangle$, where $N$ indicates that all two–point functions are extracted from the original time–ordered product $T(\psi_{I_1}\dots\psi_{I_n})$.

In consequence of the normalordering of functional states it is necessary to also transform the corresponding functional equations and the additional constraints. This can be achieved by means of the following proposition:

- **Proposition 2.13:** The commutator of the (inverse) normal–transformation (2.85) with the functional differential operator $\partial_K$ is given by

$$\left[\partial_K, \exp(-\frac{1}{2}j_{I_1}F_{I_1I_2}j_{I_2})\right]_- = -\exp\,(-\frac{1}{2}j_{I_1}F_{I_1I_2}j_{I_2})F_{KI_3}j_{I_3} \quad . \tag{2.89}$$

**Proof:** Application of (2.11) yields

$$\partial_K \exp(-\frac{1}{2} j_{I_1} F_{I_1 I_2} j_{I_2}) \tag{2.90}$$
$$= \sum_{n=0}^{\infty} \frac{(-1)^n}{n! 2^n} F_{I_1 I_2} F_{I_3 I_4} \dots F_{I_{2-1} I_{2n}} \partial_K j_{I_1} \dots j_{I_{2n}}$$
$$= \sum_{n=0}^{\infty} \frac{(-1)^n}{n! 2^n} F_{I_1 I_2} \dots F_{I_{2-1} I_{2n}} \left[ j_{I_1} \dots j_{I_{2n}} \partial_K + \sum_{l=1}^{2n} (-1)^{l+1} \delta_{I_l K} j_{I_1} \dots \overset{o}{j}_{I_l} \dots j_{I_{2n}} \right]$$
$$= \exp(-\frac{1}{2} j_{I_1} F_{I_1 I_2} j_{I_2}) \partial_K$$
$$+ \sum_{n=0}^{\infty} \frac{(-1)^n}{n! 2^n} F_{I_1 I_2} \dots F_{I_{2-1} I_{2n}} \left[ \sum_{l=1}^{n} (-1)^{2l} \delta_{K I_{2l-1}} j_{I_1} \dots \overset{o}{j}_{I_{2l-1}} \dots j_{I_{2n}} \right.$$
$$\left. + \sum_{l=1}^{n} (-1)^{2l+1} \delta_{K I_{2l}} j_{I_1} \dots \overset{o}{j}_{I_{2l}} \dots j_{I_{2n}} \right]$$
$$= \exp(-\frac{1}{2} j_{I_1} F_{I_1 I_2} j_{I_2}) \partial_K$$
$$+ \sum_{n=0}^{\infty} \frac{(-1)^n}{n! 2^n} \sum_{l=1}^{n} j_{I_1} F_{I_1 I_2} j_{I_2} j_{I_3} F_{I_3 I_4} j_{I_4} \dots j_{I_{2l-2}} F_{K I_{2l}} j_{I_{2l}} \dots j_{I_{2n-1}} F_{I_{2n-1} I_{2n}} j_{I_{2n}}$$
$$+ \sum_{n=0}^{\infty} \frac{(-1)^n}{n! 2^n} \sum_{l=1}^{n} j_{I_1} F_{I_1 I_2} \dots j_{I_{2l-2}} F_{K I_{2l-1}} j_{I_{2l-1}} j_{I_{2l+1}} F_{I_{2l+1} I_{2l+2}} j_{I_{2l+2}} \dots j_{I_{2n}} \quad ,$$

where in the last step the antisymmetry of $F_{I_1 I_2}$ was used.

By reindexing $(2l - 1 \rightarrow 2l)$ of the last term, interchanging of $F_{K I_{2l}} j_{I_{2l}}$ and renewed indexing one obtains from (2.90)

$$\partial_K \exp(-\frac{1}{2} j_{I_1} F_{I_1 I_2} j_{I_2}) \tag{2.91}$$
$$= \exp(-\frac{1}{2} j_{I_1} F_{I_1 I_2} j_{I_2}) \partial_K$$
$$+ \sum_{n=0}^{\infty} \frac{(-1)^n}{n! 2^n} 2 \sum_{l=1}^{n} j_{I_1} F_{I_1 I_2} j_{I_2} \dots j_{I_{2l-2}} F_{K I_{2l}} j_{I_{2l}} j_{I_{2l+1}} \dots F_{I_{2n-1} I_{2n}} j_{I_{2n}}$$
$$= \exp(-\frac{1}{2} j_{I_1} F_{I_1 I_2} j_{I_2}) \partial_K$$
$$+ \sum_{n=0}^{\infty} \frac{(-1)^n}{n! 2^{n-1}} n j_{I_1} F_{I_1 I_2} j_{I_2} \dots j_{I_{2n-3}} F_{I_{2n-3} I_{2n-2}} j_{I_{2n-2}} F_{K K'} j_{K'}$$
$$= \exp(-\frac{1}{2} j_{I_1} F_{I_1 I_2} j_{I_2}) (\partial_K - F_{K K'} j_{K'}) \quad .$$

◇

By means of Proposition 2.13 we can derive the normaltransformed functional equation (2.77).

- **Proposition 2.14:** The normaltransformed functional equation (2.77) reads

$$\left[ (D^\mu \partial_\mu - m)_{Z_1 Z_2} d_{Z_2}(x) + D^0_{Z_1 Z_2} A_{Z_2 Z_3} j_{Z_3}(x) \right] |\mathcal{F}(j;a)\rangle \tag{2.92}$$
$$= -\lim_U U_{Z_1 Z_2 Z_3 Z_4} d_{Z_4}(\mathbf{r}, t_4) d_{Z_3}(\mathbf{r}, t_3) d_{Z_2}(\mathbf{r}, t_2) |\mathcal{F}(j;a)\rangle$$

with

$$d_Z(x) \equiv d_I := \partial_I - F_{IK} j_K \quad . \tag{2.93}$$

**Proof:** The relation $e^{-z}e^z = 1$ can be proven by a power series expansion of the exponentials and thus it holds also for $z = j_{I_1} F_{I_1 I_2} j_{I_2}$. Therefore relation (2.85) can be resolved with respect to $|\mathcal{T}(j;a)\rangle$ and equivalently reads

$$|\mathcal{T}(j;a)\rangle = \exp(-\frac{1}{2} j_{I_1} F_{I_1 I_2} j_{I_2}) |\mathcal{F}(j;a)\rangle \quad . \tag{2.94}$$

We substitute (2.94) into equation (2.77) and interchange $\exp(-\frac{1}{2} j_{I_1} F_{I_1 I_2} j_{I_2})$ to the left–hand side of the equation. This can be achieved by use of (2.89) and by means of

$$\left[ j_K, \exp(-\frac{1}{2} j_{I_1} F_{I_1 I_2} j_{I_2}) \right]_- = 0 \quad . \tag{2.95}$$

The resulting equation is given by

$$\begin{aligned}
&\exp(-\frac{1}{2} j_{K_1} F_{K_1 K_2} j_{K_2}) \times \\
&\quad \left[ (D^\mu \partial_\mu - m)_{Z_1 Z_2} d_{Z_2}(x) + D^0_{Z_1 Z_2} A_{Z_2 Z_3} j_{Z_3}(x) \right] |\mathcal{F}(j;a)\rangle \\
= &\quad - \exp(-\frac{1}{2} j_{K_1} F_{K_1 K_2} j_{K_2}) \times \\
&\quad \lim_U U_{Z_1 Z_2 Z_3 Z_4} d_{Z_4}(\mathbf{r}, t_4) d_{Z_3}(\mathbf{r}, t_3) d_{Z_2}(\mathbf{r}, t_2) |\mathcal{F}(j;a)\rangle \quad ,
\end{aligned} \tag{2.96}$$

and as the exponential function is nondegenerate, equation (2.92) must hold. ◇

We add that in (2.92) the $\lim_U$ has to be slightly modified in order to provide for a well–defined limit $t_2, t_2, t_4 \to t$ in all expressions contained in the vertex term of (2.92). In the same way we can transform the symmetry constraints:

- **Proposition 2.15:** The constraint equations (2.42)–(2.45) are invariant under normal–transformation, i.e. it holds

$$\begin{aligned}
\mathcal{P}_\mu |\mathcal{F}(j;p)\rangle &= p_\mu |\mathcal{F}(j;p)\rangle \, , & (2.97)\\
\mathcal{P}_\mu \mathcal{P}^\mu |\mathcal{F}(j;p)\rangle &= (mc)^2 |\mathcal{F}(j;p)\rangle \, , & (2.98)\\
\mathcal{W}_\mu \mathcal{W}^\mu |\mathcal{F}(j;p)\rangle &= \hbar^2 s(s+1) |\mathcal{F}(j;p)\rangle \, , & (2.99)\\
\mathcal{W}_3 |\mathcal{F}(j;p)\rangle &= \hbar s_3 |\mathcal{F}(j;p)\rangle \, , & (2.100)
\end{aligned}$$

  where $\mathcal{P}_\mu$ and $\mathcal{W}_\mu$ are defined by (2.46) and (2.47).

**Proof:** Substitution of (2.94) into (2.42) gives

$$\mathcal{P}_\mu \exp(-\frac{1}{2} j_{I_1} F_{I_1 I_2} j_{I_2}) |\mathcal{F}(j;p)\rangle = \exp(-\frac{1}{2} j_{I_1} F_{I_1 I_2} j_{I_2}) p_\mu |\mathcal{F}(j;p)\rangle \quad . \tag{2.101}$$

Observing (2.46) and using (2.89) and (2.95) we can commute the exponential function with $\mathcal{P}_\mu$ and obtain from (2.101)

$$p_\mu|\mathcal{F}(j;p)\rangle \tag{2.102}$$

$$= -\int d^4x_1 d^4x_2 j_{Z_1}(x_1) P_\mu(x_1) \Big[\partial_{Z_1}(x_1) - F_{Z_1 Z_2}(x_1,x_2) j_{Z_2}(x_2)\Big] |\mathcal{F}(j;p)\rangle \quad . \tag{2.103}$$

Due to the translational invariance of $F_{I_1 I_2}$ it is

$$F_{Z_1 Z_2}(x_1,x_2) = F_{Z_1 Z_2}(x_1 - x_2) \tag{2.104}$$

and therefore

$$\partial_\mu(x_1) F_{Z_1 Z_2}(x_1,x_2) = -\partial_\mu(x_2) F_{Z_1 Z_2}(x_1,x_2) \quad . \tag{2.105}$$

Furthermore, $F_{I_1 I_2}$ is antisymmetric, i.e.

$$F_{Z_1 Z_2}(x_1,x_2) = -F_{Z_2 Z_1}(x_2,x_1) \quad . \tag{2.106}$$

Combining these properties and commuting the $j_I$–sources, we obtain

$$\int d^4x_1 d^4x_2 j_{Z_1}(x_1) \partial_\mu(x_1) F_{Z_1 Z_2}(x_1,x_2) j_{Z_2}(x_2) \tag{2.107}$$

$$= -\int d^4x_1 d^4x_2 j_{Z_2}(x_2) \partial_\mu(x_2) F_{Z_2 Z_1}(x_2,x_1) j_{Z_1}(x_1) \quad .$$

If we interchange $(Z_1, x_1)$ and $(Z_2, x_2)$ on the right–hand side of (2.107), the right–hand side equals the negative of the left–hand side and thus has to vanish. Hence from (2.102) it follows (2.97). By the same arguments (2.98), (2.99) and (2.100) can be proven. ◇

Equation (2.92) can be simplified if for $F_{I_1 I_2}$ the free fermion propagator is used. This choice corresponds to the definition of composite particles by hard core states and will be discussed in Section 4.3. Here we only draw formal conclusions. The free fermion propagator is defined by

$$F^f_{I_1 I_2} := \langle 0|T\left[\psi^f_{Z_1}(x_1)\psi^f_{Z_2}(x_2)\right]|0\rangle \quad , \tag{2.108}$$

where $\psi^f$ is a solution of the fermion equation

$$\left(D^\mu_{Z_1 Z_2}\partial_\mu - m_{Z_1 Z_2}\right)\psi^f_{Z_2}(x) = 0 \quad . \tag{2.109}$$

As we work with superspinors, we cannot directly take over the propagator equation for Dirac spinors. We thus give a derivation of the properties of (2.108).

- **Proposition 2.16:** The free superspinor propagator $F^f_{I_1 I_2}$ satisfies the equation

$$[D^\mu \partial_\mu(x_1) - m]_{Z_0 Z_1} F^f_{Z_1 Z_2}(x_1,x_2) = D^0_{Z_0 Z_1} A_{Z_1 Z_2}\delta(x_1 - x_2) \tag{2.110}$$

with

$$\left[\psi^f_{Z_1}(x_1), \psi^f_{Z_2}(x_2)\right]_+^{t_1=t_2} = A_{Z_1 Z_2}\delta(\mathbf{r}_1 - \mathbf{r}_2) \tag{2.111}$$

and $A_{Z_1 Z_2}$ given by (1.54).

**Proof:** With (2.63) it is

$$\partial_{t_1} T\left[\psi^f_{Z_1}(x_1)\psi^f_{Z_2}(x_2)\right] = T\left[\dot{\psi}^f_{Z_1}(x_1)\psi^f_{Z_2}(x_2)\right] + A_{Z_1Z_2}\delta(x_1 - x_2) \quad . \tag{2.112}$$

Multiplication of (2.112) from the left with $D^0_{Z_0Z_1}$ and use of (2.109) yields

$$\begin{aligned} & D^0_{Z_0Z_1}\partial_{t_1} T\left[\psi^f_{Z_1}(x_1)\psi^f_{Z_2}(x_2)\right] \qquad (2.113)\\ = \; & T\left[\left(-D^k\partial_k(x_1) + m\right)_{Z_0Z_1}\psi^f_{Z_1}(x_1)\psi^f_{Z_2}(x_2)\right] + D^0_{Z_0Z_1}A_{Z_1Z_2}\delta(x_1 - x_2) \quad . \end{aligned}$$

Formation of the vacuum expectation value of (2.113) yields (2.110). ◊

Hence, choosing $F_{I_1I_2} \equiv F^f_{I_1I_2}$ and taking into account (2.110) we get for equation (2.92)

$$\begin{aligned} & (D^\mu\partial_\mu - m)_{Z_1Z_2}\partial_{Z_2}(x)|\mathcal{F}(j;a)\rangle \qquad (2.114)\\ = \; & -\lim_U U_{Z_1Z_2Z_3Z_4} d_{Z_4}(\mathbf{r}, t_4) d_{Z_3}(\mathbf{r}, t_3) d_{Z_2}(\mathbf{r}, t_2)|\mathcal{F}(j;a)\rangle \quad . \end{aligned}$$

Resolution of (2.114) with respect to the time derivative by multiplication with $D^0_{Z_0Z_1}$, subsequent multiplication with $j_{Z_0}(x)$ and integration over $x$ in combination with (2.97) for $\mu = 0$ gives the normal–transformed energy equation

$$\begin{aligned} E_{p0}|F(j;p)\rangle = \; & \frac{ic}{\hbar}\int d^4x j_{Z_0}(x) D^0_{Z_0Z_1}\Big[(D^k\partial_k - m)_{Z_1Z_2}\partial_{Z_2}(x) \qquad (2.115)\\ & + \lim_U U_{Z_1Z_2Z_3Z_4} d_{Z_4}(\mathbf{r}, t_4) d_{Z_3}(\mathbf{r}, t_3) d_{Z_2}(\mathbf{r}, t_2)\Big]|\mathcal{F}(j;p)\rangle \quad . \end{aligned}$$

However, equation (2.92) as well as equations (2.114) and (2.115) need a further discussion, as in general the introduction of $F$–functions exposes singularities in the vertex.

For later applications we explicitly give a solution of (2.110) by

$$\begin{aligned} & F_{Z_1Z_2}(x_1x_2) = \qquad (2.116)\\ & -i\lambda_{i_1}\delta_{i_1i_2}(\gamma_5)_{\kappa_1\kappa_2}\frac{1}{(2\pi)^4}\int d^4p \exp[ip(x_1 - x_2)]\left(\frac{\gamma^\mu p_\mu - m_{i_1}}{m^2_{i_1} - i\epsilon - p^2}C\right)_{\alpha_1\alpha_2} \quad . \end{aligned}$$

The use of these Feynman propagators is correlated with the assumption of the Dirac vacuum renormalization, i.e. with a subfermionic analogue of the electron–positron theory, see Schweber, Bethe and de Hoffmann [Schwe 55]. Thus with respect to our calculation of bound states (cf. Chapter 5) the positivity of energy is secured and the subfermions $\psi$ are referred to the renormalized Dirac vacuum.

## 2.6 Vertex Renormalization

To reveal the difficulties which are connected with the definition of the interaction term in (2.114) and (2.115) we explicitly evaluate it. As it stands, in this term — due to the definiton (2.93) of $d_K$ — the sources $j_I$ and their duals $\partial_K$ appear in mixed polynomial expressions. We disentangle these terms by giving them a "functional normalordered" form with all duals shifted to the right–hand side of the sources. This gives

$$
\begin{aligned}
d_{I_4} d_{I_3} d_{I_2} \quad = \quad & \partial_{I_4} \partial_{I_3} \partial_{I_2} \\
& - F_{I_2 I_3} \partial_{I_4} + F_{I_2 I_4} \partial_{I_3} - F_{I_3 I_4} \partial_{I_2} \\
& - F_{I_2 J_2} j_{J_2} \partial_{I_4} \partial_{I_3} - F_{I_3 J_3} j_{J_3} \partial_{I_2} \partial_{I_4} - F_{I_4 J_4} j_{J_4} \partial_{I_3} \partial_{I_2} \\
& + F_{I_2 I_3} F_{I_4 J_4} j_{J_4} + F_{I_3 I_4} F_{I_2 J_2} j_{J_2} - F_{I_2 I_4} F_{I_3 J_3} j_{J_3} \\
& + F_{I_4 J_4} F_{I_2 J_2} j_{J_2} j_{J_4} \partial_{I_3} + F_{I_3 J_3} F_{I_2 J_2} j_{J_3} j_{J_2} \partial_{I_4} + F_{I_4 J_4} F_{I_3 J_3} j_{J_4} j_{J_3} \partial_{I_2} \\
& - F_{I_4 J_4} F_{I_3 J_3} F_{I_2 J_2} j_{J_4} j_{J_3} j_{J_2} \quad .
\end{aligned}
\tag{2.117}
$$

We consider the second and the fourth line of (2.117). If these lines are inserted in the vertex term of (2.114) and (or) (2.115) the $\lim_U$, together with the spatial locality of the vertex, leads to terms with $F_{Z_2 Z_3}(0)$, $F_{Z_2 Z_4}(0)$ and $F_{Z_3 Z_4}(0)$. At the origin these propagators are singular terms and it is necessary, or at least convenient, to remove these terms by additive renormalization. This additive renormalization is commonly referred to as vertex renormalization, and it is defined by

$$
\begin{aligned}
& : d_{Z_4}(x_4) d_{Z_3}(x_3) d_{Z_2}(x_2) : \\
& := \quad d_{Z_4}(x_4) d_{Z_3}(x_3) d_{Z_2}(x_2) + F_{Z_3 Z_4}(x_3, x_4) d_{Z_2}(x_2) \\
& \qquad - F_{Z_2 Z_4}(x_2, x_4) d_{Z_3}(x_2) + F_{Z_2 Z_3}(x_2, x_3) d_{Z_4}(x_4) \quad .
\end{aligned}
\tag{2.118}
$$

If this operation is applied to (2.117), the unwanted terms of (2.117) are compensated by the additive renormalization terms of (2.118) and thus drop out.

The vertex renormalization can be transferred to the dynamical equation (1.50) itself and leads to

$$
(D^\mu \partial_\mu - m)_{Z_1 Z_2} \psi_{Z_2}(x) = U_{Z_1 Z_2 Z_3 Z_4} : \psi_{Z_2}(x) \psi_{Z_3}(x) \psi_{Z_4}(x) : \tag{2.119}
$$

with

$$
\begin{aligned}
: \psi_{Z_2}(x) \psi_{Z_3}(x) \psi_{Z_4}(x) : \quad := \quad & \psi_{Z_2}(x) \psi_{Z_3}(x) \psi_{Z_4}(x) + F_{Z_3 Z_4}(0) \psi_{Z_2}(x) \\
& - F_{Z_2 Z_4}(0) \psi_{Z_3}(x) + F_{Z_2 Z_3}(0) \psi_{Z_4}(x) \quad .
\end{aligned}
\tag{2.120}
$$

If we had started our formulation of the quantum field dynamics in functional space with (2.119) we would have arrived at the corresponding functional equations (2.114) and (2.115) with the renormalized vertex (2.118). For instance, we would have obtained for (2.115)

$$
\begin{aligned}
E_{p0} | F(j;p) \rangle \quad = \quad & \frac{ic}{\hbar} \int d^4 x j_{Z_0}(x) D^0_{Z_0 Z_1} \Big[ (D^k \partial_k - m)_{Z_1 Z_2} \partial_{Z_2}(x) \\
& + \lim_U U_{Z_1 Z_2 Z_3 Z_4} : d_{Z_4}(\mathbf{r}, t_4) d_{Z_3}(\mathbf{r}, t_3) d_{Z_2}(\mathbf{r}, t_2) : \Big] | \mathcal{F}(j;p) \rangle \quad .
\end{aligned}
\tag{2.121}
$$

The latter equation is used for a further investigation in Chapter 3.

## 2.7 Limits of Covariant Formalism

Time–ordered products (2.1) and normalordering of time–ordered products originally stem from perturbation theory. In perturbation theory the representation of the field operator algebra is fixed to be the interaction picture with free–field representation, and as its evaluation is carried out according to fixed rules, there is little room for further investigations. Beyond perturbation theory the situation is quite different: the genuine field representation is influenced, or even completely determined, by the field dynamics and its constraints themselves, and one needs the full apparatus of formal quantum theory for the investigation of nonperturbative phenomena. Thus one is faced with the question: Is the covariant formalism the suitable starting point for such investigations?

Apart from the covariant formalism formal quantum theory is governed by the concept of state evolution in time. This means algebraically that the algebra of a quantum system is completely determined already on a spacelike hyperplane and that time evolution is a continuous group whose action generates a sequence of inner automorphisms of the algebra in the course of time. With respect to representations this means that the corresponding quantum state space must be complete on any spacelike hyperplane. In a preliminary way the projection states $\langle 0|T(\psi_{I_1}\dots\psi_{I_n})$ in (2.1) were referred to as basis states of the covariant representation of a state vector $|a\rangle$. These states are assumed to exist for all values of their time coordinates $t_1 \neq \dots \neq t_n$. However, the aspect of time evolution gives rise to doubts about this concept because already for any fixed $t = t_1 = \dots = t_n$ there must exist a complete set of basis vectors. Hence we are forced to suppose that the set of time–ordered "basis" states $\langle 0|T(\psi_{I_1}\dots\psi_{I_n})$ is overcomplete.

Mathematically overcompleteness is an unpleasant property because it prevents the application of non–ambigous orthonormality and completeness relations, and with respect to the time–ordered basis states their overcompleteness mixes past and future in an obscure way which admits physical interpretation only by asymptotic behaviour of the $S$–matrix.

Apart from this difficulty the time–ordered products are unpleasant in yet another respect. If one wants to perform explicit calculations one is faced with the difficulty that these products are themselves products of numerical and operator distributions which, for instance, are not well defined for equal times, see Bogoliubov, Logunov and Todorov [Bogo 75]. As a consequence the Schwinger–Dyson–Freese equations cannot be considered as differential equations for continuous functions. These difficulties already occur in field theoretic perturbation theory and enforce a very careful distributional analysis, cf. Scharf [Schar 89]. But while in perturbation theory one only has to deal with given distributions, in the general case of nonperturbative phenomena one would have to solve Schwinger–Dyson–Freese equations by a careful distributional analysis for yet unknown distributions — at present a nearly unsurmountable task.

As a way out of these difficulties path integrals were derived as formal solutions of the Schwinger–Dyson–Freese equations. At present, path integrals are commonly considered as an ideal means for the treatment of quantum field theories. However, a closer inspection reveals many weak points concerning the definition and evaluation of such integrals. In particular their evaluation with respect to composite particle dynamics shows deficiencies which will be discussed in Chapter 10.

Furthermore in Section 2.4 we have already drawn the attention to the difficulties in connection with the covariant functional formulation of a boson–fermion coupling theory.

Summarizing this discussion we conclude that there are compelling reasons to sacrify explicit covariance and to consider the covariant formalism not as a suitable starting point for the treatment of nonperturbative phenomena requiring thorough attention of the algebraic features of quantum theory. Instead of the covariant formalism we shall prepare "algebraic" Hamilton dynamics for the derivation of effective composite particle theories, and finally we will try to re–establish explicit covariance as far as possible. In this connection we refer to papers which relate noncovariant gauges with covariant ones, see, e.g., Bleuler [Bleu 50], or which discuss this question with respect to the $S$–matrix, cf. Halpern [Halp 67]; see also the discussion in Section 8.1. The covariant formalism will be used in the course of these investigations in a very limited way.

# 3 Algebraic Schrödinger Representation

## 3.1 Introduction

In the preceding chapter we pointed out our aim to develop a functional quantum theory which allows a successful treatment of quantum fields and which is governed by two ideas: that of taking into account inequivalent representations and that of explicit state space construction. To this end we proposed a map from a state space $\mathbb{H}$ onto a functional space $\mathbb{K}_F$ which has to leave metric and quantum numbers invariant.

In Chapter 2 we discussed such a map within a covariant formalism. With respect to the quantum numbers we succeeded in constructing a functional space which carries a representation of the Poincaré group and in formulating symmetry conditions in this functional space. But with respect to the construction of the state space we found the assumed time–ordered basis states $T(\psi_{I_1} \dots \psi_{I_n})|0\rangle$ to be overcomplete. This prevented us from examining in detail the metrical structure of the state space. As the overcompleteness stems from the fact that this choice of the basis states ignores the concept of causal time evolution, it suggests itself to abandon explicit covariance and to look for a Hamilton formalism where time–ordered, i.e. multi–time basis vectors are avoided in favour of a one–time formulation. We follow this line and start with a more detailed analysis of the state spaces of quantum fields.

Compared with quantum mechanics, for quantum fields the discussion of the state space, i.e. the representation space of the corresponding field operator algebra, is more complicated in a twofold way: On the one hand quantum fields as systems with an infinite number of degrees of freedom possess an infinity of inequivalent representations, see Friedrichs [Frie 53], Bratteli and Robinson [Brat 79], Haag [Haag 92]. On the other hand, nearly all interesting physical field theories, as, for instance, gauge theories, gravitation theories or the super–renormalizable spinor theory of Chapter 1, enforce the introduction of indefinite state spaces which require a careful analysis in order to guarantee the probability interpretation of the theory.

In the subsequent discussion of a Hamilton formalism with respect to inequivalent representations we will follow a general line of algebraic quantum theory, and guided by the GNS–construction we consider antisymmetrized basis states $\mathcal{A}(\psi_{I_1} \dots \psi_{I_n})_t|0\rangle$, where the arguments $I_1 \dots I_n$ are taken for equal times $t_1 = \dots = t_n = t$. With respect to the indefiniteness of the representation spaces we assume that these spaces are Krein spaces, see Bognar [Bogn 74].

This assumption is in accordance with the evaluation of axiomatic quantum field theory in indefinite metric spaces. In order to enlarge the formalism to include such theories, in particular gauge theories, Morchio and Strocchi [Mor 80] abandoned the positive definiteness of the Wightman functions, i.e. of algebraic states. Within this context Jakobczyk [Jak 84] proved that (algebraic) Strocchi–Wightman states lead to a reconstruction theorem in Krein–space and Jakobczyk and Strocchi [Jak 88] studied the Euclidean continuation of Strocchi–Wightman states in Krein space. Krein spaces exclude the more difficult cases of dipole fields (ghosts) or dipole regularizations which, however, were also studied by some authors, e.g.

by Moschella and Strocchi [Mos 90]. A group theoretical and algebraic classification of the various possibilities of indefinite metric was aimed at by Saller [Sall 89,91,92,93].

In conventional algebraic theory, cf., e.g., Bratteli and Robinson [Brat 79], as well as in original axiomatic quantum field theory, cf. Völkel [Völ 77], Glimm and Jaffe [Gli 87], Haag [Haag 92], one deals only with positive definite metrical spaces, i.e. with Hilbert spaces. The extension of these treatments of quantum fields with positive metric to the case of quantum fields with indefinite metric was performed by the above–mentioned authors and led to general statements about the structure of such theories. However, as with Hilbert space theories also in this case these efforts did not lead to a final result allowing a straightforward treatment of specific realistic models. Hence we shall be content to take over general features of both approaches, in particular of algebraic representation theory to the problem of constructing a suitable Hamilton formalism and develop our own approach which is directly concerned with the foundation of a composite particle dynamics.

In the following we will give a more detailed analysis of the metrical structure of generalized state spaces and we will demonstrate that a one–time formalism is the suitable framework to formulate our functional theory. As this formalism yields the field theoretic analogue to ordinary Schrödinger states we may call it Algebraic Schrödinger Representation.

The essential of the Algebraic Schrödinger Representation is the explicit state space construction which is, however, different from the usual approach in quantum mechanics. Projection of the above–mentioned basis states onto physical states $|a\rangle$ leads to transition matrix elements, and these matrix elements are considered as representatives of this physical state $|a\rangle$. Thus, for the construction of the state space one has to work with these quantities.

As far as the calculation of transition matrix elements for equal times is concerned, the New Tamm–Dancoff method was proposed by Dyson [Dys 53a,b]. Dyson replaced the quantum mechanical Schrödinger equation by Heisenberg's time evolution equation for products of field operators, and projecting with states from both sides of this equation leads to equations for transition matrix elements. But only an insufficient attempt was made to derive an appropriate normalization of these transition matrix elements, and no attempt at all to formulate a corresponding theory of states. Apart from some applications of the one–time New Tamm–Dancoff method in solid state physics, cf. Wahl, Duscher, Göbel and Maichle [Wahl 84], and in nuclear physics, cf. Friedrich, Gerling and Bleuler [Frie 75], Friedrich and Gerling [Frie 76], Swift and Marrero [Swi 84], Fetter and Walecka [Fett 71], this formalism was not further pursued owing to difficulties with renormalization, see Bogoliubov and Shirkov [Bogo 59]. Hence there was no further progress with respect to its relation to an explicit state space construction.

Apart from Heisenberg's fundamental evolution equation the treatment given in the following has nothing in common with the New Tamm–Dancoff method. Rather this treatment is the fieldtheoretic counterpart of quantum mechanical state representations. However, it will be seen in the following that the transfer of the method of state representations from quantum mechanics to quantum field theory is absolutely nontrivial and leads to new specific problems which are connected with the infinite number of degrees of freedom of a quantum field.

## 3.2 Indefinite State Spaces

In general a quantum theory is assumed to possess a representation in a Hilbert space, i.e. a vector space with a positive definite metric. As already mentioned in the introduction there are, however, prominent examples of theories which force us to enlarge the concept of Hilbert space.

The properties of representations and corresponding representation spaces of CAR– or CCR– algebras, respectively, depend on the sign of the anticommutators or commutators defining the algebra. If all basic anticommutators or commutators have the same sign, then there exist representation spaces with positive definite metric. If these quantities have alternating signs, then the corresponding representation spaces are indefinite. For instance, vector fields in covariant quantization possess basic commutators with alternating sign due to their dependence on the metrical tensor $\eta_{\mu\nu}$. Another example are the spinorial anticommutators (1.41), and, as announced in the introduction, we will just treat these kinds of systems. Hence the state spaces under consideration must be vector spaces $\mathbb{H}$ with indefinite metric, i.e. vector spaces with a scalarproduct $\langle\cdot,\cdot\rangle : \mathbb{H}\times\mathbb{H}\to\mathbb{C}$, and with the properties

i) $\langle\alpha_1 x_1+\alpha_2 x_2, y\rangle = \alpha_1\langle x_1, y\rangle + \alpha_2\langle x_2, y\rangle$ for all $\alpha_1,\alpha_2\in\mathbb{C},\ x_1,x_2,y\in\mathbb{H}$,

ii) $\langle y, x\rangle = \langle x, y\rangle^*$ for all $x,y\in\mathbb{H}$,

but with $\langle x, x,\rangle \leq 0$ for at least one $x\in\mathbb{H}$.

Furthermore to be able to make any nontrivial statements about the state space, in particular to introduce any basis in $\mathbb{H}$, we assume $\mathbb{H}$ to be a Krein space, i.e., $\mathbb{H}$ is nondegenerate and has the decomposition

$$\mathbb{H} = \mathbb{H}_+ \oplus \mathbb{H}_- \tag{3.1}$$

with $\mathbb{H}_+ \subset P^{++}$, $\mathbb{H}_- \subset P^{--}$, where

$$P^{++} := \{x\in\mathbb{H}\,, \langle x,x\rangle > 0 \text{ or } x=0\} \quad ,$$

$$P^{--} := \{x\in\mathbb{H}\,, \langle x,x\rangle < 0 \text{ or } x=0\} \quad ,$$

and where $\mathbb{H}_+$ and $\mathbb{H}_-$ are assumed to be intrinsic complete, while $\oplus$ denotes the orthogonal direct sum. The decomposition is referred to as fundamental decomposition.

We shortly explain the notions used in the definition of the Krein space: $\mathbb{H}$ is nondegenerate if for $\langle x,x\rangle = 0$ it is $x=0$. Intrinsic completeness of the spaces $\mathbb{H}_+$ and $\mathbb{H}_-$ means the completeness with respect to their topologies which are generated by $\| x \|_+ := \langle x,x\rangle^{\frac{1}{2}}$ for $\mathbb{H}_+$ and $\| x \|_- := |\langle x,x\rangle|^{\frac{1}{2}}$ for $\mathbb{H}_-$ and for all $x\in\mathbb{H}_+$ or all $x\in\mathbb{H}_-$, respectively. A metrical space is complete if any Cauchy sequence of elements in the space has a limit which is contained in this space. The direct orthogonal sum implies that for all $x\in\mathbb{H}_+$ and for all $y\in\mathbb{H}_-$ it is $\langle x,y\rangle = 0$.

We assume that the Krein spaces we are dealing with are spanned by a set of generalized eigenstates for a set of commuting selfadjoint observables. With respect to the latter the following statements can be verified for Krein spaces: The adjoint of an operator with dense domain of definition in $\mathbb{H}$ exists and is unique; the adjunction is an involution on the representation of the algebra, cf. Bognar [Bogn 74]. Furthermore in such spaces suitable representations of the physical transformation groups exist, see, e.g., Bracci, Morchio and Strocchi [Brac 76] or Schlieder [Schli 60a,b,c]. However, we do not go into details in order to concentrate on the evaluation of functional quantum theory.

## 3.3 Probability Interpretation

Working in indefinite state spaces entails several complications compared with ordinary Hilbert space theory. The main problem is to prepare the physical states and to show that a probability interpretation is possible.

As the eigenstates of a set of commuting observables on an indefinite state space need not constitute a basis of this space, with respect to spinor theory we are forced to modify our provisional assumption that the eigenstates of the symmetry equations (1.93) can serve as a basis of $\mathbb{H}$. However, in Krein space one can find generalized eigenstates enlarging the set of eigenstates to a set of basis states of $\mathbb{H}$, cf. Bognar [Bogn 74]. Hence all the physics and "non–physics" can be covered by this construction and we can base our probability interpretation on it.

As the physical subspace $\mathbb{H}_{\mathrm{ph}}$ of $\mathbb{H}$ we define the closed linear hull of all eigenstates with positive norm in $\mathbb{H}$ with respect to a suitable topology, see Bognar [Bogn 74]. For simplicity we consider only the Hamiltonian $\mathbf{H}$. Then we define

$$\begin{aligned} \mathbb{H}_{\mathrm{ph}} &:= \overline{LH}\mathbb{B}_{\mathrm{ph}} \quad , \\ \mathbb{B}_{\mathrm{ph}} &:= \{|p\rangle \in \mathbb{H} : \mathbf{H}|p\rangle = E_p|p\rangle \ , \ \| p \| = 1\} \quad . \end{aligned} \tag{3.2}$$

For a continuous spectrum of $\mathbf{H}$ one has to introduce wavepackets, but we will not discuss such details. Furthermore we mention that $\mathbb{H}_{\mathrm{ph}}$ in general may not be identified with $\mathbb{H}_+$ from the fundamental decomposition (3.1).

By definition $\mathbb{H}_{\mathrm{ph}}$ possesses the structure of a Hilbert space, in particular a positive definite metric. To assure the possibility of probability interpretation in $\mathbb{H}_{\mathrm{ph}}$ it remains to show the invariance of $\mathbb{H}_{\mathrm{ph}}$ under time translations. This is done by the following proposition:

- **Proposition 3.1:** For any normalized state $|x\rangle \in \mathbb{H}_{\mathrm{ph}}$ the time evolution $|x,t\rangle = \exp(-i\mathbf{H}t)|x\rangle$ conserves normalization.

**Proof:** Due to the selfadjointness of $\mathbf{H}$ we have $|x,t\rangle^+ = \langle x|\exp(i\mathbf{H}t)$ and from that it follows $\langle x,t|x,t\rangle = \langle x|x\rangle = 1$. ◊

Another complication in indefinite state spaces is the fact that the eigenvalues of selfadjoint operators, in our case the momentum and spin parameters $p$ and $s$ from (1.93), may take complex values. However, it is easy to show that in Krein space the eigenvalues of selfadjoint operators for eigenstates with a nonvanishing norm are real. Due to the non–degeneracy of the Krein space in particular the eigenvalues of the states spanning the physical subspace $\mathbb{H}_{\mathrm{ph}}$ are real. Hence quantum theory in $\mathbb{H}_{\mathrm{ph}}$ can be developed as usual.

According to definition (3.2) any physical state that can be realized must be a superposition of eigenstates of $\mathbf{H}$ with positive norm. This statement is trivial in conventional quantum mechanics. But in Krein space the elements of $\mathbb{H}_{\mathrm{ph}}$ in general form no complete basis set and due to the restricition to the physical part of the state space one has to accept that some information from the whole state space gets lost. One cannot say what is lost without considering special models. In any case this loss will effect restrictions of measurement. Then one can decide by experiment whether such restrictions can be found or not. From a theoretical point of view this definition of $\mathbb{H}_{\mathrm{ph}}$ is sufficient to allow a probability interpretation of the theory, cf. also Feynman [Feyn 87].

So far, working with Krein spaces seems not to be a drastic deviation from common quantum theory. However, up to now our discussion was only formal and the question arises: How can one explicitly calculate the eigenstates of $\mathbf{H}$ in quantum field theory and in Krein space? In the following it will be demonstrated: The essential of eigenstate construction in quantum field theory is the choice of suitable basis systems, while the application of Krein spaces requires norm calculations. Hence if we develop the quantum theory of fields along the lines of conventional quantum theory, we are forced in particular to thoroughly study the embedding state space. In the conventional covariant formalism this problem is circumvented by the derivation of the $S$–matrix where due to asymptotic theorems (and postulates) only the state spaces of free particles are needed. It is the question of state space construction which makes Hamiltonian theory of quantum fields so different from the conventional covariant formalism.

Of course this formalism should also apply to gauge theories. In this case the most simple way to solve the metrical problem consists in eliminating the gauge constraints from the beginning in order to work only in the physically admitted subspace. This leads in Quantum Electrodynamics to the Coulomb gauge, see Chapter 8, while for Quantum Chromodynamics the subsidiary conditions are properly formulated in temporal gauge, see Chapter 8, but have not been solved so far. In the super–renormalizable spinorfield theory, however, the indefinite states cannot be characterized by subsidiary conditions and one has to work with the above given criterion.

## 3.4 Nonorthogonal Basis Sets

Although we will work along the lines and with the concept of conventional quantum theory, in some respect quantum fields and Krein spaces need an extension. While in conventional quantum theory in principle orthonormal basis sets are used and nonorthogonal basis sets merely appear in applications, for quantum fields nonorthogonal basis sets have to be assumed from the outset. In this section we shall treat this extension which also covers Krein spaces. For brevity we use a symbolic notation.

Let $\mathcal{J}$ be a set of indices and

$$\mathbb{B} := \{|e_n\rangle \, , \; n \in \mathcal{J}\} \tag{3.3}$$

be a basis of a linear space $\mathbb{H}$ with an indefinite inner product $\langle . | . \rangle$. The metric of $\mathbb{H}$ is defined by the scalar product

$$G_{mn} := \langle e_m | e_n \rangle \quad . \tag{3.4}$$

Due to the assumed nonorthonormality of the basis vectors and the indefiniteness of $\mathbb{H}$ it is $G_{mn} \neq \delta_{mn}$.

For a suitable domain of definition a linear operator $\mathbf{T}$ on $\mathbb{H}$ is hermitian (selfadjoint) if

$$\langle e_n | \mathbf{T} | e_m \rangle = \langle e_m | \mathbf{T} | e_n \rangle^* \; , \quad \forall \, n, m \in \mathcal{J} \quad . \tag{3.5}$$

In particular in accordance with Section 3.2, $G_{mn}$ is hermitian. The nondiagonal form of $G_{mn}$ forces us to introduce a complete set of dual basis vectors $|e^m\rangle$ which are defined by the relation

$$\langle e^m | e_n \rangle = \delta^m_n \tag{3.6}$$

and aditionally fulfill the completeness relation

$$|e^m\rangle\langle e_m| = \mathbf{1} . \tag{3.7}$$

The dual basis vectors are collected in a dual basis

$$\mathbb{B}^d := \{|e^m\rangle \, , \ m \in \mathcal{J}\} \quad . \tag{3.8}$$

Provided that $\mathbb{B}$ is nondegenerate, i.e. that $(G_{nm})^{-1}$ exists, the dual basis vectors are determined unambiguously.

- **Proposition 3.2:** Provided that $G_{mn}$ is invertible, then $G^{mn} := \langle e^m | e^n \rangle$ equals $\left(G_{mn}\right)^{-1}$, i.e. it holds

$$G^{mk} G_{kn} = \delta^m_n , \quad G_{mk} G^{kn} = \delta^m_n . \tag{3.9}$$

**Proof:** We expand the dual basis vectors in the original basis $\mathbb{B}$ by means of (3.7):

$$|e^m\rangle = (G^{mk})^* |e_k\rangle \quad . \tag{3.10}$$

From (3.6) it follows $\langle e_n | e^m \rangle = \delta^m_n$. Multiplying (3.10) with $\langle e_n|$ we obtain

$$\langle e_n | e^m \rangle = (G^{mk})^* \langle e_n | e_k \rangle = (G^{mk})^* G^*_{kn} = \delta^m_n \quad , \tag{3.11}$$

thus it holds the first part of (3.9). The second part of (3.9) can be proved along the same line. $\diamondsuit$

For the following we assume that $\mathbb{B}$ is nondegenerate. Then any element $|a\rangle \in \mathbb{H}$ can be represented by a linear combination of basis vectors. We assume that a Krein space is isomorphic to its dual, so either the set $\mathbb{B}$ or the set $\mathbb{B}^d$ can be used for this representation of $|a\rangle$. This gives

$$|a\rangle = \sum_{n \in \mathcal{J}} \sigma^n(a) |e_n\rangle \tag{3.12}$$

or

$$|a\rangle = \sum_{n \in \mathcal{J}} \tau_n(a) |e^n\rangle \, , \tag{3.13}$$

where the summation over $n \in \mathcal{J}$ includes summation over the discrete parts of $\mathcal{J}$ and integration over the continuous parts of $\mathcal{J}$.

Projection of (3.12) and (3.13) with $\langle e^m|$ or $\langle e_m|$ yields

$$\sigma^m(a) = \langle e^m | a \rangle \, , \ m \in \mathcal{J}, \tag{3.14}$$

and

$$\tau_m(a) = \langle e_m | a \rangle \, , \ m \in \mathcal{J}. \tag{3.15}$$

By (3.14) and (3.15) two bijective maps are defined:

$$|a\rangle \to \{\tau_n(a)\,,\ n \in \mathcal{J}\} \tag{3.16}$$

and

$$|a\rangle \to \{\sigma^n(a)\,,\ n \in \mathcal{J}\} \tag{3.17}$$

and the set $\{\tau_n(a),\ n \in \mathcal{J}\}$ is referred to as the covariant representation of $|a\rangle$, while $\{\sigma^n(a),\ n \in \mathcal{J}\}$ is referred to as the contravariant representation of $|a\rangle$, see Lagally and Franz [Lag 64].

In the same way as vectors also linear operators $\mathbf{T}$ on $\mathbb{H}$ can be represented in the basis $\mathbb{B}$ or (and) $\mathbb{B}^d$.

- **Proposition 3.3:** For a linear operator $\mathbf{T}$ on $\mathbb{H}$ the following representations exist:

$$\begin{aligned} T_{mn} &:= \langle e_n|\mathbf{T}|e_m\rangle & (3.18)\\ T^{mn} &:= \langle e^n|\mathbf{T}|e^m\rangle & (3.19)\\ T_{n}^{\cdot\,m} &:= \langle e_n|\mathbf{T}|e^m\rangle & (3.20)\\ T^{n}_{\cdot\,m} &:= \langle e^n|\mathbf{T}|e_m\rangle \quad . & (3.21)\end{aligned}$$

**Proof:** Due to (3.4), (3.7) and (3.9) the unit operator $\mathbf{1}$ on $\mathbb{H}$ can be represented by

$$\begin{aligned} \mathbf{1} &= \sum_{m,n} |e^m\rangle G_{mn}\langle e^n| = \sum_{m,n} |e_m\rangle G^{mn}\langle e_n| \qquad (3.22)\\ &= \sum_{m,n} |e^m\rangle G_{m}^{\cdot\,n}\langle e_n| = \sum_{m} |e^m\rangle\langle e_m| \quad . \end{aligned}$$

With these representations one obtains

$$\begin{aligned} \mathbf{T} &= \sum_{m,n} |e^m\rangle T_{mn}\langle e^n| = \sum_{m,n} |e_m\rangle T^{mn}\langle e_n| \qquad (3.23)\\ &= \sum_{m,n} |e^m\rangle T_{m}^{\cdot\,n}\langle e^n| = \sum_{m,n} |e_n\rangle T^{n}_{\cdot\,m}\langle e^m| \quad , \end{aligned}$$

where the coefficients are given by (3.18)–(3.21). So by (3.18)–(3.21) a map from linear operators onto $\mathbb{H}$ on the space of matrix representations is defined. One easily verifies the representation properties of (3.18)–(3.21). ◊

The matrix representations (3.18)–(3.21) are referred to as double covariant, double contravariant or mixed co–contravariant representations.

By means of the representations (3.22) the scalarproduct for any elements $|a\rangle, |b\rangle \in \mathbb{H}$ can be reduced to the scalarproducts of the basis vectors of $\mathbb{B}$ and $\mathbb{B}^d$, and we obtain with (3.12), (3.13) and (3.22)

$$\begin{aligned} g_{ab} &:= \langle a|b\rangle \qquad (3.24)\\ &= \sum_{n} \sigma^n(a)^*\tau_n(b)\\ &= \sum_{m,n} \tau_m(a)^* G^{mn}\tau_n(b)\\ &= \sum_{m,n} \sigma^m(a)^* G_{mn}\sigma^n(b) \quad . \end{aligned}$$

Of special interest are eigenvalue equations for selfadjoint operators $\mathbf{S}$ on $\mathbb{H}$ in the matrix representations (3.18)–(3.21).

- **Proposition 3.4:** Let $\mathbf{S}$ be a selfadjoint operator on $\mathbb{H}$ and $|p\rangle \in \mathbb{H}$ an eigenstate of $\mathbf{S}$ with $\mathbf{S}|p\rangle = s_p|p\rangle$. Then there exists a mixed co–contravariant matrix representation of $\mathbf{S}$, $S_m^{\cdot\, n} := \langle e_m|\mathbf{S}|e^n\rangle$, and the co– and contravariant components $\tau_n(p)$ and $\sigma^n(p)$ of $|p\rangle$ are right–hand side or left–hand side solutions of the eigenvalue equations

$$\sum_n S_m^{\cdot\, n}\tau_n(p) = s_p\tau_m(p) \tag{3.25}$$

and

$$\sum_m \sigma^m(p)^* S_m^{\cdot\, n} = s_p^*\sigma^n(p)^* \ . \tag{3.26}$$

**Proof:** We substitute (3.13) into the eigenvalue equation of $\mathbf{S}$ and project from the left with $\langle e_m|$. This gives

$$\sum_n S_m^{\cdot\, n}\tau_n(p) = s_p\tau_m(p)$$

with $S_m^{\cdot\, n} = \langle e_m|\mathbf{S}|e^n\rangle$. Because of the selfadjointness of $\mathbf{S}$ from $\mathbf{S}|p\rangle = s_p|p\rangle$ it follows by hermitian conjugation

$$\langle p|\mathbf{S} = s_p^*\langle p| \quad . \tag{3.27}$$

Substitution of the hermitian conjugate of (3.12) into (3.27) and projection with $|e^n\rangle$ yields equation (3.26). ◊

Finally we observe that in spite of the selfadjointness of $\mathbf{S}$ the mixed co–contravariant matrix representation $\mathbf{S}_m^{\cdot\, n}$ in general is non–hermitian. This can be seen from matrix conjugation:

$$(S_n^{\cdot\, m})^+ = (\langle e_m|\mathbf{S}|e^n\rangle)^* = \langle e^n|\mathbf{S}|e_m\rangle = S_{\cdot\, m}^{n} \ .$$

But $\langle e^n|\mathbf{S}|e_m\rangle$ is unequal $\langle e_n|\mathbf{S}|e^m\rangle$ as long as $\langle e^m| \neq \langle e_m|$. Hence $(S_n^{\cdot\, m})^+ \neq S_n^{\cdot\, m}$. However, in any case we have $S_{mn} = S_{mn}^+$ etc., i.e. the double co– and contravariant matrix representations are hermitian.

## 3.5 Cyclic Basis Vector Representations

The explicit evaluation of eigenvalue equations etc. depends on the ability to define appropriate basis vector systems. We emphasized that in quantum field theory this must be done in accordance with general ideas of algebraic representation theory and axiomatic quantum field theory. The corresponding construction of basis vector systems in Algebraic Quantum Theory is the GNS–construction (Gelfand–Naimark–Segal construction, cf. Section 2.1), a generalization of the Fock space construction.

The GNS–construction provides genuine representations of the basic field operator algebra. As already mentioned this algebra is already complete (or even overcomplete) at any spacelike hyperplane. In our case this is expressed by the equal time anticommutation relations (1.42). For simplicity we will only consider spacelike hyperplanes defined by $t = \text{const}$.

We now again refer to the spinorfield of Chapter 1 and, according to Section 2.1, take over the concept of GNS–construction as far as possible.

The GNS–construction is based on the existence of cyclic vectors. Excluding the groundstates of thermofield dynamics we assume that a suitable cyclic vector is given by the physical groundstate $|0\rangle \in \mathbb{H}$. Then the set of basis vectors is generated by the application of products of field operators to the groundstate $|0\rangle$, and according to the completeness, or even overcompleteness of the algebra on any spacelike hyperplane we confine these products to equal times in their coordinates.

In the following we use the abbreviation

$$(\psi_{I_1} \dots \psi_{I_n})_t := \psi_{Z_1}(\mathbf{r}_1, t) \dots \psi_{Z_n}(\mathbf{r}_n, t) \tag{3.28}$$

for field operator products at equal times which are the generating elements of our adapted GNS–construction.

However, these operator products have to be subjected to further restrictions with respect to their domain of definition. Due to the equal time anticommutators

$$[\psi_{I_1}, \psi_{I_2}]_+^{t_1=t_2} = A_{I_1 I_2} \tag{3.29}$$

we can rearrange any product (3.28) into a series of products obeying some ordering relation, see Grimm, Hailer and Stumpf [Gri 91]. We choose a lexicographic ordering relation which enables us to decide for two general indices $I_1, I_2$ whether $I_1 < I_2$ or $I_2 < I_1$ apart from $I_1 = I_2$. Then any product (3.28) can be expressed by a series of products in lexicographic order, or equivalently: an arbitrary product (3.28) is linear dependent on a set of representative products with a definite denumeration. For brevity we do not discuss this rearrangement in detail. Rather we immediately define the generating GNS–products by the set

$$\left\{ (\psi_{I_1} \dots \psi_{I_n})_t,\ I_1 < I_2 < \dots < I_n,\ n \in \mathbb{N} \right\} , \tag{3.30}$$

where here and in the following we have $I_m \in \mathcal{I}$, $\forall\, 1 \le m \le n$, and where a lexicographic ordering relation for $I_1, \dots I_n$ is assumed. In this set, due to their continuum part the points $I_k = I_l$ etc. have measure zero and thus can be omitted. On this domain of the indices we now can define a set of GNS–basis vectors

$$\mathbb{B}_1 := \left\{ (\psi_{I_1} \dots \psi_{I_n})_t |0\rangle,\ I_1 < I_2 < \dots < I_n,\ n \in \mathbb{N} \right\} \quad , \tag{3.31}$$

and this set is assumed to be complete and linear independent. As a consequence of this any state vector $|a\rangle \in \mathbb{H}$ can uniquely be represented by the expansion

$$|a\rangle = \sum_n \sum_{I_1 < \dots < I_n} \rho_t^{(n)}(I_1 \dots I_n|a)\, (\psi_{I_1} \dots \psi_{I_n})_t\, |0\rangle \quad , \tag{3.32}$$

where the coefficient functions $\rho_t^{(n)}(I_1 \dots I_n|a)$ are only defined for ordered sequences $I_1 \dots I_n$ at equal times $t_1 = \dots = t_n = t$. In this way we have obtained the starting point for field theoretic state space theory. The only flaw in this representation is the restriction of the summation (integration) imposed by the ordering relation. As is familiar from statistical mechanics, one can avoid this restriction by the introduction of symmetric or antisymmetric basis vectors. In that case the restricted summation can be enlarged to comprise all equivalence classes of the original domain, and this allows integration over the whole "phase space" of $I_1 \dots I_n$ (for equal times).

For spinorfields the symmetric products can be excluded and we define for antisymmetric products the corresponding GNS–basis vectors by the set

$$\mathbb{B}_2 := \left\{ \mathcal{A}\left(\psi_{I_1} \dots \psi_{I_n}\right)_t |0\rangle,\ n \in \mathbb{N} \right\} \quad , \tag{3.33}$$

where $\mathcal{A}$ means antisymmetrization in $I_1 \dots I_n$, i.e.

$$\mathcal{A}\left(\psi_{I_1} \dots \psi_{I_n}\right)_t := \sum_{p \in S_n} \frac{1}{n!} \mathrm{sgn}(p) \psi_{I_{p_1}} \dots \psi_{I_{p_n}} \quad .$$

With respect to the set $\mathbb{B}_2$ any state $|a\rangle \in \mathbb{H}$ can be represented by

$$|a\rangle = \sum_n \sum_{I_1 \dots I_n} \frac{1}{n!} \hat{\rho}_t^{(n)}(I_1 \dots I_n|a) \mathcal{A}\left(\psi_{I_1} \dots \psi_{I_n}\right)_t |0\rangle \tag{3.34}$$

where $\hat{\rho}_n^t(I_1 \dots I_n|a)$ is antisymmetric in $I_1 \dots I_n$ for $t_1 = \dots = t_n = t$.

The proof of equivalence of (3.32) and (3.34) makes use of equivalence classes with respect to the various domains of $I_1 \dots I_n$ and the antisymmetry properties of the constituents of (3.34). For its explicit discussion we refer to statistical mechanics.

For technical reasons we still slightly modify our basis set $\mathbb{B}_2$. As the adjunction is an inner automorphism of the field algebra we can equivalently use the adjoints of the vectors of $\mathbb{B}_2$ defined by

$$\mathbb{B}_0 := \left\{ \left[\mathcal{A}\left(\psi_{I_1} \dots \psi_{I_n}\right)_t\right]^+ |0\rangle,\ n \in \mathbb{N} \right\} \quad , \tag{3.35}$$

and with respect to $\mathbb{B}_0$ the expansion of $|a\rangle$ reads

$$|a\rangle = \sum_n \sum_{I_1 \dots I_n} \sigma_t^{(n)}(I_1 \dots I_n|a) \left[\mathcal{A}\left(\psi_{I_1} \dots \psi_{I_n}\right)_t\right]^+ |0\rangle \quad , \tag{3.36}$$

where we have incorporated the factor $(n!)^{-1}$ in the definition of $\sigma_t^{(n)}$.

We can now relate expansion (3.36) to our formalism of Section 3.4 by defining the operator $\mathbf{A}_n$ by

$$\mathbf{A}_n := \mathcal{A}\left(\psi_{I_1} \dots \psi_{I_n}\right)_t \tag{3.37}$$

and the basis vectors $|e_n\rangle$ by

$$|e_n\rangle := \mathbf{A}_n^\dagger |0\rangle \ , \tag{3.38}$$

where in this symbolic notation we have $n \in \mathcal{J}$, $\mathcal{J} = \{I_1, \dots I_n\}$. Thus we can write

$$|a\rangle = \sum_n \sigma^n(a)|e_n\rangle = \sum_n \tau_n(a)|e^n\rangle \tag{3.39}$$

and according to (3.15)

$$\tau_n(a) = \langle e_n|a\rangle = \langle 0|\mathcal{A}\left(\psi_{I_1}\dots\psi_{I_n}\right)_t|a\rangle \quad . \tag{3.40}$$

In Fock space such objects like (3.40) are referred to as "wavefunctions". In GNS–spaces we can maintain this designation, but have to pay attention to the fact that the state norms are given by (3.24), i.e., they are not as simple as in Fock space.

In order to complete our discussion of the construction of state space representations by cyclic basis vector systems, we have to derive the metrical tensor with respect to these basis vectors. As already mentioned we consider the wavefunctions (3.40), or closely related transformed functions, as the central quantities of the theory. Hence we express the metrical tensor in terms of vacuum wavefunctions which will be calculable along the same lines as the transition matrix elements (3.40). This can be performed by the following proposition, Stumpf [Stu 89]:

- **Proposition 3.5:** For $G_{mn} := \langle e_m|e_n\rangle$ with $|e_m\rangle, |e_n\rangle \in \mathbb{B}_0$ the representation

$$G_{mn} = \sum_{l,k} T^l_m C^k_{ln}\tau_k(0) \tag{3.41}$$

holds, where the expansion coefficients $T^l_m$ and $C^k_{ln}$ are well–defined.

**Proof:** With (3.37) and (3.38) it is

$$\langle e_m|e_n\rangle = \langle 0|\mathcal{A}\left(\psi_{I_1}\dots\psi_{I_m}\right)_t\left[\mathcal{A}\left(\psi_{K_1}\dots\psi_{K_n}\right)_t\right]^+|0\rangle \quad . \tag{3.42}$$

As the adjunction is an automorphism of the field operator algebra, the relation

$$\mathbf{A}_n^+ \equiv \mathcal{A}\left(\psi^+_{K_n}\dots\psi^+_{K_1}\right)_t = \sum_m \sum_{I_1\dots I_m} T^{I_1\dots I_m}_{K_1\dots K_n}\mathcal{A}\left(\psi_{I_1}\dots\psi_{I_m}\right)_t \tag{3.43}$$

holds with well–defined expansion coefficients. In symbolic notation, with (3.43) equation (3.42) goes over into

$$\langle e_m|e_n\rangle = \sum_l T^l_n\langle 0|\mathbf{A}_m\mathbf{A}_l|0\rangle \quad . \tag{3.44}$$

By means of the anticommutator (3.29) the product $\mathbf{A}_m\mathbf{A}_l$ can be rearranged into a series of antisymmetric products, see Proposition 3.11:

$$\mathbf{A}_m\mathbf{A}_l = \sum_k C^k_{ml}\mathbf{A}_k \quad . \tag{3.45}$$

If we substitute (3.45) into (3.44) and observe that due to the definiton (3.40) it is

$$\langle 0|\mathbf{A}_k|0\rangle = \langle e_k|0\rangle = \tau_k(0) \, ,$$

we obtain equation (3.41). ◇

For eigenstates $|p\rangle$ of a set of selfadjoint operators the wavefunctions $\tau_n(p)$ are solutions of homogeneous eigenvalue equations as, for instance, equation (3.25). As such they are determined up to a common complex factor $\lambda(p)$. This factor can be used to transform $g_{pp'}$, given by (3.24), into its normal form, which is given by $|g_{pp}| = 1$ and $g_{pp'} = 0$ for $p \neq p'$. If we replace $\tau_n(p)$ by $\lambda(p)\tau_n(p)$ then (3.24) goes over into

$$\begin{aligned} \| p \|^2 &= g'_{pp} = \sum_{m,n} \lambda(p)^* \tau_m(p)^* G^{mn} \tau_n(p) \lambda(p) \\ &= |\lambda(p)|^2 \sum_{m,n} \tau_m(p)^* G^{mn} \tau_n(p) \\ &= |\lambda(p)|^2 g_{pp} \quad , \end{aligned} \tag{3.46}$$

and thus the eigenstate $|p\rangle$ can be normalized to $+1$ or $-1$, provided $\| p \| \neq 0$.

Furthermore this factor is transferred to $\sigma^n(p)$, as from (3.24) we can read off

$$\sigma^n(p)^* = \sum_m \tau_m(p)^* G^{mn} \quad . \tag{3.47}$$

This shows that the calculation of left–hand side solutions $\sigma^{(n)}$ does not lead to a new freedom of normalization.

## 3.6 Renormalized Eigenvalue Equations

Poincaré invariant field theories describe systems with infinite spatial and temporal extension. This implies that at least some of the absolute eigenvalues of extensive operators as, for instance, total energy, particle number etc., assume an infinite value, and if these systems are to describe an idealized "world", their absolute eigenvalues are not measurable. One can only observe differences between a state of reference (groundstate) and excited states of the system. Mathematically this leads to an additive renormalization replacing the absolute values of observables by difference values. We elaborate this concept in the cyclic representation referring to Grimm, Hailer and Stumpf, [Gri 91].

Let $\mathrm{S}$ be a selfadjoint operator on $\mathbb{H}$ with $\mathrm{S}|p\rangle = s_p|p\rangle$. Then we introduce the operator $\mathrm{S}_0$ on $\mathbb{H}$ by

$$\mathrm{S}_0|a\rangle = s_0|a\rangle \ , \quad \forall \ |a\rangle \in \mathbb{H} \tag{3.48}$$

and define the renormalized operator $\mathrm{S}^{\mathrm{r}}$ by

$$\mathrm{S}^{\mathrm{r}} := \mathrm{S} - \mathrm{S}_0 \ , \tag{3.49}$$

where renormalization refers to the physical groundstate $|0\rangle$. The explicit form of $\mathrm{S}_0$ is given by $\mathrm{S}_0 = s_0\mathbf{1}$, from which we gather that $\mathrm{S}_0$ is selfadjoint, too, and hence also $\mathrm{S}^{\mathrm{r}}$ is selfadjoint.

- **Proposition 3.6:** Let $\mathrm{S}^{\mathrm{r}}$ be the corresponding renormalized operator (3.49) of a selfadjoint operator $\mathrm{S}$ on $\mathbb{H}$ with

$$\mathrm{S}^{\mathrm{r}}|p\rangle = (s_p - s_0)|p\rangle \tag{3.50}$$

for all elements $|p\rangle \in \mathbb{B}_{\text{ph}}$, cf. (3.2). Then there exist "matrix" representations of equation (3.50) in the cyclic basis $\mathbb{B}_0$ given by

$$\sum_n (S^{\text{r}})_m^{\cdot\, n} \tau_n(p) = (s_p - s_0)\tau_m(p) \tag{3.51}$$

$$\sum_m \sigma^m(p)^* (S^{\text{r}})_m^{\cdot\, n} = (s_p^* - s_0^*)\sigma^n(p)^* \tag{3.52}$$

with

$$(S^{\text{r}})_m^{\cdot\, n} := \langle e_m|\mathbf{S}^{\text{r}}|e^n\rangle = \langle e_m|\mathbf{S} - \mathbf{A}_m^{-1}\mathbf{S}\mathbf{A}_m|e^n\rangle \quad . \tag{3.53}$$

**Proof:** We substitute (3.13) into the eigenvalue equation (3.48) and project from the left by $\langle e_m|$. This gives

$$\begin{aligned} \sum_n \langle e_m|\mathbf{S}_0|e^n\rangle \tau_n(p) &= s_0 \sum_n \langle e_m|e^n\rangle \tau_n(p) \\ &= s_0 \tau_m(p) \quad . \end{aligned} \tag{3.54}$$

On the other hand for selfadjoint $\mathbf{S}$ we have $\langle 0|\mathbf{S} = s_0\langle 0|$ and therefore

$$\begin{aligned} s_0 \sum_n \langle e_m|e^n\rangle \tau_n(p) &= \sum_n \langle 0|\mathbf{S}\mathbf{A}_m|e^n\rangle \tau_n(p) \\ &= s_0 \tau_m(p) \quad . \end{aligned} \tag{3.55}$$

Comparison of equations (3.54) and (3.55) yields

$$\sum_n \langle e_m|\mathbf{S}_0|e^n\rangle \tau_n(p) = \sum_n \langle 0|\mathbf{S}\mathbf{A}_m|e^n\rangle \tau_n(p) \quad . \tag{3.56}$$

If we formally write $\langle 0| \equiv \langle 0|\, \mathbf{A}_m \mathbf{A}_m^{-1}$ this is equivalent to

$$\sum_n \langle e_m|\mathbf{S}_0|e^n\rangle \tau_n(p) = \sum_n \langle e_m|\mathbf{A}_m^{-1}\mathbf{S}\mathbf{A}_m|e^n\rangle \tau_n(p) \tag{3.57}$$

which has to be satisfied for all $\langle e_m| \in \mathbb{B}_0^d$. Hence, if we represent (3.50) in the cyclic basis $\mathbb{B}_0$ and observe (3.57) we get equation (3.51) with (3.53). On the same way (3.52) can be proved.

To assure ourselves that the replacement of $\langle 0|$ by $\langle e_m|\mathbf{A}_m^{-1}$ is possible, according to Proposition 3.3 we represent $\mathbf{A}_m$ and $\mathbf{A}_m^{-1}$ in the cyclic basis $\mathbb{B}_0$ and its dual by

$$\mathbf{A}_m = \sum_{k,l} |e^k\rangle (A_m)_k^{\cdot\, l} \langle e_l| \quad , \tag{3.58}$$

$$\mathbf{A}_m^{-1} = \sum_{h,j} |e^h\rangle \left(A_m^{-1}\right)_h^{\cdot\, j} \langle e_j| \quad . \tag{3.59}$$

In this representation the condition $\langle 0| = \langle e_m|\mathbf{A}_m^{-1}$ goes over into

$$\sum_l (A_m)_0^{\cdot\, l} \left(A_m^{-1}\right)_l^{\cdot\, j} = \delta_0^j \quad , \tag{3.60}$$

where due to the duality we used $\langle 0|e_j\rangle = \delta_j^0$. If we consider the matrices $\left(A_m^{-1}\right)_l^{\cdot\, j}$ as a set of $j$ vectors with components $l \in \mathcal{J}$, then (3.60) means that all vectors $j \neq 0$ must be orthogonal to a given vector $(A_m)_0^{\cdot\, l}$ with components $l$. This condition can be fulfilled by direct construction. ◇

For the practical evaluation of renormalized eigenvalue equations Proposition 3.6 is not very well suited; so according to Grimm, Hailer and Stumpf [Gri 91] we give another version of this proposition which later on allows an elegant functional evaluation method.

- **Proposition 3.7:** If the operator $\mathbf{S}$ is a power series functional of the field operators on a spacelike hyperplane $t$ =const., then the commutator of $\mathbf{S}$ with $\mathbf{A}_n$ can be expressed by the expansion

$$[\mathbf{A}_n, \mathbf{S}]_- = \sum_m C_n^{\cdot\, m} \mathbf{A}_m \tag{3.61}$$

with

$$C_n^{\cdot\, m} := (S^{\mathrm{r}})_n^{\cdot\, m} \ , \tag{3.62}$$

where $(S^{\mathrm{r}})_n^{\cdot\, m}$ is given by (3.53).

**Proof:** Due to the assumption about $\mathbf{S}$ the commutator $[\mathbf{A}_n, \mathbf{S}]_-$ can be directly evaluated by means of (3.29) and leads to a power series expression in the field operators. If some of these power series terms are not antisymmetrized by means of (3.29) they can be rearranged into an antisymmetrized product expansion. Hence (3.61) holds true.

To prove (3.62) we project (3.61) from the left with $\langle 0|$ and obtain

$$\langle 0|\, [\mathbf{A}_n, \mathbf{S}]_- = \sum_k C_n^{\cdot\, k} \langle e_k| \quad . \tag{3.63}$$

Furthermore, we have with $\langle 0| \equiv \langle 0|\mathbf{A}_m \mathbf{A}_m^{-1}$ , $m \in \mathcal{J}$,

$$\langle 0|\, (\mathbf{A}_n \mathbf{S} - \mathbf{S}\mathbf{A}_n) = \langle e_n|\left(\mathbf{S} - \mathbf{A}_n^{-1}\mathbf{S}\mathbf{A}_n\right) \ , \tag{3.64}$$

and combination of (3.63) with (3.64) yields

$$\sum_k C_n^{\cdot\, k} \langle e_k| = \langle e_n|\left(\mathbf{S} - \mathbf{A}_n^{-1}\mathbf{S}\mathbf{A}_n\right) \tag{3.65}$$

which by projection with $|e^m\rangle$ from the right leads to (3.62). ◇

If we project equation (3.64) from the right with an eigenvector $|p\rangle$ to $\mathbf{S}$ and use expansion (3.39), it follows

$$\langle 0|\, [\mathbf{A}_n, \mathbf{S}]_-\, |p\rangle = \sum_k \langle e_n|\left(\mathbf{S} - \mathbf{A}_n^{-1}\mathbf{S}\mathbf{A}_n\right)|e^k\rangle \tau_k(p) \ , \tag{3.66}$$

and with (3.53) and (3.51) this yields

$$\langle 0|\, [\mathbf{A}_n, \mathbf{S}]_-\, |p\rangle = (s_p - s_0)\tau_n(p) \quad . \tag{3.67}$$

In particular for $\mathbf{S} \equiv \mathbf{H}$ we obtain

$$\langle 0| \left[\mathbf{A}_n, \mathbf{H}\right]_- |p\rangle = (E_p - E_0)\, \tau_n(p) \, , \tag{3.68}$$

the transition matrix element of Heisenberg's formula $i\hbar\dot{\mathbf{O}} = [\mathbf{O}, \mathbf{H}]_-$ for $\mathbf{O} \equiv \mathbf{A}_n$ and energy eigenstates $\langle 0|$ and $|p\rangle$. The latter equation is the starting point of the New Tamm–Dancoff method, see Silin and Fainberg [Si 56].

Summarizing our statements we see that they end up with a renormalized energy equation, but in addition this approach establishes a relation of this equation with an explicit state construction which so far has not been taken into account in conventional quantum field theory. It is just this relation which is crucial for composite particle theory.

## 3.7 Functional Eigenvalue Equations

In order to derive the renormalized eigenvalue equation (3.67) for a selfadjoint operator $\mathbf{S}$ or $\mathbf{S}^{\mathrm{r}}$, respectively, we assumed $\mathbf{S}$ to be a power series functional of the field operators on a spacelike hyperplane $t = \text{const}$. As far as the spinorfield is concerned, according to (1.79) and (1.89) this assumption applies to the generators of the Poincaré group $\mathbf{P}_\mu, \mathbf{M}_{\mu\nu}$ and therefore, by means of (3.67) the eigenvalue equations (1.93) can be formulated in the cyclic representation. In the following we concentrate on the evaluation of (3.67) for $\mathbf{P}_0 \equiv \mathbf{H}$, because the resulting renormalized energy eigenvalue equation is fundamental for composite particle theory. The evaluation of (3.67) for the other Poincaré generators appearing in (1.93) runs along the same lines, and for brevity we do not explicitly discuss their eigenvalue equations.

In analogy to the covariant formalism the appropriate tool for the evaluation of (3.67) is the mapping into functional space. But in contrast to the covariant formalism, in this case we do not define the functional CAR–algebra on the whole $\mathbb{M}^4$, rather we confine this algebra to a spacelike hyperplane $t = \text{const}$. To this end we introduce the generators of a CAR–algebra on the hyperplane $t = \text{const}$ by

$$j_I^t := j_Z^t(\mathbf{r}) \quad \text{and} \quad \partial_I^t := \partial_Z^t(\mathbf{r}) \quad , \tag{3.69}$$

and from now on for quantities with the suffix "$t$" we understand the symbol $I$ to be an abbreviation for $I = (Z, \mathbf{r}, t)$ with the fixed parameter $t$. In particular the summation convention does not include the parameter $t$ in the following.

The anticommutators of the CAR–generators on the hyperplane are then given by

$$\left[j_I^t, j_{I'}^t\right]_+ = \left[\partial_I^t, \partial_{I'}^t\right]_+ = 0 \, , \tag{3.70}$$

$$\left[j_I^t, \partial_{I'}^t\right]_+ = \delta_{ZZ'}\delta(\mathbf{r} - \mathbf{r}') \quad , \tag{3.71}$$

and their Fock space representation results from $\left(j_I^t\right)^+ |0\rangle_F = \partial_I^t |0\rangle_F = 0$. The state $|0\rangle_F$ is again assumed to be a cyclic state of a corresponding functional space $\mathbb{K}_F^t$. As apart from $\mathbf{P}_0$ we do not discuss the additional symmetry conditions of (1.93) we also do not explicitly introduce the transformation properties of the generators (3.69).

We define functional states on the hyperplane by

$$|\mathcal{A}(j^t;a)\rangle := \sum_{n=0}^{\infty} \frac{i^n}{n!} \sum_{I_1 \dots I_n} \tau_t^{(n)}(I_1 \dots I_n|a) j_{I_1}^t \dots j_{I_n}^t |0\rangle_F \tag{3.72}$$

with

$$\tau_t^{(n)}(I_1 \dots I_n|a) := \langle 0|\mathcal{A}(\psi_{I_1} \dots \psi_{I_n})_t|a\rangle \quad , \tag{3.73}$$

and in this way we associate to any state $|a\rangle \in \mathbb{H}$ a functional state $|\mathcal{A}(j^t;a)\rangle \in \mathbb{K}_F^t$.

The discussion of the "physical" metric in $\mathbb{H}$ which is based on the cyclic representation and which was given in the preceding sections enables us to establish an isometric map between $\mathbb{H}$ and $\mathbb{K}_F^t$ by postulating the isometry condition

$$\langle \mathcal{D}(\partial^t;a)|\mathcal{A}(j^t;b)\rangle = \langle a|b\rangle \tag{3.74}$$

where the dual state $\langle \mathcal{D}(\partial^t;a)|$ is defined by

$$\langle \mathcal{D}(\partial^t,a)| := \sum_{n=0}^{\infty} (-i)^n \sigma_t^{(n)}(I_1 \dots I_n|a)^* {}_F\langle 0|\partial_{I_n}^t \dots \partial_{I_1}^t \quad . \tag{3.75}$$

Then the following proposition holds:

- **Proposition 3.8:** If the expansion coefficients $\sigma_t^{(n)}(I_1 \dots I_n|a)$ of the dual functional state (3.75) are the contravariant components of the state $|a\rangle$, cf. (3.14), then (3.74) is valid and thus an isometric map from $\mathbb{H}$ to $\mathbb{K}_F$ is established.

**Proof:** Due to the defining commutator relations (3.70) and (3.71) dual basis vectors $(-i)_F^n\langle 0|\partial_{I_n}^t \dots \partial_{I_1}^t$ are orthonormal to the original basis vectors $i^n(n!)^{-1} j_{I_1}^t \dots j_{I_n}^t|0\rangle_F$. By means of this property we can calculate the left–hand side of (3.74). This gives

$$\begin{aligned}
&\langle \mathcal{D}(\partial^t;a)|\mathcal{A}(j^t;b)\rangle \\
&= \sum_{m,n=0}^{\infty} (-i)^m \frac{i^n}{n!} \delta_{mn} \sigma_t^{(m)}(K_1 \dots K_m|a)^* \tau_t^{(n)}(I_1 \dots I_n|b) \times \\
&\quad \sum_{p \in S_m} \operatorname{sgn}(p) \delta_{I_1 K_{p1}} \dots \delta_{I_n K_{pm}} \\
&= \sum_{p \in S_n} \sum_{n=0}^{\infty} \operatorname{sgn}(p) \frac{1}{n!} \sigma_t^{(n)}(K_1 \dots K_n|a)^* \tau_t^{(n)}(K_{p_1} \dots K_{p_n}|b) \\
&= \sum_{n=0}^{\infty} \sigma_t^{(n)}(K_1 \dots K_n|a)^* \tau_t^{(n)}(K_1 \dots K_n|b) \\
&= \langle a|b\rangle \, ,
\end{aligned} \tag{3.76}$$

if one observes (3.12), (3.13) and (3.24). ◇

Having provided the map from the state space $\mathbb{H}$ into functional space $\mathbb{K}_F^t$ we can now proceed to the derivation of eigenvalue equations in $\mathbb{K}_F^t$. We concentrate on equation (3.68). There are several possibilities to perform its explicit evaluation:

i) direct or indirect calculation of the commutator in (3.68);

ii) performance of the equal time limit for the multitime covariant energy equation (2.81);

iii) application of the Hausdorff formula.

We shall treat the methods i) and ii) but omit method iii) which needs a too specialized mathematical preparation. In order to gain confidence into these methods they were first tested for the anharmonic oscillator by Maison and Stumpf [Mai 66], and later on extended to the treatment of spinorfields by Pfister [Pfi 87]. In this section we apply method i) with an indirect calculation of the commutator in (3.68) which seems to be the simplest way to solve our problem.

We observe that the Heisenberg formula $i\hbar\dot{\mathbf{O}} = [\mathbf{O}, \mathbf{H}]_-$ can be applied to any observable $\mathbf{O}$, which is given as a power series of field operators, in particular to the antisymmetrized products $\mathbf{A}_n$. This leads to

$$i\hbar\frac{\partial}{\partial t}\mathcal{A}(\psi_{I_1}\dots\psi_{I_n})_t = \left[\mathcal{A}(\psi_{I_1}\dots\psi_{I_n})_t, \mathbf{H}\right]_- , \tag{3.77}$$

or in combination with equation (3.68) to

$$i\hbar\frac{\partial}{\partial t}\langle 0|\mathcal{A}(\psi_{I_1}\dots\psi_{I_n})_t|p\rangle = (E_p - E_0)\langle 0|\mathcal{A}(\psi_{I_1}\dots\psi_{I_n})_t|p\rangle \quad . \tag{3.78}$$

The evaluation of (3.78) is formulated by the following proposition, bringing the field dynamics into the game:

- **Proposition 3.9:** Equation (3.78) is equivalent to the energy equation

$$\begin{aligned}(E_p - E_0)\langle 0|\mathcal{A}(\psi_{I_1}\dots\psi_{I_n})_t|p\rangle &= \\ = \frac{c}{\hbar}\sum_{\ell=1}^{n}\Big\{\langle 0|\mathcal{A}\left[\psi_{I_1}\dots(\hat{K}_{I_\ell J}\psi_J)\dots\psi_{I_n}\right]_t|p\rangle \\ +\langle 0|\mathcal{A}\left[\psi_{I_1}\dots(\hat{W}_{I_\ell J_1J_2J_3}\psi_{J_1}\psi_{J_2}\psi_{J_3})\dots\psi_{I_n}\right]_t|p\rangle\Big\} \quad ,\end{aligned} \tag{3.79}$$

where the antisymmetrization refers to the indices $I_1, \dots I_n$.

**Proof:** We perform the time differentiation in (3.78) and obtain with $E_{p0} := (E_p - E_0)$

$$\begin{aligned}E_{p0}\langle 0|\mathcal{A}(\psi_{I_1}\dots\psi_{I_n})_t|p\rangle &= \\ = \frac{i\hbar}{n!}\sum_{p\in S_n}\mathrm{sgn}(p)\sum_{l=1}^{n}\langle 0|\left[\psi_{p_1}\dots(\frac{\partial}{\partial t}\psi_{I_{p_l}})\dots\psi_{I_{p_n}}\right]_t|p\rangle \quad .\end{aligned} \tag{3.80}$$

The time derivative of $\psi_{I_{p_l}}$ can be expressed by the field equation (1.50). We multiply (1.50) from the left with $D^0_{ZZ_1}$ and resolve it with respect to $i\dot{\psi}_I$. The result can be formulated in an abridged version by (1.62) which is given by

$$i\frac{\hbar^2}{c}\frac{\partial}{\partial t}\psi_{I_1} = \hat{K}_{I_1I_2}\psi_{I_2} + \hat{W}_{I_1I_2I_3I_4}\psi_{I_2}\psi_{I_3}\psi_{I_4} \quad . \tag{3.81}$$

We replace $\dot{\psi}_{I_{p_l}}$ in (3.80) by (3.81) and obtain (3.79). ◇

With this proposition we only made a first step in the evaluation of (3.78) because in both terms on the right–hand side of (3.79) the operators $\psi_J$ or $(\psi_{J_1}\psi_{J_2}\psi_{J_3})$, respectively, are not antisymmetrized with respect to the operators $\psi_{I_1}\dots\psi_{I_{l-1}}\psi_{I_{l+1}}\dots\psi_{I_n}$.

In Section 3.5 we demonstrated that the GNS–basis vectors for the spinorfields $\psi_I$ have to be antisymmetrized eigenstates of the operators of the permutation group with respect to the arguments $I_1,\dots I_n$, $n = 1\dots\infty$. This includes that due to the anticommutation relations of the spinorfield any vector in the corresponding state space can be expanded in terms of these antisymmetrized GNS–basis vectors. In particular the vectors $\langle 0|\mathcal{A}\Big[\psi_{I_1}\dots(\hat{K}_{I_\ell J}\psi_J)\dots\psi_{I_n}\Big]_t$ and $\langle 0|\mathcal{A}\Big[\psi_{I_1}\dots(\hat{W}_{I_\ell J_1J_2J_3}\psi_{J_1}\psi_{J_2}\psi_{J_3})\dots\psi_{I_n}\Big]_t$ appearing in (3.79) can be expanded in terms of antisymmetrized GNS–basis states, and the same holds true for the corresponding transition matrix elements which allows an expansion in terms of antisymmetric functions. Hence equations (3.79) can be rearranged in such a way that a closed system of equations for antisymmetrized transition matrix elements results.

The rearrangement is performed in two steps: First we transform the right–hand side of (3.79) into a non–antisymmetrical standard form, second we decompose this form into a series of transition matrix elements for antisymmetrical operator products which can be represented in functional space.

In the following, for abbreviation we again write instead of $\psi_I$ simply the argument $I$, and omit the explicit notation of the states $|0\rangle$ and $|p\rangle$. Furthermore, we set $\hbar = c = 1$. In this abridged version (3.79) reads

$$
\begin{aligned}
E_{p0}\sum_{p\in S_n}\frac{1}{n!}\mathrm{sgn}(p)\langle I_{p_1}\dots I_{p_n}\rangle \qquad (3.82)\\
= \sum_{p\in S_n}\frac{1}{n!}\mathrm{sgn}(p)\sum_{\ell=1}^{n}\Big[\hat{K}_{I_{p_\ell}J}\langle I_{p_1}\dots I_{p_{\ell-1}}JI_{p_{\ell+1}}\dots I_{p_n}\rangle\\
+\sum_{\ell=1}^{n}\hat{W}_{I_{p_\ell}J_1J_2J_3}\langle I_{p_1}\dots I_{p_{\ell-1}}J_1J_2J_3I_{p_{\ell+1}}\dots I_{p_n}\rangle\Big] \quad .
\end{aligned}
$$

The reduction of (3.82) to the standard form can be achieved by the following proposition by Stumpf, Fauser and Pfister [Stu 93a]:

- **Proposition 3.10:** Equation (3.82) can be equivalently written as

$$
\begin{aligned}
E_{p0}\sum_{p\in S_n}\frac{1}{n!}\mathrm{sgn}(p)\langle I_{p_1}\dots I_{p_n}\rangle \qquad (3.83)\\
= \sum_{p\in S_n}\frac{1}{n!}\mathrm{sgn}(p)\hat{K}_{I_{p_1}J}\Big[n\langle JI_{p_2}\dots I_{p_n}\rangle\\
-\frac{1}{2}n(n-1)A_{JI_{p_2}}\langle I_{p_3}\dots I_{p_n}\rangle\Big]\\
+\sum_{p\in S_n}\frac{1}{n!}\mathrm{sgn}(p)\hat{W}_{I_{p_1}J_1J_2J_3}\Big[n\langle J_1J_2J_3I_{p_2}\dots I_{p_n}\rangle\\
-\frac{3}{2}\frac{n!}{(n-2)!}A_{J_3I_{p_2}}\langle J_1J_2I_{p_3}\dots I_{p_n}\rangle
\end{aligned}
$$

$$-\frac{n!}{(n-3)!}A_{J_2 I_{p2}}A_{J_3 I_{p3}}\langle J_1 I_{p_4}\dots I_{p_n}\rangle$$
$$\left.+\frac{1}{4}\frac{n!}{(n-4)!}A_{J_1 I_{p2}}A_{J_2 I_{p3}}A_{J_3 I_{p_4}}\langle I_{p5}\dots I_{p_n}\rangle\right]$$

with the anticommutator (1.56).

**Proof:** We consider the expression

$$\sum_{p\in S_n}\frac{1}{n!}\mathrm{sgn}(p)\langle I_{p_1}\dots I_{p_{\ell-1}}JI_{p_{\ell+1}}\dots I_{p_n}\rangle \quad . \tag{3.84}$$

By repeated anticommutations using (3.29) with (1.56) we shift $J$ to the first place and, after reindexing, obtain the recursion formula

$$\begin{aligned}(3.85) \quad = \quad & \sum_{p\in S_n}\frac{1}{n!}\mathrm{sgn}(p)\Big[(-1)^{\ell-1}\langle JI_{p_1}\dots \overset{o}{I}_{p_\ell}\dots I_{p_n}\rangle \\ & +(\ell-1)A_{JI_{p_{\ell-1}}}\langle I_{p_1}\dots \overset{o}{I}_{p_{\ell-1}}\overset{o}{I}_{p_\ell}\dots I_{p_n}\rangle\Big] \quad .\end{aligned} \tag{3.85}$$

Repeated application of (3.85) to the terms on the right–hand side of (3.82) leads to the following equation:

$$\begin{aligned} & E_{p0}\sum_{p\in S_n}\frac{1}{n!}\mathrm{sgn}(p)\langle I_{p_1}\dots I_{p_n}\rangle \\ = \; & \sum_{p\in S_n}\frac{1}{n!}\mathrm{sgn}(p)\sum_{\ell=1}^{n}\hat{K}_{I_{p_\ell}J}\Big[(-1)^{\ell-1}\langle JI_{p_1}\dots \overset{o}{I}_{p_\ell}\dots I_{p_n}\rangle \\ & +(\ell-1)A_{JI_{p_{\ell-1}}}\langle I_{p_1}\dots \overset{o}{I}_{p_{\ell-1}}\overset{o}{I}_{p_\ell}\dots I_{p_n}\rangle\Big] \\ & +\sum_{p\in S_n}\frac{1}{n!}\mathrm{sgn}(p)\sum_{\ell=1}^{n}\hat{W}_{I_{p_\ell}J_1J_2J_3}\Big[(-1)^{\ell-1}\langle J_1J_2J_3I_{p_1}\dots \overset{o}{I}_{p_\ell}\dots I_{p_n}\rangle \\ & +3(\ell-1)A_{J_3I_{p_{\ell-1}}}\langle J_1J_2I_{p_1}\dots \overset{o}{I}_{p_{\ell-1}}\overset{o}{I}_{p_\ell}\dots I_{p_n}\rangle \\ & +3(-1)^{\ell-1}(\ell-1)(\ell-2)A_{J_2I_{p_{\ell-1}}}A_{J_3I_{p_{\ell-2}}}\times \\ & \quad \langle J_1I_{p_1}\dots \overset{o}{I}_{p_{\ell-2}}\overset{o}{I}_{p_{\ell-1}}\overset{o}{I}_{p_\ell}\dots I_{p_n}\rangle \\ & +(\ell-1)(\ell-2)(\ell-3)A_{J_1I_{p_{\ell-1}}}A_{J_2,I_{p_{\ell-2}}}A_{J_3,I_{p_{\ell-3}}}\times \\ & \quad \langle I_{p_1}\dots \overset{o}{I}_{p_{\ell-3}}\overset{o}{I}_{p_{\ell-2}}\overset{o}{I}_{p_{\ell-1}}\overset{o}{I}_{p_\ell}\dots I_{p_n}\rangle\Big] \quad ,\end{aligned} \tag{3.86}$$

where use was made of the antisymmetry of $\hat{W}_{I_1I_2I_3I_4}$.

A further reindexing with respect to the arguments $I_1\dots I_n$ and application of the formula

$$\sum_{\ell=1}^{n}(\ell-1)(\ell-2)\dots(\ell-r)=\frac{1}{(r+1)}n(n-1)\dots(n-r) \tag{3.87}$$

yields equation (3.83). ◊

For the rearrangment of the standard form (3.83) into the final form with antisymmetrized transition matrix elements we need a further proposition which was proved by Pfister [Pfi 87], Stumpf, Fauser and Pfister [Stu 93a]:

- **Proposition 3.11:** The recursion formula

$$\sum_{p\in S_n} \frac{1}{n!}\mathrm{sgn}(p)\langle JI_{p_1}\dots I_{p_n}\rangle \tag{3.88}$$
$$= \sum_{p\in S_n} \frac{1}{(n+1)!}\mathrm{sgn}(p)\Big[\sum_{\ell=1}^{n}(-1)^\ell\langle I_{p_1}\dots I_{p_\ell}JI_{p_{\ell+1}}\dots I_{p_n}\rangle$$
$$+\frac{1}{2}n(n+1)A_{JI_{p_1}}\langle I_{p_2}\dots I_{p_n}\rangle\Big]\ ,\ n\in\mathbb{N},$$

generates the relation between non–antisymmetric standard forms and antisymmetric matrix elements of field operators. With the definition

$$|\mathcal{A}_{J_1\dots J_m}(j^t;a)\rangle := \sum_{n=0}^{\infty}\frac{i^n}{n!}\langle 0|\psi_{J_1}\dots\psi_{J_m}\mathcal{A}(\psi_{I_1}\dots\psi_{I_n})|a\rangle j^t_{I_1}\dots j^t_{I_n}|0\rangle_F \tag{3.89}$$

equation (3.88) can be expressed by the functional relation

$$|\mathcal{A}_I(j^t;a)\rangle = \left(\frac{1}{i}\partial^t_I + \frac{i}{2}A_{IJ}j^t_J\right)|\mathcal{A}(j^t;a)\rangle \quad . \tag{3.90}$$

**Proof:** We use the identity

$$\langle JI_{p_1}\dots I_{p_n}\rangle \equiv \frac{1}{(n+1)}\sum_{k=0}^{n}\langle JI_{p_1}\dots I_{p_n}\rangle \quad , \tag{3.91}$$

and in the summand $k=\ell$ on the right–hand side of (3.91) we shift the operator $\psi_J\equiv J$ by repeated anticommutations to the position $\ell$, i.e., to $\langle I_{p_1}\dots I_{p_\ell}JI_{p_{\ell+1}}\dots I_{p_n}\rangle$. Reindexing of the permutations in the terms with the anticommutator $A_{JI}$ yields (3.88).

Next we multiply (3.88) from the right with $i^n(n!)^{-1}j^t_{I_1}\dots j^t_{I_n}|0\rangle_F$, sum over $n$ and obtain

$$\sum_n\frac{i^n}{n!}\sum_{p\in S_n}\frac{1}{n!}\mathrm{sgn}(p)\langle JI_{p_1}\dots I_{p_n}\rangle j^t_{I_1}\dots j^t_{I_n}|0\rangle_F \tag{3.92}$$
$$= \sum_n\frac{i^n}{n!}\Big[\sum_{p\in S_n}\frac{1}{(n+1)!}\mathrm{sgn}(p)\sum_{\ell=0}^{n}(-1)^\ell\langle I_{p_1}\dots I_{p_\ell}JI_{p_{\ell+1}}\dots I_{p_n}\rangle$$
$$+\frac{1}{2}\sum_{p\in S_n}\frac{1}{(n-1)!}\mathrm{sgn}(p)A_{JI_{p_1}}\langle I_{p_2}\dots I_{p_n}\rangle\Big]j^t_{I_1}\dots j^t_{I_n}|0\rangle_F \quad .$$

By means of the one–time version of (2.18) the first term on the right–hand side of (3.92) can equivalently be written in the form

$$\sum_n \frac{i^n}{n!}\langle \mathcal{A}(JI_1 \dots I_n)\rangle j^t_{I_1} \cdots j^t_{I_n}|0\rangle_F = \frac{1}{i}\partial^t_J|\mathcal{A}(j^t;a)\rangle \tag{3.93}$$

while the second term can be rearranged into

$$\begin{aligned} &\frac{1}{2}\sum_n \frac{i^n}{(n-1)!}A_{JI_1}\langle I_2 \dots I_n\rangle j^t_{I_1} \cdots j^t_{I_n}|0\rangle_F \\ =& \frac{i}{2}A_{JI}j^t_I \sum_n \frac{i^{n-1}}{(n-1)!}\langle \mathcal{A}(I_2 \dots I_n)\rangle j^t_{I_1} \cdots j^t_{I_n}|0\rangle_F \\ =& \frac{i}{2}A_{JI}j^t_I|\mathcal{A}(j^t;a)\rangle \quad . \end{aligned} \tag{3.94}$$

Substitution of (3.93) and (3.94) into (3.92) and observation of (3.89) for $m = 1$ results in (3.90). ◇

Repeated application of (3.90) leads to the general formula

$$|\mathcal{A}_{J_m \dots J_1}(j^t;a)\rangle = \prod_{k=1}^{m}\left(\frac{1}{i}\partial^t_{J_k} + \frac{i}{2}A_{J_k I_k}j^t_{I_k}\right)|\mathcal{A}(j^t;a)\rangle \quad . \tag{3.95}$$

We now proceed to the derivation of the functional energy equation for equal times.

- **Proposition 3.12:** The set of equations (3.79) for $n = 1, 2, \dots \infty$ can be equivalently expressed by the functional equation

$$\begin{aligned} &E_{p0}|\mathcal{A}(j^t;p)\rangle \\ =& \left[\hat{K}_{I_1 I_2}j^t_{I_1}\partial^t_{I_2} - \hat{W}_{I_1 I_2 I_3 I_4}j^t_{I_1}\left(\partial^t_{I_4}\partial^t_{I_3}\partial^t_{I_2} + \frac{1}{4}A_{I_4 J_1}A_{I_3 J_2}j^t_{J_1}j^t_{J_2}\partial^t_{I_2}\right)\right]|\mathcal{A}(j^t;p)\rangle \, . \end{aligned} \tag{3.96}$$

**Proof:** We multiply (3.83) with $i^n(n!)^{-1}j^t_{I_1} \cdots j^t_{I_n}|0\rangle_F$ from the right and sum over $n$. In all terms we remove the sum over $p \in S_n$ by rearrangements of $j^t_{I_1} \cdots j^t_{I_n}$ and subsequent reindexing in analogy to (3.94). With definition (3.89) this results in

$$\begin{aligned} E_{p0}|\mathcal{A}(j^t;p)\rangle =& \hat{K}_{I_1 J}j^t_{I_1}\left[i|\mathcal{A}_J(j^t;p)\rangle + \frac{1}{2}A_{JI_2}j^t_{I_2}|\mathcal{A}(j^t;p)\rangle\right] \\ &+\hat{W}_{I_1 J_1 J_2 J_3}j^t_{I_1}\Big[i|\mathcal{A}_{J_1 J_2 J_3}(j^t;p)\rangle \\ &+\frac{3}{2}A_{J_3 I_2}j^t_{I_2}|\mathcal{A}_{J_1 J_2}(j^t;p)\rangle + iA_{J_2 I_2}j^t_{I_2}A_{J_3 I_3}j^t_{I_3}|\mathcal{A}_{J_1}(j^t;p)\rangle \\ &+\frac{1}{4}A_{J_1 I_2}j^t_{I_2}A_{J_2 I_3}j^t_{I_3}A_{J_3 I_4}j^t_{I_4}|\mathcal{A}(j^t;p)\rangle\Big] \quad . \end{aligned} \tag{3.97}$$

For the terms $|\mathcal{A}_{J_1 J_2 J_3}(j^t;p)\rangle$, etc., we substitute the corresponding expressions from (3.95) and evaluate the products. Then in (3.91) various terms cancel each other and we obtain (3.96) if the symmetry of the $A_{IJ}$ and the antisymmetry of $\hat{W}_{I_1 I_2 I_3 I_4}$, namely $\hat{W}_{I_1 I_2 I_3 I_4} = \hat{W}_{I_1\{I_2 I_3 I_4\}_{as}}$, is observed. ◇

Equation (3.96) can be rewritten from the symbolic notation into the original notation. With $I_k \equiv (Z_k, \mathbf{r}_k, t)$ and equations (1.63), (1.64), equation (3.96) reads with reconstructed $\hbar$ and $c$:

$$\begin{aligned} E_{p0}|\mathcal{A}(j^t;p)\rangle &= \Big\{\frac{ic}{\hbar}\int d^3r j^t_{Z_0}(\mathbf{r}) D^0_{Z_0Z_1}(D^k\partial_k - m)_{Z_1Z_2}\partial^t_{Z_2}(\mathbf{r}) \qquad (3.98)\\ &+\frac{ic}{\hbar}\int d^3r j^t_{Z_0}(\mathbf{r}) D^0_{Z_0Z_1} U_{Z_1Z_2Z_3Z_4}\Big[\partial^t_{Z_4}(\mathbf{r})\partial^t_{Z_3}(\mathbf{r})\partial^t_{Z_2}(\mathbf{r})\\ &+\frac{1}{4}A_{Z_4Z_5}j^t_{Z_5}(\mathbf{r})A_{Z_3Z_6}j^t_{Z_6}(\mathbf{r})\partial^t_{Z_2}(\mathbf{r})\Big]\Big\}|\mathcal{A}(j^t;p)\rangle \end{aligned}$$

or concisely

$$E_{p0}|\mathcal{A}(j^t;p)\rangle = \mathcal{H}_{\mathrm{F}}(j^t,\partial^t)|\mathcal{A}(j^t;p)\rangle \tag{3.99}$$

where $\mathcal{H}_{\mathrm{F}}$ is defined by the functional operator on the right–hand side of (3.98).

In this way we have obtained a functional map from the renormalized energy equation (3.68) onto the functional equation (3.99). If we want to regain equation (3.68) from equation (3.99), we write in a symbolic notation for (3.72)

$$|\mathcal{A}(j^t;p)\rangle = \sum_n \tau_n(p)|d^n\rangle \quad . \tag{3.100}$$

Then, by the definition of the projection states

$$\langle d_m| := {}_F\langle 0|i^{-m}\partial^t_{I_m}\dots\partial^t_{I_1} \tag{3.101}$$

from (3.99) we can derive the matrix equation

$$\sum_n \langle d_m|\mathcal{H}_{\mathrm{F}}|d^n\rangle\tau_n(p) = E_{p0}\tau_m(p) \tag{3.102}$$

and by comparison with (3.66) and (3.68) it follows that

$$\sum_n \langle d_m|\mathcal{H}_{\mathrm{F}}|d^n\rangle\tau_n(p) = \sum_n \langle e_m|\mathbf{H} - \mathbf{A}_m^{-1}\mathbf{H}\mathbf{A}_m|e^n\rangle\tau_n(p) \tag{3.103}$$

for $m = 0, 1, 2\dots\infty$. Hence with (3.102) and (3.103) we have reproduced the renormalized energy equation (3.68). In all following chapters we shall use the functional equation (3.99) instead of equation (3.68) as the starting point of our discussions.

## 3.8 Normalordering

For the covariant theory the need for normalordering was discussed in full lenght in Section 2.5. The same arguments apply to the formulation of the theory on a spacelike hyperplane, and therefore normalordering must be introduced in this case, too. This will be done in this section.

In Section 2.5 we applied a free fermion propagator for normalordering. This choice will be justified in Chapter 5. However, in a preliminary way one can explain it by the following argument: The weak mapping theorems (see Chapter 4) induce decompositions of matrix elements (3.40) which simulate algebraic decompositions of matrix elements by intermediate states. These decompositions depend on the system of basis functions which are used for the weak mapping. So if one interprets weak mapping as an algebraic decomposition of matrix elements the uncorrelated parts of this decomposition must be compatible with the system of basis functions. It is this condition which for our choice of basis functions leads to the free fermion propagator, although the whole mapping procedure is strictly nonperturbative.

For any state $|a\rangle \in \mathbb{H}$ we define normalordered functional states on a spacelike hyperplane $t = \text{const}$ by

$$
\begin{aligned}
|\mathcal{F}(j^t;a)\rangle &:= \exp\left[\frac{1}{2}\int j^t_{Z_1}(\mathbf{r}_1)F^t_{Z_1Z_2}(\mathbf{r}_1,\mathbf{r}_2)j^t_{Z_2}(\mathbf{r}_2)d^3r_1d^3r_2\right]|\mathcal{A}(j^t;a)\rangle \\
&=: \sum_{n=1}^{\infty}\frac{i^n}{n!}\varphi^{(n)}_t(I_1\dots I_n|a)j^t_{I_1}\dots j^t_{I_n}|0\rangle_F
\end{aligned}
\tag{3.104}
$$

with the propagator

$$
F^t_{Z_1Z_2}(\mathbf{r}_1,\mathbf{r}_2) := \langle 0|\mathcal{A}\left[\psi^f_{Z_1}(\mathbf{r}_1,t)\psi^f_{Z_2}(\mathbf{r}_2,t)\right]|0\rangle \tag{3.105}
$$

where $\psi^f_Z(\mathbf{r},t)$ is a solution of equation (2.109).

- **Proposition 3.13:** The normaltransformed of equation (3.98) reads

$$
\begin{aligned}
&E_{p0}|\mathcal{F}(j^t;p)\rangle \\
&= \frac{ic}{\hbar}\int d^3r j^t_{Z_0}(\mathbf{r})D^0_{Z_0Z_1}\Big\{(D^k\partial_k - m)_{Z_1Z_2}\partial^t_{Z_2}(\mathbf{r}) \\
&\quad + U_{Z_1Z_2Z_3Z_4}\Big[d^t_{Z_4}(\mathbf{r})d^t_{Z_3}(\mathbf{r})d^t_{Z_2}(\mathbf{r}) \\
&\quad + \frac{1}{4}A_{Z_4Z_5}j^t_{Z_5}(\mathbf{r})A_{Z_3Z_6}j^t_{Z_6}(\mathbf{r})d^t_{Z_2}(\mathbf{r})\Big]\Big\}|\mathcal{F}(j^t;p)\rangle
\end{aligned}
\tag{3.106}
$$

with

$$
d^t_{Z_1}(\mathbf{r}_1) := \partial^t_{Z_1}(\mathbf{r}_1) - \int F^t_{Z_1Z_2}(\mathbf{r}_1,\mathbf{r}_2)j^t_{Z_2}(\mathbf{r}_2)d^3r_2 \quad . \tag{3.107}
$$

**Proof:** From (3.104) we obtain by inversion of the exponential function

$$
|\mathcal{A}(j^t;a)\rangle = \exp(-\frac{1}{2}j^t_{I_1}F^t_{I_1I_2}j^t_{I_2})|\mathcal{F}(j^t;a)\rangle \quad . \tag{3.108}
$$

We substitute (3.108) into equation (3.98) and commute the exponential factor with $\mathcal{H}_F$. Afterwards we cancel it by division. The commutation can be achieved by means of formula (2.89) which holds for functionals on the hyperplane, too, and for which with (3.107) we simply write

$$\partial^t_K \exp(-\frac{1}{2} j^t_{I_1} F^t_{I_1 I_2} j^t_{I_2}) = \exp(-\frac{1}{2} j^t_{I_1} F^t_{I_1 I_2} j^t_{I_2}) d^t_K \quad . \tag{3.109}$$

By repeated application of (3.109) we can derive the interaction terms of (3.106) from those of (3.98). With respect to the kinetic energy term we observe that due to (2.109) and (3.105) it is

$$\begin{aligned} &\left[D^k \partial_k(\mathbf{r}_1) - m\right]_{Z_0 Z_1} F^t_{Z_1 Z_2}(\mathbf{r}_1, \mathbf{r}_2) \\ &= \frac{1}{2} D^0_{Z_0 Z_1} \langle 0 | \psi^f_{Z_2}(\mathbf{r}_2, t) \dot{\psi}^f_{Z_1}(\mathbf{r}_1, t) - \dot{\psi}^f_{Z_1}(\mathbf{r}_1, t) \psi^f_{Z_2}(\mathbf{r}_2, t) | 0 \rangle \quad . \end{aligned} \tag{3.110}$$

If $\mathbf{H}^0$ is the Hamiltonian of the free field we have $\dot{\psi}^f_Z(\mathbf{r}, t) = \frac{i}{\hbar}[\mathbf{H}^0, \psi^f_Z(\mathbf{r}, t)]_-$ and (3.110) can be rewritten as

$$\left[D^k \partial_k(\mathbf{r}_1) - m\right]_{Z_0 Z_1} F^t_{Z_1 Z_2}(\mathbf{r}_1, \mathbf{r}_2) = \frac{i}{2} D^0_{Z_0 Z_1} O_{Z_1 Z_2}(\mathbf{r}_1, \mathbf{r}_2, t) \quad , \tag{3.111}$$

where

$$O_{Z_1 Z_2}(\mathbf{r}_1, \mathbf{r}_2, t) := \langle 0 | \psi^f_{Z_1}(\mathbf{r}_1, t) \mathbf{H}^0 \psi^f_{Z_2}(\mathbf{r}_2, t) + \psi^f_{Z_2}(\mathbf{r}_2, t) \mathbf{H}^0 \psi^f_{Z_1}(\mathbf{r}_1, t) | 0 \rangle \tag{3.112}$$

and thus

$$O_{Z_1 Z_2}(\mathbf{r}_1, \mathbf{r}_2, t) = O_{Z_2 Z_1}(\mathbf{r}_2, \mathbf{r}_1, t) \quad . \tag{3.113}$$

If (3.111) is substituted in the transformed kinetic energy term, due to (3.113) the $F$–term vanishes and hence only the kinetic energy term in (3.106) remains. ◇

Finally we mention that the vertex renormalization as it was discussed in Section 2.6 for the covariant formalism, can be transferred to the energy equation on a spacelike hyperplane without any alteration. In particular equation (2.118), which describes a normalordering of the sources, can be taken over to the latter case; i.e., by the effect of the counterterms the sources and their duals formally anticommute within the vertex expression, and in this way divergencies by $F(0)$–terms are avoided.

## 3.9 Covariant Equations on the Hyperplane

We shortly discuss the question whether in the limit of equal times the covariant formalism leads to the Hamilton formalism on the hyperplane, referring to Maison and Stumpf [Mai 66], Pfister [Pfi 87], Borne [Bor 91].

The starting point of our investigation is the multitime energy equation (2.81). By means of formulas (2.18)–(2.20) we project (2.81) into configuration space and obtain a system of equations between various time–ordered products. This system reads

$$\begin{aligned} &E_{p0} \langle 0 | T\left(\psi_{I_1} \dots \psi_{I_n}\right) | p \rangle \\ &= \frac{ic}{\hbar} \sum_{k=1}^{n} \Big\{ \langle 0 | T\left(\psi_{I_1} \dots \left[D^0_{Z_k X_1} \left(D^m \partial_m(x_k) - m\right)_{X_1 X_2} \psi_{X_2}(x_k)\right] \dots \psi_{I_n}\right) | p \rangle \\ &\quad - \langle 0 | T\left(\psi_{I_1} \dots \left[D^0_{Z_k X_1} U_{X_1 X_2 X_3 X_4} \psi_{X_2}(x_k) \psi_{X_3}(x_k) \psi_{X_4}(x_k)\right] \dots \psi_{I_n}\right) | p \rangle \Big\} \, , \end{aligned} \tag{3.114}$$

where we have to observe $I_k = (Z_k, x_k)$ for $1 \leq k \leq n$. Equation (3.114) corresponds to equations (3.79) with the only difference that time–ordered products are used instead of antisymmetrized one–time expressions. The connection between both kinds of products can be established by a limit procedure.

We define the following limiting process for time–ordered products:

$$\lim_T T(\psi_{I_1} \dots \psi_{I_n}) := \frac{1}{n!} \sum_{p \in S_n} \lim_{\substack{t_{p_1} > \dots > t_{p_n} \\ t_{p_1} \dots t_{p_n} \to t}} T(\psi_{I_{p_1}} \dots \psi_{I_{p_n}}) \quad . \tag{3.115}$$

By means of this limiting process $T$–products can be related to $A$–products.

- **Proposition 3.14:** The limiting process of (3.115) yields

$$\lim_T T(\psi_{I_1} \dots \psi_{I_n}) = A(\psi_{I_1} \dots \psi_{I_n})_t \quad . \tag{3.116}$$

**Proof:** By definition we have

$$T(\psi_{I_1} \dots \psi_{I_n}) = \sum_{p \in S_n} \mathrm{sgn}(p) \psi_{p_1} \dots \psi_{p_n} \Theta(t_{p_1} \dots t_{p_n}) \tag{3.117}$$

and thus

$$\lim_T T(\psi_{I_1} \dots \psi_{I_n}) = \frac{1}{n!} \sum_{p \in S_n} \lim_{\substack{t_{p_1} > \dots > t_{p_n} \\ t_{p_1} \dots t_{p_n} \to 1}} \sum_{p' \in S_n} \mathrm{sgn}(p') \psi_{p'_1} \dots \psi_{p'_n} \Theta(t_{p'_1} \dots t_{p'_n}) \,. \tag{3.118}$$

For any ordering $t_{p_1} > \dots > t_{p_n}$ there exists only one $p' \in S_n$, namely $p' = p$, for which $\Theta(t_{p'_1}, \dots t_{p'_n})$ is unequal zero, i.e. we obtain

$$\lim_T T(\psi_{I_1} \dots \psi_{I_n}) = \frac{1}{n!} \sum_{p \in S_n} \lim_{\substack{t_{p_1} > \dots > t_{p_n} \\ t_{p_1} \dots t_{p_n} \to t}} \mathrm{sgn}(p) \psi_{p_1} \dots \psi_{p_n} \Theta(t_{p_1} \dots t_{p_n}) \,; \tag{3.119}$$

and if the limiting process for $\Theta$ is performed, we obtain (3.116). ◇

As equations (3.114) are linear equations with respect to the $T$–products we can immediately apply the limiting process (3.115) to (3.114). With (3.116) this limiting process leads to equations (3.79), and the further treatment is provided by Proposition 3.12.

We showed in this way that the covariant formalism is indeed compatible with the Hamilton formalism. Nevertheless, only the latter offers a solid foundation of composite particle theory for quantum fields.

# 4 Weak Mapping Theorems

## 4.1 Introduction

What are weak mappings? Weak mappings are defined by mappings between quantum field theoretic functional states which were introduced in Chapter 3 and we regard weak mapping as a suitable means for deriving effective quantum field dynamics. The attribute "weak" is borrowed from the topological concept of the weak topology and means that these mappings are concerned with matrix elements of operators. In contrast to this are the methods which use operator products for the description of composite particles and which will be discussed in Chapter 11.

For illustrating the role of weak mappings we shortly summarize the steps we have already done towards a composite particle theory: In Chapter 3 we pointed out that antisymmetrized expectation values of field operators on a spacelike hyperplane are regarded as the basic quantities of the algebraic theory. From the algebraic point of view these matrix elements are algebraic "states". Guided by the GNS–construction of Algebraic Quantum Theory we assumed that for each algebraic state on the field algebra there exists at least one regular cyclic representation in Hilbert space. The "GNS"–basis vectors of this representation were obtained by the application of symmetrized or antisymmetrized monomials of field operators on this cyclic Hilbert space vector. We then defined functional states generating the set of projections of an arbitrary but fixed state in Hilbert space onto these basis vectors, see (3.72), (3.73). Furthermore, from the original Heisenberg dynamics of the quantum fields for these functional states corresponding functional equations could be derived, see (3.98), (3.106).

Weak mapping is now introduced by a reordering of such generating functional states according to certain bound state structures. As a result one obtains another functional state which by definition is referred to these bound states. The functional field equations may be viewed as being generated by operators acting on the set of generating functional states. In a second step by weak mapping these operators are simultaneously transformed with the states, and this yields new functional field equations which are considered as the functional description of the effective dynamics of composite particles in quantum field theory.

The interesting applications in physics are studies of maps between functional equations for basic particles and functional equations for composite particles, where the latter may be called "effective" functional equations, in analogy to the frequently used designation "effective" Lagrangian.

In this chapter we will study the formal evaluation of exact weak mapping theorems which we will derive for the case of the super–renormalizable spinorfield model of Chapter 1. But weak mapping is not restricted to this model. Rather it can also be applied to renormalizable theories like quantum electrodynamics, see Chapter 8, provided a canonical quantization is uniquely defined, and from these considerations it will become obvious that also nonrelativistic quantum field theories can be treated in this way. Physical applications and interpretations will be given in the following chapters.

Originally weak mappings were performed by means of functional "chain rules" in the transformation of elementary particle functional states to composite particle functional states, see Stumpf [Stu 85a]. This method was the starting point for the investigation of exact weak mapping theorems, and it can be justified as a borderline case for negligible exchange forces of the exact theory. As the exact mapping theorems are rather unwieldy, we will also discuss the justification of the chain rule mappings allowing a very elegant (and manageable) treatment of composite particle dynamics.

Before we start on our derivation of the weak mapping theorems it is necessary to give some general comments upon composite particles, i.e. bound states of various elementary or basic particles.

Bound states are the elements and the starting point of any composite particle dynamics. However, in contrast to quantum mechanics in quantum field theory no unique definition of bound states is possible. Rather one has to distinguish two kinds of bound states: hard core bound states and dressed particle states. The latter consist of a hard core surrounded by a polarization cloud, which in nontrivial quantum field theories is generated by the interaction of the physical vacuum with the bound state constituents. While the hard core state is the analogue to the quantum mechanical bound states which in general exist for themselves, the polarization cloud is a specific quantum field theoretic contribution to the formation of bound states, and sometimes it even stabilizes hard core bound states which otherwise would be unstable. Therefore, apart from the latter possibility, for the formulation of a composite particle dynamics one can either consider only the hard core of a bound state as the basic quantity, or one considers the hard core with inclusion of its polarization cloud as the basic quantity, what then yields a "dressed" composite particle.

However, the weak mapping of dressed particle states is much more complicated than that of hard core states and so far it has not yet been satisfactorily explored. Thus for deriving the exact mapping theorems we concentrate on the mappings of hard core states and finally add some remarks about dressed composite particle states in Sections 4.8 and 4.9.

A peculiarity of quantum field theory is the occurence of uncorrelated parts in the definition of the original antisymmetrized functional states. In order to prepare composite particle states and their dynamics these uncorrelated parts must be removed from the theory. This can be achieved by applying normal transformations to the original functional states (see Chapter 2). For the case of hard core bound states of two, three, and four subfermions it is sufficient to apply the normal transform (3.104). For the treatment of bound states of more than four constituting subfermions it is necessary to extract higher order vacuum expectation values. This also results in considerable complications which are not fully explored.

Thus in the following we restrict ourselves to the normal transform (3.104) which allows the bound state calculation for two, three and four subfermions and the treatment of the full dynamics of two– and three–fermion bound states including special cases of four–fermion bound state dynamics, as will be shown in subsequent sections. As will be demonstrated this comprises a wide range of applications in elementary particle and in solid state physics.

In solid state physics quasiparticles and their interactions can be treated by means of the weak mapping method. In this field many investigations were already performed in Fock space where in most cases hard core states were used, but sometimes dressed particle states were considered as well; for a review of such calculations, for instance with respect to semiconductors (polarons, excitons, etc.) see Stumpf [Stu 83a], or for a more general review about Fock space methods see Haken [Hak 73]. In nuclear physics, statistical mechanics and other branches of solid states physics Greenfunction techniques were applied, see, e.g., Fetter and Walecka [Fet 71]. But until now no systematic discussion of the relation of effective quasiparticle theories with the algebraic formalism exists.

## 4.2 Hard Core States

As just mentioned we first consider weak mapping on the basis of hard core bound states. Hard core states are related to the minimal number of constituents which bear the correct quantum numbers of the corresponding physical particle, apart from its mass and charge which undergo renormalization. In this way they are not different from dressed particles, because in general also dressed particles need this final renormalization. The only difference between hard core and dressed particle states is the exclusion or inclusion of a polarization cloud. Weak mapping with hard core states, however, does not mean omission of the polarization cloud. As the exact mapping theorems will provide a biunique map between the basic and the effective theory, in the case of weak mapping with hard core states the polarization cloud manifests itself in effective selfinteractions and mutual interactions of the effective hard core states. Hence in order to work with hard core states we need an appropriate definition for them.

To provide such a definition we consider the normalordered functional energy equation (3.106). If we return to the symbolic notation of (1.62)–(1.64) and observe the vertex renormalization (2.118), then equation (3.106) can be equivalently written in the following form:

$$
\begin{aligned}
E_{p0}|\mathcal{F}(j^t;p)\rangle &= j^t_{I_1}\hat{K}_{I_1I_2}\partial^t_{I_2}|\mathcal{F}(j^t;p)\rangle \\
&+\hat{W}_{I_1I_2I_3I_4}\Big[j^t_{I_1}\partial^t_{I_4}\partial^t_{I_3}\partial^t_{I_2} - 3F^t_{I_4K}j^t_{I_1}j^t_K\partial^t_{I_3}\partial^t_{I_2} \\
&+(3F^t_{I_4K_1}F^t_{I_3K_2}+\frac{1}{4}A_{I_4K_1}A_{I_3K_2})j^t_{I_1}j^t_{K_1}j^t_{K_2}\partial^t_{I_2} \\
&-(F^t_{I_4K_1}F^t_{I_3K_2}+\frac{1}{4}A_{I_4K_1}A_{I_3K_2})F^t_{I_2K_3}j^t_{I_1}j^t_{K_1}j^t_{K_2}j^t_{K_3}\Big]|\mathcal{F}(j^t;p)\rangle \quad .
\end{aligned}
\tag{4.1}
$$

To recognize the difference between hard core and dressed particle states in quantum field theory, we consider quantum mechanical bound states, such as atoms, nuclei, etc. These bound states contain only their constituting particles and no polarization cloud. Therefore they are hard core states, and if they are described in Fock space they are characterized by the appearance of only diagonal–part operators, i.e. operators which, after normalordering, contain equal numbers of creation and annihilation operators. Although due to the GNS–construction and the embedding in functional space the basic energy equation (4.1) is not directly comparable with bound state equations in Fock space, we take over the hard core criterion from Fock space to functional space. We therefore assume that the nondiagonal part of equation (4.1) is related to the polarization cloud, and if we want to characterize hard core states, we define them as solutions of the diagonal part of equation (4.1) with corresponding diagonal functional states $|\mathcal{F}(j^t;a)\rangle^{\mathrm{d}}$, cf. Stumpf [Stu 83b].

Thus in our case the defining equation for hard core bound states is the diagonal part of (4.1) and reads

$$
E^{\mathrm{d}}_{p0}|\mathcal{F}(j^t;p)\rangle^{\mathrm{d}} = \left(j^t_{I_1}\hat{K}_{I_1I_2}\partial^t_{I_2} - 3\hat{W}_{I_1I_2I_3I_4}F^t_{I_4K}j^t_{I_1}j^t_K\partial^t_{I_3}\partial^t_{I_2}\right)|\mathcal{F}(j^t;p)\rangle^{\mathrm{d}} \quad . \tag{4.2}
$$

To solve equation (4.2) we project it into configuration space by applying $\langle d_m|$ of (3.101) from the left for $m = 1, 2 \ldots \infty$. Due to the appearance of only diagonal part operators the various sectors $m = 1, 2 \ldots \infty$ completely decouple and thus can be treated separately. We denote the solutions of (4.2) in configuration space by $C_k^{I_1 \ldots I_m}$, $m = 1, 2 \ldots \infty$, where $k$ are the quantum numbers of the state for fixed $m$, and assume that the solutions in any sector build up a complete set of functions.

For the formulation of weak mapping theorems we treat the simplest example of a bound state dynamics. We restrict ourselves to the two–particle sector given by the set of functional states $|\mathcal{F}(j^t; 2, k)\rangle^{\mathrm{d}} := -\frac{1}{2} C_k^{I_1 I_2} j_{I_1}^t j_{I_2}^t |0\rangle_F$. The corresponding equations arise from (4.2) by projection with $\langle d_2| \equiv -_F\langle 0|\partial_{K_2}\partial_{K_1}$ and by substitution of $|\mathcal{F}(j^t; 2, k)\rangle^{\mathrm{d}}$; they read

$$E_{k0}^{\mathrm{d}} C_k^{K_1 K_2} = \left[ \hat{K}_{K_1 I_1} C_k^{I_1 K_2} - 3 \hat{W}_{K_1 I_2 I_3 I_4} F_{I_4 K_2}^t C_k^{I_2 I_3} \right]_{\mathrm{as}\{K_1, K_2\}} . \tag{4.3}$$

We have to make some further comments about the set of solutions of (4.3). This set contains bound states as well as scattering states. How to deal with the latter will be discussed in Section 4.6 and in the applications of weak mapping. In spite of its assumed completeness the set of solutions is, however, in general *not* an orthogonal set, since the eigenvalue equation (4.2) or (4.3), respectively, is *not* a Schrödinger equation, and the solutions in general have *not* the meaning of quantum mechanical probability amplitudes. This non–Schrödinger property is a consequence of the quantum field theoretic state construction given in Chapter 3, and it also manifests itself in the hard core equations (4.2). How to proceed in this case was discussed in detail in Chapter 3. But for the weak mapping theorems the non–orthogonality of the $C_k^{I_1 \ldots I_m}$ is of no relevance. It suffices to introduce in any sector dual functions $R_{I_1 \ldots I_m}^k$, $m = 1, 2 \ldots \infty$, which are defined by orthogonality and completeness relations. For the two–particle sector we introduce the duals $R_{I_1 I_2}^k$ by the orthogonality conditions

$$\sum_{I_1, I_2} R_{I_1 I_2}^k C_{k'}^{I_1 I_2} = \delta_{kk'} , \tag{4.4}$$

and assuming the completeness of the original set, the completeness relations

$$\sum_k R_{I_1 I_2}^k C_k^{K_1 K_2} = \frac{1}{2} \left( \delta_{I_1 K_1} \delta_{I_2 K_2} - \delta_{I_2 K_1} \delta_{I_1 K_2} \right) \tag{4.5}$$

have to hold, where in (4.5) the antisymmetries $C_k^{I_1 I_2} = -C_k^{I_2 I_1}$ and $R_{I_1 I_2}^k = -R_{I_2 I_1}^k$ were taken into account. In Section 4.8 we will give a method for calculating the duals explicitly.

## 4.3 Selfconsistent Propagators

The hard core state equation (4.2) shows that the hard core states themselves essentially depend on the one–time propagator $F_{I_1 I_2}^t$ which is the relativistic analogue to the quantum mechanical interaction potential. Hence a thorough investigation of $F_{I_1 I_2}^t$ is necessary.

From a formal point of view $F^t_{I_1 I_2}$ is a vacuum expectation value and hence an uncorrelated part of the state functional. It was emphasized in Section 3.8 that weak mapping theorems are based on an expansion of matrix elements and that uncorrelated parts of these matrix elements must be compatible with the basis functions of the expansion. In particular if the basis functions themselves are (exact) solutions of an (unperturbed) functional Hamiltonian the corresponding uncorrelated parts must be vacuum expectation values of the same functional equations. For hard core states the functional Hamiltonian is defined by the operator equation (4.2):

$$\mathcal{H}^{\mathbf{d}}_{\mathrm{F}} := j^t_{I_1} \hat{K}_{I_1 I_2} \partial^t_{I_2} - 3\hat{W}_{I_1 I_2 I_3 I_4} F^t_{I_4 K} j^t_{I_1} j^t_K \partial^t_{I_3} \partial^t_{I_2} \quad . \tag{4.6}$$

The propagator $F^t_{I_1 I_2}$ can be embedded into this formalism if one observes that vacuum expectation values have vanishing energy values, as one can easily verify by means of equation (3.78) for $|p\rangle \equiv |0\rangle$. Therefore, if we interpret the functions $C_k^{I_1 I_2}$ as two–particle matrix elements, the solution $C_0^{I_1 I_2}$ of equation (4.3) for $E_{p0} = 0$ must be a vacuum expectation value, i.e. we have $C_0^{I_1 I_2} \equiv F^t_{I_1 I_2}$. Thus $F^t_{I_1 I_2}$ has to satisfy the equation

$$\hat{K}_{K_1 I} F^t_{I K_2} - \hat{K}_{K_2 I} F^t_{I K_1} = 3\hat{W}_{K_1 I_2 I_3 I_4} F^t_{I_4 K_2} F^t_{I_3 I_2} - 3\hat{W}_{K_2 I_2 I_3 I_4} F^t_{I_4 K_1} F^t_{I_3 I_2} \quad , \tag{4.7}$$

which can be considered as a selfconsistency condition for the determination of $F^t_{I_1 I_2}$. We will discuss this feature in more detail in connection with microscopic superconductivity theory in Chapter 9. In any case from these considerations it follows that the determination of $F^t_{I_1 I_2}$ is closely related to the dynamical fixing of the quantum field theoretic representations, and further investigations are necessary.

For our investigations in *relativistic* quantum field theory we use an elementary argument to derive $F^t_{I_1 I_2}$. According to the Lehmann–Källen spectral theorem propagators can be decomposed into products of one–particle matrix elements $\langle 0|\psi_I|a\rangle$. For equal times $t = t'$ one obtains for $F^t_{I_1 I_2}$

$$F^t_{I_1 I_2} = \langle 0|\mathcal{A}(\psi_{I_1}\psi_{I_2})_t|0\rangle = \sum_{a,b} \Big\{ \langle 0|\psi_{I_1}|a\rangle g^{ab} \langle b|\psi_{I_2}|0\rangle \Big\}_{\text{as } \{I_1, I_2\}, t_1 = t_2} \quad , \tag{4.8}$$

where $g^{ab}$ is the dual to the metrical tensor $g_{ab} = \langle a|b\rangle$. Hence the properties of $F^t_{I_1 I_2}$ depend on the properties of these matrix elements. If the states $|a\rangle$ are dressed particle states apart from symmetry considerations no general statements can be made about the resulting one–particle matrix elements. For hard core states $|k\rangle$, however, we can directly calculate the corresponding one–particle matrix elements from equation (4.2) by substitution of $|\mathcal{F}(j^t;1,k)\rangle^{\mathbf{d}} = iC^I_k j^t_I |0\rangle_F$ into (4.2). Projecting (4.2) with $\langle d_1|$ from the left, we obtain the sector equation

$$E^{\mathbf{d}}_{k0} C^{I_1}_k = \hat{K}_{I_1 I_2} C^{I_2}_k \ . \tag{4.9}$$

From this equation it follows that the matrix elements $\langle 0|\psi_I|k\rangle \equiv C^I_k$ for hard core states $|k\rangle$ must be free–field matrix elements, as (4.9) is the free Dirac equation, i.e. we have $C^I_k \equiv \langle 0|\psi^f_I|k\rangle$. For free fields the metrical tensor $g^{kk'}$ can be directly calculated from the corresponding anticommutation relations (1.41), see Section 5.6. Therefore, in this case we obtain for (4.8)

$$F^t_{I_1 I_2} = \sum_{k_1, k_2} \Big\{ \langle 0|\psi^f_{I_1}|k_1\rangle g^{k_1 k_2} \langle k_2|\psi^f_{I_2}|0\rangle \Big\}_{\text{as}\{I_1, I_2\}, t_1 = t_2} \quad , \tag{4.10}$$

and this expression is equal to the spectral decomposition on a spacelike hyperplane of $F^f_{I_1 I_2}$ given by Theorem 2.16. Hence from (4.9), together with (1.41) we have $F^t_{I_1 I_2} = F^f_{I_1 I_2}$ on the hyperplane.

The statement $F^t_{I_1 I_2} = F^f_{I_1 I_2}$ is, however, in general not compatible with equation (4.7). The reason for this discrepancy is the fact that $\mathcal{H}^{\mathrm{d}}_{\mathrm{F}}$ of (4.6) does not really arise from a corresponding field theory with state space Hamiltonian $\mathbf{H}^{\mathrm{d}}$ and corresponding field operators $\psi^{\mathrm{d}}_I$. As a consequence in the decomposition (4.8) the one–particle functions cannot be regarded as solutions of a corresponding eigenvalue equation resulting from a $\mathbf{H}^{\mathrm{d}}$. In spite of this discrepancy we use $F^t_{I_1 I_2} = F^f_{I_1 I_2}$ as a first approximation for $F^t_{I_1 I_2}$.

## 4.4 Effective Boson Dynamics

A general proof of a weak mapping theorem which does not rely upon the kind of basic quantum fields and their bound states cannot be given. Although in any case the general reasoning runs along the same lines, each quantum field or quantum field combination and their bound states need an individual treatment.

In order to make the deduction as transparent as possible we demonstrate such mappings for the spinorfield of Chapter 1, i.e. for a single type of basic fields. Furthermore, we confine ourselves to two–fermion bound states, i.e. to the special mapping of the spinorfield equation onto effective boson field equations. Other cases of weak mapping will be treated in the following chapters.

For convenience we list various symbols which will be used in the following:

| | |
|---|---|
| $j^t_I$ : | fermion sources on the hyperplane $t = \text{const}$, |
| $b_k$ : | boson sources on the hyperplane $t = \text{const}$, |
| $\partial^t_I := \delta/\delta j^t_I$ : | fermion "duals" on the hyperplane $t = \text{const}$, |
| $\partial_k := \delta/\delta b_k$: | boson "duals" on the hyperplane $t = \text{const}$, |
| $\partial_\mu := \partial/\partial x_\mu$: | for Greek small letter index. |

In particular capital Latin letters refer to fermionic functional operators while small Latin letters refer to bosonic quantities. Furthermore, as long as no misunderstanding is to be expected we omit the index $t$ which indicates the hyperplane $t = \text{const}$.

We assume that the boson dynamics can be formulated by functional boson states and corresponding functional boson equations. To that end we define a functional space $\mathbb{K}_B$ analogously to the fermion case in Chapter 2 as a representation space of the bosonic functional Fock states which are generated by the boson sources $b_k$ and supplied by their "duals" $\partial_k$, where $k$ is the collection of quantum numbers characterizing the corresponding boson states. The Fock representation of the corresponding algebra is then generated by

$$\partial_k|0\rangle_B = 0\,, \qquad b^+_k|0\rangle = 0\,, \tag{4.11}$$

where $|0\rangle_B$ is the bosonic functional Fock "vacuum".

As the functional spaces $\mathbb{K}_F$ and $\mathbb{K}_B$ are in principle only auxiliary spaces allowing a compact formulation of the field dynamics, they have to be independent of each other. Therefore sources and their "duals" for fermions and bosons are assumed to commute completely. Later on, by working with the chain rule we shall abandon this postulate in favour of an approximate treatment, but for the present investigation it strictly holds. The commutation rules for the boson operators are

$$\begin{aligned} \left[\partial_k, \partial_{k'}\right]_- &= \left[b_k, b_{k'}\right]_- = 0\,, \\ \left[\partial_k, b_{k'}\right]_- &= \delta_{kk'}\,, \end{aligned} \tag{4.12}$$

and due to our assumption any boson field state $|a\rangle$ can be represented by a functional state $|\mathcal{B}(b;a)\rangle \in \mathbb{K}_B$ with the expansion

$$|\mathcal{B}(b;a)\rangle = \sum_{n=0}^{\infty} \frac{1}{n!} \rho^{(n)}(k_1 \dots k_n|a) b_{k_1} \dots b_{k_n} |0\rangle_B \tag{4.13}$$

where $\rho^{(n)}(k_1 \dots k_n|a)$ is symmetric in $k_1 \dots k_n$.

The calculation rules for boson functionals are analogous to the fermion case (2.18)–(2.20) and are given by

- **Proposition 4.1:** For $|\mathcal{B}(b;a)\rangle \in \mathbb{K}_B$ with representation (4.13) the following relations hold:

(i) $$_B\langle 0|\partial_{k_m} \dots \partial_{k_1}|\mathcal{B}(b;a)\rangle = \rho^{(m)}(k_1 \dots k_m|a)\,, \tag{4.14}$$

(ii) $$\partial_j|\mathcal{B}(b;a)\rangle = \sum_{n=0}^{\infty} \frac{1}{n!} \rho^{(n+1)}(j, k_1 \dots k_n|a) b_{k_1} \dots b_{k_n}|0\rangle_B\,, \tag{4.15}$$

(iii) $$b_j|\mathcal{B}(b;a)\rangle = \sum_{n=1}^{\infty} \frac{1}{n!} \sum_{l=1}^{n} \delta_{jk_l} \rho^{(n-1)}(k_1 \dots \overset{o}{k_l} \dots k_n|a) b_{k_1} \dots b_{k_n}|0\rangle_B \tag{4.16}$$

**Proof:** With (4.11) and (4.12) it is

$$_B\langle 0|\partial_{k_m} \dots \partial_{k_1} b_{j_1} \dots b_{j_n}|0\rangle_B = \delta_{mn} \sum_{p \in S_n} \delta_{j_1 k_{p_1}} \dots \delta_{j_n k_{p_n}}\,. \tag{4.17}$$

With (4.17) and the symmetry of the coefficients $\rho^{(n)}(k_1 \dots k_n|a)$ one obtains

$$\begin{aligned} _B\langle 0|\partial_{k_m} \dots \partial_{k_1}|\mathcal{B}(b;a)\rangle &= \frac{1}{m!} \rho^{(m)}(k_1 \dots k_m|a) \sum_{p \in S_n} \delta_{j_1 k_{p_1}} \dots \delta_{j_m k_{p_m}} \\ &= \rho^{(m)}(k_1 \dots k_m|a)\,. \end{aligned} \tag{4.18}$$

The proof of ii) and iii) runs along the same lines as that for formulas (2.18) and (2.19). ◇

The description of bosons as bound states of subfermions and the derivation of a corresponding effective dynamics has to start from the normalordered energy equation on the hyperplane (3.106). This was justified in detail in Section 2.7. Hence the intended map has to be applied to (3.106). An inspection of this equation shows that the set of solutions can be subdivided into two classes, which completely decouple: those with all coefficent functions of $|\mathcal{F}(j^t;a)\rangle$ belonging to an even number of arguments, and those with an odd number of arguments. With respect to the mapping onto a boson theory for two–particle bound boson states only the first class of solutions is of interest. Thus in Sections 4.4–4.7 we confine our considerations to the functional states

$$|\mathcal{F}(j;a)\rangle = \sum_{n=0}^{\infty} \frac{(-1)^n}{(2n)!} \varphi^{(2n)}(I_1 \dots I_{2n}|a) j_{I_1} \dots j_{I_{2n}} |0\rangle_F \tag{4.19}$$

on the hyperplane $t = \text{const}$.

Without any regard to functional space we define the weak mapping of a fermion theory onto a boson theory only by means of the coefficient functions $\varphi^{(2n)}(I_1 \dots I_{2n})$. In particular with respect to the map by two–particle boson functions we assume the set of solutions $C_k^{I_1 I_2}$ of equation (4.3) to be complete. Then we define the antisymmetrized generalized determinants by

$$D_{2n}\begin{pmatrix} I_1 \dots I_{2n} \\ k_1 \dots k_n \end{pmatrix} := C_{k_1}^{\{I_1 I_2} \dots C_{k_n}^{I_{2n-1} I_{2n}\}} \quad , \tag{4.20}$$

where $\{I_1 \dots I_{2n}\}$ means antisymmetrization in $I_1 \dots I_{2n}$ with a corresponding factor $\frac{1}{(2n)!}$. By means of the completeness relations (4.5) it can be easily shown that the set of all $D_{2n}$, where $k_1 \dots k_n$ run through all quantum numbers of the two–particle states, is complete in the space of antisymmetrized functions $f(I_1 \dots I_{2n})$. Therefore we can expand any coefficient function $\varphi^{(2n)}(I_1 \dots I_{2n})$ in terms of the determinantal functions (4.20); i.e.,

$$\varphi^{(2n)}(I_1 \dots I_{2n}|a) = \sum_{k_1 \dots k_n} D_{2n}\begin{pmatrix} I_1 \dots I_{2n} \\ k_1 \dots k_n \end{pmatrix} d(k_1 \dots k_n|a) \tag{4.21}$$

with the expansion coefficients $d(k_1 \dots k_n|a)$.

We now introduce weak mapping of a fermion theory onto a boson theory by the following definition, see Stumpf [Stu 85a], Kerschner and Stumpf [Ker 91]:

- **Definition 4.1:** Weak mapping is defined by the identification of the expansion coefficient functions $d(k_1 \dots k_n|a)$ of (4.21) with the coefficient functions $\rho^{(n)}$ of the boson functional state (4.13), i.e.,

$$d(k_1 \dots k_n|a) \equiv \rho^{(n)}(k_1 \dots k_n|a) \tag{4.22}$$

or

$$\varphi^{(2n)}(I_1 \dots I_{2n}|a) = C_{k_1}^{\{I_1 I_2} \dots C_{k_n}^{I_{2n-1} I_{2n}\}} \rho^{(n)}(k_1 \dots k_n|a) \quad . \tag{4.23}$$

Equation (4.23) establishes a map $\rho^{(n)} \to \varphi^{(2n)}$. If the set (4.20) is complete the inverse mapping $\varphi^{(2n)} \to \rho^{(n)}$ exists, i.e., the map is bijective, provided only symmetrized $\rho^{(n)}(k_1 \dots k_n)$ are admitted. The connection to effective field equations is established by interpreting the boson coefficient functions $\rho^{(n)}$ as symmetrized expectation values of effective boson field operators; for a detailed discussion we refer to the applications in the following chapters.

Equation (4.23) provides additional evidence for the use of equation (3.106) as the starting point of weak mapping. One can expect the set of functions $C_k^{I_1 I_2}$ to be complete only on a hyperplane $t = \text{const}$, if one rejects Euclidean field theories as unphysical and compares the functions $C_k^{I_1 I_2}$ with ordinary quantum mechanical wave functions. Only for the existence of an inverse mapping, (4.23) allows a faithful mapping and prevents an uncontrollable loss of information of the basic theory. Thus the need for completeness enforces the use of the noncovariant formalism for the meaningfull definition of weak mapping.

In spite of the independence of weak mapping from the functional space, we want to formulate the *results* of weak mapping by functional equations. This was done in the following proposition derived by Pfister and Stumpf [Pfi 91]. This proposition is an alternative version to a theorem of Kerschner and Stumpf [Ker 91].

- **Proposition 4.2:** By the transformation (4.23) the functional energy equation (3.107) of the fermionic spinorfield (1.49) is mapped onto the following functional boson energy equation:

$$
\begin{aligned}
&E_{p0}\mathcal{U}|\mathcal{B}(b;p)\rangle = \qquad (4.24)\\
&\quad 2R^k_{I_1K}K_{I_1I_2}b_k\mathcal{V}\begin{pmatrix} I_2K \\ k' \end{pmatrix}\partial_{k'}|\mathcal{B}(b;p)\rangle\\
&\quad -W_{K_1I_2I_3I_4}\Big\{6F_{I_4K_2}R^k_{K_1K_2}b_k\mathcal{V}\begin{pmatrix} I_2I_3 \\ k' \end{pmatrix}\partial_{k'}\\
&\quad +12F_{I_4K_2}R^{k_1}_{K_1K_3}R^{k_2}_{K_2K_4}b_{k_1}b_{k_2}\mathcal{W}\left(\begin{array}{c|c} I_2K_3 & I_3K_4 \\ k_1' & k_2' \end{array}\right)\partial_{k_1'}\partial_{k_2'}\\
&\quad -2R^k_{K_1K_2}b_k\mathcal{W}\left(\begin{array}{c|c} I_2I_3 & I_4K_2 \\ k_1' & k_2' \end{array}\right)\partial_{k_1'}\partial_{k_2'}\\
&\quad -\left(3F_{I_4K_2}F_{I_3K_3}+\frac{1}{4}A_{I_4K_2}A_{I_3K_3}\right)\times\\
&\quad \sum_{p\in S_3}\mathrm{sgn}(p)\frac{1}{3!}\Big[12R^{k_1}_{K_{p_1}K_{p_2}}R^{k_2}_{K_{p_3}K_4}b_{k_1}b_{k_2}\mathcal{V}\begin{pmatrix} I_2K_4 \\ k_1' \end{pmatrix}\partial_{k'}\\
&\quad +8R^{k_1}_{K_{p_1}K_4}R^{k_2}_{K_{p_2}K_5}R^{k_3}_{K_{p_3}K_6}b_{k_1}b_{k_2}b_{k_3}\mathcal{W}\left(\begin{array}{c|c} I_2K_6 & K_5K_4 \\ k_1' & k_2' \end{array}\right)\partial_{k_1'}\partial_{k_2'}\Big]\\
&\quad +\left(F_{I_4K_2}F_{I_3K_3}+\frac{1}{4}A_{I_4K_2}A_{I_3K_3}\right)F_{I_2K_4}\times\\
&\quad \sum_{p\in S_4}\mathrm{sgn}(p)\frac{1}{4!}\Big[12R^{k_1}_{K_{p_1}K_{p_2}}R^{k_2}_{K_{p_3}K_{p_4}}b_{k_1}b_{k_2}\mathcal{U}\\
&\quad +48R^{k_1}_{K_{p_1}K_5}R^{k_2}_{K_{p_2}K_6}R^{k_3}_{K_{p_3}K_{p_4}}b_{k_1}b_{k_2}b_{k_3}\mathcal{V}\begin{pmatrix} K_5K_6 \\ k_1' \end{pmatrix}\partial_{k'}\\
&\quad +16R^{k_1}_{K_{p_1}K_5}\ldots R^{k_4}_{K_{p_4}K_8}b_{k_1}\ldots b_{k_4}\mathcal{W}\left(\begin{array}{c|c} K_5K_6 & K_7K_8 \\ k_1' & k_2' \end{array}\right)\partial_{k_1'}\partial_{k_2'}\Big]\Big\}|\mathcal{B}(b;p)\rangle
\end{aligned}
$$

with

$$
\mathcal{U} := \sum_{n=0}^{\infty}\frac{1}{n!}b_{k_1'}\ldots b_{k_n'}|0\rangle_B R^{k_1'}_{K_1K_2}\ldots R^{k_n'}_{K_{2n-1}K_{2n}}\times \qquad (4.25)
$$

$$C_{k_1}^{\{K_1K_2} \dots C_{k_n}^{K_{2n-1}K_{2n}\}} {}_B\langle 0|\partial_{k_n} \dots \partial_{k_1} ,$$

$$\mathcal{V}\begin{pmatrix} I_1I_2 \\ k \end{pmatrix} := \sum_{n=0}^{\infty} \frac{1}{n!} b_{k'_1} \dots b_{k'_n} |0\rangle_B R_{K_1K_2}^{k'_1} \dots R_{K_{2n-1}K_{2n}}^{k'_n} \times \tag{4.26}$$

$$C_k^{\{I_1I_2} C_{k_1}^{K_1K_2} \dots C_{k_n}^{K_{2n-1}K_{2n}\}} {}_B\langle 0|\partial_{k_n} \dots \partial_{k_1} ,$$

$$\mathcal{W}\begin{pmatrix} I_1I_2 & I_3I_4 \\ k & k' \end{pmatrix} := \sum_{n=0}^{\infty} \frac{1}{n!} b_{k'_1} \dots b_{k'_n} |0\rangle_B R_{K_1K_2}^{k'_1} \dots R_{K_{2n-1}K_{2n}}^{k'_n} \times \tag{4.27}$$

$$C_k^{\{I_1I_2} C_{k'}^{I_3I_4} C_{k_1}^{K_1K_2} \dots C_{k_n}^{K_{2n-1}K_{2n}\}} {}_B\langle 0|\partial_{k_n} \dots \partial_{k_1} .$$

**Proof:** We write equation (4.1), which is equivalent to (3.106), in the form

$$E_{p0}|\mathcal{F}(j;p)\rangle = \mathcal{H}_\mathrm{F}(j,\partial)|\mathcal{F}(j;p)\rangle \tag{4.28}$$

with

$$\begin{aligned} \mathcal{H}_\mathrm{F}(j,\partial) := {} & j_{I_1} \hat{K}_{I_1I_2} \partial_{I_2} \\ & + \hat{W}_{I_1I_2I_3I_4} \Big[ j_{I_1} \partial_{I_4} \partial_{I_3} \partial_{I_2} - 3F_{I_4K} j_{I_1} j_K \partial_{I_3} \partial_{I_2} \\ & + (3F_{I_4K_1}F_{I_3K_2} + \frac{1}{4} A_{I_4K_1} A_{I_3K_2}) j_{I_1} j_{K_1} j_{K_2} \partial_{I_2} \\ & - (F_{I_4K_1}F_{I_3K_2} + \frac{1}{4} A_{I_4K_1} A_{I_3K_2}) F_{I_2K_3} j_{I_1} j_{K_1} j_{K_2} j_{K_3} \Big] \quad . \end{aligned} \tag{4.29}$$

We remember that (4.28), (4.29) and thus also (4.24)–(4.27) are defined on the hyperplane $t = \text{const}$. To transform equation (4.28) into a bosonic functional energy equation we observe that weak mapping is defined in configuration space. Hence we first project equation (4.28) into configuration space, transform the coefficient functions, and afterwards we reformulate the result of the transformation in terms of bosonic functional states and operators.

By projection with ${}_F\langle 0|(\partial_{K_{2n}} \dots \partial_{K_1})$ from the left we obtain from (4.28) for $n = 0, 1, 2 \dots \infty$

$$(-1)^n E_{p0} \varphi^{(2n)}(K_1 \dots K_{2n}|a) = {}_F\langle 0|\partial_{K_{2n}} \dots \partial_{K_1} \mathcal{H}_\mathrm{F}(j,\partial)|\mathcal{F}(j;p)\rangle \quad . \tag{4.30}$$

Substitution of (4.23) into the left–hand side of (4.30) and subsequent multiplication with $R_{K_1K_2}^{k_1} \dots R_{K_{2n-1}K_{2n}}^{k_n}$ as well as summation over $K_1 \dots K_{2n}$ yields the equation

$$\begin{aligned} & E_{p0\,B}\langle 0|\partial_{k_n} \dots \partial_{k_1} \mathcal{U}|\mathcal{B}(b;p)\rangle \\ = {} & (-1)^n R_{K_1K_2}^{k_1} \dots R_{K_{2n-1}K_{2n}}^{k_n} {}_F\langle 0|\partial_{K_{2n}} \dots \partial_{K_1} \mathcal{H}_\mathrm{F}(j,\partial)|\mathcal{F}(j;p)\rangle \end{aligned} \tag{4.31}$$

which due to definition (4.25) is equivalent to (4.30). Substitution of (4.29) into (4.31) gives

$$\begin{aligned} & E_{p0\,B}\langle 0|\partial_{k_n} \dots \partial_{k_1} \mathcal{U}|\mathcal{B}(b;p)\rangle \\ = {} & \hat{K}_{I_1,I_2} T_1(I_1,I_2) \\ & + \hat{W}_{I_1I_2I_3I_4} \Big[ T_2(I_1,I_4,I_3,I_2) - 3F_{I_4K} T_3(I_1,K,I_3,I_2) \\ & + (3F_{I_4K_1}F_{I_3K_2} + \frac{1}{4} A_{I_4K_1} A_{I_3K_2}) T_4(I_1,K_1,K_2,I_2) \\ & - (F_{I_4K_1}F_{I_3K_2} + \frac{1}{4} A_{I_4K_1} A_{I_3K_2}) F_{I_2K_3} T_5(I_1,K_1,K_2,K_3) \Big] \end{aligned} \tag{4.32}$$

with the definitions

$$T_1 := (-1)^n R^{k_1}_{K_1K_2}\dots R^{k_n}_{K_{2n-1}K_{2n}}\, {}_F\langle 0|\partial_{K_{2n}}\dots\partial_{K_1} j_{I_1}\partial_{I_2}|\mathcal{F}(j;p)\rangle\,, \tag{4.33}$$

$$T_2 := (-1)^n R^{k_1}_{K_1K_2}\dots R^{k_n}_{K_{2n-1}K_{2n}}\, {}_F\langle 0|\partial_{K_{2n}}\dots\partial_{K_1} j_{I_1}\partial_{I_4}\partial_{I_3}\partial_{I_2}|\mathcal{F}(j;p)\rangle\,, \tag{4.34}$$

$$T_3 := (-1)^n R^{k_1}_{K_1K_2}\dots R^{k_n}_{K_{2n-1}K_{2n}}\, {}_F\langle 0|\partial_{K_{2n}}\dots\partial_{K_1} j_{I_1}j_{I_2}\partial_{I_3}\partial_{I_4}|\mathcal{F}(j;p)\rangle\,, \tag{4.35}$$

$$T_4 := (-1)^n R^{k_1}_{K_1K_2}\dots R^{k_n}_{K_{2n-1}K_{2n}}\, {}_F\langle 0|\partial_{K_{2n}}\dots\partial_{K_1} j_{I_1}j_{I_2}j_{I_3}\partial_{I_4}|\mathcal{F}(j;p)\rangle\,, \tag{4.36}$$

$$T_5 := (-1)^n R^{k_1}_{K_1K_2}\dots R^{k_n}_{K_{2n-1}K_{2n}}\, {}_F\langle 0|\partial_{K_{2n}}\dots\partial_{K_1} j_{I_1}j_{I_2}j_{I_3}j_{I_4}|\mathcal{F}(j;p)\rangle\,. \tag{4.37}$$

We transform the various terms on the right–hand side of (4.32) in turn. For the evaluation we use (4.23) and the functional calculation rules for fermionic and bosonic sources without further comment.

i) For exemplification we treat the first term of $\mathcal{H}_\mathrm{F}$ in detail and evaluate $T_1$ given by (4.33). It holds

$$\begin{aligned} T_1 &= R^{k_1}_{K_1K_2}\dots R^{k_n}_{K_{2n-1}K_{2n}}\sum_{i=1}^{2n}\delta_{I_1K_i}\varphi^{(2n)}(K_1\dots K_{i-1}I_2K_{i+1}\dots K_{2n}|a) \\ &= R^{k_1}_{K_1K_2}\dots R^{k_n}_{K_{2n-1}K_{2n}}\sum_{i=1}^{2n}\delta_{I_1K_i}C^{\{K_1K_2\dots K_{i-1}I_2K_{i+1}\dots}_{k'_1}C^{K_{2n-1}K_{2n}\}}_{k'_n}\times \\ &\quad {}_B\langle 0|\partial_{k'_n}\dots\partial_{k'_1}|\mathcal{B}(b;p)\rangle\quad . \end{aligned} \tag{4.38}$$

Furthermore we define for bosonic functional states the expression

$$\tilde{T}_1 := {}_B\langle 0|\partial_{k_n}\dots\partial_{k_1}2R^{k'}_{I_1J}b_{k'}\mathcal{V}\begin{pmatrix}I_2J\\k''\end{pmatrix}\partial_{k''}|\mathcal{B}(b;p)\rangle \tag{4.39}$$

by using definition (4.26). Then it is

$$\begin{aligned} \tilde{T}_1 &= 2R^{k'}_{I_1J}\sum_{i=1}^{n}\delta_{k'k_i}R^{k_1}_{K_1K_2}\dots R^{k_{i-1}}_{K_{2i-3}K_{2i-2}}R^{k_{i+1}}_{K_{2i-1}K_{2i}}\dots R^{k_n}_{K_{2n-3}K_{2n-2}}\times \\ &\quad C^{\{I_2J}_{k'_0}C^{K_1K_2}_{k'_1}\dots C^{K_{2n-3}K_{2n-2}\}}_{k'_{n-1}}\,{}_B\langle 0|\partial_{k'_{n-1}}\dots\partial_{k'_0}|\mathcal{B}(b;p)\rangle \\ &= 2\sum_{i=1}^{n}R^{k_1}_{K_1K_2}\dots R^{k_{i-1}}_{K_{2i-3}K_{2i-2}}R^{k_i}_{I_1J}R^{k_{i+1}}_{K_{2i-1}K_{2i}}\dots R^{k_n}_{K_{2n-3}K_{2n-2}}\times \\ &\quad C^{\{I_2J}_{k'_1}C^{K_1K_2}_{k'_2}\dots C^{K_{2n-3}K_{2n-2}\}}_{k'_n}\,{}_B\langle 0|\partial_{k'_n}\dots\partial_{k'_1}|\mathcal{B}(b;p)\rangle \\ &= 2\sum_{i=1}^{n}\delta_{I_1K_{2i-1}}R^{k_1}_{K_1K_2}\dots R^{k_i}_{K_{2i-1}K_{2i}}\dots R^{k_n}_{K_{2n-1}K_{2n}}\times \\ &\quad C^{\{K_1K_2\dots K_{2i-2}I_2K_{2i}K_{2i+1}\dots}_{k'_1\dots}C^{K_{2n-1}K_{2n}\}}_{k'_n}\,{}_B\langle 0|\partial_{k'_n}\dots\partial_{k'_1}|\mathcal{B}(b;p)\rangle \\ &- \sum_{i=1}^{n}\delta_{I_1K_{2i-1}}R^{k_1}_{K_1K_2}\dots R^{k_i}_{K_{2i-1}K_{2i}}\dots R^{k_n}_{K_{2n-1}K_{2n}}\times \\ &\quad C^{\{K_1K_2\dots K_{2i-2}I_2K_{2i}K_{2i+1}\dots}_{k'_1\dots}C^{K_{2n-1}K_{2n}\}}_{k'_n}\,{}_B\langle 0|\partial_{k'_n}\dots\partial_{k'_1}|\mathcal{B}(b;p)\rangle \end{aligned} \tag{4.40}$$

$$
\begin{aligned}
&+\sum_{j=1}^{n}\delta_{I_1K_{2j-1}}R^{k_1}_{K_1K_2}\dots R^{k_j}_{K_{2j}K_{2j-1}}\dots R^{k_n}_{K_{2n-1}K_{2n}}\times\\
&\quad C^{\{K_1K_2\dots K_{2j-2}K_{2j}I_2K_{2j+1}\dots}_{k'_1\dots}C^{K_{2n-1}K_{2n}\}}_{k'_n}\,{}_B\langle 0|\partial_{k'_n}\dots\partial_{k'_1}|\mathcal{B}(b;p)\rangle\\
= &\; R^{k_1}_{K_1K_2}\dots R^{k_n}_{K_{2n-1}K_{2n}}\sum_{i=1}^{2n}\delta_{I_1K_i}C^{\{K_1K_2\dots K_{i-1}I_2K_{i+1}\dots}_{k'_1\dots}C^{K_{2n-1}K_{2n}\}}_{k'_n}\times\\
&\quad {}_B\langle 0|\partial_{k'_n}\dots\partial_{k'_1}|\mathcal{B}(b;p)\rangle\\
= &\; T_1 \quad .
\end{aligned}
$$

Hence we have

$$
\tilde{T}_1 = T_1 = {}_B\langle 0|\partial_{k_n}\dots\partial_{k_1}2R^{k'}_{I_1J}b_{k'}\mathcal{V}\begin{pmatrix} I_2J \\ k'' \end{pmatrix}\partial_{k''}|\mathcal{B}(b;p)\rangle \quad . \tag{4.41}
$$

The treatment of the other terms runs along the same lines as $T_1$. Thus we do not demonstrate their evaluation in full lenght. ii) For the second term we have

$$
\begin{aligned}
T_2 = &\; R^{k_1}_{K_1K_2}\dots R^{k_n}_{K_{2n-1}K_{2n}}\times \qquad (4.42)\\
&\sum_{i=1}^{2n}\delta_{I_1K_i}C^{\{I_2I_3}_{k}C^{K_1K_2\dots K_{i-1}I_4K_{i+1}\dots}_{k'_1\dots}C^{K_{2n-1}K_{2n}\}}_{k'_n}\,{}_B\langle 0|\partial_{k'_n}\dots\partial_{k'_1}\partial_k|\mathcal{B}(b;p)\rangle\\
= &\; 2R^{k'}_{I_1K}\,{}_B\langle 0|\partial_{k_n}\dots\partial_{k_1}b_{k'}\mathcal{W}\left(\begin{matrix} I_2I_3 \\ k''_1 \end{matrix}\middle|\begin{matrix} I_4K \\ k''_2 \end{matrix}\right)\partial_{k''_1}\partial_{k''_2}|\mathcal{B}(b;p)\rangle \qquad (4.43)
\end{aligned}
$$

with definition (4.27).

iii) The third term is given by

$$
\begin{aligned}
T_3 = &\sum_{\substack{i,j=1\\ i\neq j}}^{2n}R^{k_1\dots}_{K_1K_2\dots K_{i-1}I_1K_{i+1}\dots K_{j-1}I_2K_{j+1}\dots}R^{k_n}_{K_{2n-1}K_{2n}}\times \qquad (4.44)\\
&C^{\{K_1\dots K_{i-1}I_3K_{i+1}\dots K_{j-1}I_4K_{j+1}\dots}_{k'_1\dots}C^{K_{2n-1}K_{2n}\}}_{k'_n}\,{}_B\langle 0|\partial_{k'_n}\dots\partial_{k'_1}|\mathcal{B}(b;p)\rangle\,.
\end{aligned}
$$

We decompose the summation over $i, j$ into

$$
\sum_{\substack{i,j=1\\ i\neq j}}^{2n} = \sum_{\substack{i=2k,1\leq k\leq n\\ j=i-1}} + \sum_{\substack{j=2k,1\leq k\leq n\\ i=j-1}} + \sum{}' \quad , \tag{4.45}
$$

where $\sum'$ covers all combinations not contained in the two other summations. The first two terms can be represented by means of (4.26), while $\sum'$ has to be represented by (4.27). This gives

$$
\begin{aligned}
T_3 = &\; {}_B\langle 0|\partial_{k_n}\dots\partial_{k_1}\Bigg[2R^{k'}_{I_1I_2}b_{k'}\mathcal{V}\begin{pmatrix} I_4I_3 \\ k'' \end{pmatrix}\partial_{k''} \qquad (4.46)\\
&+4R^{k'_1}_{I_1K_1}R^{k'_2}_{I_2K_2}b_{k'_1}b_{k'_2}\mathcal{W}\left(\begin{matrix} I_4K_1 \\ k''_1 \end{matrix}\middle|\begin{matrix} I_3K_2 \\ k''_2 \end{matrix}\right)\partial_{k''_1}\partial_{k''_2}\Bigg]|\mathcal{B}(b;p)\rangle \quad .
\end{aligned}
$$

iv) For the fourth term we obtain

$$T_4 = \sum_{p\in S_3} \mathrm{sgn}(p) \sum_{\substack{i_{p_1},i_{p_2},i_{p_3}=1\\ i_{p_1}>i_{p_2}>i_{p_3}}}^{2n} (-1)^{(1+\sum_{j=1}^{3} i_{p_j})} \delta_{I_1 K_{i_1}} \delta_{I_2 K_{i_2}} \delta_{I_3 K_{i_3}} R^{k_1}_{K_1 K_2} \cdots \times \tag{4.47}$$

$$R^{k_n}_{K_{2n-1}K_{2n}} C^{\{I_4 K_1 \ldots \overset{0}{K}_{i_{p_3}} \ldots \overset{0}{K}_{i_{p_2}} \ldots \overset{0}{K}_{i_{p_1}} \cdots}_{k'_1 \cdots} C^{K_{2n-1}K_{2n}\}}_{k'_{n-1}} {}_B\langle 0|\partial_{k'_{n-1}} \ldots \partial_{k'_1} |\mathcal{B}(b;p)\rangle .$$

With the decomposition

$$\sum_{\substack{i_{p_1},i_{p_2},i_{p_3}=1\\ i_{p_1}>i_{p_2}>i_{p_3}}}^{2n} = \sum_{\substack{i_{p_3}=2j,1\le j\le n\\ i_{p_2}=i_{p_3}-1\\ i_{p_1}>i_{p_2}>i_{p_3}}}^{2n} + \sum_{\substack{i_{p_2}=2j,1\le j\le n\\ i_{p_1}=i_{p_2}-1\\ i_{p_1}>i_{p_2}>i_{p_3}}}^{2n} + \sum{}' \tag{4.48}$$

in the same way as for iii) the following expression results:

$$T_4 = {}_B\langle 0|\partial_{k_n} \ldots \partial_{k_1} \sum_{p\in S_3} \mathrm{sgn}(p) \frac{1}{3!} \Big[ 12 R^{k'_1}_{I_{p_1} I_{p_2}} R^{k'_2}_{I_{p_3} K} b_{k'_1} b_{k'_2} \mathcal{V}\begin{pmatrix} I_4 K \\ k'' \end{pmatrix} \partial_{k''} \tag{4.49}$$

$$+ 8 R^{k'_1}_{I_{p_1} K_1} R^{k'_2}_{I_{p_2} K_2} R^{k'_3}_{I_{p_3} K_3} b_{k'_1} b_{k'_2} b_{k'_3} \mathcal{W}\begin{pmatrix} I_4 K_3 & K_2 K_1 \\ k''_1 & k''_2 \end{pmatrix} \partial_{k''_1} \partial_{k''_2} \Big] |\mathcal{B}(b;p)\rangle .$$

v) For the last term it holds

$$T_5 = \sum_{p\in S_4} \mathrm{sgn}(p) \sum_{\substack{i_{p_1},\ldots,i_{p_4}=1\\ i_{p_1}>\ldots>i_{p_4}}}^{2n} (-1)^{(\sum_{j=1}^{4} i_{p_j})} \delta_{I_1 K_{i_1}} \ldots \delta_{I_4 K_{i_4}} R^{k_1}_{K_1 K_2} \cdots R^{k_n}_{K_{2n-1}K_{2n}} \times$$

$$C^{\{K_1 K_2 \ldots \overset{0}{K}_{i_{p_4}} \ldots \overset{0}{K}_{i_{p_3}} \ldots \overset{0}{K}_{i_{p_2}} \ldots \overset{0}{K}_{i_{p_1}} \cdots}_{k'_1 \cdots} C^{K_{2n-1}K_{2n}\}}_{k'_{n-1}} {}_B\langle 0|\partial_{k'_{n-2}} \ldots \partial_{k'_1} |\mathcal{B}(b;p)\rangle \tag{4.50}$$

By a decomposition analogous to iii) and iv) we obtain

$$T_5 = {}_B\langle 0|\partial_{k_n} \ldots \partial_{k_1} \sum_{p\in S_4} \mathrm{sgn}(p) \frac{1}{4!} \Big[ 12 R^{k'_1}_{I_{p_1} I_{p_2}} R^{k'_2}_{I_{p_3} I_{p_4}} b_{k'_1} b_{k'_2} \mathcal{U} \tag{4.51}$$

$$+ 48 R^{k'_1}_{I_{p_1} K_1} R^{k'_2}_{I_{p_2} K_2} R^{k'_3}_{I_{p_3} I_{p_4}} b_{k'_1} b_{k'_2} b_{k'_3} \mathcal{V}\begin{pmatrix} K_1 K_2 \\ k'' \end{pmatrix} \partial_{k''}$$

$$+ 16 R^{k'_1}_{I_{p_1} K_1} \cdots R^{k'_4}_{I_{p_4} K_4} b_{k'_1} b_{k'_4} \mathcal{W}\begin{pmatrix} K_1 K_2 & K_3 K_4 \\ k''_1 & k''_2 \end{pmatrix} \partial_{k''_1} \partial_{k''_2} \Big] |\mathcal{B}(b;p)\rangle .$$

Collecting the results and replacement of $T_1$–$T_5$ in (4.32) by (4.41), (4.42) last line, (4.46), (4.49), (4.51) eventually leads to equation (4.24). ◇

Equation (4.24) represents a faithful mapping of the spinorfield dynamics onto a boson field dynamics and it contains all subtleties of the interactions of the fermionic constituents of the boson states under consideration. In order to extract the main physical effects from this equation we have to simplify and to approximate it in a suitable way.

## 4.5 Direct and Exchange Forces

The most suitable way to reduce the complicated structure of equation (4.24) and to get an insight into its physical meaning is the decomposition of the interactions into direct and exchange force interactions.

In quantum mechanics direct forces are characterized by the exclusive occurrence of probability densities in the interaction terms, while all other bilinear combinations of probability amplitudes in the interaction terms are attributed to exchange forces. In the quantum field theoretic energy equation (4.24) the expressions $R^k_{I_1 I_2} C^{I_1 I_2}_k$ (no summation) take over the part of quantum mechanical probability densities and thus are connected with direct forces, while all other $R$–$C$–combinations have to be attributed to exchange forces. We expect the exchange forces to be dominated by the direct forces; thus we separate the latter from the exchange forces by rewriting the expressions (4.25)–(4.27). The corresponding Propositions 4.3, 4.4 and 4.5 were derived by Pfister and Stumpf [Pfi 91]. We begin with (4.25).

- **Proposition 4.3:** Let $\lambda := (\lambda_1 \dots \lambda_j)$ be a partition of $n$ with

$$\lambda_i \in \mathrm{N}\,;\;\; \lambda_1 \geq \lambda_2 \geq \dots \geq \lambda_j\,;\;\; \sum_{i=1}^{j} \lambda_i = n$$

and $s_1, \dots s_l, 1 \leq l \leq j$, the multiplicities of identical values of the $\lambda_1 \dots \lambda_j$. Then the following relation holds:

$$R^{k'_1}_{K_1 K_2} \cdots R^{k'_n}_{K_{2n-1} K_{2n}} C^{\{K_1 K_2}_{k_1} \dots C^{K_{2n-1} K_{2n}\}}_{k_n} \tag{4.52}$$

$$= \frac{(n!)^2}{(2n)!} \sum_{\text{part } \lambda} (-1)^j \frac{2^{2n-j}}{s_1! \dots s_l!\, \lambda_1 \dots \lambda_j} \left\{ \begin{matrix} k'_1 & \dots & k'_n \\ k_1 & \dots & k_n \\ \lambda_1 & \dots & \lambda_j \end{matrix} \right\}_{\text{sym}}$$

where

$$\left\{ \begin{matrix} k'_1 & \dots & k'_n \\ k_1 & \dots & k_n \\ \lambda_1 & \dots & \lambda_j \end{matrix} \right\} := \tag{4.53}$$

$$\mathrm{tr}\Big[R^{k'_1} C_{k_1} \dots R^{k'_{\lambda_1}} C_{k_{\lambda_1}}\Big] \dots \mathrm{tr}\Big[R^{k'_{n-\lambda_j+1}} C_{k_{n-\lambda_j+1}} \dots R^{k'_n} C_{k_n}\Big]$$

and the suffix "sym" means separate symmetrization in $k_1, \dots k_n$ and $k'_1, \dots k'_n$, i.e.

$$\left\{ \begin{matrix} k'_1 & \dots & k'_n \\ k_1 & \dots & k_n \\ \lambda_1 & \dots & \lambda_j \end{matrix} \right\}_{\text{sym}} := \frac{1}{n!} \sum_{p \in S_n} \frac{1}{n!} \sum_{p' \in S_n} \left\{ \begin{matrix} k'_{p_1} & \dots & k'_{p_n} \\ k_{p'_1} & \dots & k_{p'_n} \\ \lambda_1 & \dots & \lambda_j \end{matrix} \right\} \quad . \tag{4.54}$$

**Proof:** Explicitly written, the left–hand side of (4.52) fulfills the following identity:

$$\frac{1}{(2n)!} \sum_{p \in S_{2n}} \mathrm{sgn}(p) R^{k'_1}_{I_1 I_2} \dots R^{k'_n}_{I_{2n-1} I_{2n}} C^{I_{p1} I_{p2}}_{k_1} \dots C^{I_{p2n-1} I_{p2n}}_{k_n} \tag{4.55}$$

$$= \frac{1}{(2n)!} \sum_{p \in S_{2n}} \mathrm{sgn}(p) R^{k'_1}_{I_{p1} I_{p2}} \dots R^{k'_n}_{I_{p2n-1} I_{p2n}} C^{I_1 I_2}_{k_1} \dots C^{I_{2n-1} I_{2n}}_{k_n} \quad .$$

Due to the sum over permutations the l.h.s. of (4.55) is invariant under any exchange of index pairs $(I_{p_{2i-1}}, I_{p_{2i}})$ and $(I_{p_{2j-1}}, I_{p_{2j}})$, $1 \leq i,j \leq n$. As these index pairs are associated with the quantum numbers $k_i$ and $k_j$, the l.h.s. of (4.55) is invariant under any permutation of $k_i$ and $k_j$, $1 \leq i,j \leq n$ for fixed $k'_1, \ldots k'_n$. Considering the r.h.s. of (4.55) by the same arguments one concludes that it is invariant under any permutation of $k'_i$ and $k'_j$, $1 \leq i,j \leq n$ for fixed $k_1, \ldots k_n$.

If all permutations of $k_1, \ldots k_n$ and $k'_1, \ldots k'_n$ are taken into account, with (4.54) these invariances lead to the identity

$$
\begin{aligned}
& R^{k'_1}_{I_1 I_2} \ldots R^{k'_n}_{I_{2n-1} I_{2n}} C^{\{I_1 I_2 \ldots}_{k_1} C^{I_{2n-1} I_{2n}\}}_{k_n} \\
&= \left\{ R^{k'_1}_{I_1 I_2} \ldots R^{k'_n}_{I_{2n-1} I_{2n}} C^{\{I_1 I_2 \ldots}_{k_1} C^{I_{2n-1} I_{2n}\}}_{k_n} \right\}_{\text{sym}} \quad .
\end{aligned}
\tag{4.56}
$$

Each term on the left–hand side of this equation can be expressed as a product of traces:

$$
\begin{aligned}
& R^{k'_1}_{I_1 I_2} \ldots R^{k'_n}_{I_{2n-1} I_{2n}} C^{I_{p_1} I_{p_2}}_{k_{\kappa_1}} \ldots C^{I_{p_{2n-1}} I_{p_{2n}}}_{k_{\kappa_n}} \\
&= \operatorname{tr}\left[ R^{k'_1} C_{k_{\nu_1}} \ldots R^{k'_{\lambda_1}} C_{k_{\nu_{\lambda_1}}} \right] \ldots \operatorname{tr}\left[ R^{k'_{s+1}} C_{k_{\nu_{s+1}}} \ldots R^{k'_n} C_{k_{\nu_n}} \right]
\end{aligned}
\tag{4.57}
$$

where the traces have to be taken over the fermion indices $I$, while $\kappa_1 \ldots \kappa_n$ and $\nu_1 \ldots \nu_n$ describe definite distributions of the $k$–values and where we used the abbreviation $s := \sum_{i=1}^{j-1} \lambda_i$.

We now sum over all fermion permutations $p \in S_{2n}$ in (4.57) and expand it in terms of trace expressions. Due to the symmetry properties expressed by (4.56) the result must be symmetric in $k_1, \ldots k_n$ for fixed $k'_1, \ldots k'_n$ and symmetric in $k'_1, \ldots k'_n$ for fixed $k_1, \ldots k_n$. So we obtain the following expansion for the left–hand sides of (4.56) or (4.55), respectively:

$$
\begin{aligned}
\text{l.h.s. (4.56)} &= \frac{1}{(2n)!} \sum_{\text{part } \lambda} \sum_{q \in S_n} \sum_{q' \in S_n} m_{\lambda_1 \ldots \lambda_j} \times \\
& \operatorname{tr}\left[ R^{k'_{q'_1}} C_{k_{q_1}} \ldots R^{k'_{q'_{\lambda_1}}} C_{k_{q_{\lambda_1}}} \right] \ldots \operatorname{tr}\left[ R^{k'_{q'_{s+1}}} C_{k_{q_{s+1}}} \ldots R^{k'_{q'_{s+\lambda_j}}} C_{k_{q_{s+\lambda_j}}} \right]
\end{aligned}
\tag{4.58}
$$

where $\lambda := (\lambda_1, \ldots \lambda_j)$ is a partition of $n$ with $\lambda_i \in \mathbb{N}$, $\sum_{i=1}^{j} \lambda_i = n$. The multiplicity factors $m_{\lambda_1 \ldots \lambda_j}$ have been introduced in order to take into account the fact that there is no one–to–one correspondence between the single trace terms from the r.h.s. of (4.58) and terms with a definite fermionic permutation $p \in S_{2n}$ from the l.h.s. of (4.58).

Furthermore in (4.58) we have expressed the permutation invariance with respect to $k_1, \ldots k_n$ and $k'_1, \ldots k'_n$ simply by unnormalized sums over $q, q' \in S_n$. This is due to the fact that the permutations $q$ and $q'$ are a subgroup of the fermion permutations $p \in S_{2n}$, i.e. that the fermion permutations can be decomposed into the subgroups of corresponding $k$– and $k'$–permutations and a complement of remaining exchange permutations within the subgroup members.

Now we apply definition (4.54) and obtain for the expansion (4.58):

$$
\text{l.h.s. (4.56)} = \sum_{\text{part } \lambda} \frac{(n!)^2}{(2n)!} m_{\lambda_1 \ldots \lambda_j} \left\{ \begin{array}{c} k'_1 \ldots k'_n \\ k_1 \ldots k_n \\ \lambda_1 \ldots \lambda_j \end{array} \right\}_{\text{sym}} \quad . \tag{4.59}
$$

For fixed values $k'_1, \dots k'_n$ and $k_1, \dots k_n$ the $m_{\lambda_1 \dots \lambda_n}$ are determined by the connections of the fermion indices $I$ in the various permutations. These relations are not altered if $k'_1, \dots k'_n$ and $k_1, \dots k_n$ are varied. Hence the multiplicity factors $m_{\lambda_1 \dots \lambda_n}$ cannot depend on the distributions of $k'_1, \dots k'_n$ and $k_1, \dots k_n$. The problem is now to determine these multiplicity factors $m_{\lambda_1 \dots \lambda_n}$. This will be done in two steps.

In a first step we consider the multiplicities emerging from single trace expressions. We take as an example the expression

$$\begin{aligned} & \mathrm{tr}\left[R^{k'_1} C_{k_1} \dots R^{k'_{\lambda_1}} C_{k_{\lambda_1}}\right] \\ \equiv \;& R^{k'_1}_{K_1 K_2} C^{K_2 K_3}_{k_1} R^{k'_2}_{K_3 K_4} C^{K_4 K_5}_{k_2} \dots R^{k'_{\lambda_1}}_{K_{2\lambda_1 - 1} K_{2\lambda_1}} C^{K_{2\lambda_1} K_1}_{k_{\lambda_1}} \end{aligned} \tag{4.60}$$

where $K_1, \dots K_{2\lambda_1}$ is a subset of $I_{p_1}, \dots I_{p_{2n}}$ with a definite $p \in S_{2n}$ which leads to the trace (4.60). Simultaneously with this permutation there exists a further permutation $p' \in S_{2n}$ where only $C^{K_2 K_3}_{k_1}$ is replaced by $-C^{K_3 K_2}_{k_1}$. But due to the antisymmetry $C^{K_3 K_2}_{k_1} = -C^{K_2 K_3}_{k_1}$ the original trace can be recovered with the same sign. If these considerations are extended to all $C_{k_i}$ appearing in (4.60), one obtains a multiplicity factor $2^{\lambda_1}$ for all permutations connected with internal commutation of the fermion coordinates of the $C_{k_i}$. In other words: the trace expression (4.60) results from exactly $2^{\lambda_1}$ terms of the l.h.s. of (4.56).

Another fermionic permutation in (4.56) which leaves the trace (4.60) invariant is the interchange of arguments between two successive $C$–functions. In this case, for instance the sequence $C^{K_2 K_3}_{k_1} R^{k'_2}_{K_3 K_4} C^{K_4 K_5}_{k_2}$ is mapped into the sequence $-C^{K_2 K_4}_{k_1} R^{k'_2}_{K_3 K_4} \times C^{K_3 K_5}_{k_2}$. Due to the antisymmetry $R^{k'_2}_{K_3 K_4} = -R^{k'_2}_{K_4 K_3}$ this interchange can be compensated by an interchange in the arguments of $R^{k'_2}$ and thus also this permutation yields the same trace expression (4.60). If these considerations are applied to all $R^{k_i}$ in (4.60), apart from the first one, this leads to a multiplicity factor $2^{\lambda_1 - 1}$ and the combination of inner interchanges of $C$– as well as of $R$–function arguments yields the multiplicity $2^{2\lambda_1 - 1}$. Hence we see that there is a class $P_{\lambda_1}(k, k')$ of $2^{2\lambda_1 - 1}$ fermionic permutations from the r.h.s. of (4.56) which contribute to a single trace of the kind (4.60).

Extending this result to a product of traces contained in a partition $\lambda$ means that there is a class $P_\lambda(k, k')$ of fermionic permutations contributing to a definite partition of traces with fixed $k_1, \dots k_n$ and $k'_1, \dots k'_n$. Thus we obtain

$$m_{\lambda_1 \dots \lambda_j} = \prod_{i=1}^{j} 2^{2\lambda_i - 1} m^{(1)}_{\lambda_1 \dots \lambda_j} = 2^n 2^{n-j} m^{(1)}_{\lambda_1 \dots \lambda_j} \quad . \tag{4.61}$$

The next step is the investigation of symmetries connected with products of traces. In this case we are concerned with fermionic permutations $p \in S_{2n}$ which generate permutations of the $k_1, \dots k_n$ and of the $k'_1, \dots k'_n$.

For fixed $k'_1, \dots k'_n$ it is obvious that any permutation of the $k_1, \dots k_n$ is generated by one corresponding class $P_\lambda(k, k')$ of fermion permutations. Thus the symmetrization in $k_1, \dots k_n$ on the r.h.s. of (4.58) does not give rise for an additional factor in $m^{(1)}_{\lambda_1 \dots \lambda_j}$.

However, things are more complicated for the symmetrization in $k'_1, \dots k'_n$ for fixed $k_1, \dots k_n$. We know from (4.56) that each permutation of the $k'_1, \dots k'_n$ in (4.58) can be written as a term from the l.h.s. of (4.56), i.e. as a term with the ordering $(k'_1 \dots k'_n)$ and corresponding permutations of the fermionic indices. This leads to a redundance of the explicit symmetrization in $k'_1, \dots k'_n$ with respect to the symmetry properties of the various trace terms.

First we consider the invariance of each trace under cylic permutation of the factors $(R^{k'_i} C_{k_j})$. For fixed $k_1, \dots k_n$ on the r.h.s. of (4.58) there are $\lambda_1$ traces of the kind (4.60), corresponding to $\lambda_1$ permutations of the $k'_1 \dots k'_{\lambda_1}$, which by cyclic permutations of the factors $(R^{k'_i} C_{k_j})$ are equivalent to or are generated by the one term (4.60), i.e. with the ordering $(k'_1 \dots k'_{\lambda_1})$. Extension of this result to a given partition $\lambda$ yields the fact that $\lambda_1 \dots \lambda_j$ terms from the r.h.s. of (4.58) correspond to one term from the l.h.s. of (4.58). Thus we have

$$m^{(1)}_{\lambda_1 \dots \lambda_j} = \frac{1}{\lambda_1 \dots \lambda_j} m^{(2)}_{\lambda_1 \dots \lambda_j} \quad . \tag{4.62}$$

Next we consider partitions which contain $s_k$ traces with equal length $\lambda_i$. There are $s_k!$ permutations of the involved $k'_l$ which by means of permutations of the $s_k$ traces are generated by permutations of corresponding $k_{l'}$. Because these terms are already contained in the explicit symmetrization in $k_1, \dots k_n$, we obtain

$$m^{(2)}_{\lambda_1 \dots \lambda_j} = \frac{1}{s_1! \dots s_l!} m^{(3)}_{\lambda_1 \dots \lambda_j} \tag{4.63}$$

where $s_1, \dots s_l$ are the multiplicities of traces of equal length contained in a definite partition $\lambda$.

Finally the permutation of traces with distinct length leads to another redundance which can be removed by the condition $\lambda_1 \geq \dots \geq \lambda_j$.

As all possibilities for the determination of multiplicities were fully utilized, we have $m^{(3)}_{\lambda_1 \dots \lambda_j} \equiv 1$.

The sign of each partition can be determined from permutations of the fermionic indices to give $(-1)^j$. So we eventually obtain

$$\text{l.h.s. } (4.56) = \sum_{\text{part } \lambda} (-1)^j \frac{(n!)^2}{(2n)!} \frac{2^n 2^{(n-j)}}{\lambda_1 \dots \lambda_j s_1! \dots s_l!} \left\{ \begin{array}{c} k'_1 \dots k'_n \\ k_1 \dots k_n \\ \lambda_1 \dots \lambda_j \end{array} \right\}_{\text{sym}} \tag{4.64}$$

with $\lambda_1 \geq \dots \geq \lambda_j$, which is equivalent to (4.52).

To convince ourselves from the correctness of counting the various terms, we sum the terms on both sides of (4.64) and get the formula

$$(2n)! = 2^n n! \sum_{\text{part } \lambda} 2^{n-j} \frac{n!}{s_1! \dots s_l! \lambda_1 \dots \lambda_j}$$

which can be explicitly verified for small $n$. ◇

We extend these techniques to the derivation of a suitable representation of (4.26). With respect to (4.26) the following proposition holds:

- **Proposition 4.4:** Using the definitions of Proposition 4.3 the following relation holds:

$$R^{k'_1}_{K_1K_2}\dots R^{k'_n}_{K_{2n-1}K_{2n}} C^{\{I_1I_2}_{k_1} C^{K_1K_2}_{k_2}\dots C^{K_{2n-1}K_{2n}\}}_{k_{n+1}} \tag{4.65}$$
$$= \frac{1}{(2n+1)} R^{k'_1}_{K_1K_2}\dots R^{k'_n}_{K_{2n-1}K_{2n}} C^{I_1I_2}_{[k_1} C^{\{K_1K_2}\dots C^{K_{2n-1}K_{2n}\}}_{k_{n+1}]}$$
$$+\frac{(n+1)!2^{n+1}}{(2(n+1))!}\sum_{\text{part }\lambda}(-1)^{j+1}\frac{2^{n-j+1}n!}{s_1!\dots s_l!\,\lambda_1\dots\lambda_j}\left\{\begin{array}{ccc} k'_1 & \dots & k'_n, I_1I_2 \\ k_1 & \dots & k_nk_{n+1} \\ \lambda_1 & \dots & \lambda_j\end{array}\right\}_{\text{sym}}$$

with

$$\left\{\begin{array}{ccc} k'_1 & \dots & k'_n, I_1I_2 \\ k_1 & \dots & k_nk_{n+1} \\ \lambda_1 & \dots & \lambda_j\end{array}\right\} := \sum_{l=1}^{j}\lambda_l\,\mathrm{tr}\left[R^{k'_1}C_{k_1}\dots R^{k'_{\lambda_1}}C_{k_{\lambda_1}}\right]\dots\times \tag{4.66}$$
$$\left[C_{k_{s+1}}R^{k'_{s+1}}\dots R^{k'_{s+\lambda_l}}C_{k_{s+\lambda_l+1}}\right]^{\{I_1I_2\}}\dots\mathrm{tr}\left[R^{k'_{n-\lambda_j+1}}C_{k_{n-\lambda_j+2}}\dots R^{k'_n}C_{k_{n+1}}\right]$$

where $s := \sum_{i=1}^{l-1}\lambda_i$ and the symmetrization (4.54) is generalized to $k_1\dots k_{n+1}$ and $k'_1\dots k'_n$.

**Proof:** Due to the permutations of $I_1, I_2, K_1\dots K_{2n}$ the left–hand side of (4.65) is symmetric with respect to $k_1\dots k_{n+1}$. Thus we have

$$\text{l.h.s. (4.50)} = (n+1)!R^{k'_1}_{K_1K_2}\dots R^{k'_n}_{K_{2n-1}K_{2n}} C^{\{I_1I_2}_{[k_1} C^{K_1K_2}_{k_2}\dots C^{K_{2n-1}K_{2n}\}}_{k_{n+1}]}\,. \tag{4.67}$$

If in $C^{I_1I_2}_{k_1}\dots C^{K_{2n-1}K_{2n}}_{k_{n+1}}$ we separate those terms from the others where $(I_1I_2)$ is exactly placed over one $k_i$, the symmetry in $k_1\dots k_{n+1}$ can be used to shift $C^{I_1I_2}_{k_i}$ to the first position and we obtain the first term on the right–hand side of (4.65). In the remaining terms the indices $I_1, I_2$ are distributed in such a way that two values $k_i, k_j$ are involved. If the product on the left–hand side of (4.65) is factorized into a product of traces, this yields definition (4.66) (with omission of those terms already separated from (4.65)). Then the multiplicity factors have to be determined. Their calculations run along the same lines as those of Proposition 4.3. We thus do not discuss this in detail. ◇

The treatment of the last expression (4.27) is completely similar. Therefore, without further proof we formulate the corresponding proposition:

- **Proposition 4.5:** Using the definitions of Proposition 4.3 the following relation holds:

$$R^{k'_1}_{K_1K_2}\dots R^{k'_n}_{K_{2n-1}K_{2n}} C^{\{I_1I_2}_{k_1} C^{I_3I_4}_{k_2} C^{K_1K_2}_{k_3}\dots C^{K_{2n-1}K_{2n}\}}_{k_{n+2}} \tag{4.68}$$
$$= \frac{3}{(2n+3)(2n+1)} R^{k'_1}_{K_1K_2}\dots R^{k'_n}_{K_{2n-1}K_{2n}} C^{\{I_1I_2}_{[k_1} C^{I_3I_4\}}_{k_2} C^{\{K_1K_2}_{k_3}\dots C^{K_{2n-1}K_{2n}\}}_{k_{n+2}]}$$
$$+\frac{(n+2)!2^{n+2}}{(2n+4)!}\sum_{\text{part }\lambda}(-1)^{j+2}\times$$
$$\frac{2^{n-j+2}n!}{s_1!\dots s_r!\,\lambda_1\dots\lambda_j}\left\{\begin{array}{ccc} k'_1 & \dots & k'_n, I_1I_2I_3I_4 \\ k_1 & \dots & k_nk_{n+1}k_{n+2} \\ \lambda_1 & \dots & \lambda_j\end{array}\right\}_{\text{sym}}$$

where the bracket with $I_1 \dots I_4$ is the corresponding generalization of definition (4.66) and the symmetrization (4.54) is generalized to $k_1 \dots k_{n+2}$ and $k'_1 \dots k'_n$.

By means of these propositions the direct forces can be separated from the exchange forces. According to our definition the direct forces are characterized by the exclusive appearance of density expressions, which in our case are given by $\mathrm{tr}(RC)$. Obviously these expressions result from all partitions with $\lambda_1 = \dots = \lambda_j = 1$. Hence only these partitions lead to direct forces while all other partitions have to be identified with the exchange forces.

## 4.6 Estimate of Exchange Forces

So far the resolution of (4.25), (4.26), (4.27) into terms describing direct forces and terms describing exchange forces is purely formal and means only a rearrangement of the original expressions. A simplification of equation (4.24) can be achieved if for instance exchange terms can be neglected. As will be demonstrated in the following an estimate of exchange forces can only be performed for bound states. Hence as a further step we have to subdivide the boson states $C_k^{I_1 I_2}$ into a subset of bound states and a subset of scattering states. The latter subset can be neglected in all calculations if the masses $m_0$, $m_1$, $m_2$ of the original spinorfields are very large. In this case one can apply the decoupling theorem, see for instance Symanzik [Sym 73], Appelquist and Carrazone [App 75], Collins [Coll 87]), which justifies the neglect of the contribution of the scattering states to the low energy dynamics. Without further comment we will make use of this theorem in all further considerations; cf. also the discussion in Section 6.3. Thus, as far as the simplification of (4.24) is concerned we can omit the contribution of scattering states and for an estimate of exchange forces we will have to deal only with bound states in the low energy limit. Under these conditions such an estimate was given by Pfister and Stumpf in [Pfi 91]. In this section we will follow the expositions given in that paper.

For a first draft we will not use the realistic (i.e. complicated) boson bound state functions $C_k^{I_1 I_2}$. Rather, with respect to exchange forces we assume that the essential property of wave functions is their spatial extension. In discussing only this aspect we restrict ourselves to a simple class of wave functions which are used in nuclear physics, namely oscillator functions. We thus consider the set of functions

$$\tilde{C}_k^{r_1 r_2} := \frac{m^{\frac{3}{2}}}{(2\pi)^{\frac{3}{2}} \pi^{\frac{3}{4}}} \exp\left( i\mathbf{kz} - \frac{m^2}{2}\mathbf{u}^2 \right) \;, \; \mathbf{k} \in \mathbb{R}^3 \tag{4.69}$$

with $\mathbf{z} = \frac{1}{2}(\mathbf{r}_1 + \mathbf{r}_2)$, $\mathbf{u} = (\mathbf{r}_1 - \mathbf{r}_2)$ and $m = \sigma^{-1}$ for $\hbar = c = 1$, where $\sigma$ is the length scale expressed by the mass scale $m$. The dual $\tilde{R}^k_{r_1 r_2}$ of $\tilde{C}_k^{r_1 r_2}$ is given by its conjugate complex function.

We generalize this nonrelativistic ansatz by assuming the wave functions (4.69) to describe moving boson bound states with momentum $\mathbf{k}$ relative to the general frame of reference of the functional equation (4.24). The scale factor $m = \sigma^{-1}$ in (4.69) describes the spatial extension of this function. According to special relativity, length scales $\sigma$ of moving objects are contracted with respect to observers in a rest frame. Thus $m$ cannot be a constant but has to be transformed inversely to $\sigma$, i.e., $m = m(\mathbf{k}) = m/(1-\beta^2)^{\frac{1}{2}}$. To determine $m = m(\mathbf{k})$

we assume that the composite bosons have a rest mass $\mu$. Then $k^2 = \mu^2$ or $k_0^2 = \mu^2 + \mathbf{k}^2$. Furthermore it is $k_0^2 = E^2 = \mu^2(1-\beta^2)^{-1}$ with $\beta^2 = v^2$ for $\hbar = c = 1$, and therefore by comparison $(1-\beta^2)^{-1} = (1+\mathbf{k}^2/\mu^2)$. From this it follows

$$m(\mathbf{k}) = m\left(1+\mathbf{k}^2/\mu^2\right)^{\frac{1}{2}} \tag{4.70}$$

which has to replace $m$ in (4.69).

With the set (4.69) for any $m$ any trace expression can be exactly calculated. However, further evaluation of the exchange terms can be considerably simplified if we consider large $m$. In this case we have

$$\lim_{m\to\infty} m^3(2\pi)^{-\frac{3}{2}} \exp\left(-\frac{m^2}{2}\mathbf{u}^2\right) = \delta(\mathbf{u}) \quad ,$$

and thus we can approximately replace the oscillator functions by $\delta$–distributions. This yields

$$\operatorname{tr}\left[\tilde{R}^{k_1'}\tilde{C}_{k_1}\ldots\tilde{R}^{k_l'}\tilde{C}_{k_l}\right] \approx 2^{\frac{3}{2}}\left(\pi^{-\frac{3}{2}}\right)^{l-1}\delta\left(\sum_{i=1}^{l}(\mathbf{k}_i-\mathbf{k}_i')\right)\prod_{i=1}^{l}\frac{m^3(\mathbf{k}_l)}{m^{-\frac{3}{2}}(\mathbf{k}_i)m^{-\frac{3}{2}}(\mathbf{k}_i')} \tag{4.71}$$

and

$$\begin{aligned}&\operatorname{tr}\left[\tilde{C}_k\tilde{R}^{k_1'}\tilde{C}_{k_1}\ldots\tilde{R}^{k_l'}\tilde{C}_{k_l}\right]^{\{r_1r_2\}} \\ &\approx \left(\pi^{-\frac{3}{2}}\right)^{l}\tilde{C}_k^{r_1r_2}\exp\left[i\mathbf{z}\left(\sum_{i=1}^{l}(\mathbf{k}_i-\mathbf{k}_i')\right)\right]\prod_{i=1}^{l}m^{-\frac{3}{2}}(\mathbf{k}_i)m^{-\frac{3}{2}}(\mathbf{k}_i')\end{aligned} \tag{4.72}$$

with $\mathbf{z} = \frac{1}{2}(\mathbf{r}_1+\mathbf{r}_2)$.

In the appendix of the paper quoted above the following inequalities are proven:

$$\begin{aligned}&\Bigg|\int\left\{\operatorname{tr}\left[\tilde{R}^{k_1'}\tilde{C}_{k_1}\ldots\tilde{R}^{k_{\lambda_1}'}\tilde{C}_{k_{\lambda_1}}\right]\ldots\operatorname{tr}\left[\tilde{R}^{k_{n-\lambda_j+1}'}\tilde{C}_{k_{n-\lambda_j+1}}\ldots\tilde{R}^{k_n'}\tilde{C}_{k_n}\right]\right\}_{\text{sym}}\times \\ &\qquad \rho^{(n)}(k_1\ldots k_n)dk_1\ldots dk_n\Bigg| \leq \left(a\frac{\mu}{m}\right)^{3(n-j)}|\rho|_{\max}\end{aligned} \tag{4.73}$$

and

$$\begin{aligned}&\Bigg|\int\left\{\operatorname{tr}\left[\tilde{R}^{k_1'}\tilde{C}_{k_1}\ldots\tilde{R}^{k_{\lambda_1}'}\tilde{C}_{k_{\lambda_1}}\right]\ldots\left[\tilde{C}_k\tilde{R}^{k_{s+1}'}\tilde{C}_{k_{s+1}}\ldots\tilde{R}^{k_{s+\lambda_l}'}\tilde{C}_{k_{s+\lambda_l}}\right]^{\{r_1r_2\}}\times\right. \\ &\qquad \left.\ldots\operatorname{tr}\left[\tilde{R}^{k_{n-\lambda_j+1}'}\tilde{C}_{k_{n-\lambda_j+1}}\ldots\tilde{R}^{k_n'}\tilde{C}_{k_n}\right]\right\}_{\text{sym}}\rho^{(n)}(k_1\ldots k_n)dk_1\ldots dk_n\Bigg| \\ &\leq \left|\tilde{C}_k^{r_1r_2}\right|\left(a\frac{\mu}{m}\right)^{3(n-j+1)}|\rho|_{\max}\end{aligned} \tag{4.74}$$

where $\rho^{(n)}$ is the corresponding simplified boson wavefunction, $s := \sum_{i=1}^{l-1}\lambda_i$ and $a \approx 4$.

By means of these inequalities an estimate for (4.52) can be given. We decompose (4.52) into the partition $\lambda_i = 1, i = 1\ldots n$ and the rest. This yields

$$\tilde{R}^{k'_1}_{r'_1 r'_2} \dots \tilde{R}^{k'_n}_{r'_{2n-1} r'_{2n}} \tilde{C}^{\{r'_1 r'_2}_{k_1} \dots \tilde{C}^{r'_{2n-1} r'_{2n}\}}_{k_n} \tag{4.75}$$
$$= \frac{n!}{(2n)!} 2^n \left( \sum_{p \in S_n} \frac{1}{n!} \delta_{k'_1 k_{p1}} \dots \delta_{k'_n k_{pn}} + R_1 \right) .$$

If we consider the left–hand side of (4.75) as an integral kernel which acts on $\rho^{(n)}$, we obtain by means of (4.73)

$$|R_1| \le \sum_{j=1}^{n-1} \sum_{\text{part } \lambda} \frac{2^{n-j} n!}{s_1! \dots s_l! \, \lambda_1 \dots \lambda_j} \left( a \frac{\mu}{m} \right)^{3(n-j)} . \tag{4.76}$$

For a further estimate of (4.76) we apply the following inequality without proof:

$$\frac{2^{n-j} n!}{s_1! \dots s_r! \, \lambda_1 \dots \lambda_j} \le (2n^3)^{n-j} . \tag{4.77}$$

This gives

$$|R_1| \le \sum_{j=1}^{n-1} \Delta_j(n) \left( \frac{a' n \mu}{m} \right)^{3(n-j)} \tag{4.78}$$

where $a' = 2^{\frac{1}{3}} a$ and $\Delta_j(n)$ denotes the number of partitions with exactly $j$ trace factors in (4.52). If $P_j(n)$ is defined by

$$P_j(n) := \sum_{k=1}^{j} \Delta_k(n) \tag{4.79}$$

then the relation

$$\Delta_j(n) = P_j(n-j) \tag{4.80}$$

holds, cf. Niven and Zuckerman [Niv 72]. Thus we can rewrite (4.78) as

$$|R_1| \le \sum_{k=1}^{n-1} P(k) \left( \frac{a' n \mu}{m} \right)^{3k} \tag{4.81}$$

and by means of the generating function for the $P(k)$, for $(an\mu/m)^3 \ll 1$ we get the inequality

$$|R_1| \le n^3 \left( \frac{a' \mu}{m} \right)^3 . \tag{4.82}$$

In an analogous way (4.65) can be estimated. This expression can be written in the form

$$\tilde{R}^{k'_1}_{r'_1 r'_2} \dots \tilde{R}^{k'_n}_{r'_{2n-1} r'_{2n}} \tilde{C}^{\{r_1 r_2}_{k_1} \tilde{C}^{r'_1 r'_2}_{k_2} \dots \tilde{C}^{r'_{2n-1} r'_{2n}\}}_{k_{n+1}} \tag{4.83}$$
$$= \frac{1}{(2n+1)} \tilde{R}^{k'_1}_{r'_1 r'_2} \dots \tilde{R}^{k'_n}_{r'_{2n-1} r'_{2n}} \tilde{C}^{r_1 r_2}_{[k_1} \tilde{C}^{\{r'_1 r'_2}_{k_2} \dots \tilde{C}^{r'_{2n-1} r'_{2n}\}}_{k_{n+1}]} + R_2$$

with

$$\begin{aligned}
&\frac{1}{(2n+1)}\tilde{R}^{k'_1}_{r'_1 r'_2}\dots\tilde{R}^{k'_n}_{r'_{2n-1}r'_{2n}}\tilde{C}^{r_1 r_2}_{[k_1}\tilde{C}^{\{r'_1 r'_2}_{k_2}\dots\tilde{C}^{r'_{2n-1}r'_{2n}\}}_{k_{n+1}]} \\
&= \frac{2(n+1)}{(2(n+1))!}\,2^n n!\sum_{p\in S_{n+1}}\frac{1}{(n+1)!}\tilde{C}^{r_1 r_2}_{k_{p_1}}(\delta_{k_{p_2}k'_1}\dots\delta_{k_{p_{n+1}}k'_n}+R_1)
\end{aligned} \tag{4.84}$$

and

$$|R_2| \le \frac{(n+1)!2^{n+1}}{(2(n+1))!}\,|\tilde{C}^{r_1 r_2}_k|\,n\left(\frac{a'\mu}{m}\right)^3 . \tag{4.85}$$

Finally we obtain for the first term on the right–hand side of (4.68)

$$\begin{aligned}
&\frac{3}{(2n+3)(2n+1)}\tilde{R}^{k'_1}_{r'_1 r'_2}\dots\tilde{R}^{k'_n}_{r'_{2n-1}r'_{2n}}\tilde{C}^{\{r_1 r_2}_{[k_1}\tilde{C}^{r_3 r_4\}}_{k_2}\tilde{C}^{\{r'_1 r'_2}_{k_3}\dots\tilde{C}^{r'_{2n-1}r'_{2n}\}}_{k_{n+2}]} \\
&= \frac{3(n+2)!}{(2(n+2))!}\,2^n\sum_{p\in S_{n+2}}\frac{1}{(n+2)!}\tilde{C}^{\{r_1 r_2}_{k_{p_1}}\tilde{C}^{r_3 r_4\}}_{k_{p_2}}(\delta_{k'_1 k_{\lambda_3}}\dots\delta_{k'_n k_{p_{n+2}}}+R_1) .
\end{aligned} \tag{4.86}$$

For brevity we do not cite the inequality of the rest on the right–hand side of (4.68).

Now we assume that in the case of realistic wavefunctions $C^{I_1 I_2}_k$ the rest terms corresponding to $R_1$, $R_2$ and $R_3$ also can be estimated by powers of $\frac{\mu}{m}$; then because of very large fermion masses $m$ and finite boson masses $\mu$ these rest terms are negligible, too.

We examine the consequences of this estimate for the functional equation (4.24). Toward this end we define a new state functional by

$$\begin{aligned}
|\hat{\mathcal{B}}(b;a)\rangle &:= \sum_{n=0}^{\infty}\frac{1}{n!}\frac{2^n n!}{(2n)!}\,\rho^{(n)}(k_1\dots k_n|a)b_{k_1}\dots b_{k_n}|0\rangle_B \\
&=: \sum_{n=0}^{\infty}\frac{1}{n!}\,\hat{\rho}^{(n)}(k_1\dots k_n|a)b_{k_1}\dots b_{k_n}|0\rangle_B \quad ,
\end{aligned} \tag{4.87}$$

where $\rho^{(n)}(k_1\dots k_n|a)$ are the coefficient functions of the functional state $|\mathcal{B}(b;a)\rangle$ from (4.13); i.e. $|\hat{\mathcal{B}}(b;a)\rangle$ is modified by some statistical factors. Then from (4.75), (4.84) and (4.86) we obtain the equations

$$\langle 0|\partial_{k_n}\dots\partial_{k_1}\mathcal{U}|\mathcal{B}(b;a)\rangle = \langle 0|\partial_{k_n}\dots\partial_{k_1}|\hat{\mathcal{B}}(b;a)\rangle , \tag{4.88}$$

$$\langle 0|\partial_{k_n}\dots\partial_{k_1}\mathcal{V}\begin{pmatrix} I_1 I_2 \\ k \end{pmatrix}\partial_k|\mathcal{B}(b;a)\rangle = \langle 0|\partial_{k_n}\dots\partial_{k_1}C^{I_1 I_2}_k\partial_k|\hat{\mathcal{B}}(b;a)\rangle , \tag{4.89}$$

$$\begin{aligned}
&\langle 0|\partial_{k_n}\dots\partial_{k_1}\mathcal{W}\begin{pmatrix} I_1 I_2 & | & I_3 I_4 \\ k & | & k' \end{pmatrix}\partial_k\partial_{k'}|\mathcal{B}(b;a)\rangle \\
&= 3C^{\{I_1 I_2}_k C^{I_3 I_4\}}_{k'}\langle 0|\partial_{k_n}\dots\partial_{k_1}\partial_k\partial_{k'}|\hat{\mathcal{B}}(b;a)\rangle
\end{aligned} \tag{4.90}$$

if $R_1$, $R_2$ and $R_3$ are neglected.

If (4.88)–(4.90) are substituted into (4.24) the following boson functional equation results:

$$
\begin{aligned}
E_{p0}|\hat{\mathcal{B}}(b;p)\rangle &= 2R^k_{I_1K}\hat{K}_{I_1I_2}C^{I_2K}_{k'}b_k\partial_{k'}|\hat{\mathcal{B}}(b;p)\rangle \\
&-\hat{W}_{K_1I_2I_3I_4}\Big\{6F_{I_4K_2}R^k_{K_1K_2}C^{I_2I_3}_{k'}b_k\partial_{k'} \\
&+36F_{I_4K_2}R^{k_1}_{K_1K_3}R^{k_2}_{K_2K_4}C^{\{I_2K_3}_{k'_1}C^{I_3K_4\}}_{k'_2}b_{k_1}b_{k_2}\partial_{k'_1}\partial_{k'_2} \\
&-6R^k_{K_1K_2}C^{\{I_2I_3}_{k'_1}C^{I_4K_2\}}_{k'_2}b_k\partial_{k'_1}\partial_{k'_2} \\
&-\left(3F_{I_4K_2}F_{I_3K_3}+\frac{1}{4}A_{I_4K_2}A_{I_3K_3}\right)\sum_{p\in S_3}\frac{1}{3!}\mathrm{sgn}(p) \quad\times \\
&\Big[12R^{k_1}_{K_{p1}K_{p2}}R^{k_2}_{K_{p3}K_4}C^{I_2K_4}_{k'}b_{k_1}b_{k_2}\partial_{k'} \\
&+24R^{k_1}_{K_{p1}K_4}R^{k_2}_{K_{p2}K_5}R^{k_3}_{K_{p3}K_6}C^{\{I_2K_6}_{k'_1}C^{K_5K_4\}}_{k'_2}b_{k_1}b_{k_2}b_{k_3}\partial_{k'_1}\partial_{k'_2}\Big] \\
&+\left(F_{I_4K_2}F_{I_3K_3}F_{I_2K_4}+\frac{1}{4}A_{I_4K_2}A_{I_3K_3}A_{I_2K_4}\right)\sum_{p\in S_4}\frac{1}{4!}\mathrm{sgn}(p) \quad\times \\
&\Big[12R^{k_1}_{K_{p1}K_{p2}}R^{k_2}_{K_{p3}K_{p4}}b_{k_1}b_{k_2}+48R^{k_1}_{K_{p1}K_5}R^{k_2}_{K_{p2}K_6}R^{k_3}_{K_{p3}K_{p4}}C^{K_6K_5}_{k'}b_{k_1}b_{k_2}b_{k_3}\partial_{k'} \\
&+48R^{k_1}_{K_{p1}K_5}R^{k_2}_{K_{p2}K_6}R^{k_3}_{K_{p3}K_7}R^{k_4}_{K_{p4}K_8}C^{\{K_5K_6}_{k'_1}C^{K_7K_8\}}_{k'_2}b_{k_1}b_{k_2}b_{k_3}b_{k_4}\partial_{k'_1}\partial_{k'_2}\Big]\Big\}|\hat{\mathcal{B}}(b;p)\rangle\,.
\end{aligned}
\tag{4.91}
$$

Within the $(\mu/m)$–approximation, see also Stumpf [Stu 85a], equation (4.91) can be further simplified. If (4.91) is projected into configuration space, terms which differ by the factor $b_k\partial_{k'}$ act in configuration space on the same amplitude $\rho^{(n)}(k_1\dots k_n)$. Thus between such terms an estimate can be made with respect to their relative magnitude. It can be shown that the term with $b_{k_1}b_{k_2}b_{k_3}\partial_{k'_1}\partial_{k'_2}$ is proportional to $(\mu/m)^3b_{k_1}b_{k_2}\partial_{k'}$; for a more detailed discussion of such estimates we refer to Section 6.2. Thus with respect to the term $b_{k_1}b_{k_2}\partial_{k'}$ the term $b_{k_1}b_{k_2}b_{k_3}\partial_{k'_1}\partial_{k'_2}$ can be neglected.

Analogous relations hold for the terms with $b_{k_1}b_{k_2}b_{k_3}b_{k_4}\partial_{k'_1}\partial_{k'_2}$, $b_{k_1}b_{k_2}b_{k_3}\partial_{k'}$ and $b_{k_1}b_{k_2}$ and for $b_{k_1}b_{k_2}\partial_{k'_1}\partial_{k'_2}$ and $b_k\partial_{k'}$, respectively. From the latter terms only the terms $b_{k_1}b_{k_2}$ and $b_k\partial_{k'}$ are left within this approximation.

All these estimates were confirmed by means of alternative calculations by Kerschner [Ker 94]. Taking into account these estimates, from (4.91) we obtain the functional equation

$$
\begin{aligned}
E_{p0}|\hat{\mathcal{B}}(b;p)\rangle &= 2R^k_{I_1K}\hat{K}_{I_1I_2}C^{I_2K}_{k}b_k\partial_{k'}|\hat{\mathcal{B}}(b;p)\rangle \\
&-\hat{W}_{I_1I_2I_3I_4}\Big\{6F_{I_4K}R^k_{I_1K}C^{I_2I_3}_{k'}b_k\partial_{k'} \\
&-6R^k_{I_1K}C^{\{I_2I_3}_{k'_1}C^{I_4K\}}_{k'_2}b_k\partial_{k'_1}\partial_{k'_2} \\
&-12\left(3F_{I_4K_1}F_{I_3K_2}+\frac{1}{4}A_{I_4K_1}A_{I_3K_2}\right)R^{k_1}_{\{I_1K_1}R^{k_2}_{K_2\}K_3}C^{I_2K_3}_{k'}b_{k_1}b_{k_2}\partial_{k'} \\
&+12\left(F_{I_4K_1}F_{I_3K_2}F_{I_2K_3}+\frac{1}{4}A_{I_4K_1}A_{I_3K_2}F_{I_2K_3}\right)R^{k_1}_{\{I_1K_1}R^{k_2}_{K_2K_3\}}b_{k_1}b_{k_2}\Big\}|\hat{\mathcal{B}}(b;p)\rangle\,.
\end{aligned}
\tag{4.92}
$$

Equation (4.92) was first calculated in papers of Stumpf [Stu 85a], [Stu 86a,b] by means of a short–cut calculation with subsequent omission of those terms which are neglected within the $(\mu/m)$–approximation. Thus this technique, which will be presented in the next section, is justified provided exchange forces can be neglected. The physical content of (4.92) will be discussed in Chapter 6.

## 4.7 Weak Mapping in Functional Space

In order to exhibit the physical and mathematical foundations of weak mappings we strictly separated the definition of the latter from the embedding of the results of weak mappings into functional space. In this section we soften up this point of view in favour of an elegant functional treatment of weak mappings which neglects the exchange forces contained in (4.25), (4.26) and (4.27) from the beginning and which was developed by Stumpf [Stu 85a].

To introduce this method we define bosonic source operators by operator products in functional space:

$$b_k := \frac{1}{2} C_k^{I_1 I_2} j_{I_1} j_{I_2} , \tag{4.93}$$

where $C_k^{I_1 I_2}$ are the two–particle solutions of (4.3), $k$ denotes the boson quantum numbers and where in contrast to Section 4.4 the boson operators $b_k$ depend on the fermionic sources. It is just this property which now allows to directly transform equation (4.1) in functional space.

Let $|\mathcal{F}(j;p)\rangle$ be the functional eigenstate (4.19) of equation (4.1) or (4.28), respectively. Then we postulate the invariance of the functional state $|\mathcal{F}(j;p)\rangle$ under the boson transformation (4.93) by

$$|\mathcal{F}(j;p)\rangle = |\tilde{\mathcal{B}}(b;p)\rangle \tag{4.94}$$

where according to (4.93) $|\tilde{\mathcal{B}}(b;p)\rangle$ is an abbreviation for $|\tilde{\mathcal{B}}(b[j];p)\rangle$ and denotes the transformed functional state which is assumed to have a power series expansion in the $b$–operators. With (4.94) equation (4.28) can be equivalently written as

$$E_{p0}|\tilde{\mathcal{B}}(b;p)\rangle = \mathcal{H}_{\mathrm{F}}(j,\partial)|\tilde{\mathcal{B}}(b;p)\rangle \quad . \tag{4.95}$$

If we introduce functional boson "duals" $\partial^b$ and transform $\mathcal{H}_{\mathrm{F}}(j,\partial)$ into $\mathcal{H}_{\mathrm{B}}(b,\partial^b) \equiv \mathcal{H}_{\mathrm{F}}(j[b],\partial[\partial^b])$ and subsequently take the $b$– and $\partial^b$–operators as independent from the $j$– and $\partial$–operators we will arrive at an effective boson theory given by

$$E_{p0}|\tilde{\mathcal{B}}(b;p)\rangle = \mathcal{H}_{\mathrm{B}}(b,\partial^b)|\tilde{\mathcal{B}}(b;p)\rangle \quad . \tag{4.96}$$

Thus we first have to calculate the effect of fermion duals $\partial_I$ on $|\tilde{\mathcal{B}}(b[j];p)\rangle$. Analogously to Section 4.4 we introduce the dual $\partial_k^b$ to $b_k$ with the commutation relation

$$\left[\partial_{k'}^b, b_k\right]_- = \delta_{kk'} \tag{4.97}$$

and apply the functional chain rule by

$$\partial_I|\tilde{\mathcal{B}}(b[j];a)\rangle = \sum_k (\partial_I b_k)\partial_k^b|\tilde{\mathcal{B}}(b;a)\rangle \equiv \sum_k \frac{\delta b_k[j]}{\delta j_I}\frac{\delta}{\delta b_k}|\tilde{\mathcal{B}}(b[j];a)\rangle \quad . \tag{4.98}$$

With (4.93) we obtain $\partial_{I_1} b_k = C_k^{I_1 I_2} j_{I_2}$ and therefore

$$\partial_{I_1} |\tilde{\mathcal{B}}(b;a)\rangle = \sum_k C_k^{I_1 I_2} j_{I_2} \partial_k^b |\tilde{\mathcal{B}}(b;a)\rangle \quad . \tag{4.99}$$

Repeated application of the chain rule yields

$$\partial_{I_1}\partial_{I_2} |\tilde{\mathcal{B}}(b;a)\rangle = \sum_k C_k^{I_2 I_1} \partial_k^b |\tilde{\mathcal{B}}(b;a)\rangle - \sum_{k_1,k_2} 2 C_{k_1}^{I_2 K_1} C_{k_2}^{I_1 K_2} j_{K_1} j_{K_2} \partial_{k_2}^b \partial_{k_1}^b |\tilde{\mathcal{B}}(b;a)\rangle \tag{4.100}$$

and

$$\begin{aligned} \partial_{I_1}\partial_{I_2}\partial_{I_3} |\tilde{\mathcal{B}}(b;a)\rangle &= 3 \sum_{k_1,k_2} C_{k_1}^{I_2 I_1} C_{k_2}^{I_3 K_1} j_{K_1} \partial_{k_2}^b \partial_{k_1}^b |\tilde{\mathcal{B}}(b;a)\rangle \\ &+ \sum_{k_1,k_2,k_3} C_{k_1}^{I_1 K_1} C_{k_2}^{I_2 K_2} C_{k_3}^{I_3 K_3} j_{K_1} j_{K_2} j_{K_3} \partial_{k_1}^b \partial_{k_2}^b \partial_{k_3}^b |\tilde{\mathcal{B}}(b;a)\rangle_{|\mathrm{as}\{I_1,I_2,I_3\}} \;. \end{aligned} \tag{4.101}$$

In this way the fermionic duals can be transformed into bosonic duals. As is obvious from $\mathcal{H}_{\mathrm{F}}$ and from (4.99), (4.100) and (4.101) the transformation has to be completed by giving a transformation rule for the fermionic sources themselves. By observing (4.4) and (4.5) we can resolve (4.93) with respect to the fermionic sources in the form

$$j_{I_1} j_{I_2} = 2 R_{I_1 I_2}^k b_k \quad . \tag{4.102}$$

The transformation of products of more than two fermion sources can be consistently performed only in connection with dressed particle states. This will be shown in Section 4.8. For the moment we mention that it is necessary to observe antisymmetries of these products; in particular we assume

$$j_{I_1} j_{I_2} j_{I_3} j_{I_4} = 4 R_{\{I_1 I_2}^{k_1} R_{I_3 I_4\}}^{k_2} b_{k_1} b_{k_2} \quad . \tag{4.103}$$

The five equations (4.99)–(4.103) allow a direct transformation of $\mathcal{H}_{\mathrm{F}}$ which results in

$$\begin{aligned} E_{p0} |\tilde{\mathcal{B}}(b;p)\rangle \;=\;& 2\hat{K}_{I_1 I_2} R_{I_1 K}^k C_{k'}^{I_2 K} b_k \partial_{k'}^b |\tilde{\mathcal{B}}(b;p)\rangle \\ &+ \hat{W}_{I_1 I_2 I_3 I_4} \Big\{ -6 F_{I_4 K} R_{I_1 K}^k C_{k'}^{I_2 I_3} b_k \partial_{k'}^b \\ &+ 6 R_{I_1 K}^k C_{k'_1}^{\{I_2 I_3} C_{k'_2}^{I_4 K\}} b_k \partial_{k_1}^b \partial_{k_2}^b \\ &+ 4 R_{I_1 K_1}^{k_1} R_{K_2 K_3}^{k_2} C_{k'_1}^{I_2 K_1} C_{k'_2}^{I_3 K_2} C_{k'_3}^{I_4 K_3} b_{k_1} b_{k_2} \partial_{k'_1}^b \partial_{k'_2}^b \partial_{k'_3}^b \\ &+ 12 F_{I_4 K} R_{\{I_1 K_1}^{k_1} R_{K_2 K_3\}}^{k_2} C_{k'_1}^{I_2 K_2} C_{k'_2}^{I_3 K_3} b_{k_1} b_{k_2} \partial_{k'_1}^b \partial_{k'_2}^b \Big) \\ &+ 12 \Big( 3 F_{I_4 K_1} F_{I_3 K_2} + \frac{1}{4} A_{I_4 K_1} A_{I_3 K_2} \Big) \times \\ &\qquad R_{\{I_1 K_1}^{k_1} R_{K_2 K_3\}}^{k_2} C_{k'}^{I_2 K_3} b_{k_1} b_{k_2} \partial_{k'}^b \\ &- 12 \Big( F_{I_4 K_1} F_{I_3 K_2} + \frac{1}{4} A_{I_4 K_1} A_{I_3 K_2} \Big) F_{I_2 K_3} \times \\ &\qquad R_{\{I_1 K_1}^{k_1} R_{K_2 K_3\}}^{k_2} b_{k_1} b_{k_2} \Big\} |\tilde{\mathcal{B}}(b;p)\rangle \quad . \end{aligned} \tag{4.104}$$

We add that in (4.104) we can give up the relation (4.93) between boson and fermion sources as in (4.104) no reference is made to (4.93) and as the duals $\partial_k^b$ were introduced without reference to fermionic duals. Thus we have obtained a genuine boson functional equation.

Analogously to Section 4.6 the fourth and the fifth term can be estimated relative to the second and the third term on the right–hand side of (4.104). Neglecting these terms and identifying $|\tilde{\mathcal{B}}(b;p)\rangle \equiv |\hat{\mathcal{B}}(b;p)\rangle$, we see that equation (4.104) is equivalent to equation (4.92). So by the estimates in Section 4.6 we have justified the method of the functional chain rule which yields equation (4.104).

The successful use of the chain rule suggests to formulate weak mapping completely in a functional version. This was performed by Kerschner [Ker 94]. For brevity we do not explicitly discuss this approach in full length. Rather we sketch only the general line of arguments for the case of the boson map.

In both approaches weak mapping is defined by equation (4.23). In order to guarantee the existence of an inverse map we have to investigate the inverse relation of (4.23).

If we apply (4.4) we obtain from (4.23) the equation

$$R^{k_1}_{I_1 I_2} \dots R^{k_n}_{I_{2n-1} I_{2n}} \varphi^{(2n)}(I_1 \dots I_{2n}|a) = S^{k_1 \dots k_n}_{k'_1 \dots k'_n} \rho^{(n)}(k'_1 \dots k'_n|a) \tag{4.105}$$

with

$$S^{k_1 \dots k_n}_{k'_1 \dots k'_n} := R^{k_1}_{I_1 I_2} \dots R^{k_n}_{I_{2n-1} I_{2n}} C^{\{I_1 I_2}_{k'_1} \dots C^{I_{2n-1} I_{2n}\}}_{k'_n} \quad . \tag{4.106}$$

From this it follows that equation (4.23) is not a suitable definition with respect to the formation of an inverse mapping, i.e. to the resolution with respect to $\rho^{(n)}(k_1 \dots k_n|a)$. However, we are free to modify (4.23) in such a way that this resolution can be performed. Instead of (4.23) we use the following definition of weak mapping (for bosons!)

$$\varphi^{(2n)}(I_1 \dots I_{2n}|a) = C^{\{I_1 I_2}_{k_1} \dots C^{I_{2n-1} I_{2n}\}}_{k_n} S^{k'_1 \dots k'_n}_{k_1 \dots k_n} \tilde{\rho}^{(n)}(k'_1 \dots k'_n|a) \quad . \tag{4.107}$$

By means of equations (4.4) and (4.5) one can show, see Kerschner and Stumpf [Ker 91], that $S$ has projector properties:

$$S^{k_1 \dots k_n}_{k'_1 \dots k'_n} S^{k'_1 \dots k'_n}_{l_1 \dots l_n} = S^{k_1 \dots k_n}_{l_1 \dots l_n} \quad , \tag{4.108}$$

and if we apply the duals $R^k_{I_1 I_2}$ to (4.107), due to (4.108) we obtain

$$R^{k_1}_{I_1 I_2} \dots R^{k_n}_{I_{2n-1} I_{2n}} \varphi^{(2n)}(I_1 \dots I_{2n}|a) = S^{k_1 \dots k_n}_{k'_1 \dots k'_n} \tilde{\rho}^{(n)}(k'_1 \dots k'_n|a) \, ; \tag{4.109}$$

i.e., if we represent $\rho^{(n)}(k_1 \dots k_n|a)$ in the form

$$\rho^{(n)}(k_1 \dots k_n|a) \equiv S^{k'_1 \dots k'_n}_{k_1 \dots k_n} \tilde{\rho}^{(n)}(k'_1 \dots k'_n|a) \tag{4.110}$$

we are able to invert equation (4.23). Then, instead of (4.13) we introduce the boson functional state

$$|\mathcal{B}'(b;a)\rangle = \sum_{n=0}^{\infty} \frac{1}{n!} \tilde{\rho}^{(n)}(k'_1 \dots k'_n|a) S^{k'_1 \dots k'_n}_{k_1 \dots k_n} b_{k_1} \dots b_{k_n} |0\rangle_B \, , \tag{4.111}$$

and with definition (4.106) the analogue to the operator $\mathcal{U}$ of (4.25) is now given by

$$\tilde{\mathcal{U}} = \sum_{n=0}^{\infty} \frac{(-1)^n}{n!} b_{k_1} \dots b_{k_n} |0\rangle_B S^{k_1 \dots k_n}_{k'_1 \dots k'_n} {}_B\langle 0| \partial^b_{k'_n} \dots \partial^b_{k'_1} \quad . \tag{4.112}$$

Then one easily verifies that

$$\tilde{\mathcal{U}} |\mathcal{B}'(b; a)\rangle = |\mathcal{B}'(b; a)\rangle \tag{4.113}$$

holds. So far, the functionals (4.111) and (4.19) are related by (4.107) or (4.109), respectively. According to our program (4.107) and (4.109) are the basic source–free relations. In order to transfer these relations into functional space one has to be cautious, because in any case the independence of the boson space basis vectors from the fermion space basis vectors has to be respected. A suitable definition was proposed by Kerschner [Ker 94]. Instead of (4.93), (4.94) the relation

$$|\mathcal{B}'(b; a)\rangle = \mathcal{T}_1(b, \partial) |\mathcal{F}(j; a)\rangle \tag{4.114}$$

is postulated with

$$\mathcal{T}_1(b, \partial) := \sum_{n=0}^{\infty} \frac{1}{n!} b_{k_1} \dots b_{k_n} |0\rangle_B R^{k_1}_{I_1 I_2} \dots R^{k_n}_{I_{2n-1} I_{2n}} {}_F\langle 0| \partial_{I_{2n}} \dots \partial_{I_1} \tag{4.115}$$

and $\partial_I := \delta / \delta_{j_I}$, which is the functional version of equation (4.109). In the same way (4.107) can be transferred into the functional equation

$$|\mathcal{F}(j; a)\rangle = \mathcal{T}_2(j, \partial^b) |\mathcal{B}'(b; a)\rangle \tag{4.116}$$

with

$$\mathcal{T}_2(j, \partial^b) := \sum_{n=0}^{\infty} \frac{1}{(2n)!} j_{I_1} \dots j_{I_{2n}} |0\rangle_F C^{I_1 I_2}_{k_1} \dots C^{I_{2n-1} I_{2n}}_{k_n} {}_B\langle 0| \partial^b_{k_n} \dots \partial^b_{k_1} \quad . \tag{4.117}$$

The operators $\mathcal{T}_1$ and $\mathcal{T}_2$ are nonsingular and fulfill the relations

$$\mathcal{T}_1(b, \partial) \mathcal{T}_2(j, \partial^b) = \tilde{\mathcal{U}}(b, \partial^b) \quad , \tag{4.118}$$

$$\mathcal{T}_2(j, \partial^b) \mathcal{T}_1(b, \partial) = \mathbb{1}(j, \partial) \quad . \tag{4.119}$$

Due to the existence of inverse mappings and their functional formulation it is possible to directly transform the functional energy equation (4.28) for fermions

$$E_{p0} |\mathcal{F}(j; p)\rangle = \mathcal{H}_\mathrm{F}(j, \partial) |\mathcal{F}(j; p)\rangle \tag{4.120}$$

into a corresponding energy equation for bosons. Multiplying (4.120) by $\mathcal{T}_1$ and inserting (4.119) we obtain

$$E_{p0} \mathcal{T}_1 |\mathcal{F}(j; p)\rangle = \mathcal{T}_1 \mathcal{H}_\mathrm{F}(j, \partial) \mathcal{T}_2 \mathcal{T}_1 |\mathcal{F}(j; p)\rangle \tag{4.121}$$

and with (4.114)

$$E_{p0} |\mathcal{B}'(b; p)\rangle = \mathcal{T}_1 \mathcal{H}_\mathrm{F}(j, \partial) \mathcal{T}_2 |\mathcal{B}'(b; p)\rangle \quad . \tag{4.122}$$

Therefore the exact effective boson energy operator is defined by

$$\mathcal{H}_\mathrm{B}(b, \partial^b) := \mathcal{T}_1 \mathcal{H}_\mathrm{F}(j, \partial) \mathcal{T}_2 \quad . \tag{4.123}$$

Evaluation of the right–hand side of equation (4.123) leads to an equivalent representation of the boson dynamics of Proposition 4.2. It reads according to Kerschner [Ker 94]:

$$\begin{aligned} E_{p0}|\mathcal{B}'(b;p)\rangle &= \big[K^{kl} b_k \partial_l^b \\ &+W_1^{kl_1l_2} b_k \partial_{l_1}^b \partial_{l_2}^b \\ &+W_2^{k_1l_1}\left(b_{k_1}+\Gamma_{k_1l_2}^{k_2k_3} b_{k_2} b_{k_3}\partial_{l_2}^b\right)\partial_{l_1}^b \\ &+W_3^{k_1k_2l_1}\left(b_{k_1}+\Gamma_{k_1l_2}^{k_3k_4} b_{k_3} b_{k_4}\partial_{l_2}^b\right) b_{k_2}\partial_{l_1}^b \\ &+W_4^{k_1k_2}\left(b_{k_1}+\Gamma_{k_1l_1}^{k_3k_4} b_{k_3} b_{k_4}\partial_{l_1}^b\right)\left(b_{k_2}+\Gamma_{k_2l_2}^{k_4k_5} b_{k_4} b_{k_5}\partial_{l_2}^b\right)\big]|\mathcal{B}'(b;p)\rangle \end{aligned} \tag{4.124}$$

with

$$\begin{aligned} K^{kl} &:= 2\hat{K}_{I_1I_2} R_{I_1K}^k C_l^{I_2K} \\ W_1^{kl_1l_2} &:= 2\hat{W}_{I_1I_2I_3I_4} R_{I_1K}^k C_{l_1}^{I_4K} C_{l_2}^{I_2I_3} \\ W_2^{kl} &:= -6\hat{W}_{I_1I_2I_3I_4} F_{I_4K} R_{I_1K}^k C_l^{I_2I_3} \\ W_3^{k_1k_2l} &:= 4\hat{W}_{I_1I_2I_3I_4}\left(3F_{I_4K_1}F_{I_3K_2}+\frac{1}{4}A_{I_4K_1}A_{I_3K_2}\right) R_{I_1K_1}^{k_1} R_{K_2K_3}^{k_2} C_l^{I_2K_3} \\ W_4^{k_1k_2} &:= -4\hat{W}_{I_1I_2I_3I_4}\left(F_{I_4K_1}F_{I_3K_2}+\frac{1}{4}A_{I_4K_1}A_{I_3K_2}\right) F_{I_2K_3} R_{I_1K_1}^{k_1} R_{K_2K_3}^{k_2} \\ \Gamma_{l_1l_2}^{k_1k_2} &:= -2C_{l_1}^{I_1I_2} R_{I_2I_3}^{k_1} C_{l_2}^{I_3I_4} R_{I_4I_1}^{k_2} \quad . \end{aligned} \tag{4.125}$$

In addition, by neglection of the exchange terms $\Gamma_{kk'}^{ll'}$ the results of the chain rule–method can be independently reproduced.

## 4.8 Dressed Particle States

Hard core bound states are characterized by a definite number of coordinates in their wave functions, the minimal number of subfermions, and they are described by the state functionals of Section 4.2. Corresponding dressed particle states start with the same number of coordinates in their wave functions, but in addition they contain an infinite set of wave functions with higher coordinate numbers than the minimal hard core number. Thus we can describe dressed particle states built up of subfermions by the functional states

$$|\mathcal{F}(j;k)\rangle^{dr} = \sum_{n=0}^{\infty} \frac{i^{2n+m}}{(2n+m)!} C_{m,k}^{I_1\ldots I_{2n+m}} j_{I_1}\cdots j_{I_{2n+m}}|0\rangle_F \tag{4.126}$$

where $k$ denotes the full set of quantum numbers which characterize the dressed particle, while $m$ is the minimal number of fermionic subcomponents which is needed for the formation of the hard core of a specific dressed particle state. In contrast to the hard core states which are assumed to be solutions of the diagonal part (4.2) of the energy equation, the dressed particle states are assumed to be solutions of the full energy equation (4.1). As in (4.1) the sectors of even and odd fermion numbers completely decouple, the specific form of (4.126) is justified. However, it should be noted that although (4.126) is assumed to be a solution of (4.1), the direct product of (4.126) with itself, i.e. a state of two dressed particles, is in general not a solution of (4.1). Rather equation (4.1) enforces an interaction between two (or more) dressed particles, i.e. a dressed particle dynamics, a situation which is completely analogous to the hard core bound state dynamics.

In order to derive an effective dressed particle dynamics we need orthonormality and completeness relations analogously to those for hard core states, cf. Stumpf [Stu 87]. By definition we introduce dual sets $\left\{ R^{m,k}_{I_1 \ldots I_{2n+m}}, n = 0, 1, \ldots, \infty \right\}$ where $m$ and $k$ run through all dressed particle state quantum numbers. These dual sets are characterized by the orthonormality relations

$$\sum_{I_1 \ldots I_{2n+m}} \sum_{n,n'} R^{m,k}_{I_1 \ldots I_{2n+m}} C^{I_1 \ldots I_{2n'+m'}}_{m',k'} \delta_{2n+m,2n'+m'} = \delta_{kk'} \delta_{mm'} \quad . \tag{4.127}$$

These relations are an appropriate extension of the hard core state orthonormality relations (4.4).

From (4.127) it follows that orthonormality or duality, respectively, is connected with a selection rule. Due to (4.126) either even or odd numbers $2n+m$ appear in (4.127), i.e., boson states with even numbers and fermion states with odd numbers are completely separated. Therefore, if we generalize weak mapping to dressed particle states, it is convenient to distinguish between fermions and bosons from the beginning. In full generality the analogue of the hard core boson transformation (4.93) for dressed particles then have to be defined by the transformations

$$f_{2j+1,l} = \sum_{n=0}^{\infty} C^{I_1 \ldots I_{2j+1+2n}}_{2j+1,l} j_{I_1} \cdots j_{I_{2j+1+2n}} \,, \quad j = 0, 1, \ldots \infty, \tag{4.128}$$

$$b_{2j,k} = \sum_{n=0}^{\infty} C^{I_1 \ldots I_{2j+2n}}_{2j,k} j_{I_1} \cdots j_{I_{2j+2n}} \,, \quad j = 1, 2 \ldots \infty, \tag{4.129}$$

for dressed fermion operators $f_{2j+1,l}$ and dressed boson operators $b_{2j,k}$. With respect to weak mapping the inverse relations of (4.128), (4.129) are needed.

- **Proposition 4.6:** The right–inverses of (4.128) and (4.129) are given by the relations

$$j_{I_1} \cdots j_{I_{2r+1}} = \sum_l \sum_{j=0}^{r} R^{2j+1,l}_{I_1 \ldots I_{2r+1}} f_{2j+1,l} \,, \tag{4.130}$$

$$j_{I_1} \cdots j_{I_{2r}} = \sum_k \sum_{j=1}^{r} R^{2j,k}_{I_1 \ldots I_{2r}} b_{2j,k} \,, \tag{4.131}$$

  where $k \neq k'$ for all $R^{m,k}$ and $R^{n,k'}$ with $m \neq n$.

**Proof:** We assume $j < j'$, but $k = k'$. Then we have a state $(2j+1, k)$ with $C^{I_1 \ldots I_{2j+1}}_{2j+1,k} \neq 0$ and a state $(2j'+1, k)$ with $C^{I_1 \ldots I_{2j+1}}_{2j+1,k} \equiv 0$, as by definition $C^{I_1 \ldots I_{2j'+1}}_{2j'+1,k}$ is the first nonvanishing amplitude for the state $(2j'+1, k)$. This is a contradiction, and thus it follows $k \neq k'$.

To prove (4.130) we write it in the form

$$j_{I_1} \cdots j_{I_{2j+1+2n}} = \sum_{l'} \sum_{j'=0}^{j+n} R^{2j'+1,l'}_{I_1 \dots I_{2j+1+2n}} f_{2j'+1,l'} \tag{4.132}$$

and substitute (4.132) into (4.128). This gives

$$f_{2j+1,l} = \sum_{n=0}^{\infty} C^{I_1 \dots I_{2j+1+2n}}_{2j+1,l} \sum_{l'} \sum_{j'=0}^{j+n} R^{2j'+1,l'}_{I_1 \dots I_{2j+1+2n}} f_{2j'+1,l'} \tag{4.133}$$

and if we apply (4.127) we get

$$f_{2j+1,l} = \sum_{l'} \sum_{j'=0}^{j+n} \delta_{2j+1,2j'+1} \delta_{l,l'} f_{2j'+1,l'} = f_{2j+1,l} \quad . \tag{4.134}$$

The same considerations can be applied to (4.131). If we symbolically write $f = U \circ j$ we have $j = U^{-1} \circ f$. By relation (4.134) we have thus shown $U \circ U^{-1} = \mathbf{1}$, i.e. that $U^{-1}$ given by (4.130) and (4.131) is a right–inverse to $U$. ◊

On the other hand one should also show $j = U^{-1} \circ Uj = j$, i.e. that $U^{-1}$ is simultaneously a left– and a right–inverse. The proof of the latter property depends on completeness relations for dressed particle states. For the simple applications of dressed particle states which we will discuss in the following this proof is not needed.

Finally we have to derive the duals $R^{m,k}_{I_1 \dots I_{2n+m}}$ for themselves. There is no way to simply guess these functions; rather they have to be calculated. We assume to have calculated a complete set of left–hand side dressed particle solutions of equation (4.1) which are given by their coefficient functions $L^{m,k}_{I_1 \dots I_{2n+m}}$. Then according to (3.24) we obtain for the scalar product of right–hand and left–hand side solutions

$$\sum_{n} \sum_{I_1 \dots I_{2n+m}} \left( L^{m,k}_{I_1 \dots I_{2n+m}} \right)^* C^{I_1 \dots I_{2n+m}}_{m,k} = g_{kk'} \quad . \tag{4.135}$$

If we assume that $g_{kk'}$ is non–degenerate, it follows from (4.135)

$$\sum_{n,k} \sum_{I_1 \dots I_{2n+m}} \left( g^{-1} \right)_{k''k} \left( L^{m,k}_{I_1 \dots I_{2n+m}} \right)^* C^{I_1 \dots I_{2n+m}}_{m,k'} = \delta_{k''k'} \quad , \tag{4.136}$$

and by comparison with (4.127) we obtain

$$R^{m,k''}_{I_1 \dots I_{2n+m}} \equiv \sum_{k} \left( g^{-1} \right)_{k''k} \left( L^{m,k}_{I_1 \dots I_{2n+m}} \right)^* \quad . \tag{4.137}$$

The left–hand side solutions are not directly calculated by means of equation (4.1), rather they are derived by means of Proposition 3.8. We will illustrate this procedure in Chapter 7 by calculating dressed one–fermion states.

## 4.9 Effective Boson– and Composite Fermion–Dynamics

In Section 4.4 the weak mapping of quantum fields was examplified by the proof of a weak mapping theorem for bosons as two–fermion composites. In physical reality, however, bosons and fermions occur simultaneously, and often both kinds of particles are assumed to be non–elementary. The most prominent example is offered by quantum chromodynamics where mesons and nucleons are considered to be composed of two or three quarks, respectively. Another example is provided by the standard model. This model contains at least 37 elementary bosons and fermions and more than 20 parameters, cf., for instance, Nachtmann [Nach 86]. This strongly supports the assumption that this model describes composite particles and effective forces rather than elementary ones. In a subfermion model for the explanation of the standard model the standard fermions should be composed of three subfermions, while the weak bosons should be composed of two subfermions. Hence in both cases one has to study the effective dynamics for two–fermion and three–fermion bound states.

With respect to the quark model this problem is more involved than for the subfermion model, as the quark model is based on a complicated nonabelian coupling theory while the subfermion model simply rests on a spinorfield with nonlinear selfinteraction. Therefore, in a first step to treat these problems we concentrate on the spinorfield model.

Within this model Kerschner [Ker 94] has derived a weak mapping theorem for the combined dynamics of two–fermion and three–fermion hard core states. While this theorem demonstrates the ability of the weak mapping technique to derive effective dynamics for complicated systems, there are indications that the effective dynamics for combined boson–fermion hard core states does not correctly reflect the physical situation. Without going into details we merely state that there are compelling reasons to treat, instead of a hard core dynamics, a dressed particle dynamics, because only in the latter case one can achieve a correct coupling of the bosons to the composite fermion current. An exact treatment for the combined dynamics of two–fermion and three–fermion *dressed* particle states has not yet been given. Hence for a first exploration we discuss this problem by means of the chain rule formalism. This formalism was already applied by Stumpf [Stu 87], [Stu 88] to the problem under consideration, but the improved discussion of dressed particle states in Section 4.8 also allows an improved application of the chain rule which we will formally discuss in this section.

The relation of the chain rule to the exact mapping theorems was discussed in Section 4.7. Of course, these arguments are still valid if we generalize the chain rule to comprise dressed particles. So without further comments we rewrite the formalism of Section 4.7 for dressed particles, i.e., in the case under consideration, for dressed two–fermion and dressed three–fermion states. In accordance with (4.128) and (4.129) we define boson and fermion source operators by

$$b_{2,k} := \sum_{n=0}^{\infty} C_{2,k}^{I_1 \ldots I_{2n+2}} j_{I_1} \cdots j_{I_{2n+2}} , \tag{4.138}$$

$$f_{3,l} := \sum_{n=0}^{\infty} C_{3,l}^{I_1 \ldots I_{2n+3}} j_{I_1} \cdots j_{I_{2n+3}} , \tag{4.139}$$

while the reciprocal relations are given by (4.130) and (4.131). In defining (4.138) and (4.139) we have restricted ourselves to $j = 1$ for fermions in (4.128) and $j = 1$ for bosons in (4.129).

In order to study the consequences of this restriction we consider the reciprocal relations (4.130) and (4.131). From these formulae it follows that $j_{I_1} \dots j_{I_{2r+1}}$ is correlated with $f_{1,l}, \dots f_{2r+1,l}$ and $j_{I_1} \dots j_{I_{2r}}$ is correlated with $b_{2,k}, \dots b_{2r,k}$. So if we suppress $f_{2r+1,l}$, $r = 2, \dots \infty$, and $b_{2r,k}$, $r = 2, \dots \infty$, we are not able to correctly express higher order products of the $j$–sources. On the other hand, due to the chain rule such higher order products appear in the transformation formulae if we insist on the full expansion in (4.138) and (4.139). From this we conclude that full expansion in (4.138) and (4.139) and restriction to $b_{2,k}$ and $f_{3,l}$ are not compatible. Therefore, if we want to exclusively study the $(b_{2,k}, f_{3,l})$–dynamics we are forced to start with the lowest order of the expansions (4.138) and (4.139), i.e. with the corresponding hard core states. The dressing property of these states then solely appears in the reciprocal relations (4.130) and (4.131). Thus we define the hard core sources (without using new letters)

$$b_{2,k} := C_{2,k}^{I_1 I_2} j_{I_1} j_{I_2} \tag{4.140}$$

$$f_{3,l} := C_{3,l}^{I_1 I_2 I_3} j_{I_1} j_{I_2} j_{I_3} \tag{4.141}$$

and assume that the original functional state $|\mathcal{F}(j;a)\rangle$, which is an eigenstate of equation (4.28), is transformed into a state $|\mathcal{G}(b,f;a)\rangle$, which is defined by the invariance relation $|\mathcal{F}(j;a)\rangle = |\mathcal{G}(b,f;a)\rangle$. Thus equation (4.28) can equivalently be written as

$$E_{p0}|\mathcal{G}(b,f;p)\rangle = \mathcal{H}_{\mathrm{F}}(j,\partial)|\mathcal{G}(b,f;p)\rangle \tag{4.142}$$

and the chain rule now reads

$$\begin{aligned} \partial_I|\mathcal{G}\rangle &= \left[(\partial_I b_{2,k})\partial_{2,k}^b + (\partial_I f_{3,l})\partial_{3,l}^f\right]|\mathcal{G}\rangle \\ &= \left[2C_{IK}^{2,k} j_K \partial_{2,k}^b + 3C_{IK_1K_2}^{3,l} j_{K_1} j_{K_2} \partial_{3,l}^f\right]|\mathcal{G}\rangle \quad . \end{aligned} \tag{4.143}$$

By repeated application of this chain rule all derivatives $\partial_I \equiv \delta/\delta j_I$ of $\mathcal{H}_{\mathrm{F}}$ can be expressed by $\partial^b$– and $\partial^f$–derivatives together with products of sources $j_I$. The latter can be replaced by linear combinations of $b$– and $f$–sources if one substitutes for these products the relations (4.130) and (4.131). If we restrict ourselves to $b_{2,k}$– and $f_{3,l}$–sources, these relations can be expressed by

$$j_{I_1} \dots j_{I_{2r+1}} = \sum_l R_{I_1 \dots I_{2r+1}}^{3,l} f_{3,l} \,, \tag{4.144}$$

$$j_{I_1} \dots j_{I_{2r}} = \sum_k R_{I_1 \dots I_{2r}}^{2,k} b_{2,k} \,, \tag{4.145}$$

what means that we exclude from our calculation any admixture of hard core states (and their dressing) with fermion number $n \geq 4$. Furthermore, from the beginning we suppress all contributions of the elementary subfermions with fermion number $n = 1$, because due to the decoupling theorem discussed in Section 4.6 these heavy particles do not contribute to the low energy dynamics.

The corresponding calculations are straightforward, so we can directly give the result. We decompose the resulting effective functional Hamilton operator $\mathcal{H}_{\mathrm{C}}$ into

$$\mathcal{H}_{\mathrm{C}}(b,f,\partial^b,\partial^f) = \mathcal{H}_{\mathrm{b}} + \mathcal{H}_{\mathrm{f}} + \text{quantization terms} \tag{4.146}$$

and obtain

$$\mathcal{H}_\mathrm{b} = b_k\Big(K^{kk'}\partial^b_{k'} + M^{kk'}\partial^b_{k'} + N^{kk'_1k'_2}\partial^b_{k'_1}\partial^b_{k'_2} + S^{k\ell'_1\ell'_2}\partial^f_{\ell'_1}\partial^f_{\ell'_2} + H^{kk'_1k'_2k'_3}\partial^b_{k'_1}\partial^b_{k'_2}\partial^b_{k'_3} + V^{kk'\ell'_1\ell'_2}\partial^b_{k'}\partial^f_{\ell'_1}\partial^f_{\ell'_2}\Big) \tag{4.147}$$

$$\mathcal{H}_\mathrm{f} = f_\ell\Big(\hat{K}^{\ell\ell'}\partial^f_{\ell'} + \hat{M}^{\ell\ell'}\partial^f_{\ell'} + \hat{N}^{\ell k\ell'}\partial^b_k\partial^f_{\ell'} + \hat{H}^{\ell\ell'_1\ell'_2\ell'_3}\partial^f_{\ell'_1}\partial^f_{\ell'_2}\partial^f_{\ell'_3} + \hat{V}^{\ell\ell'k'_1k'_2}\partial^f_{\ell'}\partial^b_{k'_1}\partial^b_{k'_2}\Big) \tag{4.148}$$

with

$$K^{kk'} := 2K_{I_1I_2}C^{2,k'}_{I_2I_3}R^{2,k}_{I_1I_3} \tag{4.149}$$

$$M^{kk'} := W_{I_4I_1I_2I_3}\Big[-6F_{I_3K}C^{2,k'}_{I_1I_2}R^{2,k}_{I_4K} + 2\left(3F_{I_1K_1}F_{I_4K_2} + \frac{1}{4}A_{I_1K_1}A_{I_4K_2}\right)C^{2,k'}_{I_3K_3}R^{2,k}_{I_2K_1K_2K_3}\Big] \tag{4.150}$$

$$N^{kk'_1k'_2} := W_{I_4I_1I_2I_3}\Big[12F_{I_3K_1}C^{2,k'_1}_{I_1K_2}C^{2,k'_2}_{I_2K_3}R^{2,k}_{I_4K_1K_2K_3} + 12C^{2,k'_1}_{I_1I_2}C^{2,k'_2}_{I_3K}R^{2,k}_{I_4K}\Big] \tag{4.151}$$

$$S^{k\ell'_1\ell'_2} := -W_{I_4I_1I_2I_3}\Big[-27F_{I_3K_1}C^{3,\ell'_1}_{I_1K_2K_3}C^{3,\ell'_2}_{I_2K_4K_5}R^{2,k}_{I_4K_1K_2K_3K_4K_5} + 36C^{3,\ell'_1}_{I_1I_3K_1}C^{3,\ell'_2}_{I_2K_2K_3}R^{2,k}_{I_4K_1K_2K_3}\Big] \tag{4.152}$$

$$H^{kk'_1k'_2k'_3} := -8W_{I_4I_1I_2I_3}C^{2,k'_1}_{I_1K_1}C^{2,k'_2}_{I_2K_2}C^{2,k'_3}_{I_3K_3}R^{2,k}_{I_4K_1K_2K_3} \tag{4.153}$$

and

$$\hat{K}^{\ell\ell'} := 3K_{I_1I_2}C^{3,\ell'}_{I_2K_1K_2}R^{3,\ell}_{I_1K_1K_2} \tag{4.154}$$

$$\hat{M}^{\ell\ell'} := W_{I_4I_1I_2I_3}\Big[-18F_{I_3K_1}C^{3,\ell'}_{I_1I_2K_2}R^{3,\ell}_{I_4K_1K_2} + 3\left(3F_{I_1K_1}F_{I_2K_2} + \frac{1}{4}A_{I_1K_1}A_{I_2K_2}\right)C^{3,\ell'}_{I_3K_3K_4}R^{3,\ell}_{I_4K_1K_2K_3K_4}\Big] \tag{4.155}$$

$$\hat{N}^{\ell k\ell'} := W_{I_4I_1I_2I_3}\Big[36F_{I_3K_1}C^{3,\ell'}_{I_2K_3K_4}C^{2,k}_{I_1K_2}R^{3,\ell}_{I_4K_1K_2K_3K_4} + \Big(18C^{3,\ell'}_{I_3K_1K_2}C^{2,k}_{I_1I_2} + 36C^{3,\ell'}_{I_2I_3K_2}C^{2,k}_{I_1K_1}\Big)R^{3,\ell}_{I_4K_1K_2}\Big] \tag{4.156}$$

$$\hat{H}^{\ell\ell'_1\ell'_2\ell'_3} := 27W_{I_4I_1I_2I_3}C^{3,\ell'_1}_{I_1K_1K_2}C^{3,\ell'_2}_{I_2K_3K_4}C^{3,\ell'_3}_{I_3K_5K_6}R^{3,\ell}_{I_4K_1K_2K_3K_4K_5K_6}, \tag{4.157}$$

while we do not explicitly write out the terms $V^{\ell k\ell'_1\ell'_2}$ and $\hat{V}^{\ell k\ell'_1\ell'_2}$.

Although so far this map was only partially discussed by explicit calculation of the various terms in (4.147) and (4.148) with realistic wavefunctions, we can use the results from Stumpf [Stu 85a], [Stu 86a] and Chapter 6 in order to draw some general conclusions.

First of all, the various terms in (4.147) and (4.148) allow a distinct physical interpretation. This interpretation results from the derivation of the functional energy equation of quantum fields. Apart from time derivatives the expressions in the brackets of (4.147) and (4.148) are the functional version of the field equations: $K^{kk'}$ and $\hat{K}^{\ell\ell'}$ contain the kinetic terms of the boson and fermion field equations, respectively. $M^{kk'}$ and $\hat{M}^{\ell\ell'}$ yield bosonic and fermionic mass corrections in the field equations. $N^{kk'_1k'_2}$ represents the nonabelian bosonic selfcoupling and $S^{k\ell'_1\ell'_2}$ the coupling of the boson fields to the fermionic current in the field equations, while $H^{kk'_1k'_2k'_3}$ describes a Higgs selfcoupling term. The term $\hat{N}^{\ell k\ell'}$ leads to the covariant coupling of fermions to bosons and $\hat{H}^{\ell\ell'_1\ell'_2\ell'_3}$ to a selfcoupling of the fermion fields in the field equations. Examples justifying such interpretations can be found in Chapter 6.

A further insight into the peculiarities of this map can be gained by estimates. Such estimates were performed in [Stu 85a] for the problem under consideration with respect to hard core states. If for a first exploration we extend these results to our case an estimate can be given for the boson–fermion coupling terms. One obtains that $V^{kk'\ell'_1\ell'_2}$ and $\hat{V}^{kk'\ell'_1\ell'_2}$ are small if they are compared with $\hat{N}^{\ell k\ell'}$ in the case of large mean subfermion mass. Hence the leading coupling terms are $\hat{N}$ and $S$, and in the limit $m \to \infty$ we can neglect $V$ and $\hat{V}$. In this limit the map yields

$$\mathcal{H}_{\mathrm{F}}(j,\partial)|\mathcal{F}(j;p)\rangle = \mathcal{H}_{\mathrm{C}}(b,f,\partial^b,\partial^f)|\mathcal{G}(b,f;p)\rangle \tag{4.158}$$

with

$$\begin{aligned}\mathcal{H}_{\mathrm{C}}(b,f,\partial^b,\partial^f) \;:=\;& b_k\Big(K^{kk'}\partial^b_{k'} + M^{kk'}\partial^b_{k'} + N^{kk'_1k'_2}\partial^b_{k'_1}\partial^b_{k'_2} \\ &+ S^{k\ell'_1\ell'_2}\partial^f_{\ell'_1}\partial^f_{\ell'_2} + H^{kk'_1k'_2k'_3}\partial^b_{k'_1}\partial^b_{k'_2}\partial^b_{k'_3}\Big) \\ &+ f_\ell\Big(\hat{K}^{\ell\ell'}\partial^f_{\ell'} + \hat{M}^{\ell\ell'}\partial^f_{\ell'} + \hat{N}^{\ell k'\ell'}\partial^b_{k'}\partial^f_{\ell'} + \hat{H}^{\ell\ell'_1\ell'_2\ell'_3}\partial^f_{\ell'_1}\partial^f_{\ell'_2}\partial^f_{\ell'_3}\Big) \\ &+ \text{quantization terms} \quad .\end{aligned} \tag{4.159}$$

The quantization does not reflect the field dynamics. Therefore, in this general discussion we did not evaluate the corresponding terms. In Chapter 6 an explicit calculation and interpretation of such terms is given.

# 5 Bound State Calculations

## 5.1 Introduction

The aim of our investigations is the derivation of a composite particle dynamics by means of weak mapping. As this procedure is referred to corresponding bound states we have to determine them.

According to Section 4.1 there are two possibilities of formulating a composite particle theory: either in terms of hard core bound states or in terms of dressed composite particle states. Because of the complications connected with the latter we concentrate on the calculation of hard core bound states. We will discuss a two–particle vector boson solution which is generated by the fusion of two subfermions and an approximated spin 2–tensor boson solution resulting as a bound state from four subfermions. Such states will be the essential elements of corresponding effective theories. Finally we will make some comments on dressed particle states, in particular we will prepare the calculation of an approximated subfermion polarization cloud.

For decades in quantum field theory the calculation of bound states was dominated by the use of Bethe–Salpeter equations, see Salpeter and Bethe [Sal 51]. However, with increasing familiarity with this type of equations, strange and unphysical properties of their solutions were discovered, see, e.g., Itzykson and Zuber [Itz 80]. Bethe–Salpeter equations were derived within the covariant formalism. Therefore, with respect to our discussion in Section 2.7 the strange properties of the solutions of such equations seem to be a further hint that the covariant formalism may lead to wrong results even in a non–perturbative treatment. As we work in the algebraic Schrödinger representation (cf. Chapter 3) the (hard core) bound state equations (5.1) are of another type than Bethe–Salpeter equations. Structurally they can be considered as the energy representation of generalized Bargmann–Wigner equations, cf. [Barg 48].

Indeed, beyond quantum field theory there was a continuous interest in the discussion of Bargmann–Wigner equations. A review concerning this topic is contained in the introduction of a paper of Pfister, Rosa and Stumpf [Pfi 89]. It should be noted that Bargmann–Wigner equations are also primarily covariant equations. But in contrast to the Bethe–Salpeter equations, Bargmann–Wigner equations possess an energy representation and admit in their covariant generalized form the transition from the multi–time version to equations for equal times. Obviously this difference is responsible for the fact that in the latter case the covariance does not lead to difficulties.

In this chapter we concentrate on the general way of deriving bound state solutions of equation (5.1) for the case of the subfermion numbers $n = 2$ and $n = 4$. The explicit evaluation of the corresponding wavefunctions is partly postponed to subsequent chapters where these functions are used for weak mapping calculations. In deriving solutions for $n = 2$ and $n = 4$ we use as an intermediate step covariant equations, but the close connection of these equations with the energy representation of corresponding one–time equations is explicitly demonstrated. The calculations for vector bosons were performed by Pfister, Rosa and Stumpf [Pfi 89], and for spin–2 bosons by Stumpf, Borne and Kaus [Stu 93b], as well as by Kaus [Kaus 94]. In Sections 5.3 and 5.4 we review the content of these papers.

As we will calculate massless vector bosons, we have to make a comment with respect to these states: The so–called Weinberg–Witten theorem [Wein 80] excludes massless charged vector bosons as being composite. However, this statement does not apply to our case, because its presuppositions are too restrictive. Hence from this theorem there cannot be established a contradiction to our approach.

Finally we mention that the solution method for (5.1) via covariant equations is not imperative. The definition of the hard core equation (5.1) and a direct integration method for $n = 2$ was proposed by Stumpf [Stu 83b] and further investigated by Hailer [Hai 85], Hornung [Hor 86], and for $n = 3$ by Lauxmann [Lau 86]. For an illustrative treatment of bound states see also Grosser [Gro 87]. As the extension to $n = 4$ has not been performed so far we will not discuss this method.

Furthermore, for abbreviation from now on we work with "natural units", i.e. we set $\hbar = c = 1$.

## 5.2 Covariant Bound State Equations

According to Section 4.2 we define hard core bound states for fermion numbers $n = 2, 3, 4$ as solutions of equation (4.2):

$$E_{p0}|\mathcal{F}(j^t;p)\rangle^{\mathrm{d}} = (j^t_{I_1}\hat{K}_{I_1I_2}\partial^t_{I_2} - 3\hat{W}_{I_1I_2I_3I_4}F^t_{I_4K}j^t_{I_1}j^t_{K}\partial^t_{I_3}\partial^t_{I_2})|\mathcal{F}(j^t;p)\rangle^{\mathrm{d}} \,. \tag{5.1}$$

Due to the algebraic degrees of freedom which are contained in the symbolic indices $I$, equation (5.1) is rather complicated even for the low fermion numbers $n = 2, 3, 4$. However, these complications can be reduced if (5.1) is related to its covariant counterpart equation. This is a special feature of quantum field theory and it may be used to obtain solutions of (5.1) via the solution of the associated covariant equation

$$K_{I_1I_2}\partial_{I_2}|\mathcal{F}(j;a)\rangle^{\mathrm{d}} = 3W_{I_1I_2I_3I_4}F_{I_4K}j_K\partial_{I_3}\partial_{I_2}|\mathcal{F}(j;a)\rangle^{\mathrm{d}} \tag{5.2}$$

where according to (1.58)–(1.61) and (2.116) it is

$$\begin{aligned}
K_{I_1I_2} &= \left[D^{\mu}_{Z_1Z_2}\partial_\mu(x_1) - m_{Z_1Z_2}\right]\delta(x_1-x_2) \quad , && (5.3)\\
W_{I_1I_2I_3I_4} &= U_{Z_1\{Z_2Z_3Z_4\}_{\mathrm{as}}}\delta(x_1-x_2)\delta(x_1-x_3)\delta(x_1-x_4) \quad , && (5.4)\\
F_{I_1I_2} &= i\lambda_{i_1}\delta_{i_1i_2}(\gamma_5)_{\kappa_1\kappa_2}\left[(i\gamma^\mu\partial_\mu(x_1)+m_{i_1})C\right]_{\alpha_1\alpha_2} \times && (5.5)\\
&\quad \Delta(x_1-x_2,m_{i_1}) \quad ,
\end{aligned}$$

and where $\Delta(x_1 - x_2, m_{i_1})$ is the Feynman propagator

$$\Delta(x_1-x_2,m_{i_1}) := (2\pi)^{-4}\int d^4p\frac{e^{-ip(x_1-x_2)}}{p^2-m^2+i\epsilon} \quad .$$

According to Sections 2.4 and 2.5 in (5.4) the limit $t_2, t_3, t_4 \to t_1$ has to be performed in a definite way in order to obtain well–defined one–time quantities.

Equation (5.2) is the diagonal part of equation (2.92) with the regularization (2.118). We will explicitly discuss the relation between equations (5.2) and (5.1) for the case $n = 2$.

Equation (5.2) admits separate solutions in any fermion number sector, so for the bound state functionals we apply the following ansatz:

$$|\,\mathcal{F}(j;a,n)\rangle^{\mathrm{d}} := \frac{i^n}{n!}\varphi^{(n)}\,(I_1\ldots I_n|a)\,j_{I_1}\ldots j_{I_n}|0\rangle_F \quad . \tag{5.6}$$

The bound state equation (5.2) can now be reduced by a technique developed for the "covariant" treatment of the anharmonic oscillator by Schuler and Stumpf [Schu 67]. For this purpose we introduce some useful definitions.

First we decompose the collective index $I$ (cf. Section 1.5) into $I = (i,a)$ with $a := (\kappa,\alpha,x)$ where the index $\kappa$ is given by the following assignment:

$$\kappa = \begin{cases} 1 & \text{for} \quad \Lambda = 1, A = 1 \\ 2 & \text{for} \quad \Lambda = 1, A = 2 \\ 3 & \text{for} \quad \Lambda = 2, A = 1 \\ 4 & \text{for} \quad \Lambda = 2, A = 2\,. \end{cases} \tag{5.7}$$

Then we have

$$W_{I_1I_2I_3I_4} = \lambda_{i_1}B_{i_2i_3i_4}W_{a_1a_2a_3a_4} \tag{5.8}$$

and

$$W_{a_1a_2a_3a_4} := V^{\kappa_1\kappa_2\kappa_3\kappa_4}_{\alpha_1\alpha_2\alpha_3\alpha_4}\,\delta(x_1-x_2)\delta(x_1-x_3)\delta(x_1-x_4) \tag{5.9}$$

with

$$V^{\kappa_1\kappa_2\kappa_3\kappa_4}_{\alpha_1\alpha_2\alpha_3\alpha_4} := \frac{g}{2}\sum_{h=1}^{2}\left\{v^h_{\alpha_1\alpha_2}(v^hC)_{\alpha_3\alpha_4}\delta_{\kappa_1\kappa_2}\left[\gamma_5(1-\gamma^0)\right]_{\kappa_3\kappa_4}\right\}_{\mathrm{as}\{2,3,4\}} \quad . \tag{5.10}$$

The Green's function $G$ is defined by $GK\psi = KG\psi = \psi$, the averages over the auxiliary fields are defined by

$$\varphi^{(n)}_{a_1\ldots a_n} := \sum_{i_1\ldots i_n}\varphi^{(n)}_{\substack{i_1\ldots i_n\\ a_1\ldots a_n}} \quad , \tag{5.11}$$

$$F_{a_1a_2} := \sum_{i_1,i_2}F_{\substack{i_1i_2\\ a_1a_2}} \quad , \tag{5.12}$$

$$G_{a_1a_2} := \sum_{i_1,i_2}\lambda_{i_1}G_{\substack{i_1i_2\\ a_1a_2}} \quad , \tag{5.13}$$

and the Fouriertransforms of $\varphi^{(n)}$, $F$ and $G$ are denoted by $\tilde{\varphi}^{(n)}$, $\tilde{F}$ and $\tilde{G}$ with coordinates $q,p$ instead of $x,y$. (For brevity here and in the following we sometimes omit quantum numbers of the wavefunctions.)

For the derivation of $n = 2$ bound state solutions we first mention that the two–particle wave functions have to be of the form

$$\varphi^{(2)}_{Z_1Z_2}(x_1,x_2|a,k) = e^{-i\frac{k}{2}(x_1+x_2)}\chi_{Z_1Z_2}(x_1-x_2|a,k) \quad , \tag{5.14}$$

where $a$ is a collection of quantum numbers besides the four–momentum $k$.

One easily verifies the translational invariance of the diagonal equation (5.2). Therefore $|\mathcal{F}(j;a,k,2)\rangle^{\mathrm{d}}$ has to fulfill (2.97) with the momentum quantum number $k$; this is equivalent to (5.14) for the state function $\varphi^{(2)}_{Z_1Z_2}(x_1,x_2|a,k)$. We interpret $k$ as the center of mass momentum of the two–particle state.

The Fouriertransform of (5.14) reads

$$\tilde{\varphi}^{(2)}_{Z_1Z_2}(q_1,q_2|a,k) = \delta(q_1+q_2-k)\tilde{\chi}_{Z_1Z_2}\left(\frac{q_1-q_2}{2}\middle| a,k\right) \quad . \tag{5.15}$$

Then for $n=2$ equation (5.2) can be reduced to a purely algebraic problem by

- **Proposition 5.1:** Let $\varphi^{(2)}_{I_1I_2}$ be the wavefunction of the hard core functional state (5.6) for $n=2$. Then $\varphi^{(2)}_{I_1I_2}$ is a solution of equation (5.2), if it is given by

$$\begin{aligned}
&\varphi^{(2)}_{\substack{i_1 i_2\\ \kappa_1\kappa_2\\ \alpha_1\alpha_2}}(x_1,x_2) \\
&= 3\int G_{\substack{i_1 j_1\\ \kappa_1\nu_1\\ \alpha_1\beta_1}}(x_1-y)\lambda_{j_1}V_{\substack{\nu_1\nu_2\nu_3\nu_4\\ \beta_1\beta_2\beta_3\beta_4}}\sum_{j_2}F_{\substack{j_2 i_2\\ \nu_2\kappa_2\\ \beta_2\alpha_2}}(y-x_2)\varphi^{(2)}_{\substack{\nu_3\nu_4\\ \beta_3\beta_4}}(x,x)d^4x
\end{aligned}\tag{5.16}$$

with

$$\varphi^{(2)}_{\substack{\kappa_1\kappa_2\\ \alpha_1\alpha_2}}(x_1,x_2) \equiv \varphi^{(2)}_{a_1a_2} = \sum_{i_1,i_2}\varphi^{(2)}_{\substack{i_1 i_2\\ \kappa_1\kappa_2\\ \alpha_1\alpha_2}}(x_1,x_2) \quad .$$

The Fouriertransform of $\varphi^{(2)}_{a_1a_2}$ is determined by

$$\begin{aligned}
&\tilde{\varphi}^{(2)}_{\substack{\kappa_1\kappa_2\\ \alpha_1\alpha_2}}(q_1,q_2) \\
&= -3(2\pi)^4\tilde{G}_{\substack{\kappa_1\nu_1\\ \alpha_1\beta_1}}(q_1)V_{\substack{\nu_1\nu_2\nu_3\nu_4\\ \beta_1\beta_2\beta_3\beta_4}}\tilde{F}_{\substack{\kappa_2\nu_2\\ \alpha_2\beta_2}}(q_2)\delta(q_1+q_2-k)\tilde{\chi}_{\substack{\nu_3\nu_4\\ \beta_3\beta_4}} \quad ,
\end{aligned}\tag{5.17}$$

where $\tilde{\chi}$ has to fulfill the eigenvalue equation

$$\tilde{\chi}_{\substack{\kappa_1\kappa_2\\ \alpha_1\alpha_2}} = -3(2\pi)^4\int d^4\eta\,\tilde{G}_{\substack{\kappa_1\nu_1\\ \alpha_1\beta_1}}(\eta)V_{\substack{\nu_1\nu_2\nu_3\nu_4\\ \beta_1\beta_2\beta_3\beta_4}}\tilde{F}_{\substack{\kappa_2\nu_2\\ \alpha_2\beta_2}}(k-\eta)\tilde{\chi}_{\substack{\nu_3\nu_4\\ \beta_3\beta_4}} \tag{5.18}$$

and the eigenvalue $k$ is the four–momentum of the state.

**Proof:** We project equation (5.2) for $n=2$ into coordinate space and apply the Green's function $G$. This gives (5.16), where $\varphi^{(2)}_{I_1I_2}$ has to be antisymmetric in $I_1, I_2$. Averaging (5.16) over the auxiliary indices and Fourier transforming the resulting equation we obtain

$$\tilde{\varphi}^{(2)}_{\substack{\kappa_1\kappa_2\\ \alpha_1\alpha_2}}(q_1,q_2) = -3(2\pi)^4\int d^4u\,\tilde{G}_{\substack{\kappa_1\nu_1\\ \alpha_1\beta_1}}(q_1)V_{\substack{\nu_1\nu_2\nu_3\nu_4\\ \beta_1\beta_2\beta_3\beta_4}}\tilde{F}_{\substack{\kappa_2\nu_2\\ \alpha_2\beta_2}}(q_2)\tilde{\varphi}^{(2)}_{\substack{\nu_3\nu_4\\ \beta_3\beta_4}}(q_1+q_2-u,u)\,. \tag{5.19}$$

We define the contracted function

$$\tilde{\varphi}^{(2)}(q_1|) := \int \tilde{\varphi}^{(2)}(\eta,q_1-\eta)d^4\eta\,. \tag{5.20}$$

Contraction of (5.19) according to (5.20) yields

$$\tilde{\varphi}^{(2)}_{\substack{\kappa_1\kappa_2\\ \alpha_1\alpha_2}}(q_1|) = -3(2\pi)^4 \int d^4\eta\, \tilde{G}_{\substack{\kappa_1\nu_1\\ \alpha_1\beta_1}}(\eta) V_{\substack{\nu_1\nu_2\nu_3\nu_4\\ \beta_1\beta_2\beta_3\beta_4}} \tilde{F}_{\substack{\kappa_2\nu_2\\ \alpha_2\beta_2}}(q_1-\eta)\tilde{\varphi}^{(2)}_{\substack{\nu_3\nu_4\\ \beta_3\beta_4}}(q_1|)\,. \tag{5.21}$$

Substituting the ansatz (5.15) into (5.20) gives

$$\tilde{\varphi}^{(2)}_{\substack{\kappa_1\kappa_2\\ \alpha_1\alpha_2}}(q_1|) = \delta(q_1-k)\tilde{\chi}_{\substack{\kappa_1\kappa_2\\ \alpha_1\alpha_2}} \tag{5.22}$$

with

$$\tilde{\chi}_{\substack{\kappa_1\kappa_2\\ \alpha_1\alpha_2}} := \int d^4u \tilde{\chi}_{\substack{\kappa_1\kappa_2\\ \alpha_1\alpha_2}}(u)\;; \tag{5.23}$$

inserting (5.22) into (5.21) gives (5.18) and inserting (5.22) into (5.19) gives (5.17). ◇

Thus, by means of Proposition 5.1 we have reduced the solution of equation (5.2) for $n=2$ to the solution of an algebraic equation (5.18). In a similar way we can proceed with (5.2) for $n=3$ and $n=4$. For brevity we do not cite the corresponding propositions, but later on discuss only the explicit calculations which were performed for the cases $n=2$ and $n=4$.

With respect to the eigenvalue $k$ we add that due to the relativistic invariance of (5.2) it is sufficient to calculate the solutions of (5.2) in the rest frame, while the solution for arbitrary $k$ can be obtained by an appropriate Lorentz transformation of the corresponding wave function. The case $m=0$ needs an extra discussion.

For illustration we now explicitly discuss the relationship between the many–time solutions of (5.2) and the equal–time solutions of (5.1). For $n=2$ this is done by the following two propositions:

- **Proposition 5.2:** Let $\varphi^{(2)}_{Z_1Z_2}(x_1,x_2|a,k)$ be the wave function of the hard core state functional (5.6) for $n=2$ and (5.6) be a solution of equation (5.2). Then for

$$\varphi^{(2)}_{Z_1Z_2}(\mathbf{r}_1,\mathbf{r}_2|a,k) := \frac{1}{2}\left(\lim_{t_1\to t_2+0} + \lim_{t_1\to t_2-0}\right)\varphi^{(2)}_{Z_1Z_2}(x_1,x_2|a,k) \tag{5.24}$$

the corresponding state functional

$$|\mathcal{F}(j^t;a,k,2)\rangle^{\mathrm{d}} := -\frac{1}{2}\int d^3r_1 d^3r_2\, \varphi^{(2)}_{Z_1Z_2}(\mathbf{r}_1,\mathbf{r}_2|a,k) j^t_{Z_1}(\mathbf{r}_1) j^t_{Z_2}(\mathbf{r}_2)|0\rangle_F \tag{5.25}$$

is a solution of equation (5.1).

**Proof:** We consider the function $\varphi^{(2)}_{Z_1Z_2}(x_1,x_2|a,k)$. By projection of equation (5.2) into configuration space due to the antisymmetry of $\varphi^{(2)}(I_1,I_2)$ we obtain the following equations:

$$K_{I_1K_1}\varphi^{(2)}(K_1,I_2) = 3W_{I_1K_2K_3K_4}F_{K_4I_2}\varphi^{(2)}(K_2,K_3)\;, \tag{5.26}$$

$$K_{I_2K_1}\varphi^{(2)}(I_1,K_1) = -3W_{I_2K_2K_3K_4}F_{K_4I_1}\varphi^{(2)}(K_2,K_3)\;. \tag{5.27}$$

With $I=(Z,x)$, (5.3) and (5.4), equations (5.26) and (5.27) can be rewritten as

$$\Big[-\delta_{Z_1X_1}\partial_0(t_1) + (D^0D^k)_{Z_1X_1}\partial_k(x_1) - (D^0m)_{Z_1X_1}\Big]\varphi^{(?)}_{X_1Z_2}(x_1,x_2) \tag{5.28}$$
$$= 3D^0_{Z_1X_1}U_{X_1X_2X_3X_4}F_{X_4Z_2}(x_1-x_2)\varphi^{(2)}_{X_2X_3}(x_1,x_1)$$

and

$$\Big[-\delta_{Z_2X_1}\partial_0(t_2)+(D^0D^k)_{Z_2X_1}\partial_k(x_2)-(D^0m)_{Z_2X_1}\Big]\varphi^{(2)}_{Z_1X_1}(x_1,x_2) \tag{5.29}$$
$$= -3D^0_{Z_2X_1}U_{X_1X_2X_3X_4}F_{X_4Z_1}(x_2-x_1)\varphi^{(2)}_{X_2X_3}(x_2,x_2) \quad .$$

We introduce new coordinates

$$x_s := \frac{1}{2}(x_1+x_2)\,, \quad x_r := (x_1-x_2) \tag{5.30}$$

with

$$\partial_{x_1} = \frac{1}{2}\partial_{x_s}+\partial_{x_r}\,, \quad \partial_{x_2} = \frac{1}{2}\partial_{x_s}-\partial_{x_r} \tag{5.31}$$

and observe

$$\varphi^{(2)}_{Z_1Z_2}(x_1,x_2) = \exp(-ikx_s)\,\chi_{Z_1Z_2}(x_r|a,k) \quad . \tag{5.32}$$

Substitution of (5.31), (5.32) into (5.28), (5.29) and subsequent addition of the resulting equations yields after reintroduction of $\varphi^{(2)}$

$$\begin{aligned} k_0\varphi^{(2)}_{Z_1Z_2}(x_1,x_2) \;=\;& iD^0_{Z_1X_1}\Big[D^k_{X_1X_2}\partial_k(x_1)-m_{X_1X_2}\Big]\varphi^{(2)}_{X_2Z_2}(x_1,x_2) \\ &+iD^0_{Z_2X_1}\Big[D^k_{X_1X_2}\partial_k(x_2)-m_{X_1X_2}\Big]\varphi^{(2)}_{Z_1X_2}(x_1,x_2) \\ &-3i\Big[D^0_{Z_1X_1}U_{X_1X_2X_3X_4}F_{X_4Z_2}(x_1-x_2)\varphi^{(2)}_{X_2X_3}(x_1,x_1) \\ &-D^0_{Z_2X_1}U_{X_1X_2X_3X_4}F_{X_4Z_1}(x_2-x_1)\varphi^{(2)}_{X_2X_3}(x_2,x_2)\Big] \,. \end{aligned} \tag{5.33}$$

We mention that apart from energy renormalization (5.33) is the diagonal part of the multi–time energy equation (2.121), projected into coordinate space for n=2.

Due to the normal transform (2.85) we have with $\langle k|0\rangle = 0$

$$\varphi^{(2)}_{Z_1Z_2}(x_1,x_2|a,k) = \tau^{(2)}_{Z_1Z_2}(x_1,x_2|a,k) - \delta_{0ak}\delta_{0k}F_{Z_1Z_2}(x_1,x_2) \quad , \tag{5.34}$$

for the two–point functions, i.e. for $a,k \neq 0$, $\varphi^{(2)}$ and $\tau^{(2)}$ are identical. Now according to Proposition 3.14 the limiting process (5.24) gives an equation which is identical with the coordinate space representation of (5.1) for $n=2$. ◇

The transition from the covariant representation (5.6) to the energy representation (5.25) of a state $|a,k\rangle$ does not imply any loss of information. This is secured by the following proposition, cf. Stumpf [Stu 85a]:

- **Proposition 5.3:** The covariant multi–time solution $\varphi^{(2)}_{Z_1Z_2}(x_1,x_2|a,k)$ is completely determined by the one–time solution $\varphi^{(2)}_{Z_1Z_2}(\mathbf{r}_1,\mathbf{r}_2|a,k)$.

**Proof:** We again consider the function $\varphi^{(2)}_{Z_1Z_2}(x_1, x_2|a)$. Substitution of (5.31), (5.32) into (5.28), (5.29) and subsequent subtraction of the resulting equations yields with $x_r \equiv (\mathbf{u}, t)$:

$$
\begin{aligned}
-2\partial_0(t)\chi_{Z_1Z_2}(\mathbf{u}, t|a, k) = & \qquad\qquad (5.35)\\
& -D^0_{Z_1X_1}\left[D^j_{X_1X_2}\big(-\frac{i}{2}k + \partial_j(\mathbf{u})\big) - m_{X_1X_2}\right]\chi_{X_2Z_2}(\mathbf{u}, t|a, k)\\
& +D^0_{Z_2X_1}\left[D^j_{X_1X_2}\big(-\frac{i}{2}k - \partial_j(\mathbf{u})\big) - m_{X_1X_2}\right]\chi_{Z_1X_2}(\mathbf{u}, t|a, k)\\
& +3D^0_{Z_1X_1}U_{X_1X_2X_3X_4}F_{X_4Z_2}(\mathbf{u}, t)\chi_{X_2X_3}(0, 0|a, k)\\
& +3D^0_{Z_2X_1}U_{X_1X_2X_3X_4}F_{X_4Z_1}(-\mathbf{u}, -t)\chi_{X_2X_3}(0, 0|a, k)\\
=: \; & H_1(\nabla_{\mathbf{u}})\chi(\mathbf{u}, t) + H_2(\mathbf{u}, t)\chi(0, 0)
\end{aligned}
$$

or

$$\chi(\mathbf{u}, t) = -\frac{1}{2}\int_0^t H_1(\nabla_{\mathbf{u}})\chi(\mathbf{u}, \tau)d\tau + f(\mathbf{u}, t) \tag{5.36}$$

with

$$f(\mathbf{u}, t) := -\frac{1}{2}\int_0^t H_2(\mathbf{u}, \tau)\chi(0, 0)d\tau + \chi(\mathbf{u}, 0) \quad .$$

Equation (5.36) is an integral equation of the Volterra type which allows convergent solutions by iteration for all values of its coupling constant $\lambda$, in particular for $\lambda = -\frac{1}{2}$. The iteration solution of (5.36) is completely fixed by the inhomogeneous term $f(\mathbf{u}, t)$. For $t = 0$ this term is given by $f(\mathbf{u}, 0) \equiv \chi(\mathbf{u}, 0)$, i.e. the one–time solution. $\diamond$

Finally it should be noted that although we will concentrate on hard core bound states, equations (5.1) and (5.2) are not restricted to the calculation of the latter, but can also be used for the calculation of hard core scattering states. For $n = 2$ such scattering states were calculated by Pfister, Rosa and Stumpf in [Pfi 89]. For brevity we do not discuss this topic, but only mention that the scattering states are essential for the construction of complete orthonormal systems of functions which are applied in the weak mapping procedure.

## 5.3 Vector Boson States

The simplest bound states are two–fermion states. As it was shown in the preceding section these states, defined as solutions of (5.1) or (5.2), can be exactly calculated. With respect to their physical interpretation we assume that spin–1 two–fermion bound states are to be identified with massless gauge vector bosons of the weak and electromagnetic interactions before symmetry breaking. Hence in the following we will concentrate on the discussion of these spin–1 states by evaluating (5.18). After a slight rearrangement we obtain for this equation

$$\tilde{\chi}^{\kappa_1\kappa_2}_{\alpha_1\alpha_2} = 3(2\pi)^4\int d^4q\,\tilde{G}^{\kappa_1\rho_1}_{\alpha_1\beta_1}(k-q)V^{\rho_1\rho_2\rho_3\rho_4}_{\beta_1\beta_2\beta_3\beta_4}\tilde{F}^{\rho_4\kappa_2}_{\beta_4\alpha_2}(-q)\tilde{\chi}^{\rho_2\rho_3}_{\beta_2\beta_3} \quad . \tag{5.37}$$

Due to Proposition 5.1 for the solution of (5.16) it is sufficient to solve (5.37). To do this we expand $\tilde{\chi}$ on the superspinor–isospinor and on the spinor algebra. As both algebras have the same dimension we can use the same basis elements. Denoting the symmetric basis elements of the Dirac algebra by $S^b$ and the antisymmetric elements by $T^a$ we may represent $\tilde{\chi}$ by

$$\tilde{\chi}_{\substack{\rho_1\rho_2\\ \beta_1\beta_2}} = \sum_{a,b} m^{ab} T^a_{\rho_1\rho_2} S^b_{\beta_1\beta_2} + \sum_{a,b} n^{ab} S^b_{\rho_1\rho_2} T^a_{\beta_1\beta_2} \tag{5.38}$$

in order to obtain a completely antisymmetric $\tilde{\chi}$. With the explicit form of $\tilde{G}$, $\tilde{F}$ and $V$, cf. (5.47), (5.48) and (5.10), one can immediately deduce from (5.37) that linear independent solutions of (5.37) are given by the following submanifolds of (5.38):

$$\tilde{\chi}^{(a,1)}_{\substack{\kappa_1\kappa_2\\ \alpha_1\alpha_2}} := \sum_b m^b T^a_{\kappa_1\kappa_2} S^b_{\alpha_1\alpha_2} \quad , \tag{5.39}$$

and

$$\tilde{\chi}^{(2)}_{\substack{\kappa_1\kappa_2\\ \alpha_1\alpha_2}} := \sum_{a,b} n^{ab} S^b_{\kappa_1\kappa_2} T^a_{\alpha_1\alpha_2} \tag{5.40}$$

with $a = 1, \ldots, 6$ and $b = 1, \ldots, 10$. By substitution of (5.39) into (5.37) it follows that the solutions (5.39) are degenerate with respect to any element of the superspinor–isospinor algebra. Differences appear with respect to the spinor algebra: while (5.39) describes vector boson states, solutions (5.40) describe scalar boson states. In the following we will discuss solutions of the type (5.39), i.e. vector boson states.

We write the vector boson ansatz (5.39) explicitly as

$$\tilde{\chi}^{(a,1)}_{\substack{\kappa_1\kappa_2\\ \alpha_1\alpha_2}} = T^a_{\kappa_1\kappa_2} \Big[ A^\mu (\gamma_\mu C)_{\alpha_1\alpha_2} + F^{\mu\nu} (\Sigma_{\mu\nu} C)_{\alpha_1\alpha_2} \Big] \tag{5.41}$$

and obtain the wave functions for vector bosons by

- **Proposition 5.4:** Let $\varphi^{(2)}_{I_1 I_2}$ be a solution of (5.16). Then $\varphi^{(2)}_{I_1 I_2}$ describes a vector boson with momentum $k$, if it is given by

$$\varphi^{(2)}_{\substack{i_1 i_2\\ \kappa_1\kappa_2\\ \alpha_1\alpha_2}}(x_1, x_2) = T^a_{\kappa_1\kappa_2} \exp\Big[-i\frac{k}{2}(x_1 + x_2)\Big] A_\mu \chi^\mu_{\substack{i_1 i_2\\ \alpha_1\alpha_2}}(x_1 - x_2 | a, k) \tag{5.42}$$

with

$$\chi^\mu_{\substack{i_1 i_2\\ \alpha_1\alpha_2}}(x_1, x_2 | k) := \frac{2ig}{(2\pi)^4} \lambda_{i_1} \lambda_{i_2} \int d^4p \exp\big[-ip(x_1 - x_2)\big] \times \Big[ S_F\Big(p + \frac{k}{2}, m_{i_1}\Big) \gamma^\mu S_F\Big(p - \frac{k}{2}, m_{i_2}\Big) C \Big]_{\alpha_1\alpha_2} \tag{5.43}$$

(no summation over $i_1, i_2$) and

$$S_F\Big(p, m\Big) := \frac{\gamma^\mu p_\mu + m}{p^2 - m^2 + i\epsilon} \quad . \tag{5.44}$$

**Proof:** The general form of the solution in coordinate space is given by (5.14). By introducing center of mass coordinates (5.30) one verifies that (5.14) is a general solution of (5.16). Hence for the determination of $\varphi^{(2)}$ in (5.14) we have to calculate only $\chi_{Z_1Z_2}(x_1 - x_2)$. It is convenient to consider instead of $\chi_{Z_1Z_2}(x_1 - x_2)$ its Fouriertransform $\tilde{\chi}_{Z_1Z_2}(\frac{q_1-q_2}{2})$. We transform (5.16) into momentum space, introduce center of mass coordinates $q = q_1 + q_2$, $p = \frac{1}{2}(q_1 - q_2)$ and substitute (5.15) into the left–hand side of the Fouriertransform of (5.16). This gives

$$\delta(q-k)\tilde{\chi}_{\substack{i_1i_2\\ \kappa_1\kappa_2\\ \alpha_1\alpha_2}}(p) = \tag{5.45}$$

$$-3(2\pi)^4\,\tilde{G}_{\substack{i_1j_1\\ \kappa_1\rho_1\\ \alpha_1\beta_1}}(q_1)\lambda_{j_1}V_{\substack{\rho_1\rho_2\rho_3\rho_4\\ \beta_1\beta_2\beta_3\beta_4}}\sum_{j_2}\tilde{F}_{\substack{j_2i_2\\ \rho_2\kappa_3\\ \beta_2\alpha_3}}(q_2)\delta(q-k)\tilde{\chi}_{\substack{\rho_3\rho_4\\ \beta_3\beta_4}}$$

where $\tilde{\chi}$ is defined by (5.23). Then due to the $\delta$–distribution on the right–hand side of (5.45) we can replace $q_1 = \frac{1}{2}q + p$, $q_2 = \frac{1}{2}q - p$ by $q_1 = \frac{1}{2}k + p$, $q_2 = \frac{1}{2}k - p$. Subsequent integration over $q$ yields

$$\tilde{\chi}_{\substack{i_1i_2\\ \kappa_1\kappa_2\\ \alpha_1\alpha_2}}(p) = \tag{5.46}$$

$$-3(2\pi)^4\,\tilde{G}_{\substack{i_1j_1\\ \kappa_1\rho_1\\ \alpha_1\beta_1}}(\frac{k}{2}+p)\lambda_{j_1}V_{\substack{\rho_1\rho_2\rho_3\rho_4\\ \beta_1\beta_2\beta_3\beta_4}}\sum_{j_2}\tilde{F}_{\substack{j_2i_2\\ \rho_2\kappa_3\\ \beta_2\alpha_3}}(\frac{k}{2}-p)\tilde{\chi}_{\substack{\rho_3\rho_4\\ \beta_3\beta_4}}\quad .$$

In the latter equation we substitute $V$ given by (5.10), $\tilde{F}$ as the Fouriertransform of (5.5), and $\tilde{\chi}$ given by (5.41). Furthermore, it is $\tilde{G} = \frac{i}{\lambda}\gamma^5\tilde{F}$. Direct evaluation of (5.46) then leads to the representation (5.42), (5.43). ◊

A detailed calculation with respect to this proposition and the subsequent discussion of eigenvalue equations is contained in the work of Sand [Sand 91].

The above given representation of the boson wave functions holds for all $k$, as the calculation was performed in a strictly relativistic invariant way. The integral in (5.43) can be evaluated by standard methods and leads, of course, to a singular behavior on the light cone. Performing the equal–time limit one explicitly sees that $\varphi^{(2)}$ vanishes for large relative distances $\mathbf{r}_1 - \mathbf{r}_2$, so (5.42), (5.43) really describe bound states. Furthermore one can verify the antisymmetry of (5.42) in the indices $I_1$,$I_2$. Finally we mention that (5.42) is only determined up to a normalization constant.

We are interested in massless vector bosons, i.e. $k^2 = m^2 = 0$. For discussing this case we will deduce a secular equation for the "vector potentials" $A^\mu$. We introduce the explicit representations

$$\tilde{G}_{\substack{\alpha_1\alpha_2\\ \kappa_1\kappa_2}}(p) \;=\; \delta_{\kappa_1\kappa_2}\left[R_\mu(p)\gamma^\mu_{\alpha_1\alpha_2} + S(p)\delta_{\alpha_1\alpha_2}\right]\quad , \tag{5.47}$$

$$\tilde{F}_{\substack{\alpha_1\alpha_2\\ \kappa_1\kappa_2}}(p) \;=\; -i\gamma^5_{\kappa_1\kappa_2}\left[R_\mu(p)(\gamma^\mu C)_{\alpha_1\alpha_2} + S(p)C_{\alpha_1\alpha_2}\right] \tag{5.48}$$

with

$$R_\mu(p) \;:=\; \sum_{i-0}^{2}\lambda_i\Delta(p,m_i)p_\mu \tag{5.49}$$

$$S(p) \;:=\; \sum_{i=0}^{2}\lambda_i m_i\Delta(p,m_i) \tag{5.50}$$

and $\Delta(p, m_i) = (2\pi)^{-4}(p_i^2 - m_i^2 + i\epsilon)^{-1}$, which follows from (5.12) and (5.13). Observing $\tilde{\chi}_{\alpha_1\alpha_2}^{\kappa_1\kappa_2} = \chi_{\alpha_1\alpha_2}^{\kappa_1\kappa_2}(0)$ and (5.41), from (5.18) we obtain the set of equations

$$\begin{aligned} A^\mu &= 2i(2\pi)^4 g \int d^4q \Big\{ \left[ S(k-q)S(q) + R^\rho(k-q)R_\rho(q) \right] A^\mu \\ &\qquad -2R^\mu(k-q)R_\rho(q)A^\rho \Big\} \end{aligned} \tag{5.51}$$

and

$$F^{\mu\nu} = 4(2\pi)^4 g \int d^4q \; S(q) \left[ R^\nu(k-q)A^\mu \right]_{\mathrm{as}\{\mu,\nu\}} \tag{5.52}$$

for the determination of the center–of–mass momentum eigenvalue $k$. Thus the "field strengths" $F^{\mu\nu}$ are determined by the $A^\mu$.

With the abbreviations

$$(2\pi)^8 \int d^4q \; S(k-q)S(q) \;=:\; J_0(k^2)\,, \tag{5.53}$$

$$(2\pi)^8 \int d^4q \; R^\mu(k-q)R^\nu(q) \;=:\; \left[ J_1(k^2) - J_2(k^2) \right] k^\mu k^\nu + J_3(k^2)\eta^{\mu\nu} \tag{5.54}$$

the set of equations (5.51) can be written in the form

$$A^\mu = 2i(2\pi)^{-4} g \Big\{ \left[ J_0 + (J_1 - J_2)k^2 + 2J_3 \right] A^\mu - 2k^\mu (J_1 - J_2) k^\rho A_\rho \Big\}\,. \tag{5.55}$$

The integrals in (5.55) can be further evaluated. Using the definitions

$$\begin{aligned} c_j &:= m_k m_l \qquad (j,k,l \text{ cyclic}) \\ \Delta &:= \prod_{i>k} (m_i - m_k)^{-1} \\ P_{il}(x) &:= k^2x^2 - (m_l^2 - m^2 + k^2) + m_i^2 \end{aligned} \tag{5.56}$$

we obtain

$$\begin{aligned} J_0 &= i\pi^2\Delta^2 \sum_{j,s} c_k c_l \int_0^1 d^4x \ln P_{ri}(x) \epsilon_{kij}\epsilon_{lrs} \\ J_1 &= i\pi^2\Delta^2 \sum_{j,s} m_k m_l \int_0^1 d^4x \; x \ln P_{ri}(x) \epsilon_{kij}\epsilon_{lrs} \\ J_2 &= i\pi^2\Delta^2 \sum_{j,s} m_k m_l \int_0^1 d^4x \; x^2 \ln P_{ri}(x) \epsilon_{kij}\epsilon_{lrs} \\ J_3 &= \frac{i}{2}\pi^2\Delta^2 \sum_{j,s} m_k m_l \int_0^1 d^4x \; P_{ri}(x) \ln P_{ri}(x) \epsilon_{kij}\epsilon_{lrs}\,, \end{aligned} \tag{5.57}$$

where the summation convention is used.

Because we are interested in massless vector bosons, i.e. $k^2 = m^2 = 0$, we formulate the conditions for the existence of those states. For $k^2 = 0$ the integrals (5.57) can be exactly calculated. With the transversality condition $k^\rho A_\rho = 0$, which is compatible with (5.55), from equation (5.55) the secular equation results

$$\frac{2ig}{(2\pi)^4}(J_0 + 2J_3) = 1 \quad . \tag{5.58}$$

It constitutes a relation between $g, m_0, m_1, m_2$ in order to fulfill $k^2 = 0$. Due to the length of the resulting expression, for brevity we do not give it here explicitly, nor do we discuss its numerical consequences.

In the following we assume the coupling constant $g$ to be determined by the masses $m_0$, $m_1$ and $m_2$ and denote it by $g_V$. We add that the parameter $g$ here is fixed only in the context of defining suitable bound states by means of hard core equations. However, this does not mean that $g$ as the coupling constant of the full spinorfield dynamics (2.92) is fixed by this adjustment. We consider $g$ in this context as an independent coupling constant; by weak mapping this $g$ will be related to parameters of effective theories.

With (5.43) we can discuss the transformation properties of $\chi^\mu(x|k)$. If $\varphi^{(2)}_{I_1 I_2}$ undergoes an orthochronous Lorentz transformation, then according to Proposition 2.1 it transforms as a direct product of Dirac spinors. Therefore $\chi^\mu(x|k)$ also transforms in this way and we have

$$\chi^{\mu}{}_{\substack{i_1 i_2\\ \alpha_1\alpha_2}}(x|k) \to S_{\alpha_1\beta_1}(\Lambda)\chi^{\mu}{}_{\substack{i_1 i_2\\ \beta_1\beta_2}}(x|k')S_{\alpha_2\beta_2}(\Lambda) \quad , \tag{5.59}$$

with $k' = \Lambda k$, where $\Lambda$ is the ordinary Lorentz matrix and $S(\Lambda)$ the representation of the Lorentz transformation in classical spinor space.

With (5.43) one can easily verify that (5.59) can be rewritten as

$$\chi^{\mu}{}_{\substack{i_1 i_2\\ \alpha_1\alpha_2}}(x|k) \to \Lambda^\mu_\rho \chi^{\rho}{}_{\substack{i_1 i_2\\ \alpha_1\alpha_2}}(x|k') \quad . \tag{5.60}$$

Equation (5.60) shows that $\chi^\mu(x|k)$ indeed transforms as a vector boson. Note that $A^\mu k_\mu = 0$ implies that $A$ transforms as a four–vector, which exhibits the form invariance property

$$A_\mu \chi^{\mu}{}_{\substack{i_1 i_2\\ \alpha_1\alpha_2}}(x|k) \to A'_\mu \chi^{\mu}{}_{\substack{i_1 i_2\\ \alpha_1\alpha_2}}(x|k') \tag{5.61}$$

with $k' = \Lambda k,\ A' = \Lambda A$.

Finally a physical interpretation of the results is needed. We concentrate on the bound state solutions for vector bosons. A comparison of (5.42) with the solutions of ordinary Bargmann–Wigner equations shows that (5.42) not only has symmetric, but also antisymmetric spin parts $C$ and $\gamma^\mu\gamma_5 C$ which are not contained in the ordinary Bargmann–Wigner solutions. Nevertheless, solution (5.42) describes spin–1 states. This is possible because in contrast to the strictly local Bargmann–Wigner solutions our solutions are nonlocal, i.e. have extended wave functions in the relative coordinate.

Thus only the superspin–isospin parts $T^a_{\kappa_1\kappa_2}$ need a further discussion. The full set of antisymmetric $T^a$–matrices explicitly reads

$$\left\{T^a, a - 1, \dots 6\right\} \tag{5.62}$$

$$= \left\{ \begin{pmatrix} \sigma^2 & 0 \\ 0 & \sigma^2 \end{pmatrix}, \begin{pmatrix} \sigma^2 & 0 \\ 0 & -\sigma^2 \end{pmatrix}, \begin{pmatrix} 0 & 1 \\ -1 & 0 \end{pmatrix}, \begin{pmatrix} 0 & \sigma^t \\ (-1)^t\sigma^t & 0 \end{pmatrix}, t = 1, 2, 3 \right\} .$$

In Chapter 6 it will be shown that in connection with a weak mapping the triplet

$$T^t_{SU(2)} \in \left\{ \begin{pmatrix} 0 & \sigma^t \\ (-1)^t\sigma^t & 0 \end{pmatrix}, \ t = 1,2,3 \right\} \tag{5.63}$$

has to be identified with $SU(2)$–gauge vector boson states, while

$$T_{U(1)} := \begin{pmatrix} 0 & 1 \\ -1 & 0 \end{pmatrix} \tag{5.64}$$

is the electroweak $U(1)$–gauge boson. States (5.63) and (5.64) lead to the effective $SU(2) \otimes U(1)$ electroweak dynamics, while the first two states in (5.62) completely decouple from this dynamics.

## 5.4 Four–Fermion Bound States

The treatment of multi–fermion bound states runs along the same lines as that of the two–fermion bound states, i.e. by contraction of the corresponding integral equations. However, in contrast to the two–fermion case, in the multi–fermion case even the contracted integral equations cannot be exactly solved. As an example we treat four–fermion bound states which are the basis for the discussion of a microscopic theory of gravitation and which were given by Stumpf in [Stu 88] and Stumpf, Borne and Kaus [Stu 93b], see Chapter 7. The detailed calculations were made by Kaus [Kaus 94]. For brevity we describe only the results, more or less omitting the intermediate steps of calculation.

The starting point of the calculation is the general bound state equation (5.2), which, as already remarked, admits separate solutions in any sector of the fermion number $n = 1, 2, \ldots, \infty$. Then the analogue to Proposition 5.2 is given by

- **Proposition 5.5:** Let $\varphi^{(4)}(I_1, I_2, I_3, I_4)$ be the wavefunction of the hard core state functional (5.6) for $n = 4$. Then if $\varphi^{(4)}(I_1, I_2, I_3, I_4)$ is a solution of equation (5.2) for $n = 4$, it can be reconstructed from the contracted functions $\tilde{\varphi}^{(4)}(q_1, q_3|)$ and $\tilde{\varphi}^{(4)}(q_1|q_3, q_4)$, which satisfy the integral equation

$$\begin{aligned} \tilde{\varphi}^{(4)}_{\substack{\kappa_1\kappa_2\kappa_3\kappa_4 \\ \alpha_1\alpha_2\alpha_3\alpha_4}}(q_1, q_3|) = 3(2\pi)^4 V_{\substack{\rho_1\rho_2\rho_3\rho_4 \\ \beta_1\beta_2\beta_3\beta_4}} \quad \times \\ \Big\{ \int d^4\eta \, \tilde{G}_{\substack{\kappa_4\rho_1 \\ \alpha_4\beta_1}}(q_3 - \eta) \tilde{F}_{\substack{\rho_2\kappa_3 \\ \beta_2\alpha_3}}(-\eta) \tilde{\varphi}^{(4)}_{\substack{\kappa_1\kappa_2\rho_3\rho_4 \\ \alpha_1\alpha_2\beta_3\beta_4}}(q_3, q_1|) \\ - \int d^4\eta d^4\xi \, \tilde{G}_{\substack{\kappa_4\rho_1 \\ \alpha_4\beta_1}}(q_3 - \eta) \tilde{F}_{\substack{\rho_2\kappa_2 \\ \beta_2\alpha_2}}(\xi - q_1) \tilde{\varphi}^{(4)}_{\substack{\kappa_1\kappa_3\rho_3\rho_4 \\ \alpha_1\alpha_3\beta_3\beta_4}}(q_1 + q_3 - \xi - \eta|\xi, \eta) \\ + \int d^4\eta d^4\xi \, \tilde{G}_{\substack{\kappa_4\rho_1 \\ \alpha_4\beta_1}}(q_3 - \eta) \tilde{F}_{\substack{\rho_2\kappa_1 \\ \beta_2\alpha_1}}(-\xi) \tilde{\varphi}^{(4)}_{\substack{\kappa_2\kappa_3\rho_3\rho_4 \\ \alpha_2\alpha_3\beta_3\beta_4}}(q_3 + \xi - \eta|q_1 - \xi, \eta) \Big\} \end{aligned} \tag{5.65}$$

  where $\tilde{\varphi}^{(4)}$ is the Fouriertransform of $\varphi^{(4)}$, the contractions are defined by

$$\tilde{\varphi}^{(4)}(q_1|q_3, q_4) := \int d^4\xi \, \tilde{\varphi}^{(4)}(\xi, q_1 - \xi, q_3, q_4) \quad , \tag{5.66}$$

$$\varphi^{(4)}(q_1, q_3|) := \int d^4\eta \, \varphi^{(4)}(q_1|\eta, q_3 - \eta) \tag{5.67}$$

and the definitions (5.11)–(5.13) have to be taken into account.

**Proof:** We assume that the bound state functional (5.6) is defined for $n = 4$. Substitution of (5.6) into (5.2), projection into coordinate space and subsequent application of the Green's function $G$ yields the equation

$$\begin{aligned} \varphi^{(4)}(I_1, I_2, I_3, I_4) &= 3G_{I_4K_1}W_{K_1K_2K_3K_4}\Big[F_{K_2I_3}\varphi^{(4)}(I_1, I_2, K_3, K_4) \\ &\quad -F_{K_2I_2}\varphi^{(4)}(I_1, I_3, K_3, K_4) + F_{K_2I_1}\varphi^{(4)}(I_2, I_3, K_3, K_4)\Big] . \end{aligned} \tag{5.68}$$

If we sum over the auxiliary field indices $i_1, i_2, i_3, i_4$, observe (5.8), (5.9) and definitions (5.11)–(5.13) and eventually apply a Fourier transformation, equation (5.68) goes over into

$$\begin{aligned} \tilde{\varphi}^{(4)}_{\substack{\kappa_1\kappa_2\kappa_3\kappa_4 \\ \alpha_1\alpha_2\alpha_3\alpha_4}}(q_1, q_2, q_3, q_4) &= 3(2\pi)^4 \tilde{G}_{\substack{\kappa_4\rho_1 \\ \alpha_4\beta_1}}(q_4) V_{\substack{\rho_1\rho_2\rho_3\rho_4 \\ \beta_1\beta_2\beta_3\beta_4}} \quad \times \\ &\Big\{\tilde{F}_{\substack{\rho_2\kappa_3 \\ \beta_2\alpha_3}}(-q_3)\tilde{\varphi}^{(4)}_{\substack{\rho_3\rho_4\kappa_1\kappa_2 \\ \beta_3\beta_4\alpha_1\alpha_2}}(q_3 + q_4|q_1, q_2) \\ &-\tilde{F}_{\substack{\rho_2\kappa_2 \\ \beta_2\alpha_2}}(-q_2)\tilde{\varphi}^{(4)}_{\substack{\rho_3\rho_4\kappa_1\kappa_3 \\ \beta_3\beta_4\alpha_1\alpha_3}}(q_2 + q_4|q_1, q_3) \\ &+\tilde{F}_{\substack{\rho_2\kappa_1 \\ \beta_2\alpha_1}}(-q_1)\tilde{\varphi}^{(4)}_{\substack{\rho_3\rho_4\kappa_2\kappa_3 \\ \beta_3\beta_4\alpha_2\alpha_3}}(q_1 + q_4|q_2, q_3)\Big\} . \end{aligned} \tag{5.69}$$

Application of the contractions (5.66), (5.67) to (5.69) gives equation (5.65).

Let $\tilde{\varphi}^{(4)}(q_1, q_3|)$ and $\tilde{\varphi}^{(4)}(q_1|q_3, q_4)$ be solutions of (5.65). Then we can substitute $\tilde{\varphi}_4(q_1|q_3, q_4)$ into equation (5.69) and obtain $\tilde{\varphi}^{(4)}(q_1, q_2, q_3, q_4)$. Hence in this way the full $\tilde{\varphi}^{(4)}$ or $\varphi^{(4)}$ can be reconstructed, and $\tilde{\varphi}^{(4)}$ is a solution of the uncontracted equation (5.69). ◇

In contrast to the secular equation (5.18) for the two–particle case the integral equation (5.65) does not allow an immediate solution. As we are only interested in approximate solutions we do not speculate about exact methods of solution of equation (5.65) in order to get four–fermion bound states.

To properly approximate the wavefunctions on the left–hand side of (5.65) or (5.68), respectively, we only consider a submanifold of solutions. As we will investigate gravitational effects we are interested in particular in four–fermion spin–2 states which can be related to the tensor expressions of relativistic gravity theory.

The restriction to spin–2 states suggests to imagine these states as being composed of two spin–1 states, i.e., we consider the four–fermion spin–2 state as a bound state of two vector bosons. In order to utilize this idea for an approximate calculation of the four–fermion states we further assume that in spite of their fusion the internal wave functions of the vector bosons are approximately conserved. In this case the remaining task is the determination of the part of the wavefunction for the center of mass motion of both bosons, and for their relative motion.

The wavefunctions of the single vector bosons are given by (5.42) and (5.43), and with $F_{\mu\nu} := A_\mu k_\nu$ the Fouriertransforms of (5.42), (5.43) can be written in the following form:

$$\begin{aligned} \tilde{\varphi}^{(2)}\begin{pmatrix} q_1 & q_2 \\ i_1 & i_2 \\ \kappa_1 & \kappa_2 \\ \alpha_1 & \alpha_2 \end{pmatrix} \equiv \tilde{\varphi}^{(2)}_{\substack{i_1 i_2 \\ \kappa_1\kappa_2 \\ \alpha_1\alpha_2}}(q_1, q_2) &= \frac{2ig_V}{(2\pi)^4} T^a_{\kappa_1\kappa_2}\delta(k - q_1 - q_2) \quad \times \\ &\Big[A_\mu u^\mu_{\substack{i_1 i_2 \\ \alpha_1\alpha_2}}(q_1 - q_2) + F_{\mu\nu} v^{\mu\nu}_{\substack{i_1 i_2 \\ \alpha_1\alpha_2}}(q_1 - q_2)\Big] \end{aligned} \tag{5.70}$$

with

$$u^{\mu}_{\substack{i_1 i_2 \\ \alpha_1\alpha_2}}(p) := \lambda_{i_1}\lambda_{i_2} \quad \times \tag{5.71}$$

$$\left\{ \frac{\frac{1}{4}p_\rho p_\lambda \gamma^\rho\gamma^\mu\gamma^\lambda + \frac{1}{2}p_\nu\gamma^\nu\gamma^\mu(m_{i_1} - m_{i_2}) - m_{i_1}(p_\nu\eta^{\nu\mu} - \gamma^\mu m_{i_2})}{\left[\left(\frac{p-k}{2}\right)^2 - m_{i_1}^2 + i\epsilon\right]\left[\left(\frac{p+k}{2}\right)^2 - m_{i_2}^2 + i\epsilon\right]} C \right\}_{\alpha_1\alpha_2}$$

and

$$v^{\mu\nu}_{\substack{i_1 i_2 \\ \alpha_1\alpha_2}}(p) := \lambda_{i_1}\lambda_{i_2} \quad \times \tag{5.72}$$

$$\left\{ \frac{-\frac{1}{2}(m_{i_1} + m_{i_2})\gamma^\mu\gamma^\nu + m_{i_2}\eta^{\mu\nu} + \frac{1}{2}p_\rho\epsilon_\nu{}^{\rho\mu\nu}\gamma^\eta\gamma_5 + \frac{1}{4}\gamma^\mu k^\nu}{\left[\left(\frac{p-k}{2}\right)^2 - m_{i_1}^2 + i\epsilon\right]\left[\left(\frac{p+k}{2}\right)^2 - m_{i_2}^2 + i\epsilon\right]} C \right\}_{\alpha_1\alpha_2} .$$

As a preliminary step we consider the direct product of two vector boson states:

$$\tilde{\varphi}^{(2)}_{\substack{i_1 i_2 \\ \alpha_1\alpha_2 \\ \kappa_1\kappa_2}}(q_1, q_2) \otimes \tilde{\varphi}^{(2)}_{\substack{i_3 i_4 \\ \alpha_3\alpha_4 \\ \kappa_3\kappa_4}}(q_3, q_4) = \tag{5.73}$$

$$N g_V^2 T^a_{\kappa_1\kappa_2}\delta(k_1 - q_1 - q_2)\left[A_\mu u^{\mu}_{\substack{i_1 i_2 \\ \alpha_1\alpha_2}}(q_1 - q_2) + F_{\mu\nu}v^{\mu\nu}_{\substack{i_1 i_2 \\ \alpha_1\alpha_2}}(q_1 - q_2)\right] \otimes$$

$$T^b_{\kappa_3\kappa_4}\delta(k_2 - q_3 - q_4)\left[A_\mu u^{\mu}_{\substack{i_3 i_4 \\ \alpha_3\alpha_4}}(q_3 - q_4) + F_{\mu\nu}v^{\mu\nu}_{\substack{i_3 i_4 \\ \alpha_3\alpha_4}}(q_3 - q_4)\right]$$

with a normalization constant $N$. This ansatz describes two bosons moving in directions $k_1$ and $k_2$ without any mutual interaction. If the bosons are assumed to form a bound state, according to our assumption their internal wavefunctions remain unchanged, but their center of mass coordinates become correlated. This correlation has to provide for a definite center of mass momentum of the four–fermion state.

We therefore modify (5.73) and make the following ansatz for the four–fermion spin–2 hard core state, which respects the symmetry conditions (2.97):

$$\tilde{\varphi}^{(4,g)}\begin{pmatrix} q_1 & q_2 & q_3 & q_4 \\ i_1 & i_2 & i_3 & i_4 \\ \kappa_1 & \kappa_2 & \kappa_3 & \kappa_4 \\ \alpha_1 & \alpha_2 & \alpha_3 & \alpha_4 \end{pmatrix} = N g_V^2 T^a_{\kappa_1\kappa_2} T^b_{\kappa_3\kappa_4}\delta(k - q_1 - q_2 - q_3 - q_4) \quad \times$$

$$\Big\{ u^{\nu}_{\substack{i_1 i_2 \\ \alpha_1\alpha_2}}(q_1 - q_2) u^{\mu}_{\substack{i_3 i_4 \\ \alpha_3\alpha_4}}(q_3 - q_4) g_{\mu\nu}\left[(q_1 + q_2) - (q_3 + q_4)\right]$$

$$+ u^{\nu}_{\substack{i_1 i_2 \\ \alpha_1\alpha_2}}(q_1 - q_2) v^{\sigma\rho}_{\substack{i_3 i_4 \\ \alpha_3\alpha_4}}(q_3 - q_4)\Gamma_{\nu\sigma\rho}\left[(q_1 + q_2) - (q_3 + q_4)\right]$$

$$+ v^{\nu\kappa}_{\substack{i_1 i_2 \\ \alpha_1\alpha_2}}(q_1 - q_2) u^{\mu}_{\substack{i_3 i_4 \\ \alpha_3\alpha_4}}(q_3 - q_4)\bar{\Gamma}_{\nu\kappa\mu}\left[(q_1 + q_2) - (q_3 + q_4)\right]$$

$$+ v^{\nu\kappa}_{\substack{i_1 i_2 \\ \alpha_1\alpha_2}}(q_1 - q_2) v^{\sigma\rho}_{\substack{i_3 i_4 \\ \alpha_3\alpha_4}}(q_3 - q_4) R_{\nu\kappa\sigma\rho}\left[(q_1 + q_2) - (q_3 + q_4)\right]\Big\} , \tag{5.74}$$

where the tensors of the boson bonding $g_{\mu\nu}, \Gamma_{\nu\sigma\rho}, \bar{\Gamma}_{\nu\kappa\mu}, R_{\nu\kappa\sigma\rho}$ are unknown functions which have to be determined by means of equation (5.65). We remark that we did not explicitly write the quantum numbers of the graviton state (5.74).

The wavefunction $\tilde{\varphi}^{(4)}$ and its approximation (5.74) should be antisymmetric in the general coordinates $I_1 \dots I_4$. But in order to simplify matters we do not insist on antisymmetrization for the calculation of the unknown tensor functions. Rather we postpone antisymmetrization until we calculate matrix elements which contain the four–fermion states. Thus we can directly substitute (5.74) into (5.65). For the contracted functions we introduce center of mass coordinates

$$s := q_1 + q_3\,, \quad p := q_1 - q_3 \tag{5.75}$$

and in some terms define new integration variables by

$$\begin{aligned} z &:= \eta + \xi - \frac{s}{2} \quad , \\ w &:= \eta - \xi + \frac{p}{2} \quad . \end{aligned} \tag{5.76}$$

After using the properties of the contracted $\delta$–function we obtain

$$\begin{aligned}
& T^a_{\kappa_1\kappa_2} T^b_{\kappa_3\kappa_4} \Big[ U^\nu_{\alpha_1\alpha_2} U^\mu_{\alpha_3\alpha_4} g_{\mu\nu}(p) + U^\nu_{\alpha_1\alpha_2} V^{\sigma\rho}_{\alpha_3\alpha_4} \Gamma_{\nu\sigma\rho}(p) \\
& \quad + V^{\nu\kappa}_{\alpha_1\alpha_2} U^\mu_{\alpha_3\alpha_4} \bar{\Gamma}_{\nu\kappa\mu}(p) + V^{\nu\kappa}_{\alpha_1\alpha_2} V^{\sigma\rho}_{\alpha_3\alpha_4} R_{\nu\kappa\sigma\rho}(p) \Big] \\
& = 3(2\pi)^4 V_{\substack{\epsilon_1\epsilon_2\epsilon_3\epsilon_4 \\ \beta_1\beta_2\beta_3\beta_4}} \Bigg\{ \int d^4\eta\, T^a_{\kappa_1\kappa_2} T^b_{\epsilon_3\epsilon_4} \tilde{G}_{\substack{\kappa_4\epsilon_1 \\ \alpha_4\beta_1}} \left( \frac{k-p}{2} - \eta \right) \tilde{F}_{\substack{\epsilon_2\kappa_3 \\ \beta_2\alpha_3}}(-\eta) \times \\
& \quad \Big[ U^\nu_{\alpha_1\alpha_2} U^\mu_{\beta_3\beta_4} g_{\mu\nu}(-p) + U^\nu_{\alpha_1\alpha_2} V^{\sigma\rho}_{\beta_3\beta_4} \Gamma_{\nu\sigma\rho}(-p) \\
& \quad + V^{\nu\kappa}_{\alpha_1\alpha_2} U^\mu_{\beta_3\beta_4} \bar{\Gamma}_{\nu\kappa\mu}(-p) + V^{\nu\kappa}_{\alpha_1\alpha_2} V^{\sigma\rho}_{\beta_3\beta_4} R_{\nu\kappa\sigma\rho}(-p) \Big] \\
& \quad -2^{-4} \int d^4z d^4w\, T^a_{\kappa_1\kappa_3} T^b_{\epsilon_3\epsilon_4} \quad \times \\
& \quad \tilde{G}_{\substack{\kappa_4\epsilon_1 \\ \alpha_4\beta_1}} \left( \frac{k-p}{4} - \frac{z+w}{2} \right) \tilde{F}_{\substack{\epsilon_2\kappa_2 \\ \beta_2\alpha_2}} \left( -\frac{k+p}{4} + \frac{z-w}{2} \right) \quad \times \\
& \quad \Big[ U^\nu_{\alpha_1\alpha_3} u^\mu_{\beta_3\beta_4} \left( \frac{p}{2} - w \right) g_{\mu\nu}(-2z) + U^\nu_{\alpha_1\alpha_3} v^{\sigma\rho}_{\beta_3\beta_4} \left( \frac{p}{2} - w \right) \Gamma_{\nu\sigma\rho}(-2z) \\
& \quad + V^{\nu\kappa}_{\alpha_1\alpha_3} u^\mu_{\beta_3\beta_4} \left( \frac{p}{2} - w \right) \bar{\Gamma}_{\nu\kappa\mu}(-2z) + V^{\nu\kappa}_{\alpha_1\alpha_3} v^{\sigma\rho}_{\beta_3\beta_4} \left( \frac{p}{2} - w \right) R_{\nu\kappa\sigma\rho}(-2z) \Big] \\
& \quad +2^{-4} \int d^4z d^4w\, T^a_{\kappa_2\kappa_3} T^b_{\epsilon_3\epsilon_4} \quad \times \\
& \quad \tilde{G}_{\substack{\kappa_4\epsilon_1 \\ \alpha_4\beta_1}} \left( \frac{k-p}{4} - \frac{z+w}{2} \right) \tilde{F}_{\substack{\epsilon_2\kappa_1 \\ \beta_2\alpha_1}} \left( -\frac{k+p}{4} - \frac{z-w}{2} \right) \\
& \quad \Big[ U^\nu_{\alpha_2\alpha_3} u^\mu_{\beta_3\beta_4} \left( \frac{p}{2} - z \right) g_{\mu\nu}(-2w) + U^\nu_{\alpha_2\alpha_3} v^{\sigma\rho}_{\beta_3\beta_4} \left( \frac{p}{2} - z \right) \Gamma_{\nu\sigma\rho}(-2w)
\end{aligned} \tag{5.77}$$

$$+V^{\nu\kappa}_{\alpha_2\alpha_3}u^{\mu}_{\beta_3\beta_4}\left(\frac{p}{2}-z\right)\bar{\Gamma}_{\nu\kappa\mu}(-2w)+V^{\nu\kappa}_{\alpha_2\alpha_3}v^{\sigma\rho}_{\beta_3\beta_4}\left(\frac{p}{2}-z\right)R_{\nu\kappa\sigma\rho}(-2w)\Big]\Big\}$$

with

$$u^{\mu}_{\alpha_1\alpha_2}(\xi) \quad := \quad \sum_{i_1,i_2} u^{\mu}{}_{\substack{i_1 i_2\\ \alpha_1\alpha_2}}(\xi) \quad , \tag{5.78}$$

$$U^{\mu}_{\alpha_1\alpha_2} \quad := \quad \int d^4\xi \; u^{\mu}_{\alpha_1\alpha_2}(2\xi) \tag{5.79}$$

and analogous definitions for $v^{\mu\nu}$ and $V^{\mu\nu}$.

In contrast to the initial integral equation (5.65) which cannot be directly integrated, equation (5.77) is a general Fredholm equation with implicit eigenvalue $k$ which is well–defined and allows the application of conventional methods of solution for the determination of the unknown tensor functions $g_{\mu\nu}(p),\ldots R_{\nu\kappa\sigma\rho}(p)$. The "eigenvalue" $k$ is the four–momentum of the bound four–fermion state, and due to Lorentz invariance of equation (5.77) it is sufficient to consider $k$ in the rest frame, i.e. $k=(k_0,0,0,0)$, while all other solutions can be obtained by a corresponding Lorentz transformation of the rest frame into a moving frame, with simultaneous transformation of the above mentioned tensor functions. Thus these tensor functions will depend on $k$, i.e. they have to be labelled by $k$ as a good quantum number. Hence in the following we write $g_{\mu\nu}(p|k),\ldots R_{\nu\kappa\sigma\rho}(p|k)$. If we assume that equation (5.77) admits only one genuine bound state for spin 2 with mass $m_0$, then the tensor functions can equally well be characterized by $g_{\mu\nu}(p|\mathbf{k}),\ldots R_{\nu\kappa\sigma\rho}(p|\mathbf{k})$, i.e. only by the three–momentum $\mathbf{k}$.

Although we have already simplified and transformed equation (5.65) by the ansatz (5.74), the solution of the resulting integral equation (5.77) is still a formidable task. Thus for a first inspection we introduce a further crude approximation. We expand all functions in the kernel of (5.77) in terms of $z$ or $w$, respectively, and neglect all terms with angular dependence, i.e. we take into account terms as functions of $z^2$ or $w^2$ only. Furthermore we assume that the tensor functions $g_{\mu\nu}(p|\mathbf{k}),\ldots R_{\nu\kappa\sigma\rho}(p|\mathbf{k})$ are concentrated around the light cone and take their (finite) maximum value at the light cone. In this case the tensor functions act like distributions, and we approximate the effect of the kernels on the tensor functions by

$$\int K(z^2)g(z)d^4z \approx K(0)\int g(z)d^4z \quad . \tag{5.80}$$

For integrals of the tensor functions themselves we have

$$\tilde{g}_{\mu\nu}(0|\mathbf{k}) = 2^4\int g_{\mu\nu}(-2z|\mathbf{k})d^4z \tag{5.81}$$

etc., where $\tilde{g}$ is the Fouriertransform of $g$ etc., and for brevity we write $\tilde{g}_{\mu\nu}(0|\mathbf{k})\equiv\tilde{g}_{\mu\nu}(\mathbf{k})$, etc. (We remark that $\tilde{g}$ is the Fouriertransform of a function in momentum space, i.e. $\tilde{g}$ is referred to coordinate space!) Then equation (5.77) can be rewritten in the following form:

$$T^a_{\kappa_1\kappa_2}T^b_{\kappa_3\kappa_4}\Big[U^{\nu}_{\alpha_1\alpha_2}U^{\mu}_{\alpha_3\alpha_4}g_{\mu\nu}(p|\mathbf{k})+U^{\nu}_{\alpha_1\alpha_2}V^{\sigma\rho}_{\alpha_3\alpha_4}\Gamma_{\nu\sigma\rho}(p|\mathbf{k}) \tag{5.82}$$

$$+V^{\nu\kappa}_{\alpha_1\alpha_2}U^{\mu}_{\alpha_3\alpha_4}\bar{\Gamma}_{\nu\kappa\mu}(p|\mathbf{k})+V^{\nu\kappa}_{\alpha_1\alpha_2}V^{\sigma\rho}_{\alpha_3\alpha_4}R_{\nu\kappa\sigma\rho}(p|\mathbf{k})\Big]$$

$$
\begin{aligned}
&= 3(2\pi)^4 V_{\substack{\epsilon_1\epsilon_2\epsilon_3\epsilon_4\\ \beta_1\beta_2\beta_3\beta_4}} \left\{ \int d^4\eta \, T^a_{\kappa_1\kappa_2} T^b_{\epsilon_3\epsilon_4} \tilde{G}_{\substack{\kappa_4\epsilon_1\\ \alpha_4\beta_1}} \left( \frac{k-p}{2} - \eta \right) \tilde{F}_{\substack{\epsilon_2\kappa_3\\ \beta_2\alpha_3}}(-\eta) \quad \times \right.\\
&\quad \left[ U^\nu_{\alpha_1\alpha_2} U^\mu_{\beta_3\beta_4} g_{\mu\nu}(-p|\mathbf{k}) + U^\nu_{\alpha_1\alpha_2} V^{\sigma\rho}_{\beta_3\beta_4} \Gamma_{\nu\sigma\rho}(-p|\mathbf{k}) \right.\\
&\quad \left. + V^{\nu\kappa}_{\alpha_1\alpha_2} U^\mu_{\beta_3\beta_4} \bar{\Gamma}_{\nu\kappa\mu}(-p|\mathbf{k}) + V^{\nu\kappa}_{\alpha_1\alpha_2} V^{\sigma\rho}_{\beta_3\beta_4} R_{\nu\kappa\sigma\rho}(-p|\mathbf{k}) \right]\\
&\quad -2^{-8} \int d^4w \, T^a_{\kappa_1\kappa_3} T^b_{\epsilon_3\epsilon_4} \quad \times\\
&\quad \tilde{G}_{\substack{\kappa_4\epsilon_1\\ \alpha_4\beta_1}} \left( \frac{k-p}{4} - \frac{w}{2} \right) \tilde{F}_{\substack{\epsilon_2\kappa_2\\ \beta_2\alpha_2}} \left( -\frac{k+p}{4} - \frac{w}{2} \right)\\
&\quad \left[ U^\nu_{\alpha_1\alpha_3} u^\mu_{\beta_3\beta_4} \left( \frac{p}{2} - w \right) \tilde{g}_{\mu\nu}(\mathbf{k}) + U^\nu_{\alpha_1\alpha_3} v^{\sigma\rho}_{\beta_3\beta_4} \left( \frac{p}{2} - w \right) \tilde{\Gamma}_{\nu\sigma\rho}(\mathbf{k}) \right.\\
&\quad \left. + V^{\nu\kappa}_{\alpha_1\alpha_3} u^\mu_{\beta_3\beta_4} \left( \frac{p}{2} - w \right) \tilde{\bar{\Gamma}}_{\nu\kappa\mu}(\mathbf{k}) + V^{\nu\kappa}_{\alpha_1\alpha_3} v^{\sigma\rho}_{\beta_3\beta_4} \left( \frac{p}{2} - w \right) \tilde{R}_{\nu\kappa\sigma\rho}(\mathbf{k}) \right]\\
&\quad +2^{-8} \int d^4z \, T^a_{\kappa_2\kappa_3} T^b_{\epsilon_3\epsilon_4} \quad \times\\
&\quad \tilde{G}_{\substack{\kappa_4\epsilon_1\\ \alpha_4\beta_1}} \left( \frac{k-p}{4} - \frac{z}{2} \right) \tilde{F}_{\substack{\epsilon_2\kappa_1\\ \beta_2\alpha_1}} \left( -\frac{k+p}{4} - \frac{z}{2} \right)\\
&\quad \left[ U^\nu_{\alpha_2\alpha_3} u^\mu_{\beta_3\beta_4} \left( \frac{p}{2} - z \right) \tilde{g}_{\mu\nu}(\mathbf{k}) + U^\nu_{\alpha_2\alpha_3} v^{\sigma\rho}_{\beta_3\beta_4} \left( \frac{p}{2} - z \right) \tilde{\Gamma}_{\nu\sigma\rho}(\mathbf{k}) \right.\\
&\quad \left.\left. + V^{\nu\kappa}_{\alpha_2\alpha_3} u^\mu_{\beta_3\beta_4} \left( \frac{p}{2} - z \right) \tilde{\bar{\Gamma}}_{\nu\kappa\mu}(\mathbf{k}) + V^{\nu\kappa}_{\alpha_2\alpha_3} v^{\sigma\rho}_{\beta_3\beta_4} \left( \frac{p}{2} - z \right) \tilde{R}_{\nu\kappa\sigma\rho}(\mathbf{k}) \right] \right\} .
\end{aligned}
$$

The further treatment of equation (5.82) is straightforward. First, equation (5.82) is projected on the direct products ${}^s\Gamma^k_{\alpha_1\alpha_2} \otimes {}^s\Gamma^l_{\alpha_3\alpha_4}$ of the symmetric elements of the Dirac algebra. By means of these projections one obtains just as many equations as there are unknown tensor functions. Second, the resulting equations are (algebraically!) resolved with respect to the tensor functions $g_{\mu\nu}(p|\mathbf{k}), \ldots R_{\nu\kappa\sigma\rho}(p|\mathbf{k})$. Finally, integration over $p$ is performed in order to obtain an algebraic secular problem for the determination of $\tilde{g}_{\mu\nu}(\mathbf{k}), \ldots \tilde{R}_{\nu\kappa\sigma\rho}(\mathbf{k})$. We assume that this problem is solved and the eigenvectors $\tilde{g}_{\mu\nu}(\mathbf{k}), \ldots \tilde{R}_{\nu\kappa\sigma\rho}(\mathbf{k})$ are known. Then, if we stop after the second step we can directly read off the explicit form of the functions $g_{\mu\nu}(p|\mathbf{k}), \ldots R_{\nu\kappa\sigma\rho}(p|\mathbf{k})$. In this way we have determined the bonding tensor functions of our two vector bosons. The explicit evaluation is performed in Section 7.4.

## 5.5 Dressed Fermion States

We illustrate the formal discussion of Section 4.8 by an explicit calculation of dressed fermion states $C^{I_1\ldots I_{1+2n}}_{1,k}$ and $C^{I_1\ldots I_{3+2n}}_{3,k}$ in lowest order of the polarization cloud. According to (4.128) we define for dressed particles the corresponding fermion operators

$$
f_{1,k} \quad := \quad C^{I_1}_{1,k} j_{I_1} + C^{I_1 I_2 I_3}_{1,k} j_{I_1} j_{I_2} j_{I_3} + \ldots \tag{5.83}
$$

$$f_{3,l} \quad := \quad C_{3,l}^{I_1 I_2 I_3} j_{I_1} j_{I_2} j_{I_3} + \ldots \tag{5.84}$$

and in a first approximation we omit all terms in the (infinite) power series of (5.83) and (5.84) with more than three sources $j_I$. Equation (5.83) defines dressed one–fermion operators, while equation (5.84) defines dressed three–fermion operators.

If the wave functions $C_{3,l}^{I_1 I_2 I_3}$ have quantum numbers $l$ which coincide with those of $C_{1,k}^{I_1}$, these wave functions cannot be genuine three–particle states. Rather, due to the fact that for coinciding quantum numbers these wave functions are contained in the solutions of the energy equation (4.1), the three–fermion wave functions with $k = l$ must necessarily be polarization cloud wave functions which are correlated with a non–vanishing $C_{1,k}^{I_1}$. Thus the state operators (5.83) and (5.84) in any case have different quantum numbers $k \neq l$. This leads to the following orthonormality relations, see (4.127):

$$R_{I_1}^{1,k} C_{1,k'}^{I_1} + R_{I_1 I_2 I_3}^{1,k} C_{1,k'}^{I_1 I_2 I_3} \quad = \quad \delta_{k'}^{k} \tag{5.85}$$

$$R_{I_1 I_2 I_3}^{3,l} C_{3,l'}^{I_1 I_2 I_3} \quad = \quad \delta_{l'}^{l} \tag{5.86}$$

$$R_{I_1 I_2 I_3}^{3,l} C_{1,k}^{I_1 I_2 I_3} \quad = \quad R_{I_1 I_2 I_3}^{1,k} C_{3,l}^{I_1 I_2 I_3} = 0 \quad . \tag{5.87}$$

The set of coefficient functions $C_{1,k}^{I_1 \ldots I_{1+2n}}$ is assumed to be a solution of equation (4.1) (in configuration space); so they have to be calculated by means of this equation. However, as in the hard core case we can alternatively use the relativistic invariant equation (2.92). Then by Proposition 3.14 we can relate the relativistic invariant multi–time normalordered functionals to the one–time normalordered functionals which later on are needed for the weak mapping procedure.

We project equation (2.92) into configuration space and obtain with the vertex renormalization (2.118) for $n = 1$ and $n = 3$

$$\left(D_{I_1 I_2}^{\mu} \partial_\mu - m_{I_1 I_2}\right) \varphi^{(1)}(I_2) = W_{I_1 I_2 I_3 I_4} \varphi^{(3)}\left(I_2, I_3, I_4\right) \tag{5.88}$$

and

$$\begin{aligned} &\left(D_{I_1 I_2}^{\mu} \partial_\mu - m_{I_1 I_2}\right) \varphi^{(3)}\left(K_1, K_2, I_2\right) = \\ &3 W_{I_1 I_2 I_3 I_4} \Big\{ \Big[ F_{I_4 K_1} F_{I_3 K_2} - F_{I_4 K_2} F_{I_3 K_1} \Big] \varphi^{(1)}(I_2) \\ &\quad - \Big[ F_{I_4 K_2} \varphi^{(3)}\left(K_1, I_2, I_3\right) - F_{I_4 K_1} \varphi^{(3)}\left(K_2, I_2, I_3\right) \Big] \Big\} \quad , \end{aligned} \tag{5.89}$$

where in (5.89) we suppressed the $\varphi^{(5)}$–term due to our three–source approximation in (5.83). In the same way, due to this approximation we also suppress all higher order equations for $\varphi^{(n)}$ with $n > 3$ which result from equation (2.92).

For a straightforward integration even these remaining equations (5.88) and (5.89) are too complicated. For a first inspection we thus apply an iteration procedure which we break off in the lowest order. We denote the lowest order functions by $\varphi_0^{(1)}$ and $\varphi_0^{(3)}$ and define the lowest order equations by

$$\left(D_{I_1 I_2}^{\mu} \partial_\mu - m_{I_1 I_2}\right) \varphi_0^{(1)}(I_2) = 0 \tag{5.90}$$

and

$$
\begin{aligned}
&\left(D^{\mu}_{I_1 I_2}\partial_\mu - m_{I_1 I_2}\right)\varphi_0^{(3)}\left(K_1, K_2, I_2\right) = \\
&\qquad 3W_{I_1 I_2 I_3 I_4}\left(F_{I_4 K_1}F_{I_3 K_2} - F_{I_4 K_2}F_{I_3 K_1}\right)\varphi_0^{(1)}(I_2) \quad .
\end{aligned} \tag{5.91}
$$

The other terms in equations (5.88) and (5.89) which do not appear in equations (5.90) and (5.91) are taken into account in the first iteration step.

Equations (5.90) and (5.91) can directly be integrated. While $\varphi_0^{(1)}$ is a Dirac superspinor, $\varphi_0^{(3)}$ is given by

$$
\begin{aligned}
&\varphi_0^{(3)}(K_1, K_2, K_3) = \\
&\left[3G_{K_3 I_1}W_{I_1 I_2 I_3 I_4}\left(F_{I_4 K_1}F_{I_3 K_2} - F_{I_4 K_2}F_{I_3 K_1}\right)\varphi_0^{(1)}(I_2)\right]_{\mathrm{as}\{K_1,K_2,K_3\}} \quad .
\end{aligned} \tag{5.92}
$$

The explicit evaluation of this expression and the transition from the multi–time to the one–time form of (5.92) is straightforward. Details of this evaluation are given in Section 7.5.

## 5.6 Metric of Dressed Fermion States

In this section we will derive the explicit expression for a consistent approximation of the metrical tensor in the dressed fermion sector. According to Proposition 3.5 the metrical tensor can be expressed in terms of vacuum expectation values of the $\psi_I$–fields. This expression holds for the exact theory and relates the metrical tensor to the definite representation of the $\psi_I$–fields. Simultaneously, if the energy eigenstates $|p\rangle$ of the spinorfield are described by state functionals, these state functionals have to be referred to the same representation. Thus state description and metrical tensor have to be compatible with respect to the common underlying algebraic representation.

However, if the state functionals are approximately calculated it is unclear what kind of approximation has to be applied to the calculation of the metric tensor in order to achieve this compatibility. It is therefore desirable to directly relate the metric tensor calculation to the state calculation in order to a priori guarantee the compatibility of the representations. We follow this program and start our investigation with an evaluation of the metric tensor for the one–fermion sector.

We observe that according to Sections 1.5 and 3.7 $\psi_I$ is defined by $\psi_I := \psi_{A\alpha i\Lambda}(\mathbf{r}) \equiv (\psi_{A\alpha i}(\mathbf{r}), \psi^C_{A\alpha i}(\mathbf{r}))$ with $\psi^C_{A\alpha i} := (C\gamma^0)_{\alpha\beta}\psi^+_{A\beta i}$, where with regard to the one–time formalism we suppressed the time parameter $t$. Expanded, (3.4) with (3.37), (3.38) reads

$$
\begin{aligned}
G_{II'} &\equiv G_{A\alpha i\Lambda, A'\alpha' i'\Lambda'}(\mathbf{r}, \mathbf{r}') \\
&= \langle 0|\psi_{A\alpha i\Lambda}(\mathbf{r})\psi^+_{A'\alpha' i'\Lambda'}(\mathbf{r}')|0\rangle \quad .
\end{aligned} \tag{5.93}
$$

To relate (5.93) to state representations we consider the one–fermion states $\varphi^{(1)}_{A\alpha i\Lambda}(\mathbf{r}|k)$ and classify their quantum numbers. By definition of our approximation these matrix elements are assumed to be solutions of the free superspinor Dirac equation (5.90). The quantum numbers of these solutions are given by the three–momentum $\mathbf{k}$, the spin $s$, superspinor–isospinor numbers $a$ and the auxiliary field number $j$. For the following we need a more detailed breakdown of $a$. With respect to $a$ we observe that the quantized fields $\psi$ and $\psi^C$ are considered as independent from each other, i.e., that there exist solutions either for $\psi$ or for $\psi^C$ which are compatible with (5.90) and lead to state (quantum) numbers $\Gamma = 1, 2$. Furthermore, the isospin quantum numbers can simply be expressed by $B = 1, 2$. Then we assume that for each solution $a \equiv (B, \Gamma)$ there exists a full set of spinor eigenstates of equation (5.90). Now the following proposition holds:

- **Proposition 5.6:** The metrical tensor (5.93) of the one–fermion sector is uniquely determined by the one–fermion transition matrix elements $\varphi^{(1)}(I|k)$ and by the anticommutators (1.42). It reads in an approximation which is consistent with the free Dirac equation (5.90)

$$G_{II'} = \frac{1}{2}\lambda_i \delta_{ii'} \delta_{AA'} \delta_{\Lambda\Lambda'} \delta_{\alpha\alpha'} \delta(\mathbf{r} - \mathbf{r}') \quad . \tag{5.94}$$

In addition the metrical tensor for the one–fermion energy eigenstates $|k\rangle$ is determined to be

$$g_{kk'} = 2\lambda_j \delta_{kk'} \tag{5.95}$$

where $k, k'$ are abbreviations for the set of quantum numbers $k = (\mathbf{k}, s, B, \Gamma, j)$.

**Proof:** We decompose (5.93) into the form

$$G_{II'} = \begin{pmatrix} \langle 0|\psi_{A\alpha i}\psi^+_{A'\alpha' i'}|0\rangle\,, & \langle 0|\psi_{A\alpha i}(\psi^C)^+_{A'\alpha' i'}|0\rangle \\ \langle 0|\psi^C_{A\alpha i}\psi^+_{A'\alpha' i'}|0\rangle\,, & \langle 0|\psi^C_{A\alpha i}(\psi^C)^+_{A'\alpha' i'}|0\rangle \end{pmatrix} \tag{5.96}$$

and successively evaluate the terms of (5.96).

Introducing intermediate states we obtain for $\langle 0|\psi_{A\alpha i}\psi^+_{A'\alpha' i'}|0\rangle$ the expression

$$\langle 0|\psi_{A\alpha i}\psi^+_{A'\alpha' i'}|0\rangle = \sum_{k,k'} \langle 0|\psi_{A\alpha i}(\mathbf{r})|k\rangle g^{kk'} \langle k'|\psi^+_{A'\alpha' i'}(\mathbf{r}')|0\rangle \quad , \tag{5.97}$$

where $g^{kk'}$ is the dual tensor to $g_{kk'} = \langle k|k'\rangle$.

For the further evaluation we observe that it holds by definition

$$\varphi^{(1)}_{A\alpha i}(\mathbf{r}|k) := \langle 0|\psi_{A\alpha i}(\mathbf{r})|k\rangle = \left(\langle k|\psi^+_{A\alpha i}(\mathbf{r})|0\rangle\right)^* \tag{5.98}$$

or

$$\langle k|\psi^+_{A\alpha i}(\mathbf{r})|0\rangle = \varphi^{(1)}_{A\alpha i}(\mathbf{r}|k)^* \quad . \tag{5.99}$$

Hence we have

$$\begin{aligned} \langle 0|\psi_{A\alpha i}\psi^+_{A'\alpha' i'}|0\rangle &= \sum_{k,k'} \langle 0|\psi_{A\alpha i}(\mathbf{r})|k\rangle g^{kk'} \langle k'|\psi^+_{A'\alpha' i'}(\mathbf{r}')|0\rangle \\ &= \sum_{k,k'} \varphi^{(1)}_{A\alpha i}(\mathbf{r}|k) g^{kk'} \varphi^{(1)}_{A'\alpha' i'}(\mathbf{r}'|k')^* \quad . \end{aligned} \tag{5.100}$$

Equation (5.100) may be further evaluated by specifying the metrical tensor $g^{kk'}$ and the quantum numbers $k$, which are given by $k = (\mathbf{k}, s, B, \Gamma, j)$.

With respect to the metrical tensor according to the assumption of an underlying Krein space the energy eigenstates have to be orthogonal. If we assume

$$g_{kk'} = \langle k|k'\rangle = N\lambda_j\delta_{jj'}\delta_{BB'}\delta_{ss'}\delta_{\Gamma\Gamma'}\delta(\mathbf{k}-\mathbf{k}') \tag{5.101}$$

it follows

$$g^{kk'} = N^{-1}\lambda_j^{-1}\delta_{jj'}\delta_{BB'}\delta_{ss'}\delta_{\Gamma\Gamma'}\delta(\mathbf{k}-\mathbf{k}') \tag{5.102}$$

with a constant $N$. Substituting (5.102) into (5.100) we obtain

$$\sum_{k,k'}\varphi_{A\alpha i}(\mathbf{r}|k)g^{kk'}\varphi_{A'\alpha'i'}(\mathbf{r}'|k')^* = \tag{5.103}$$

$$= \frac{1}{N}\sum_{s,j,\Gamma,B}\int d^3k\varphi^{(1)}_{A\alpha i}(\mathbf{r}|\mathbf{k},s,B,\Gamma,j)\lambda_j^{-1}\varphi^{(1)}_{A'\alpha'i'}(\mathbf{r}'|\mathbf{k},s,B,\Gamma,j)^* \quad .$$

With respect to the quantum numbers $k$ we observe that $\varphi^{(1)}_{A\alpha i}$ is a solution of the free Dirac equation (5.90) with quantum numbers $\mathbf{k}, s, B, \Gamma, j$ (see the discussion above); i.e. for these solutions the following represention holds:

$$\varphi^{(1)}_{A\alpha i}(\mathbf{r}|\mathbf{k},s,B,\Gamma,j) = \lambda_j\varphi_\alpha(\mathbf{r}|\mathbf{k},s)\delta_{ji}\delta_{BA}\delta_{\Gamma 1} \tag{5.104}$$

where $\varphi_\alpha(\mathbf{r}|\mathbf{k},s)$ is an ordinary Dirac spinor. The factor $\lambda_j$ in (5.104) is in accordance with the ansatz (5.101) for the metrical tensor, which for its part determines the physical units of the eigenstates $|k\rangle$. The special form of (5.104) will be justified by the consistency relation (5.109). It reflects the fact, that the wavefunctions (5.98) are referred to regularized spinorfields which in their physical units differ from the wavefunctions for ordinary (nonregularized) Dirac fields.

Substitution of (5.104) in (5.103) yields

$$\sum_{k,k'}\varphi_{A\alpha i}(\mathbf{r}|k)g^{kk'}\varphi_{A'\alpha'i'}(\mathbf{r}'|k')^* = \tag{5.105}$$

$$= \frac{1}{N}\sum_{s}\int d^3k\varphi_\alpha(\mathbf{r}|\mathbf{k},s)\lambda_i\varphi_{\alpha'}(\mathbf{r}'|\mathbf{k},s)^*\delta_{ii'}\delta_{AA'}\delta_{\Lambda 1}\delta_{\Lambda 1} \quad .$$

Application of the completeness relations for a full set of Dirac spinors leads to

$$\langle 0|\psi_{A\alpha i}\psi^+_{A'\alpha'i'}|0\rangle = \frac{1}{N}\lambda_i\delta_{\alpha\alpha'}\delta_{ii'}\delta_{AA'}\delta_{\Lambda 1}\delta_{\Lambda' 1}\delta(\mathbf{r}-\mathbf{r}') \quad . \tag{5.106}$$

Now the problem remains to determine the normalization constant $N$. Toward this end we make use of the anticommutation relations (1.42), from which it follows

$$\left[\psi_{\alpha Ai}(\mathbf{r}),\psi^+_{\alpha' A'i'}(\mathbf{r}')\right]_+ = \lambda_i\delta_{ii'}\delta_{AA'}\delta_{\alpha\alpha'}\delta(\mathbf{r}-\mathbf{r}') \tag{5.107}$$

and

$$\left[\psi^C_{\alpha Ai}(\mathbf{r}),(\psi^C)^+_{\alpha' A'i'}(\mathbf{r}')\right]_+ = \lambda_i\delta_{ii'}\delta_{AA'}\delta_{\alpha\alpha'}\delta(\mathbf{r}-\mathbf{r}') \quad . \tag{5.108}$$

Taking the vacuum expectation value of (5.107) and introducing intermediate states we obtain the condition

$$\sum_{k,k'}\Big[\langle 0|\psi_{A\alpha i}(\mathbf{r})|k\rangle g^{kk'}\langle k'|\psi^+_{A'\alpha' i'}(\mathbf{r}')|0\rangle + \langle 0|\psi^+_{A'\alpha' i'}(\mathbf{r}')|k\rangle g^{kk'}\langle k'|\psi_{A\alpha i}(\mathbf{r})|0\rangle\Big]$$
$$= \lambda_i\delta_{ii'}\delta_{AA'}\delta_{\alpha\alpha'}\delta(\mathbf{r}-\mathbf{r}') \quad . \tag{5.109}$$

In order to utilize this condition we evaluate the second term on the left–hand side analogously to the first term. We define

$$\bar{\varphi}^{(1)}_{A'\alpha' i'}(\mathbf{r}'|k) := \langle 0|\bar{\psi}_{A'\alpha' i'}(\mathbf{r}')|k\rangle \tag{5.110}$$

or

$$\langle 0|\psi^+_{A'\alpha' i'}(\mathbf{r}')|k\rangle = \gamma^0_{\alpha'\beta'}\bar{\varphi}^{(1)}_{A'\beta' i'}(\mathbf{r}'|k) \quad . \tag{5.111}$$

Furthermore it is

$$\gamma^0_{\alpha'\beta'}\bar{\varphi}^{(1)}_{A'\beta' i'}(\mathbf{r}'|k)^* = \langle k|\psi_{A'\alpha' i'}(\mathbf{r}')|0\rangle \, . \tag{5.112}$$

From this it follows

$$\begin{aligned}\langle 0|\psi^+_{A'\alpha' i'}\psi_{A\alpha i}|0\rangle &= \sum_{k,k'}\langle 0|\psi^+_{A'\alpha' i'}(\mathbf{r}')|k\rangle g^{kk'}\langle k'|\psi_{A\alpha i}(\mathbf{r})|0\rangle \\ &= \sum_{k,k'}\gamma^0_{\alpha'\beta'}\bar{\varphi}^{(1)}_{A'\beta' i'}(\mathbf{r}'|k)g^{kk'}\gamma^0_{\alpha\beta}\bar{\varphi}^{(1)}_{A\beta i}(\mathbf{r}|k')^* \, .\end{aligned} \tag{5.113}$$

Substitution of the metrical tensor (5.102) and the expression (5.104) gives

$$\begin{aligned}&\sum_{k,k'}\gamma^0_{\alpha'\beta'}\bar{\varphi}_{A'\beta' i'}(\mathbf{r}'|k)g^{kk'}\gamma^0_{\alpha\beta}\bar{\varphi}_{A\beta i}(\mathbf{r}|k')^* \\ &= \frac{1}{N}\sum_s\int d^3k\gamma^0_{\alpha'\beta'}\bar{\varphi}_{\beta'}(\mathbf{r}'|\mathbf{k},s)\lambda_i\gamma^0_{\alpha\beta}\bar{\varphi}_{\beta}(\mathbf{r}|\mathbf{k},s)\delta_{ii'}\delta_{AA'}\delta_{\Lambda 1}\delta_{\Lambda 1} \\ &= \frac{1}{N}\lambda_i\delta_{\alpha\alpha'}\delta_{ii'}\delta_{AA'}\delta_{\Lambda 1}\delta_{\Lambda' 1}\delta(\mathbf{r}-\mathbf{r}') \quad .\end{aligned} \tag{5.114}$$

With (5.114) and (5.106) equation (5.109) has to be selfconsistently satisfied; from this it follows $N = 2$. As we have

$$\langle 0|\psi_I\psi^+_{I'}|0\rangle = \sum_{k,k'}\varphi^{(1)}_{A\alpha i}(\mathbf{r}|k)g^{kk'}\varphi^{(1)}_{A'\alpha' i'}(\mathbf{r}'|k')^*$$

we have calculated the first submatrix of (5.94) or (5.96), respectively.

The same considerations can be applied to $\langle 0|\psi^C_I\left(\psi^C_{I'}\right)^+|0\rangle$ with the result

$$\langle 0|\psi^C_I\left(\psi^C_{I'}\right)^+|0\rangle = \frac{1}{2}\lambda_i\delta_{\alpha\alpha'}\delta_{ii'}\delta_{AA'}\delta_{\Lambda 2}\delta_{\Lambda' 2}\delta(\mathbf{r}-\mathbf{r}') \quad . \tag{5.115}$$

In the nondiagonal elements of (5.96) the introduction of intermediate states leads to matrix elements $\langle 0|\psi^+|k\rangle$, $\langle k|\psi^C|0\rangle$. In accordance with (5.104) $\langle 0|\psi^+|k\rangle$ contains a factor $\delta_{\Lambda 1}$, while $\langle k|\psi^C|0\rangle$ has a factor $\delta_{\Lambda' 2}$. If these states are combined with $g^{kk'}$, there results the combination $\delta_{\Lambda 1}\delta_{\Lambda\Lambda'}\delta_{\Lambda' 2} \equiv 0$. Hence the nondiagonal elements of (5.96) vanish, while the diagonal elements are given by (5.106) and (5.115). This gives (5.94). ◇

It is easy to extend these considerations to the calculation of the metrical tensor for three–fermion states. We only have to observe that on the hyperplane, i.e. in three–dimensional position space, the products of three one–(sub–)fermion transition matrix elements $\varphi^{(1)}(I|k)$ are complete. Therefore any three–fermion transition matrix element $\varphi^{(3)}(I_1, I_2, I_3|k)$ can be expanded in terms of these one–fermion matrix elements

$$\varphi^{(3)}_{Z_1 Z_2 Z_3}(\mathbf{r}_1, \mathbf{r}_2, \mathbf{r}_3|k) = \sum_{a_1,a_2,a_3} c(a_1 a_2 a_3|k)\varphi^{(1)}_{Z_1}(\mathbf{r}_1|a_1)\varphi^{(1)}_{Z_2}(\mathbf{r}_2|a_2)\varphi^{(1)}_{Z_3}(\mathbf{r}_3|a_3) \quad . \tag{5.116}$$

But this suggests the assumption that the metrical tensor $G_{I_1 I_2 I_3, I'_1 I'_2 I'_3}$ can be represented by the direct product of the corresponding one–particle tensor $G_{II'}$

$$G_{I_1 I_2 I_3, I'_1 I'_2 I'_3} = G_{I_1 I'_1} G_{I_2 I'_2} G_{I_3 I'_3}{}_{\mathrm{as}\{I_1, I_2, I_3\}} \, . \tag{5.117}$$

It should however be emphasized that (5.117) is only a suitable approximation because according to Proposition 3.5 the general expression for $G_{I_1 I_2 I_3, I'_1 I'_2 I'_3}$ is determined by the quantum field dynamics and not by the reduction of $G_{I_1 I_2 I_3, I'_1 I'_2 I'_3}$ to one–fermion norms.

# 6 Effective Yang–Mills Dynamics

## 6.1 Introduction

In Chapter 4 the weak mapping of quantum fields was examplified by the derivation of a weak mapping theorem for bosons as two–fermion composites. In this chapter we investigate applications of weak mapping in the realm of high energy physics.

The aim of experimental high energy physics is to detect the ultimate building blocks of matter. Such elementary quantities are in general not accessible to everyday experience. In order to detect them several levels of increasing decomposition have to be gone through by means of increasingly complicated experimental settings. On the other hand, for the corresponding theory this means that for the theoretical reconstruction of macroscopic matter these levels have to be passed through in reverse order, i.e. by successive composition of elementary entities. Therefore for the success of theoretical high energy physics it is of utmost importance to understand the formation and reactions of composite particles.

While in numerous papers the formation of composite particles from elementary (relativistic) constituents was described by Bethe–Salpeter equations and their solutions, the search for an appropriate description of the reactions, i.e. of the interactions of composite particles deviates from this concept. In contrast to quantum mechanics and Bethe–Salpeter equations which describe composite objects by wavefunctions, the reactions of composite particles were theoretically studied by means of field operators or field operator products. For instance, the early approach to describe the photon as a composite object was performed by operator products. A photon operator was assumed to be a product of a neutrino and an antineutrino operator, see, e.g., Pryce [Pry 38]. Similar assumptions were made with respect to other composite objects either to derive an $S$–matrix description for them or to generate effective Lagrangians. In particular the transition from a Lagrangian for elementary constituents to an effective Lagrangian for composite particles by use of field operator products can be considered as a mapping. Because this kind of mapping is concerned with operators, one can denote it as "strong" mapping, in contrast to "weak" mapping discussed in the preceding chapters. Gradually these techniques were modified and incorporated into the evaluation of path integrals. Chapter 10 is devoted to the connection of path integrals and composite particles.

As far as genuine strong mapping is concerned one has to observe that the representation of an operator product, i.e. its various matrix elements between all possible states of the representation space under consideration, describe much more physical constellations than only the formation of a composite particle, and only in $S$–matrix theory it is possible to asymptotically project out just the information about such a particle. With respect to the formation of a local effective Lagrangian such an asymptotic projection is not possible and the concept of strong mapping has to be rejected. Indeed, this concept was frequently criticized. Reviews of papers about strong mapping and their criticism can be found in Stumpf [Stu 85a,b], [Stu 86a].

At present most of the attempts of explaining the hierarchy of particles and their dynamics are related to the standard model which is the generally accepted high energy theory of matter. This model contains at least 37 "elementary" particles and more than 20 parameters. This strongly supports the assumption that this model describes composite particles and resulting forces rather than elementary ones. A great number of subcomponent models was proposed, see, e.g., Lyons [Lyo 83], claiming to be the subcomponent basis of the standard model, but without delivering convincing proofs for that claim. Without discussing these numerous models we establish the spinor–isospinor model of Heisenberg [Heis 66] on the subfermion level, which in this context we modify by another vertex and another regularization as was discussed in Chapter 1. In this fusion model of the standard theory the standard fermions should be composites of three subfermions, while the weak bosons have to be considered as composites of two subfermions.

Although first calculations were made in this direction by Stumpf in [Stu 86b] and [Stu 87], this assumption leads to a rather complicated effective boson–fermion dynamics for which at present no detailed treatment exists, i.e. no explicit evaluation of the equations of Section 4.9. Thus, in a first step we consider the effective dynamics of bosons as two–subfermion composites and their effective interactions with the elementary subfermions themselves.

For an exact treatment even this problem is rather complicated. In Chapter 4 it was demonstrated that for bound states the mutual exchange interactions of the subfermions contained in different bound states can be neglected. If one disregards these exchange interactions from the beginning, it was shown that the performance of the weak mapping procedure can be considerably simplified by applying functional chain rules. In this chapter we use these techniques for the derivation of the effective dynamics of bosons and fermions.

The pure map of the spinorfield onto an effective Yang–Mills dynamics was treated by Stumpf [Stu 86a]. With inclusion of fermion–boson coupling the same map was investigated by Pfister [Pfi 90], and with discussion of the quantum properties of the effective boson fields by Stumpf and Pfister [Stu 92]. The calculation of fermion–boson coupling constants and of the effective fermion-boson Lagrangian was done by Pfister [Pfi 93]. In Sections 6.2–6.5 we follow Stumpf and Pfister [Stu 92], while Sections 6.6 and 6.7 refer to Pfister [Pfi 93].

## 6.2 Effective Boson–Fermion Dynamics

The basic subfermion theory is defined by the spinorfield model of Chapter 1. Its quantum field theoretic energy representation is given by equations (3.106) with the subfermion field functional $|\mathcal{F}(j^t;a)\rangle$ defined by

$$|\mathcal{F}(j^t;a)\rangle := \sum_n \frac{i^n}{n!}\varphi_t^{(n)}(I_1\dots I_n|a)j^t_{I_1}\dots j^t_{I_n}|0\rangle_F \quad . \tag{6.1}$$

In the following for brevity we will suppress the one–time index $t$.

In a symbolic notation we write for (3.106) or (4.1), respectively,

$$E_{p0}|\mathcal{F}(j;p)\rangle = \mathcal{H}_F(j,\partial)|\mathcal{F}(j;p)\rangle \quad , \tag{6.2}$$

where $\mathcal{H}_F(j,\partial)$ is given by (4.29).

With respect to the boson–fermion coupling theory we define the following (effective) functional states

$$|\mathcal{G}(b,f;a)\rangle := \sum_{n,m=0}^{\infty} \frac{i^n}{n!m!} \rho^{(n+m)}(k_1 \dots k_m; I_1 \dots I_n|a) b_{k_1} \dots b_{k_m} f_{I_1} \dots f_{I_n} |0\rangle_B \otimes |0\rangle_F \tag{6.3}$$

where $b_k$ and $f_I$ are the boson or fermion sources, respectively, and $|0\rangle_B$ and $|0\rangle_F$ are the corresponding functional vacua. Here and in the following we use the suggestive abbreviations $b_k$, $f_I$ ect. for the operators $b_k \otimes \mathbf{1}$, $\mathbf{1} \otimes f_I$ ect. on the direct product space of (6.3). We denote the functional "duals" of the boson and fermion sources by $\partial_k^b$ and $\partial_I^f$ and assume the following commutation relations:

$$\begin{aligned} \left[\partial_k^b, b_{k'}\right]_- &= \delta_{kk'} \quad , \\ \left[\partial_I^f, f_{I'}\right]_+ &= \delta_{II'} \quad , \\ \left[\partial_k^b, f_{I'}\right]_- &= \left[\partial_I^f, b_k\right]_- = 0 \quad , \end{aligned} \tag{6.4}$$

while all other commutators (anticommutators) vanish.

Furthermore we assume $\partial_k^b|0\rangle_B = \partial_I^f|0\rangle_F = 0$, which enforces the functional basis states in (6.3) to be Fock states. This functional Fock space is completely independent of the subcomponent fermionic functional space used for the definition of $|\mathcal{F}(j;a)\rangle$. For a more detailed discussion of functional spaces we refer to Chapters 2 and 3.

The weak mapping theorem, see Proposition 4.2, is based on the transformation (4.23) between the fermionic and bosonic coefficient functions of the corresponding state functionals; i.e. it is completely independent of the basis states of the auxiliary functional spaces. In order to tranfer this method to the combined boson–fermion dynamics, we introduce complete sets of one–particle states $C_I^{I'}$ and two–particle states $C_{II'}^k$ and their duals $R_I^{I'}$ and $R_k^{II'}$ and define the transformation

$$\varphi^{(l)}(I_1 \dots I_l|a) = \sum_{n,m} C^{k_1}_{\{I_1 I_2} \dots C^{k_m}_{I_{2m-1} I_{2m}} C^{I'_1}_{I_{2m+1}} \dots C^{I'_n}_{I_{2m+n}\}} \rho^{(m+n)}(k_1 \dots k_m; I'_1 \dots I'_n|a) \ , \tag{6.5}$$

for $l = 2m + n$, where the transformation matrix in (6.5) has to be antisymmetrized in all indices $I_1 \dots I_{2m+n}$. As in Proposition 4.2 this definition allows an exact transformation of the original functional equation (6.2) into an effective functional equation of a combined boson–fermion dynamics which is completely independent of the auxiliary functional spaces.

The exact treatment includes exchange forces and leads to rather complicated expressions. However, for the case of a pure effective boson theory it was shown in Chapter 4 , that if one can neglect the exchange forces, then there exists an alternative transformation procedure between the subfermion theory and the effective theory which reproduces the (exchange free) results of the exact mapping theorems and is very transparent. In this case the functional auxiliary spaces are involved from the beginning. The corresponding "functional chain rule" has been treated in Section 4.7. We shortly repeat this concept and apply it to the case of an effective boson–fermion theory, expecting that also in this case the application of the functional chain rule corresponds to the neglect of exchange forces.

We assume the invariance relation

$$|\mathcal{F}(j;a)\rangle = |\mathcal{G}(b,f;a)\rangle \ . \tag{6.6}$$

This relation is based on the idea of invariant functional state vectors, which are referred to different functional basis vectors; i.e. the $\mathcal{G}$–functional states are only rearrangements of the corresponding $\mathcal{F}$–functional states.

To evaluate the consequences of this equation a relation between the various functional basis states has to be established. For the problem under consideration this can be done by defining relations between the original sources $j_I$ and the effective sources $b_K$ and $f_I$ which are given by

$$f_I = C_I^{I'} j_{I'} \quad , \tag{6.7}$$

and

$$b_k = \frac{1}{2} C_k^{I_1 I_2} j_{I_1} j_{I_2} \quad , \tag{6.8}$$

where the transformation matrices $C$ fulfill orthonormality and completeness relations like (4.4) and (4.5) with respect to their duals $R$. This allows the resolution

$$j_I = R_I^{I'} f_{I'} \tag{6.9}$$

and

$$j_{I_1} j_{I_2} = 2 R^k_{I_1 I_2} b_k \ . \tag{6.10}$$

From (6.2) and (6.6) it follows that the equation

$$E_{p0}|\mathcal{G}(b,f;p)\rangle = \mathcal{H}_{\mathrm{F}}(j,\partial)|\mathcal{G}(b,f;p)\rangle \tag{6.11}$$

has to hold where $|\mathcal{G}(b,f;p)\rangle \equiv |\mathcal{G}(b[j],f[j];p)\rangle$.

In order to derive an effective boson–fermion equation we have to transform $\mathcal{H}_{\mathrm{F}}$ into the boson–fermion representation; i.e. we have to evaluate the action of $\mathcal{H}_{\mathrm{F}}$ on $|\mathcal{G}\rangle$ in order to obtain an effective equation for $|\mathcal{G}\rangle$. This can be performed by applying the chain rule for functional differentiation

$$\partial_I|\mathcal{G}\rangle = \left[(\partial_I b_k)\,\partial^b_k + (\partial_I f_J)\,\partial^f_J\right]|\mathcal{G}\rangle \tag{6.12}$$

and the substitution rules (6.7)–(6.10) for the replacement of the subfermion sources and duals by effective sources and duals. These rules are sufficient for the transformation of $\mathcal{H}_{\mathrm{F}}$ into $\mathcal{H}_{\mathrm{eff}}$, where $\mathcal{H}_{\mathrm{eff}}$ is the effective boson–fermion functional operator.

In general the fermions will be modified by the presence of the bosons. Hence the transformation matrix $C_I^{I'}$ of (6.7) will be nontrivial. On the other hand, the subfermions are unobservable. Thus, in the following for simplicity we assume $C^I_{I'} \equiv \delta^I_{I'}$. Then for the transformation of the subfermion dynamics into an effective boson–fermion dynamics we have the following result:

- **Proposition 6.1:** If exchange forces between subfermions belonging to different composite particles are neglected, then for a combined boson–fermion effective dynamics the state equation (6.2) of the subfermion dynamics is transformed into the effective state equation

$$E_{p0}|\mathcal{G}(b,f;p)\rangle = \mathcal{H}_{\mathrm{eff}}(b,\partial^b,f,\partial^f)|\mathcal{G}(b,f;p)\rangle \tag{6.13}$$

where $\mathcal{H}_{\text{eff}}$ is given by

$$\mathcal{H}_{\text{eff}} = \mathcal{H}_{\text{f}} + \mathcal{H}_{\text{b}} + \mathcal{H}_{\text{b}}^{\text{Q}} + \mathcal{H}_{\text{bf}} + \mathcal{H}_{\text{corr}} \tag{6.14}$$

with the effective fermion part

$$\mathcal{H}_{\text{f}} := K_{I_1 I_2} f_{I_1} \partial^f_{I_2} + W_{I_1 I_2 I_3 I_4} f_{I_1} \partial^f_{I_4} \partial^f_{I_3} \partial^f_{I_2} \, , \tag{6.15}$$

the effective boson part

$$\begin{aligned} \mathcal{H}_{\text{b}} \;&:=\; 2K_{I_1 I_2} R^k_{I_1 I_3} C^{I_2 I_3}_{k'} b_k \partial^b_{k'} \\ &\quad +6W_{I_1 I_2 I_3 I_4} R^k_{I_1 I_5} C^{I_2 I_3}_{k'_1} C^{I_4 I_5}_{k'_2} b_k \partial^b_{k'_1} \partial^b_{k'_2} \\ &\quad -6W_{I_1 I_2 I_3 I_4} F_{I_4 I_5} R^k_{I_1 I_5} C^{I_2 I_3}_{k'} b_k \partial^b_{k'} \, , \end{aligned} \tag{6.16}$$

the effective boson quantization term

$$\mathcal{H}_{\text{b}}^{\text{Q}} := 12 W_{I_1 I_2 I_3 I_4} \left( 3F_{I_4 K_1} F_{I_3 K_2} + \frac{1}{4} A_{I_4 K_1} A_{I_3 K_2} \right) R^{k_1}_{\{I_1 K_1} R^{k_2}_{K_2 K_3\}} C^{I_2 K_3}_{k'} b_{k_1} b_{k_2} \partial^b_{k'} \tag{6.17}$$

and the boson–fermion effective interaction term

$$\mathcal{H}_{\text{bf}} := 3W_{I_1 I_2 I_3 I_4} \left( C^{I_2 I_3}_{k'} f_{I_1} \partial^f_{I_4} \partial^b_{k'} - 2F_{I_4 I_5} R^k_{I_1 I_5} b_k \partial^f_{I_3} \partial^f_{I_2} \right) \, . \tag{6.18}$$

Furthermore, the correction term $\mathcal{H}_{\text{corr}}$ is given by

$$\begin{aligned} \mathcal{H}_{\text{corr}} \;:=\; & W_{I_1 I_2 I_3 I_4} \Big[ 4R^{k_1}_{I_1 K_1} R^{k_2}_{K_2 K_3} C^{I_4 K_1}_{k'_1} C^{I_3 K_2}_{k'_2} C^{I_2 K_3}_{k'_3} b_{k_1} b_{k_2} \partial^b_{k'_1} \partial^b_{k'_2} \partial^b_{k'_3} \\ & +6R^k_{I_1 K_1} C^{I_4 K_1}_{k'_1} C^{I_3 K_2}_{k'_2} b_k f_{K_2} \partial^f_{I_2} \partial^b_{k'_1} \partial^b_{k'_2} \\ & -6R^k_{I_1 I_5} C^{I_4 I_5}_{k'} b_k \partial^f_{I_2} \partial^f_{I_3} \partial^b_{k'} \\ & +12F_{I_4 K_1} \Big( R^{k_1}_{I_1 K_1} R^{k_2}_{K_3 K_2} C^{I_2 K_3}_{k'_1} C^{I_3 K_2}_{k'_2} b_{k_1} b_{k_2} \partial^b_{k'_1} \partial^b_{k'_2} \\ & \quad +3R^k_{K_1 K_2} C^{I_2 K_2}_{k'} b_k f_{I_1} \partial^f_{I_3} \partial^b_{k' \,\text{as}\{I_1, K_1, K_2\}} \Big) \\ & -12\Big( F_{I_4 K_1} F_{I_3 K_2} + \frac{1}{4} A_{I_4 K_1} A_{I_3 K_2} \Big) F_{I_2 K_3} R^{k_1}_{\{I_1 K_1} R^{k_2}_{K_2 K_3\}} b_{k_1} b_{k_2} \\ & +6\Big( 3F_{I_4 K_1} F_{I_3 K_2} + \frac{1}{4} A_{I_4 K_1} A_{I_3 K_2} \Big) R^k_{I_1 K_1} b_k f_{K_2} \partial^f_{I_2 \,\text{as}\{I_1, K_1, K_2\}} \Big] \, . \end{aligned} \tag{6.19}$$

**Proof:** Repeated application of the chain rule (6.12) yields

$$\begin{aligned} \partial_{I_2} \partial_{I_1} |\mathcal{G}\rangle \;=\; & \Big[ (\partial_{I_2} \partial_{I_1} b_k) \, \partial^b_k - (\partial_{I_1} b_{k_1}) (\partial_{I_2} b_{k_2}) \, \partial^b_{k_2} \partial^b_{k_1} \\ & - (\partial_{I_1} b_k) (\partial_{I_2} f_J) \, \partial^f_J \partial^b_k + (\partial_{I_1} f_J) (\partial_{I_2} b_k) \, \partial^b_k \partial^f_J \\ & + (\partial_{I_1} f_{J_1}) (\partial_{I_2} f_{J_2}) \, \partial^f_{J_1} \partial^f_{J_2} \Big] |\mathcal{G}\rangle \end{aligned} \tag{6.20}$$

and

$$\begin{aligned}\partial_{I_3}\partial_{I_2}\partial_{I_1}|\mathcal{G}\rangle = & \Big[(\partial_{I_3}b_{k_1})(\partial_{I_2}b_{k_2})(\partial_{I_1}b_{k_3})\,\partial^b_{k_3}\partial^b_{k_2}\partial^b_{k_1} \\ & +3\,(\partial_{I_3}\partial_{I_2}b_{k_1})(\partial_{I_1}b_{k_2})\,\partial^b_{k_2}\partial^b_{k_1} \\ & +3\,(\partial_{I_3}b_{k_1})(\partial_{I_2}b_{k_2})(\partial_{I_1}f_J)\,\partial^f_J\partial^b_{k_2}\partial^b_{k_1} \\ & +3\,(\partial_{I_3}\partial_{I_2}b_k)(\partial_{I_1}f_J)\,\partial^f_J\partial^b_k \\ & -3\,(\partial_{I_3}b_k)(\partial_{I_2}f_{J_1})(\partial_{I_1}f_{J_2})\,\partial^f_{J_2}\partial^f_{J_1}\partial^b_k \\ & -(\partial_{I_3}f_{J_1})(\partial_{I_2}f_{J_2})(\partial_{I_1}f_{J_3})\,\partial^f_{J_3}\partial^f_{J_2}\partial^f_{J_1}\Big]_{\mathrm{as}\{I_1,I_2,I_3\}}|\mathcal{G}\rangle\,.\end{aligned} \tag{6.21}$$

If these expressions are evaluated by means of (6.7)–(6.10) and if they are substituted in the original equations (6.2) or (6.11), respectively, one obtains equation (6.13) with (6.14). ◇

One could expect that weak mapping for a combined boson–fermion theory in the pure fermion sector $\mathcal{H}_{\mathrm{f}}$ yields the full original subfermion dynamics, i.e. $\mathcal{H}_{\mathrm{f}} \equiv \mathcal{H}_{\mathrm{F}}$. Proposition 6.1 shows that this is not the case. The reason for this result is a certain arbitrariness of the replacement of powers of $j$ by powers of $b$ and $f$ for hard core states. It was supposed that this replacement has to be performed in accordance with a combinatorial equipartition of $b$– and $f$–sources in the representation of powers of $j$. This assumption was proved by Kerschner [Ker 94]. With respect to dressed particle states this difficulty does not exist. As this problem is connected with higher order terms, for the present purpose it is sufficient to consider the approximation of weak mapping which results from using chain rule and hard core states.

An essential point for the physical interpretation of the effective functional equation (6.13) is to neglect $\mathcal{H}_{\mathrm{corr}}$. The arguments for this neglect follow from a direct calculation of the corresponding expressions and a comparison of their magnitude with the magnitude of expressions in $\mathcal{H}_{\mathrm{b}}, \mathcal{H}_{\mathrm{f}}, \mathcal{H}_{\mathrm{bf}}$. Such direct calculations were performed by Stumpf [Stu 85a]. These estimates are based on very large subfermion masses which are needed for the application of the decoupling theorem and which lead to the neglect of unpleasant terms as well as to a low energy limit of the mapping. In the following we will show, see Sections 6.4 and 6.6, that the low energy evaluation of a term like $KR^kC_{k'}b_k\partial^b_{k'}$ yields an expression $G\int d^3r b(\mathbf{r})\partial^b(\mathbf{r})$. The parameters $G$ can be shown to be proportional to powers of the mean subfermion mass $m$; we express this fact by writing $b\partial^b \propto m^n$. Thus all terms of $\mathcal{H}_{\mathrm{eff}}$ can be associated with powers of the mean subfermion mass $m$. For large $m$ only the leading terms in these power series expansion in $m$ have to be considered, while all other terms can be neglected in comparison with the leading terms.

We begin with $\mathcal{H}_{\mathrm{b}}$ given by (6.16). The terms proportional to $b\partial^b$ are the kinetic energy of the bosons and a mass–correction term, while the term proportional to $b(\partial^b)^2$ induces a boson–boson interaction. The latter term competes with corresponding terms in $\mathcal{H}_{\mathrm{corr}}$ which are given by $b^2(\partial^b)^2$ and $b^2(\partial^b)^3$. With an appropriate normalization of the wavefunctions $C_k$ we get by direct calculation $b(\partial^b)^2 \propto m^0$; $b^2(\partial^b)^3 \propto m^{-n}, n \geq 1$; $b^2(\partial^b)^2 \propto m^{-n}, n \geq 1$. Hence these terms can be neglected in comparison with $b(\partial^b)^2$.

The terms of $\mathcal{H}_{\mathrm{bf}}$ partially occur in the projected fermion equation or in the boson equation. The coupling term of the bosons in the fermion equation is proportional to $f\partial^f\partial^b$ and yields $f\partial^f\partial^b \propto m^0$. The corresponding term in $\mathcal{H}_{\mathrm{corr}}$ is proportional to $fb\partial^f(\partial^b)^2$ and yields $fb\partial^f(\partial^b)^2 \propto m^{-n}, n \geq 2$, i.e. it can be neglected. The coupling term of the fermionic current in the boson equation is proportional to $b(\partial^f)^2$ with $b(\partial^f)^2 \propto m^2$. The corresponding term in $\mathcal{H}_{\mathrm{corr}}$ is poportional to $b\partial^b(\partial^f)^2 \propto m^n, n \leq 1$ and thus can be neglected.

Finally, in $\mathcal{H}_{\text{corr}}$ the terms proportional to $bf\partial^f\partial^b$ and $bf\partial^f$ do not allow a meaningfull interpretation in the hard core formulation. Taking into account our discussion with respect to composite particles, the correct representation of this term is given by $f\partial^f\partial^b$ or $f\partial^f$ with $f \propto j^3$, i.e. a polarization cloud term. In this way it is obvious that $f\partial^f\partial^b$ contributes to the first term in $\mathcal{H}_{\text{bf}}$, i.e., it is a correction to the coupling term of fermions to bosons. By the same arguments the term $f\partial^f$ is a mass correction in the fermion sector.

In the $bb$–term scalar propagators are coupled to the vector boson duals, so this term vanishes for reasons of symmetry.

With respect to these estimates it should be noted that a completely conclusive physical interpretation is assumed to be achieved only by use of dressed particle states, as was demonstrated in Sections 4.8 and 4.9. However, for this case so far detailed calculations are not available.

To explicitly evaluate equation (6.13) it is convenient to proceed in two steps: First we discuss a pure boson theory, then we treat the full equation (6.13).

The pure boson theory is a subsector of the solutions of equation (6.13). This subsector is obtained if (6.13) is projected with the pure boson projectors ${}_F\langle 0|{}_B\langle 0|\partial^b_{k_1}\dots\partial^b_{k_l}$. By means of the commutation relations (6.4) one easily convinces oneself that the resulting equation can be equivalently written as the bosonic equation

$$E_{p0}|\mathcal{B}(b;p)\rangle = \left[\mathcal{H}_{\text{b}} + \mathcal{H}_{\text{b}}^{\text{Q}}\right]|\mathcal{B}(b;p)\rangle \tag{6.22}$$

where the boson state functional $|\mathcal{B}(b;p)\rangle$ is given by (4.13) and where we have already neglected the terms from $\mathcal{H}_{\text{corr}}$. Of course equation (6.22) is identical with (4.92). It may be obtained from equation (6.13) by setting $f = \partial^f \equiv 0$ .

In order to evaluate (6.22) we have to investigate the boson hard core states $C_k^{I_1 I_2}$ and their duals $R^k_{I_1 I_2}$.

## 6.3 Boson States and Dual States

The basis of weak mapping is the set of boson states $C_k^{I_1 I_2}$ together with their dual states $R^k_{I_1 I_2}$, where the hard core states $C_k^{I_1 I_2}$ are assumed to be solutions of equation (5.1).

Exact solutions of (5.1) were derived in Section 5.3 following Pfister, Rosa and Stumpf [Pfi 89]. The set of these solutions contains bound state and scattering state solutions. The energy of the latter at least amounts to $2m$, where $m$ is the lowest mass of $m_0$, $m_1$ and $m_2$ in (1.5). In the following we assume $m$ to be very large, i.e. larger than any observable energy limit. If we confine the evaluation of (6.22) to an energy range $E < m$, then due to the energy conservation of the spinorfield processes no break up processes of bound fermion states can occur, i.e. the dynamics of the bound states and the dynamics of the scattering states should decouple, see also Section 4.6. This decoupling should exactly be derived from (6.22). However, a general investigation of this hypothesis has not yet been performed. But one can apply a simple argument to justify decoupling: The dynamics of scattering states of fermions contained in (6.22) can equivalently be described by the original dynamics of the spinorfield (3.106). Thus the evaluation of (6.22) with inclusion of scattering states must result in a coupling theory between composite bosons and fermions, and if one is exclusively interested in the boson dynamics one can omit the coupling terms to the fermions. But this is equivalent to calculate (6.22) from the beginning only with bound states.

Also with respect to bound states we concentrate on the discussion of the spin 1–boson–fermion dynamics, i.e. we omit spin 0–bosons. For a complete inclusion of all bound boson states into the effective dynamics see Grimm [Gri 94].

Of course, in a quantum theory of boson–fermion systems the field fluctuations of the fermion field influence the boson field and vice versa. But this mutual influence is an effect of the coupling between both kind of particles and does not prevent the independent formulation of a separate boson and (or) fermion dynamics. Thus in the following we consider only the bound states of (5.1).

Furthermore, from the beginning we restrict ourselves to the vector boson solutions of Section 5.3 and thus omit the treatment of scalar and other fields. For a combined treatment of all these fields see Grimm [Gri 94]. According to Section 5.3 the vector boson bound states are given by the equal time limit of the functions

$$C_{\substack{i_1 i_2\\ \kappa_1\kappa_2\\ \alpha_1\alpha_2}}(x_1,x_2|k,a) := NT^a_{\kappa_1\kappa_2}\exp\Big[-ik(x_1+x_2)/2\Big]A_\mu\chi^\mu_{\substack{i_1 i_2\\ \alpha_1\alpha_2}}(x_1-x_2) \tag{6.23}$$

with

$$\chi^\mu_{\substack{i_1 i_2\\ \alpha_1\alpha_2}}(x_1-x_2) := \tag{6.24}$$
$$g_V\lambda_{i_1}\lambda_{i_2}\int\left\{\frac{\gamma^\rho(p-k/2)_\rho - m_{i_1}}{m^2_{i_1}-(p-k/2)^2-i\varepsilon}\gamma^\mu\frac{\gamma^\nu(p+k/2)_\nu - m_{i_2}}{m^2_{i_2}-(p+k/2)^2-i\varepsilon}C\right\}_{\alpha_1\alpha_2}\times$$
$$\exp\Big[ip(x_1-x_2)\Big]d^4p$$

where $k$ is the four–momentum of the boson and $T^a_{\kappa_1\kappa_2}$ are superspin–isospin matrices, while $N$ is a normalization constant, which will be specified later. The four–vector $A^\mu$ fullfills the condition $A^\mu k_\mu = 0$ and the coupling constant $g_V$ is fixed to give massless vector bosons ($k^2 = 0$).

The functions (6.23) are exact solutions of generalized Bargmann–Wigner equations (5.26), (5.27), whose energy representation coincides with (5.1). As the energy representation (5.1) is referred to the state description in the equal time limit which is defined by (5.24), this holds also for the special state representation (6.23).

Before performing this limit, we apply a low energy approximation by omitting $k/2$ in the denominators of (6.24). Furthermore, as already mentioned we assume a vanishing boson mass and a small difference of the masses $m_1$, $m_2$ and $m_3$. By a straightforward calculation we then obtain for (6.24) in the equal time limit with $\mathbf{r} := \mathbf{r}_1 - \mathbf{r}_2$ and for $\mu = l = 1,2,3$:

$$\chi^l_{\substack{i_1 i_2\\ \alpha_1\alpha_2}}(\mathbf{r}) = ik_\nu\hat{\chi}^{l\nu}_{\substack{i_1 i_2\\ \alpha_1\alpha_2}}(\mathbf{r}) + \hat{\chi}^l_{\substack{i_1 i_2\\ \alpha_1\alpha_2}}(\mathbf{r}) \quad , \tag{6.25}$$

where the decomposition is defined by

$$\begin{aligned}\hat{\chi}^{l\nu}_{\substack{i_1 i_2\\ \alpha_1\alpha_2}}(\mathbf{r}) \;:=\;& 2i\pi^2\lambda_{i_1}\lambda_{i_2}\left\{(\Sigma^{l\nu}C)_{\alpha_1\alpha_2}M_{i_1 i_2}\int_0^1 dz K_0(\hat{z}r)\right.\\ &\left.+i\varepsilon_\eta{}^{jl\nu}(\gamma^\eta\gamma_5 C)_{\alpha_1\alpha_2}\int_0^1 dz x^j\frac{\hat{z}K_1(\hat{z}r)}{r}\right\}\end{aligned} \tag{6.26}$$

and

$$\hat{\chi}^{l}{}_{\substack{i_1 i_2 \\ \alpha_1\alpha_2}}(\mathbf{r}) := 2i\pi^2\lambda_{i_1}\lambda_{i_2}\int_0^1 dz \Big\{-(\gamma^l C)_{\alpha_1\alpha_2}\frac{2\hat{z}K_1(\hat{z}r)}{r} + 2(\gamma^j C)_{\alpha_1\alpha_2}x^l x^j\frac{\hat{z}^2K_2(\hat{z}r)}{r^2}$$
$$+(\gamma^l C)_{\alpha_1\alpha_2}\left[\frac{2\hat{z}K_1(\hat{z}r)}{r} - \hat{z}^2K_2(\hat{z}r) + M_{i_1i_2}^2K_0(\hat{z}r)\right] \tag{6.27}$$
$$+ 2iM_{i_1i_2}C_{\alpha_1\alpha_2}x^l\frac{\hat{z}K_1(\hat{z}r)}{r} - 2M_{i_1i_2}\tilde{M}_{i_1i_2}(\Sigma^{lj}C)_{\alpha_1\alpha_2}x^j\frac{\hat{z}K_1(\hat{z}r)}{r}\Big\}$$

with

$$M_{i_1i_2} := \frac{1}{2}(m_{i_1}+m_{i_2}), \tag{6.28}$$
$$\tilde{M}_{i_1i_2} := \frac{m_{i_1}-m_{i_2}}{2M_{i_1i_2}}, \tag{6.29}$$
$$\hat{z} := [m_{i_1}^2 + z(m_{i_2}^2 - m_{i_1}^2)]^{\frac{1}{2}} \tag{6.30}$$

and the modified Bessel functions $K_i(r)$.

We now come to an important point of weak mapping. In order to introduce the dynamical quantities of the effective theory, in the present case the Yang–Mills fields, the expansion functions of weak mapping are not the full boson bound states (6.23), but a modified set of functions. As we will consider the quantities $A_l$ and $F_{l\nu} := iA_{\{l}k_{\nu\}}$ as the new dynamical quantities of an effective boson theory in temporal gauge, we separate these factors from the equal time version of (6.23) with (6.25) and define the following basis functions for the weak mapping procedure:

$$\hat{C}_k^{I_1I_2} \equiv \left\{\hat{C}^{l}{}_{\substack{i_1 i_2\\ \kappa_1\kappa_2\\ \alpha_1\alpha_2}}(\mathbf{r}_1,\mathbf{r}_2|\mathbf{k},a),\ \hat{C}^{l\nu}{}_{\substack{i_1 i_2\\ \kappa_1\kappa_2\\ \alpha_1\alpha_2}}(\mathbf{r}_1,\mathbf{r}_2|\mathbf{k},a)\right\}, \tag{6.31}$$

where $k$ denotes the full set of quantum numbers of the state and where

$$\hat{C}^{l}{}_{\substack{i_1 i_2\\ \kappa_1\kappa_2\\ \alpha_1\alpha_2}}(\mathbf{r}_1,\mathbf{r}_2|\mathbf{k},a) := NT^a_{\kappa_1\kappa_2}\exp\left[i\mathbf{k}(\mathbf{r}_1+\mathbf{r}_2)/2\right]\hat{\chi}^{l}{}_{\substack{i_1 i_2\\ \alpha_1\alpha_2}}(\mathbf{r}_1-\mathbf{r}_2) \tag{6.32}$$

and

$$\hat{C}^{l\nu}{}_{\substack{i_1 i_2\\ \kappa_1\kappa_2\\ \alpha_1\alpha_2}}(\mathbf{r}_1,\mathbf{r}_2|\mathbf{k},a) := NT^a_{\kappa_1\kappa_2}\exp\left[i\mathbf{k}(\mathbf{r}_1+\mathbf{r}_2)/2\right]\hat{\chi}^{k\nu}{}_{\substack{i_1 i_2\\ \alpha_1\alpha_2}}(\mathbf{r}_1-\mathbf{r}_2). \tag{6.33}$$

Because $\hat{\chi}^{l\nu}$ is antisymmetric in $l,\nu$ for $\nu = 1,2,3$, we have to restrict the set of basis functions (6.31) by the condition $l > \nu$.

We consider the temporal gauge, i.e. $A_0 = 0$. Then a general solution $C_{I_1I_2}$ of equation (5.33) is given by the following linear combination:

$$C_{I_1I_2} = \sum_{l,\mathbf{k}}\left\{A_l(\mathbf{k})\hat{C}^{l}{}_{\substack{i_1 i_2\\ \kappa_1\kappa_2\\ \alpha_1\alpha_2}}(\mathbf{r}_1,\mathbf{r}_2|\mathbf{k},a) + \sum_\nu iA_l(\mathbf{k})k_\nu\hat{C}^{l\nu}{}_{\substack{i_1 i_2\\ \kappa_1\kappa_2\\ \alpha_1\alpha_2}}(\mathbf{r}_1,\mathbf{r}_2|\mathbf{k},a)\right\}. \tag{6.34}$$

Comparison of the solution (6.34) with the vector boson ansatz

$$\Lambda(x) = m\hat{A}_\mu(x)\gamma^\mu C + \frac{1}{2}\Sigma^{\mu\nu}C\hat{F}_{\mu\nu}(x) \tag{6.35}$$

for an ordinary Bargmann–Wigner equation

$$(i\gamma^\mu\partial_\mu - m)_{\alpha\alpha'}\Lambda_{\alpha'\beta}(x) = 0 \tag{6.36}$$

with

$$\hat{F}_{\mu\nu} = \partial_\mu\hat{A}_\nu - \partial_\nu\hat{A}_\mu \tag{6.37}$$

suggests to interpret $A_l$, $l = 1,2,3$, as vector potential in temporal gauge and $F_{\mu\nu} := \frac{i}{2}(k_\mu A_\nu - k_\nu A_\mu)$ as field strength tensor in momentum space. This fixed relation between $A_l$ and $F_{l\nu}$ is a consequence of the linearized field dynamics in (5.1). If by the weak mapping procedure the full field dynamics is switched on we expect that this relation between $A_l$ and $F_{l\nu}$ will be modified. Thus, with respect to the weak mapping, we consider the set of boson functions (6.31) as independent states and leave it to the full field dynamics to establish a relation between vector potentials and field strengths. As will be demonstrated this enforces a mass renormalization.

Furthermore, the representation (6.34) is based on the temporal gauge $A_0 \equiv 0$. This choice at first seems to be arbitrary. But in connection with the weak mapping of the energy representation it will turn out that the use of the temporal gauge is conclusive, i.e. there is a unique connection between the Hamilton formalism and temporal gauge, which was already emphasized in the literature, see for instance Christ and Lee [Chri 80]. For a more detailed discussion we refer to Section 8.5.

From (6.26) and (6.27) it follows that the functions $\hat{\chi}_l$ and $\hat{\chi}_{l\nu}$ are symmetric with respect to a simultaneous interchange of the index sets $(\mathbf{r}_h, \alpha_h, i_h)$, $h = 1,2$. As the complete functions (6.31) must be antisymmetric with respect to a simultaneous interchange of the index sets $(\mathbf{r}_h, \alpha_h, \kappa_h, i_h)$, $h = 1,2$, it follows that the superspin–isospin matrices $T^a_{\kappa_1\kappa_2}$ must be antisymmetric. In accordance with Stumpf [Stu 86a] and Section 5.3 we choose the three antisymmetric matrices

$$T^a := \begin{pmatrix} 0 & \sigma^a \\ (-1)^a\sigma^a & 0 \end{pmatrix}, \qquad a = 1,2,3, \tag{6.38}$$

where $\{\sigma^a\}$ is the set of Pauli–matrices. Substitution of (6.38) into the wave functions (6.23) or (6.32), (6.33), respectively, and a subsequent determination of the good quantum numbers of these states shows that these wave functions represent an isospin triplet of vector bosons with isospin 1 and fermion number 0. The investigation of the full state spectrum of two–fermion bound states of equation (5.33) and its physical interpretation can be found in the thesis of Grimm [Gri 94]. Here we will not discuss this topic.

With respect to the calculation of the dual functions $R^k_{I_1 I_2}$ we observe that in principle we need the complete set of functions $C^{I_1 I_2}_k$, i.e. bound states and scattering states. Then the duals $R^k_{I_1 I_2}$ can be explicitly calculated by applying Schmidt's orthonormalization procedure. In order to avoid this laborious procedure we use a short–cut and define only a projector on the bound states which guarantees orthonormality. With respect to the algebraic part of the set (6.31) this dual projector is unique. The non–uniqueness is solely concerned with the space function part of (6.31). But the explicit evaluation of the map will show that the algebraic part essentially determines the resulting terms, and the space part only fixes numerical values of constants. Corresponding to the decomposition (6.31) we define the following projectors:

$$
\begin{aligned}
R^k_{I_1 I_2} &:= \left\{ \frac{k_1}{N}\hat{T}^a_{\kappa_1\kappa_2}(\gamma^l C)^+_{\alpha_1\alpha_2} e^{-i\frac{1}{2}\mathbf{k}(\mathbf{r}_1+\mathbf{r}_2)} \left(e^{-m_{i_1} r} + e^{-m_{i_2} r}\right) \right., \\
&\quad \left. \frac{k_2}{N}\hat{T}^a_{\kappa_1\kappa_2}(\Sigma^{l\nu} C)^+_{\alpha_1\alpha_2} e^{-i\frac{1}{2}\mathbf{k}(\mathbf{r}_1+\mathbf{r}_2)} \left(e^{-m_{i_1} r} + e^{-m_{i_2} r}\right) \right\} \\
&=: \left\{ R^{l,a}_{I_1 I_2}(\mathbf{k}) \, , \, R^{l\nu,a}_{I_1 I_2}(\mathbf{k}) \right\} \quad ,
\end{aligned}
\tag{6.39}
$$

with $r = |\mathbf{r}_1 - \mathbf{r}_2|$ and $\hat{T}^a := \frac{1}{4}\left(T^a\right)^+$. The normalization constant $N$ stems from (6.23) and the constants $k_1$ and $k_2$ are given by

$$
\begin{aligned}
k_1 &:= \frac{5(m_{i_1} + m_{i_2})}{2^{16} 3 \pi^6 \lambda_{i_1} \lambda_{i_2}} \quad , \\
k_2 &:= \frac{5(m_{i_1} + m_{i_2})^2}{2^{15} 3 \pi^6 \lambda_{i_1} \lambda_{i_2}} \quad .
\end{aligned}
\tag{6.40}
$$

The set of these projectors is complete and satisfies orthonormality and in the algebraic submanifold also completeness relations with respect to the original set (6.31).

Finally, for the evaluation of the map we need the explicit form of the one–time limit of the propagator (2.116) of the fermion fields. This limit can be performed by standard methods and is given by

$$
\begin{aligned}
F^t_{Z_1 Z_2}(\mathbf{r}_1, \mathbf{r}_2) &= \frac{1}{(2\pi)^2} \lambda_{i_1} \delta_{i_1 i_2} \gamma^5_{\kappa_1\kappa_2} \quad \times \\
&\left\{ -m^2_{i_1} \frac{K_1(m_{i_1} r)}{r} C_{\alpha_1\alpha_2} + i m_{i_1} (\gamma^k C)_{\alpha_1\alpha_2} x^k \left[ \frac{2K_1(m_{i_1} r)}{r^3} + m_{i_1} \frac{K_0(m_{i_1} r)}{r^2} \right] \right\}
\end{aligned}
\tag{6.41}
$$

with the modified Bessel functions $K_i(r)$.

It should be noticed that after summation over the auxiliary field indices $i_1, i_2$ in (6.41) the corresponding regularized propagator is linearly divergent at the origin $\mathbf{r}_1 = \mathbf{r}_2$.

## 6.4 Evaluation of the Map

The evaluation of the map in the bosonic sector simply consists in substituting the bound states (6.31) and the dual projectors (6.39) into equation (6.22). The necessary calculations are lengthy, but elementary. Details of the calculation with simplified functions can be found in Stumpf [Stu 86a]; the full calculation with the functions (6.31) is contained in Pfister [Pfi 90]. In all calculations algebraic expressions were exactly evaluated, while space integrations were performed in low energy approximations and under the assumption $|(m_{i_1} - m_{i_2})/(m_{i_1} + m_{i_2})| \ll 1$. But we emphasize that due to the initial regularization all singularities of the wavefunctions finally cancel. Here we give only the results.

To properly express these results we have to specify the notation of (6.8) by explicitly introducing the indices of the set (6.31) instead of a general index $k$. By separating the dynamical quantities $A_\mu$ and $F_{\mu\nu}$ we have determined the expansion functions of (6.8) to be the functions (6.32) and (6.33) and we explain the set of bosonic sources $b_k$ and their corresponding "duals" $\partial_k$ to be given by

$$\{b_k\} \equiv \{b^A_{l,a}(\mathbf{k}), b^F_{l\nu,a}(\mathbf{k})\} \ , \ \{\partial^b_k\} \equiv \{\partial^A_{l,a}(\mathbf{k}), \partial^F_{l\nu,a}(\mathbf{k})\} \tag{6.42}$$

with quantum numbers corresponding to (6.31). Furthermore, the Fouriertransforms of (6.42) are denoted by $b^A_{l,a}(\mathbf{r})$, $b^F_{l\nu,a}(\mathbf{r})$ etc. We assume $b^F_{l\nu,a}$ and $\partial^F_{l\nu,a}$ to be antisymmetric in $l$ and $\nu$ for $\nu = 1,2,3$. In order to work with independent quantities we postulate the restriction $l > \nu$ for all $b^F$ and $\partial^F$.

We first consider the dynamical terms of (6.13) and write $\mathcal{H}_\mathrm{b} = \mathcal{H}^0_\mathrm{b} + \mathcal{H}^1_\mathrm{b}$ with

$$\mathcal{H}^0_\mathrm{b} := 2R^k_{I_1 I_3} K_{I_1 I_2} C^{I_2 I_3}_{k'} b_k \partial^b_{k'} - 6W_{I_1\{I_2 I_3 I_4\}} F_{I_4 K} R^k_{I_1 K} C^{I_2 I_3}_{k'} b_k \partial^b_{k'} \tag{6.43}$$

and obtain

$$\begin{aligned} \mathcal{H}^0_\mathrm{b} \;=\; & \left[\sum_{i_1,i_2}\left(\frac{2}{9}M^2_{i_1 i_2} + \frac{8}{9}M_{i_1 i_2} m_{i_1} + \frac{g}{9\pi^2}\frac{M^2_{i_1 i_2}}{m_0^2}\right)\right] \times \\ & \frac{1}{4}\int d^3r \left\{ b^A_{i,a}(\mathbf{r}) \mathrm{tr}\left[(\gamma^i C)^+ \gamma^0 \gamma^k C\right] \partial^A_{k,a}(\mathbf{r}) \right. \\ & \left. + b^F_{i\nu,a}(\mathbf{r})_{|i>\nu} \mathrm{tr}\left[(\Sigma^{i\nu} C)^+ \gamma^0 \gamma^k C\right] \partial^A_{k,a}(\mathbf{r}) \right\} \\ & -i \int d^3r b^A_{i,a}(\mathbf{r}) \partial^F_{i0,a}(\mathbf{r}) \\ & +i \int d^3r b^F_{ij,a}(\mathbf{r})_{|i>j} \varepsilon_{ijk}\varepsilon_{lmk} \partial_m(\mathbf{r}) \partial^F_{l0,a}(\mathbf{r}) \\ & +i \int d^3r b^F_{i0,a}(\mathbf{r}) \varepsilon_{ijk}\varepsilon_{lmk} \partial_j(\mathbf{r}) \partial^F_{lm,a}(\mathbf{r})_{|l>m} \quad . \end{aligned} \tag{6.44}$$

In this context we remember that the symbol $\partial_k(\mathbf{r})$ without an upper index means the partial derivative with respect to the space coordinate $x^k$.

The first term of (6.44) is the renormalized mass term of the vector bosons, and by a suitable choice of the coupling constant $g$ we can make it to vanish. This value of $g$ is given by

$$g = -10\pi^2 m_0^2 \ . \tag{6.45}$$

As the next term we consider

$$\mathcal{H}^1_\mathrm{b} := 6W_{I_1\{I_2 I_3 I_4\}} R^k_{I_1 K} C^{I_2 I_3}_{k'_1} C^{I_4 K}_{k'_2} b_k \partial^b_{k'_1} \partial^b_{k'_2} \quad . \tag{6.46}$$

Its evaluation yields

$$\begin{aligned} \mathcal{H}^1_\mathrm{b} \;=\; & iG_1 \int d^3r \varepsilon_{abc} \left[\varepsilon_{ijk}\varepsilon_{lmk} b^F_{ij,b}(\mathbf{r})_{|i>j} \partial^F_{l0,a}(\mathbf{r}) \partial^A_{m,c}(\mathbf{r}) \right. \\ & \left. + \varepsilon_{ijk}\varepsilon_{lmk} b^F_{l0,b}(\mathbf{r}) \partial^F_{ij,a}(\mathbf{r})_{|i>j} \partial^A_{m,c}(\mathbf{r})\right] \end{aligned} \tag{6.47}$$

with

$$G_1 = \frac{32\sqrt{2} g N \pi^{\frac{7}{2}}}{105} \frac{1}{m_0^4} \ . \tag{6.48}$$

The terms $\mathcal{H}^0_\mathrm{b}$ and $\mathcal{H}^1_\mathrm{b}$ represent the genuine gauge boson dynamics.

The quantization term is defined by

$$\begin{aligned}\mathcal{H}_{\mathrm{b}}^{\mathrm{Q}} \;&:=\; 12W_{I_1\{I_2I_3I_4\}}\left(3F_{I_4K_1}F_{I_3K_2}+\frac{1}{4}A_{I_4K_1}A_{I_3K_2}\right)\quad\times \\ &\quad R^{k_1}_{\{I_1K_1}R^{k_2}_{K_2\}K_3}C^{I_2K_3}_{k'}b_{k_1}b_{k_2}\partial^b_{k'}\quad .\end{aligned} \tag{6.49}$$

The $F$–terms in (6.49) describe a coupling of the Fermion vacuum to the boson fields, while the $A$–terms describe the action of the Fermion field quantization on the boson fields. Apart from the difference between $F$ and $A$, i.e. between propagator and anticommutator, both terms are structurally equivalent. The argument for this equivalence will be given in the next section. A lengthy but straightforward calculation yields in the low energy limit

$$\mathcal{H}_{\mathrm{b}}^{\mathrm{Q}}=G_2\varepsilon_{abc}\int d^3r b^F_{ik,a}(\mathbf{r})_{|i>k}b^F_{k0,b}(\mathbf{r})\partial^A_{i,c}(\mathbf{r}) \tag{6.50}$$

with

$$G_2:=-m_0\frac{855g}{56N}\quad . \tag{6.51}$$

According to our decomposition of $\mathcal{H}_{\mathrm{b}}$ we write (6.22) as

$$E_{p0}|\mathcal{B}(b;p)\rangle=\left(\mathcal{H}_{\mathrm{b}}^0+\mathcal{H}_{\mathrm{b}}^1+\mathcal{H}_{\mathrm{b}}^{\mathrm{Q}}\right)|\mathcal{B}(b;p)\rangle\quad . \tag{6.52}$$

For relating the energy equation (6.22) to dynamical boson equations which explicitly contain time derivatives we remember the symmetry condition (2.97) for the covariant counterparts of the states (6.1). This symmetry condition reads for $\mu=0$ and after energy renormalization

$$E_{p0}|\mathcal{F}(j;p)\rangle=i\int d^4x j_Z(x)\partial(x_0)\partial_Z(x)|\mathcal{F}(j;p)\rangle\,. \tag{6.53}$$

We observe that the transformation (6.7) does not destroy the time–translational invariance. Thus the symmetry condition (6.53) must also hold for the covariant counterparts of the functionals (6.3) and (6.13).

In order to perform a one–time limit to an arbitrary moment $t$, which is possible because of the time–translational invariance, we write the covariant sources $b_k(x)$ as $b_k(\mathbf{r},t')=\delta(t-t')b_k(\mathbf{r},t)$, where $t$ is the reference time of the one–time formalism and obtain for the boson functionals $|\mathcal{B}(b;p)\rangle$:

$$E_{p0}|\mathcal{B}(b;p)\rangle=i\int d^3r b_k(\mathbf{r},t)\left[\frac{\partial}{\partial t}\partial^b_k(\mathbf{r},t)\right]|\mathcal{B}(b;p)\rangle\,. \tag{6.54}$$

By shifting the time $t$ we have restored a time parameter in the one–time equations (6.54).

But we have to comment (6.54) in yet another regard: If we perform the one–time limit in the above mentioned manner, this does not lead to a non–ambigous definition of the effect of the time–derivative in (6.54). However, the application of (6.54) can be justified as the functional version of the Heisenberg equation (3.78) is identical with the projection of the one–time equations (6.54).

By substitution of (6.54) into (6.52) and with the independence of $b^A$ and $b^F$, equation (6.52) can be interpreted as the energy representation of the functional equations

$$i\frac{\partial}{\partial t}\partial^A_{i,a}(\mathbf{r},t)|\mathcal{B}(b;p)\rangle=-i\partial^F_{i0,a}(\mathbf{r})|\mathcal{B}(b;p)\rangle\quad , \tag{6.55}$$

$$i\frac{\partial}{\partial t}\partial^F_{i0,a}(\mathbf{r},t)|\mathcal{B}(b;p)\rangle = \Big[\frac{i}{2}\varepsilon_{ijk}\varepsilon_{lmk}\partial_j(\mathbf{r})\partial^F_{lm,a}(\mathbf{r},t) - \frac{i}{2}G_1\varepsilon_{abc}\varepsilon_{ijk}\varepsilon_{lmk}\partial^F_{lm,b}(\mathbf{r},t)\partial^A_{j,c}(\mathbf{r},t)\Big]|\mathcal{B}(b;p)\rangle \tag{6.56}$$

and

$$\frac{i}{2}\frac{\partial}{\partial t}\partial^F_{ij,a}(\mathbf{r},t)|\mathcal{B}(b;p)\rangle = \big[i\varepsilon_{ijk}\varepsilon_{lmk}\partial_m(\mathbf{r})\partial^F_{l0,a}(\mathbf{r},t) - iG_1\varepsilon_{abc}\varepsilon_{ijk}\varepsilon_{lmk}\partial^F_{l0,b}(\mathbf{r},t)\partial^A_{m,c}(\mathbf{r},t) + G_2\varepsilon_{abc}b^F_{j0,b}(\mathbf{r},t)\partial^A_{i,c}(\mathbf{r},t)\big]|\mathcal{B}(b;p)\rangle \quad , \tag{6.57}$$

where the time derivatives only act on the sequent functional dual $\partial$, not on the time parameter of the functionals $|\mathcal{B}\rangle$.

We now subdivide (6.55)–(6.57) into non–abelian electric and magnetic fields by introducing the definitions

$$\begin{aligned} b^E_{k,a}(\mathbf{r},t) &:= b^F_{k0,a}(\mathbf{r},t)\ , \\ b^B_{k,a}(\mathbf{r},t) &:= -\sum_{l<m}\varepsilon_{klm}b^F_{lm,a}(\mathbf{r},t) \\ &= -\frac{1}{2}\sum_{l,m}\varepsilon_{klm}b^F_{lm,a}(\mathbf{r},t) \end{aligned} \tag{6.58}$$

for the sources $b_k$.

With corresponding definitions for the duals $\partial_k$ one can easily derive the "canonical" commutation relations for these transformed functional operators

$$\Big[\partial^X_{k,a}(\mathbf{r},t), b^X_{k',a'}(\mathbf{r}',t)\Big]_- = \delta_{kk'}\delta_{aa'}\delta(\mathbf{r}-\mathbf{r}') \tag{6.59}$$

for $X = A, E, B$.

Equations (6.55)–(6.57) now can be equivalently written as

$$i\frac{\partial}{\partial t}\partial^A_{i,a}(\mathbf{r},t)|\mathcal{B}(b;p)\rangle = -i\partial^E_{i,a}(\mathbf{r},t)|\mathcal{B}(b;p)\rangle\ , \tag{6.60}$$

$$i\frac{\partial}{\partial t}\partial^B_{i,a}(\mathbf{r},t)|\mathcal{B}(b;p)\rangle = \big[-i\varepsilon_{ijk}\partial_k(\mathbf{r})\partial^E_{j,a}(\mathbf{r},t) + iG_1\varepsilon_{abc}\varepsilon_{ijk}\partial^E_{j,b}(\mathbf{r},t)\partial^A_{k,c}(\mathbf{r},t) - G_2\varepsilon_{abc}\varepsilon_{ijk}b^E_{k,b}(\mathbf{r},t)\partial^A_{j,c}(\mathbf{r},t)\big]|\mathcal{B}(b;p)\rangle \tag{6.61}$$

and

$$i\frac{\partial}{\partial t}\partial^E_{i,a}(\mathbf{r},t)|\mathcal{B}(b;p)\rangle = \big[i\varepsilon_{ijk}\partial_k(\mathbf{r})\partial^B_{j,a}(\mathbf{r},t) - iG_1\varepsilon_{abc}\varepsilon_{ijk}\partial^B_{j,b}(\mathbf{r},t)\partial^A_{k,c}(\mathbf{r},t)\big]|\mathcal{B}(b;p)\rangle \quad . \tag{6.62}$$

Equations (6.60), (6.61), (6.62) describe a quantized composite boson dynamics. Thus a complete comparison with the phenomenological Yang–Mills theory can be made only at the quantized level. This comparison will be performed in the next section. But for a first outlook we switch off the quantization terms by setting $G_2 = 0$. Then we can perform the transition from the functional version of the theory to its classical formulation by means of the ansatz, cf. Stumpf [Stu 88]:

$$|\mathcal{B}(b;p)\rangle = e^{Z_0[b;p]}|0\rangle_B \tag{6.63}$$

with

$$Z_0[b;p] := \int \sum_X \sum_{k,a} X_a^k(\mathbf{r},t) b_{k,a}^X(\mathbf{r},t) d^3r \tag{6.64}$$

for the classical fields $X = A, E, B$.

If we project the functional equations (6.60)–(6.62) onto the functional vacuum sector and observe the ansatz (6.63) we obtain the following equations for the covariant classical fields:

$$\begin{aligned}
\partial_0 A_a^i(\mathbf{r},t) &= -E_a^i(\mathbf{r},t) \\
\partial_0 B_a^i(\mathbf{r},t) &= \varepsilon_{ijk}\partial_j(\mathbf{r})E_a^k(\mathbf{r},t) + G_1\varepsilon_{abc}\varepsilon_{ijk}E_b^j(\mathbf{r},t)A_c^k(\mathbf{r},t) \\
\partial_0 E_a^i &= \varepsilon_{ijk}\partial_k(\mathbf{r})B_a^j(\mathbf{r},t) + G_1\varepsilon_{ijk}\varepsilon_{abc}B_b^k(\mathbf{r},t)A_c^j(\mathbf{r},t) \quad .
\end{aligned} \tag{6.65}$$

For comparison we consider the field equations of a classical $SU(2)$–Yang–Mills theory

$$\nabla_\mu^{ab} F_b^{\mu\nu}(x) = j_a^\nu \tag{6.66}$$

with

$$\nabla_\mu^{ab} F_b^{\mu\nu}(x) := \partial_\mu F_a^{\mu\nu}(x) + \hat{g}\varepsilon_{abc}A_c^\mu(x)F_b^{\mu\nu}(x) \tag{6.67}$$

and

$$F_a^{\mu\nu}(x) := \partial^\mu A_a^\nu(x) - \partial^\nu A_a^\mu(x) - \hat{g}\varepsilon_{abc}A_b^\mu(x)A_c^\mu(x) \quad , \tag{6.68}$$

where $a, b, c = 1, 2, 3$ refer to isospin–indices and $\mu, \nu = 0, 1, 2, 3$ to the Lorentz–vector indices, while $\hat{g}$ is the coupling constant of the theory. Equations (6.66)–(6.68) are a special case of the general $SU(N)$–equations (8.80) and (8.78).

Introducing the generalized non–abelian electric and magnetic fields

$$E_a^i(x) := F_a^{i0}(x) \tag{6.69}$$

$$B_a^i(x) := -\frac{1}{2}\varepsilon_{ijk}F_a^{jk}(x) \tag{6.70}$$

the field equations (6.66) can be rewritten as

$$\begin{aligned}
\partial_0 A_a^i &= -E_a^i - \partial_i A_a^0 - \hat{g}\varepsilon_{abc}A_b^i A_c^0 \, , \\
\partial_0 B_a^i &= \varepsilon_{ijk}\partial_j\left(-E_a^k - \partial_k A_a^0 - \hat{g}\varepsilon_{abc}A_b^k A_c^0\right) \\
&\quad +\hat{g}\varepsilon_{ijk}\varepsilon_{abc}\left(-E_b^j - \partial_j A_b^0 - \hat{g}\varepsilon_{bad}A_a^j A_d^0\right)A_c^k \, , \\
\partial_0 E_a^i &= \hat{g}\varepsilon_{abc}A_b^0 E_c^i + \varepsilon_{ijk}\partial_j B_a^k + \hat{g}\varepsilon_{ijk}\varepsilon_{abc}A_b^i B_c^k - j_a^i
\end{aligned} \tag{6.71}$$

with the Gauss law

$$\partial_i E_a^i = j_a^0 - \hat{g}\varepsilon_{abc} A_b^k E_c^k \quad . \tag{6.72}$$

For $j \equiv 0$ and in temporal gauge, i.e. for $A_a^0 = 0$, we obtain from (6.71)

$$\begin{aligned} \partial_0 A_a^i &= -E_a^i \,, \\ \partial_0 B_a^i &= -\varepsilon_{ijk}\partial_j E_a^k - \hat{g}\varepsilon_{ijk}\varepsilon_{abc} E_b^j A_c^k \,, \\ \partial_0 E_a^i &= \varepsilon_{ijk}\partial_j B_a^k + \hat{g}\varepsilon_{ijk}\varepsilon_{abc} A_b^j B_c^k \quad . \end{aligned} \tag{6.73}$$

Shifting now $B_a^i$ to $-B_a^i$ and identifying $\hat{g}$ with $G_1$, we obtain equations (6.65).

Thus with (6.65) we have derived a classical Yang–Mills dynamics in temporal gauge, except for the non–abelian Gauß law. In Stumpf [Stu 86a] the Gauß law was introduced by the condition of relativistic invariance. In the meanwhile it was shown, cf. Magg [Magg 84] and Grimm [Gri 94], that the Gauß law may be proposed without contradictions to equations (6.65). From this it follows that with the additional assumption of the Gauß law the weak mapping produces a complete Yang–Mills dynamics in the temporal gauge. We add that with a suitable choice of the normalization constant $N$ we can achieve a coupling constant $G_1$ to be dimensionless whereas the fields $A, B, E$ get the usual dimensions. This will also be discussed in Section 6.6.

## 6.5 Quantum Properties of Mapped Fields

The quantum properties of the basic spinorfields $\Psi_{A\alpha}$ are expressed by the anticommutation relations of its auxiliary fields (1.42), namely by singular distributions. Due to the nonperturbative Pauli–Villars regularization the equal time anticommutator of the original spinorfield with its "canonical" conjugate field vanishes, i.e. we have, cf. (1.45),

$$\left[\Psi_{A\alpha}(\mathbf{r},0), \bar{\Psi}_{B\beta}(\mathbf{r}',0)\right]_+ = 0 \quad . \tag{6.74}$$

As in the final formulas of the mapping procedure always the regularization, i.e. the summation over the auxiliary fields is performed, one might suspect that by this regularization the quantum properties of the $\Psi$–field have been completely destroyed. However, this is not the case. To see this we observe that the quantum properties of a fermionic field $\chi$ can also be expressed by its equal time propagator $F_{12} := \langle 0|\mathcal{A}(\chi_1\chi_2)|0\rangle$. An example for this relation between propagator and anticommutator is the propagator equation (2.110) or its equal–time version, respectively. Furthermore by this relation the simultaneous appearence of $F$– and $A$–terms in the "quantization terms " of the spinorfield equations (3.106) or (4.1) and consequently in the effective theories, for instance (4.104), can be understood.

In order to investigate the quantum properties of $\Psi_{A\alpha}(\mathbf{r},t)$ we thus have to consider its equal–time propagator $F^t_{\hat{Z}_1\hat{Z}_2}(\mathbf{r}_1,\mathbf{r}_2) := \sum_{i_1,i_2} F^t_{Z_1Z_2}(\mathbf{r}_1,\mathbf{r}_2)$ with $\hat{Z} = (A,\alpha)$ where $F^t_{Z_1Z_2}$ is (approximately) given by (6.41).

One easily convinces oneself that $F^t_{\hat{Z}_1\hat{Z}_2} \neq 0$. This means that $\Psi_{A\alpha}(\mathbf{r},t)$ is a nonclassical field whose quantum properties are expressed by $F^t_{Z_1Z_2}$ instead of (6.74).

With respect to the appearance of quantization terms in the dynamical equations it is therefore obvious to consider the normalordered version of the dynamical equations, because normalordering explicitly introduces the propagators into the dynamical equations. This has already been illustrated by the spinorfield equations (3.106) or (4.1), respectively. In particular the normalordered version has to be used if the commutators or anticommutators are deformed, as for instance is the case in (6.74). Another case where in general such deformations have to be expected are the effective commutators or anticommutators for composite particles and fields.

According to these considerations the investigation of the quantum properties of mapped fields, i.e. in the present case of the $A$–, $B$– and $E$–fields, proceeds in two steps: First we will derive the effective commutation relations for the $A$– and $E$–fields, or more precisely the vacuum expectation values of such commutators. Second we will investigate the normalordered dynamical equations resulting from weak mapping in order to compare them with the phenomenological equations and discuss the quantum properties of the effective fields.

With respect to the weak mapping procedure we expect to discover local canonical commutation relations only in the low energy limit, in particular for vacuum expectation values (VEV), while for higher energies, i.e. for arbitrary matrix elements, formfactors will appear which regularize the commutators. As for the spinorfield, also for the boson field we expect that this regularization will be reflected in the normalordered equations.

Furthermore, in the phenomenological theory in temporal gauge the **A**– and **E**–fields are the canonical conjugate fields. Assuming the same for the mapped fields which are produced by the bound state construction, we thus have to calculate the VEV

$$\hat{\rho}(k_1, k_2) := \langle 0| \left[X_{a_1}^{l_1}(\mathbf{k}_1, t), X_{a_2}^{l_2}(\mathbf{k}_2, t)\right]_- |0\rangle \tag{6.75}$$

with $X = A, E$ and $k = (l, a, \mathbf{k})$.

Toward this end we consider the commutator

$$\begin{aligned}&[\psi_{I_1}\psi_{I_2}, \psi_{I_3}\psi_{I_4}]_- = \\ &\quad A_{I_2 I_3}\psi_{I_1}\psi_{I_4} + A_{I_2 I_4}\psi_{I_3}\psi_{I_1} - A_{I_1 I_3}\psi_{I_2}\psi_{I_4} - A_{I_1 I_4}\psi_{I_3}\psi_{I_2}\end{aligned} \tag{6.76}$$

where the fermionic anticommutator $A_{I_1 I_2}$ is given by (1.56).

We define the vacuum expectation of the left–hand side of (6.76) by $\hat{a}(I_1 \dots I_4) := \langle 0| [\psi_{I_1}\psi_{I_2}, \psi_{I_3}\psi_{I_4}]_- |0\rangle$. The quantity $\hat{a}$ now can be directly calculated by means of commutator rules and the relation $\psi_{I_1}\psi_{I_2} = \frac{1}{2}A_{I_1 I_2} + \mathcal{A}(\psi_{I_1}\psi_{I_2})$ to give

$$\hat{a}(I_1 \dots I_4) = A_{I_2 I_3}F^t_{I_1 I_4} + A_{I_2 I_4}F^t_{I_3 I_1} - A_{I_1 I_3}F^t_{I_2 I_4} - A_{I_1 I_4}F^t_{I_3 I_2} \cdot \tag{6.77}$$

To relate this fermion vacuum expectation value to boson expectation values we specialize the weak mapping transformation (6.5) to a pure boson transformation, i.e. we set $n = 0$. This gives

$$\varphi^{(2m)}(I_1 \dots I_{2m}|a) = C_{k_1}^{\{I_1 I_2} \dots C_{k_m}^{I_{2m-1} I_{2m}\}} \rho^{(m)}(k_1 \dots k_m|a) \quad , \tag{6.78}$$

where the $\rho^{(2m)}$ are the coefficient functions of the pure boson functional

$$|\mathcal{B}(b, a)\rangle = \sum_{n=0}^{\infty} \frac{1}{n!} \rho^{(n)}(k_1 \dots k_n|a) b_{k_1} \dots b_{k_n} |0\rangle_B \quad . \tag{6.79}$$

If we assume $|a\rangle \equiv |0\rangle$, (6.78) is a transformation between bosonic and fermionic vacuum expectation values. We consider the boson VEV $\rho^{(2)}(k_1, k_2|0)$ which according to (6.78) corresponds to $\varphi^{(4)}(I_1 \ldots I_4|0)$ on the level of the subfermions. This term is the first nonvanishing VEV of the subfermion field after the removal of the uncorrelated parts arising from subfermion two–point functions. Thus the relation

$$\varphi^{(4)}(I_1 \ldots I_4|0) = C_{k_1}^{\{I_1 I_2} C_{k_2}^{I_3 I_4\}} \rho^{(2)}(k_1, k_2|0) \tag{6.80}$$

is well–defined and nontrivial. If we resolve (6.80) for $\rho^{(2)}(k_1, k_2|0)$ we obtain

$$\rho^{(2)}(k_1, k_2|0) = R^{k_1}_{I_1 I_2} R^{k_2}_{I_3 I_4} \varphi^{(4)}(I_1 \ldots I_4|0) \quad . \tag{6.81}$$

Due to its linearity this equation can be also applied to the transformation between commutators. The VEV of the commutator on the subfermion level is defined by (6.77). Denoting its boson transform by $\hat{\hat{\rho}}(k_1, k_2)$ we have according to (6.81)

$$\hat{\hat{\rho}}(k_1, k_2) = R^{k_1}_{I_1 I_2} R^{k_2}_{I_3 I_4} \hat{a}(I_1, I_2, I_3, I_4) \tag{6.82}$$

where $\hat{\hat{\rho}}(k_1, k_2)$ must have the form (6.75), i.e. $\hat{\hat{\rho}}(k_1, k_2) \equiv \hat{\rho}(k_1, k_2)$. This can be verified by the introduction of intermediate states in $\hat{a}(I_1 \ldots I_4)$, i.e. by considering the identity

$$\hat{a}(I_1 \ldots I_4) = \sum_a \left( \langle 0|\psi_{I_1}\psi_{I_2}|a\rangle\langle a|\psi_{I_3}\psi_{I_4}|0\rangle - \langle 0|\psi_{I_3}\psi_{I_4}|a\rangle\langle a|\psi_{I_1}\psi_{I_2}|0\rangle \right) ,$$

where we assume positive metric of the mapped field theory in the one–quantum sector, which is identical with the positive metric of the two–subfermion bound states. If one again applies $\psi_{I_1}\psi_{I_2} = \frac{1}{2}A_{I_1 I_2} + \mathcal{A}(\psi_{I_1}\psi_{I_2})$ and observes $\langle 0|\mathcal{A}(\psi_{I_1}\psi_{I_2})|a\rangle = C^a_{I_1 I_2}$, then the duals in (6.82) project out a term of the form (6.75).

On the other hand a straightforward evaluation of $\hat{\rho}(k_1, k_2)$ by means of (6.77) yields in the low energy limit

$$\hat{\rho}(k_1, k_2) = \langle 0| \left[ E^k_a(\mathbf{k}), A^{k'}_{a'}(\mathbf{k}') \right]_- |0\rangle = -i\frac{\text{const}}{N^2}\delta_{kk'}\delta_{aa'}\delta(\mathbf{k} - \mathbf{k}'), \ \text{const} > 0, \tag{6.83}$$

while the $E$–$E$– and $A$–$A$–commutators vanish. With an appropriately chosen $N$, the Fourier transformation of (6.83) leads to

$$\langle 0| \left[\mathbf{E}_a(\mathbf{r}), \mathbf{A}_{a'}(\mathbf{r}')\right]_- |0\rangle = -i\mathbf{1}\delta_{aa'}\delta(\mathbf{r} - \mathbf{r}') \quad . \tag{6.84}$$

This is exactly the canonical commutation relation for the phenomenological $E$– and $A$–fields in temporal gauge, cf., e.g., Itzykson and Zuber [Itz 80], admittedly only for vacuum expectation values.

As already mentioned above for a more profound discussion of the quantum properties of mapped fields we need the normalordered version of the dynamical equations. The normalordering of nonabelian gauge fields, however, must be done with caution. As in the Hamilton formalism the **B**–field is not a canonical variable, but only an auxiliary field, its normalordering with **A**– and **E**–fields does not yield $c$–number functions. Although such difficulties can be circumvented by repeated application of Hausdorff formulas, the conventional normalordering transformation cannot be used. Thus, to avoid these difficulties, in a first step we eliminate the **B**–field from equations (6.60)–(6.62). We use (6.60) and $P_0 \equiv -i\partial/\partial t$ to rewrite (6.61) as

$$
\begin{aligned}
P_0 \partial^B_{i,a}(\mathbf{r},t)|\mathcal{B}\rangle &= \Big\{ P_0 \varepsilon_{ijk} \partial_k(\mathbf{r}) \partial^A_{j,a}(\mathbf{r},t) - \frac{1}{2} G_1 \varepsilon_{abc} \varepsilon_{ijk} P_0 \left[ \partial^A_{j,b}(\mathbf{r},t) \partial^A_{k,c}(\mathbf{r},t) \right] \\
&\quad + G_2 \varepsilon_{abc} \varepsilon_{ijk} P_0 P_0^{-1} b^E_{k,b}(\mathbf{r},t) \partial^A_{j,c}(\mathbf{r},t) \Big\} |\mathcal{B}\rangle \quad . \qquad (6.85)
\end{aligned}
$$

With $\partial_{l,a} \equiv \partial_{l,a}(\mathbf{r},t)$ etc., we can conclude from (6.85)

$$
\begin{aligned}
\partial^B_{i,j}|\mathcal{B}\rangle &= \Big\{ \varepsilon_{ijk} \partial_k \partial^A_{j,a} - \frac{1}{2} G_1 \varepsilon_{abc} \varepsilon_{ijk} \partial^A_{j,b} \partial^A_{k,c} \qquad (6.86) \\
&\quad + G_2 \varepsilon_{abc} \varepsilon_{ijk} P_0^{-1} \left[ b^E_{k,b} \partial^A_{j,c} \right] \Big\} |\mathcal{B}\rangle
\end{aligned}
$$

with a vanishing integration constant. For $G_2 = 0$ (6.86) corresponds to the phenomenological equation $\mathbf{B} = \partial \times \mathbf{A} - \frac{\hat{g}}{2} \mathbf{A} \times \mathbf{A}$.

If we observe that the coefficients of the original state functionals (3.73) fulfill the Heisenberg equation (3.78) for $t \neq 0$, we have

$$
\tau^{(n)}_{Z_1 \ldots Z_n}(\mathbf{r}_1 \ldots \mathbf{r}_n, t|p) = \exp(-iE_{p0}t) \tau^{(n)}_{Z_1 \ldots Z_n}(\mathbf{r}_1 \ldots \mathbf{r}_n|p) \qquad (6.87)
$$

and this also holds for the coefficients of the boson state functional (6.79). But then we can substitute $P_0^{-1} \equiv E_{p0}^{-1}$ in (6.86). So (6.86) is well–defined and can be substituted into (6.62). By again using (6.54) this yields

$$
\begin{aligned}
E_{p0}|\mathcal{B}\rangle &= \int d^3r \Big\{ -ib^A_{i,a} \partial^E_{i,a} + ib^E_{i,a} \varepsilon_{ijk} \partial_j \varepsilon_{lmk} \partial_l \partial^A_{m,a} \qquad (6.88) \\
&\quad + ib^E_{i,a} \varepsilon_{abc} \varepsilon_{ijk} \varepsilon_{lmk} \partial_j \left( \frac{G_1}{2} \partial^A_{l,b} + \frac{2G_2}{E_{k0}} b^E_{l,b} \right) \partial^A_{m,c} \\
&\quad + iG_1 b^E_{i,a} \varepsilon_{abc} \varepsilon_{ij} \varepsilon_{lmk} \partial_m \partial^A_{l,b} \partial^A_{j,c} \\
&\quad - iG_1 b^E_{i,a} \varepsilon_{abc} \varepsilon_{cde} \varepsilon_{ijk} \varepsilon_{lmk} \left( \frac{G_1}{2} \partial^A_{m,e} + \frac{2G_2}{E_{k0}} b^E_{m,e} \right) \partial^A_{l,d} \partial^A_{j,b} \Big\} |\mathcal{B}\rangle
\end{aligned}
$$

where in the following we restrict ourselves to the independent boson sources $b^A$ and $b^E$, i.e. we put $b^B = 0$ in $|\mathcal{B}\rangle$.

We now introduce the normaltransformed functionals by

$$
|\mathcal{N}(b;k)\rangle := \exp\left[ -\frac{1}{2} \int b^X_{\ell,a}(\mathbf{r}) F^{XX'}_{\substack{\ell\ell' \\ aa'}}(\mathbf{r}-\mathbf{r}') b^{X'}_{\ell',a'}(\mathbf{r}') d^3r d^3r' \right] |\mathcal{B}(b;k)\rangle \qquad (6.89)
$$

with $X = E, A$.

In principle the propagators $F^{XX'}_{\substack{\ell\ell' \\ aa'}}(\mathbf{r}-\mathbf{r}') := \langle 0|\mathcal{S}\left[ X^{\ell,a}(\mathbf{r}) X^{\ell',a'}(\mathbf{r}') \right]|0\rangle$ have to be selfconsistently calculated from (6.88). As for the commutator we expect $F^{XX'}$ in a low energy limit to be given by the phenomenological propagator; this will be justified by a comparison of the dynamical equations of the mapped and the phenomenological theories.

Analogously to Proposition 2.14 equation (6.88) is transformed into the following equation:

$$
\begin{aligned}
E_{k0}|\mathcal{N}(b;k)\rangle \;=\; & \int d^3r \Big\{ -ib^A_{i,a}(\mathbf{r}) d^E_{i,a}(\mathbf{r}) - ib^E_{i,a}(\mathbf{r}) \varepsilon_{ijk} \partial_j \varepsilon_{lmk} \partial_m d^A_{n,a}(\mathbf{r}) \\
& -ib^E_{i,a}(\mathbf{r}) \varepsilon_{abc} \varepsilon_{ijk} \varepsilon_{lmk} \partial_j \frac{G_1}{2} \left[ d^A_{m,b}(\mathbf{r}) + \frac{2G_2}{G_1 E_{k0}} b^E_{m,b}(\mathbf{r}) \right] d^A_{l,c}(\mathbf{r}) \\
& +iG_1 b^E_{i,a}(\mathbf{r}) \varepsilon_{abc} \varepsilon_{ijk} \varepsilon_{lmk} \partial_m d^A_{l,b}(\mathbf{r}) d^A_{j,c}(\mathbf{r}) \\
& +iG_1 b^E_{i,a}(\mathbf{r}) \varepsilon_{abc} \varepsilon_{cde} \varepsilon_{ijk} \varepsilon_{lmk} \frac{G_1}{2} \left[ d^A_{m,e}(\mathbf{r}) + 2\frac{G_2}{G_1 E_{k0}} b^E_{m,e}(\mathbf{r}) \right] \times \\
& d^A_{l,d}(\mathbf{r}) d^A_{j,b}(\mathbf{r}) \Big\} |\mathcal{N}(b;k\rangle
\end{aligned}
\tag{6.90}
$$

with

$$
d^X_{i,a}(\mathbf{r}) := \partial^X_{i,a}(\mathbf{r}) + \int d^3r' F^{XX'}_{\substack{ii' \\ aa'}}(\mathbf{r}-\mathbf{r}') b^{X'}_{i',a'}(\mathbf{r}') \quad . \tag{6.91}
$$

If we now define

$$
\int \hat{F}^{EX'}_{\substack{ii' \\ aa'}}(\mathbf{r}-\mathbf{r}') b^{X'}_{i',a'}(\mathbf{r}') d^3r' \;:=\; \int F^{EX'}_{\substack{ii' \\ aa'}}(\mathbf{r}-\mathbf{r}') b^{X'}_{i',a'}(\mathbf{r}') d^3r' \tag{6.92}
$$

$$
\begin{aligned}
\int \hat{F}^{AX'}_{\substack{ii' \\ aa'}}(\mathbf{r}-\mathbf{r}') b^{X'}_{i',a'}(\mathbf{r}') d^3r' \;:=\; & \int d^3r' b^{X'}_{i',a'}(\mathbf{r}') \times \\
& \left[ F^{AX'}_{\substack{ii' \\ aa'}}(\mathbf{r}-\mathbf{r}') + 2\frac{G_2}{G_1 E_{k0}} \delta_{EX'} \delta_{ii'} \delta_{jj'} \delta(\mathbf{r}-\mathbf{r}') \right]
\end{aligned}
\tag{6.93}
$$

we obtain the result:

$$
\begin{aligned}
E_{k0}|\mathcal{N}(b;k)\rangle \;=\; & \int d^3r \Big\{ -ib^A_{i,a}(\mathbf{r}) \hat{d}^E_{i,a}(\mathbf{r}) - ib^E_{i,a}(\mathbf{r}) \varepsilon_{ijk} \partial_j \varepsilon_{lmk} \partial_m d^A_{l,a}(\mathbf{r}) \\
& +ib^E_{i,a}(\mathbf{r}) \varepsilon_{abc} \varepsilon_{ijk} \varepsilon_{lmk} \partial_j \frac{G_1}{2} \hat{d}^A_{m,b}(\mathbf{r}) d^A_{l,c}(\mathbf{r}) \\
& -iG_1 b^E_{i,a}(\mathbf{r}) \varepsilon_{abc} \varepsilon_{ijk} \varepsilon_{lmk} \partial_m d^A_{l,c}(\mathbf{r}) d^A_{j,b}(\mathbf{r}) \\
& -i\frac{G_1^2}{2} b^E_{i,a}(\mathbf{r}) \varepsilon_{abc} \varepsilon_{cde} \varepsilon_{ijk} \varepsilon_{lmk} \hat{d}^A_{m,e}(\mathbf{r}) d^A_{l,d}(\mathbf{r}) d^A_{j,b}(\mathbf{r}) \Big\} |\mathcal{N}(b;k)\rangle
\end{aligned}
\tag{6.94}
$$

with

$$
\hat{d}^X_{l,a}(\mathbf{r}) := \partial^X_{l,a}(\mathbf{r}) + \int \hat{F}^{XX'}_{\substack{ll' \\ aa'}}(\mathbf{r}-\mathbf{r}') b^{X'}_{l',a'}(\mathbf{r}') d^3r' \quad . \tag{6.95}
$$

As we will show in Section 8.5, apart from further quantization terms the normalordered energy equations of the phenomenological Yang–Mills theory are identical with (6.94), if in (6.94) $\hat{F}$ is replaced by the ordinary propagator $F$. But $\hat{F}$ is equal to $F$, apart from deviations at the origin of the light cone. As the origin of the light cone is in the high energy range and the additional quantization terms of the phenomenological theory vanish outside the ligt cone, it follows that the phenomenological canonical quantized Yang–Mills theory and the boson theory which results from weak mapping are identical in the low energy range.

## 6.6 Effective Boson–Fermion Lagrangian

After having discussed the pure boson theory given by equation (6.22), we consider the full boson–fermion dynamics contained in equation (6.13). In order to elaborate the dynamical structure of the theory we will consider only the classical equations following from (6.13), i.e. from the beginning we omit those terms of (6.13) which contain quantization and correction terms. The "classical" terms will turn out to be $\mathcal{H}_{\mathrm{f}}$, $\mathcal{H}_{\mathrm{b}}$ and $\mathcal{H}_{\mathrm{bf}}$ given by (6.15), (6.16) and (6.18). The boson term $\mathcal{H}_{\mathrm{b}}$ was already evaluated in Section 6.4. Thus it remains to calculate the coupling terms $\mathcal{H}_{\mathrm{bf}}$.

This is done along the same lines as the calculation of the boson terms. With the boson states $C_k^{I_1 I_2}$ and their duals $R^k_{I_1 I_2}$ given by (6.31) and (6.39), the explicit form (6.41) of the propagator $F^t_{I_1 I_2}$ and with the decomposition (6.42) one eventually obtains in the low energy limit:

$$\begin{aligned}\mathcal{H}_{\mathrm{bf}} &= -iG_3 \sum_{i_1,i_2} \int d^3r\, b^F_{k0,a}(\mathbf{r})\partial^f_{\kappa_1\beta_1 i_1}(\mathbf{r})\,(T^a)^+_{\kappa_1\kappa_2}\,(C\gamma^k)_{\beta_1\beta_2}\,\partial^f_{\kappa_2\beta_2 i_2}(\mathbf{r}) \qquad (6.96)\\ &+ G_4 \sum_{i_1,i_2} \int d^3r\, \lambda_{i_1} f_{\kappa_1\alpha_1 i_1}(\mathbf{r}) L^a_{\kappa_1\kappa_2}(\gamma^0\gamma^k)_{\alpha_1\alpha_2}\partial^f_{\kappa_2\alpha_2 i_2}(\mathbf{r})\partial^A_{k,a}(\mathbf{r})\end{aligned}$$

with

$$G_3 = \frac{1}{N}\frac{5}{(2\pi)^{\frac{11}{2}}} g m^2 \quad , \qquad (6.97)$$

$$G_4 = N\frac{2(2\pi)^{\frac{7}{2}}}{5}\frac{g}{m^2} \quad , \qquad (6.98)$$

where the matrices $L^a := \frac{1}{2}T^a\gamma_5$ are four–dimensional representations of the $SU(2)$–generators and $N$ is a normalization constant which stems from the expansion functions (6.31)–(6.33).

We now consider the first terms of the functional state $|\mathcal{G}(b,j;k)\rangle$. According to (6.3) and the decomposition (6.42) $|\mathcal{G}(b,j;k)\rangle$ may be written as

$$\begin{aligned}|\mathcal{G}(b,j;k)\rangle &= \sum_{\nu<l,a}\int d^3r\, \rho^{l\nu}_a(\mathbf{r},t|k) b^F_{l\nu,a}(\mathbf{r})|v\rangle \qquad (6.99)\\ &+\sum_{l,a}\int d^3r\, \rho^l_a(\mathbf{r},t|k) b^A_{l,a}(\mathbf{r})|v\rangle\\ &+i\sum_Z \int d^3r \rho_Z(\mathbf{r},t|k) f_Z(\mathbf{r})|v\rangle + \ldots\end{aligned}$$

with $|v\rangle := |0\rangle_F \otimes |0\rangle_B$.

To obtain the connection to *classical* phenomenological quantities we interpret the coefficient functions of (6.99) as representing classical phenomenological matrix elements, i.e. we postulate

$$\rho^{l\nu}_a(\mathbf{r},t|k) \equiv F^{l\nu}_a(\mathbf{r},t) \quad , \qquad (6.100)$$

$$\rho^l_a(\mathbf{r},t|k) \equiv A^l_a(\mathbf{r},t) \quad ,$$

$$\rho_Z(\mathbf{r},t|k) \equiv \Psi_Z(\mathbf{r},t) \qquad (6.101)$$

etc., see also (6.63), (6.64).

In this way we are prepared to derive the (classical) field equations for $F$, $A$ and $\Psi$ or $E$, $B$, $A$ and $\Psi$, respectively. We multiply (6.13) from the left with $\langle v|\partial^F_{l\nu,a}(\mathbf{r})$, $\langle v|\partial^A_{l,a}(\mathbf{r})$, and $i^{-1}\langle v|\partial^f_Z(\mathbf{r})$ and observe that these projections vanish for all quantization terms in (6.13) and (6.14).

To evaluate the projections we substitute (6.15), (6.43), (6.47) and (6.96) into (6.14), make use of the commutation relations for the boson sources and their duals and observe the relation

$$E_{p0}|\mathcal{G}(b,j;p)\rangle = i\int d^3r b^X_k(\mathbf{r},t)\left[\frac{\partial}{\partial t}\partial^X_k(\mathbf{r},t)\right]|\mathcal{G}(b,j;p)\rangle + i\int d^3r j_Z(\mathbf{r},t)\left[\frac{\partial}{\partial t}\partial_Z(\mathbf{r},t)\right]|\mathcal{G}(b,j;p)\rangle \tag{6.102}$$

for $X = A, F$ and energy eigenstates $|p\rangle$, which is analogous to (6.54).

With $E^i_a(\mathbf{r},t) := F^{i0}_a(\mathbf{r},t)$ and $B^i_a := -\sum_{j<k}\varepsilon_{ijk}F^{jk}_a(\mathbf{r},t)$ we eventually obtain the following set of equations:

$$i\partial_t A^i_a(\mathbf{r},t) = -iE^i_a(\mathbf{r},t) \quad , \tag{6.103}$$

$$i\partial_t B^i_a(\mathbf{r},t) = -i\varepsilon_{ijk}\partial_k E^j_a(\mathbf{r},t) + iG_1\varepsilon_{abc}\varepsilon_{ijk}E^j_b(\mathbf{r},t)A^k_c(\mathbf{r},t) \quad , \tag{6.104}$$

$$\begin{aligned} i\partial_t E^i_a(\mathbf{r},t)|k &= -i\varepsilon_{ijk}\partial_j B^k_a(\mathbf{r},t) \\ &-iG_1\varepsilon_{abc}\varepsilon_{ijk}B^j_b(\mathbf{r},t)A^k_c(\mathbf{r},t) \\ &-iG_3\sum_{i_1,i_2} L^a_{\kappa_1\kappa_2}\gamma^i_{\alpha_1\alpha_2}\mathcal{A}\left[\Psi_{\kappa_1\alpha_1 i_1}(\mathbf{r},t)\Psi_{\kappa_2\alpha_2 i_2}(\mathbf{r},t)\right] \quad , \end{aligned} \tag{6.105}$$

$$\begin{aligned} \partial_t\Psi_{Z_1}(\mathbf{r},t) &= D^0_{Z_1Z_2}\left(D^k_{Z_2Z_3}\partial_k(\mathbf{r}) - m_{Z_2Z_3}\right)\Psi_{Z_3}(\mathbf{r},t) \\ &-D^0_{Z_1Z_2}U_{Z_2\{Z_3Z_4Z_5\}}\Psi_{Z_5}(\mathbf{r},t)\Psi_{Z_4}(\mathbf{r},t)\Psi_{Z_3}(\mathbf{r},t) \\ &-iG_4\sum_{i_2}\lambda_{i_1}L^a_{\kappa_1\kappa_2}(\gamma^0\gamma^k)_{\alpha_1\alpha_2}\Psi_{\kappa_2\alpha_2 i_2}(\mathbf{r},t)A^k_a(\mathbf{r},t) \end{aligned} \tag{6.106}$$

with $Z = (\kappa, \alpha, i)$. In particular for $G_3 = 0$ we obtain the pure boson equations (6.65) from Section 6.4.

If we compare these field equations with the phenomenological Yang–Mills equations in temporal gauge, we see from (6.105) that the fermion current couples to the bosons with the coupling constant $G_3$ given by (6.97) and from (6.106) that the bosons couple to the fermions with the coupling constant $G_4$ given by (6.98).

In order to get a local gauge invariant Lagrangian (cf. Section 8.5) for the field equations (6.103)–(6.106) we must have $G_3 = \frac{1}{2}G_4 =: \hat{G}$. Therefore, we require $N$ to satisfy the relation

$$\hat{G} = \frac{1}{N}\frac{5}{(2\pi)^{\frac{11}{2}}}gm^2 = N\frac{(2\pi)^{\frac{7}{2}}}{5}\frac{g}{m^2} \tag{6.107}$$

which yields

$$N = m^2\frac{5}{(2\pi)^{9/2}} \quad . \tag{6.108}$$

With this choice for $N$, the following Lagrangian corresponds to the classical field dynamics:

$$\mathcal{L} = \mathcal{L}_{\text{sf}}[\Psi] - \frac{1}{2}\text{tr}\,(F^{\mu\nu}F_{\mu\nu}) - \hat{G}\mathcal{A}\sum_{i_1,i_2}[\Psi_{\kappa_1\alpha_1 i_1}(\not{A}_a)_{\alpha_1\alpha_2}(L^a)_{\kappa_1\kappa_2}\Psi_{\kappa_2\alpha_2 i_2}] \quad , \tag{6.109}$$

where $\hat{G} = \frac{g}{\pi} \approx 0.3g$ and $\mathcal{L}_{\text{sf}}$ is the Lagrangian (1.33).

From (6.109) we see, that we have a dimensional boson–fermion coupling constant $\hat{G}$ and that due to the cross terms in the summation over the auxiliary field indices, we do not have a genuine gauge covariant boson–fermion coupling. However, it should be noted that this deviation from gauge invariance is merely an effect of regularization by auxiliary fields. If the unobservable auxiliary fields are removed, the gauge invariant coupling is restored. This can be demonstrated by a transition to observable fermion states which are, for instance, constituted by bound states of three subfermions. In this case it is possible to derive a genuine gauge covariant coupling of the $SU(2)$–vector bosons to three–particle bound states, cf. Stumpf [Stu 86b], which can be interpreted as leptons and Han–Nambu quarks. The boson–fermion coupling constant turns out to be dimensionless but a numerical value of this coupling constant can only be given with improved three–particle wavefunctions.

We turn to the coupling constant $G_1$ of the boson–boson interaction, which is given by (6.48). If we substitute the value of $N$, which is given by (6.108) and the value of $g$, which is given by (6.45) into (6.48), we get for the boson–fermion coupling constant $G_1$:

$$G_1 = \frac{40\pi}{21} \approx 6.0 \quad . \tag{6.110}$$

Obviously, our value for $G_1$ does not coincide with the experimental one which is 0.6. In fact we cannot expect to get the correct experimental value for $G_1$ if we fix $N$ in the unobservable elementary fermion sector. But we expect that if $N$ is fixed in the three–particle sector, not only the correct gauge covariant coupling appears, but also the coupling constant $G_1$ will be modified such that it will be in accordance with experiment.

Finally we want to discuss the units of the phenomenological fields and of the effective fields. In particular we want to show that there is no inconsistency in the identifications (6.100).

In the cgs–unit system we have three independent units, namely length, mass and time. Using "natural units" ($\hbar = 1,\ c = 1$) there is only one independent unit which we choose to be the mass unit. The units of all physical quantities thus can be described by one single number, the power of the mass unit. We denote it by $[m] = 1$, $[x_\mu] = -1$ etc.

In the phenomenological Yang–Mills theory we have

$$\Big[\mathcal{L}(x)\Big] = \Big[-\frac{1}{2}\text{tr}(F_{\mu\nu}F^{\mu\nu})\Big] = 4 \tag{6.111}$$

with $F^{\mu\nu} = \partial^\mu A^\nu - \partial^\nu A^\mu + i\hat{g}[A^\mu, A^\nu]_-$, from which we infer

$$[F_{\mu\nu}] = 2, \quad [A_\mu] = 1, \quad [\hat{g}] = 0 \quad . \tag{6.112}$$

In the weak mapping equations, the units of the field operators which correspond to the sources $b_k$ are identical with those of the phenomenological Yang–Mills theory if we appropriately normalize the wavefunctions $C_k^{I_1 I_2}$. Therefore we require the boson state $|k\rangle = |n, \mathbf{k}\rangle$ to be normalized in accordance with the covariant normalization prescription, cf., e.g., Lurié [Luri 68],

$$\langle n, \mathbf{k} | n', \mathbf{k}' \rangle = \delta_{nn'} \, 2\omega \, \delta(\mathbf{k} - \mathbf{k}') \quad , \tag{6.113}$$

with $\omega := (\mathbf{k}^2 + \mu^2)^{\frac{1}{2}}$, where $\mu$ is the boson mass and where we assume positive definiteness without explicit calculation of the norm. From (6.113) we have $[|k\rangle] = -1$ and therefore we can conclude from $C_k^{I_1 I_2} = \langle 0 | \mathcal{N}(\psi_{I_1} \psi_{I_2}) | k \rangle$ with $\big[|0\rangle\big] = 0$:

$$\left[C_{l,a}^{I_1 I_2}(\mathbf{k})\right] = 0, \quad \left[C_{l\nu,a}^{I_1 I_2}(\mathbf{k})\right] = -1 \quad . \tag{6.114}$$

The spinorfields $\psi_I$ have the units $[\psi] = \frac{1}{2}$ and from this and the definition of the one–time functional states it follows

$$[j_I] = 5/2 \, , \tag{6.115}$$

and with (6.115) and (6.114) we infer from (6.7):

$$\left[b_{l,a}^{A}(\mathbf{r})\right] = 2 \, , \quad \left[b_{l\nu,a}^{F}(\mathbf{r})\right] = 1 \quad . \tag{6.116}$$

From (6.116), (6.99), $[d^3 r] = -3$ and

$$\left[\,|\mathcal{G}(b, j; k)\rangle\,\right] = \left[|k\rangle\right] = -1 \quad , \tag{6.117}$$

we get

$$[F_{\mu\nu}] = 2 \, , \quad [A_\mu] = 1 \, , \quad [G] = 0 \tag{6.118}$$

i.e., the phenomenological correct dimensions.

Finally from $[G] = 0$, (6.48) and $[g] = 2$, which follows from (6.45), we get $[N] = 2$.

# 7 Fermions and Gravitation

## 7.1 Introduction

The gravitational force acts indiscriminately on any kind of matter or energy. Thus it also affects the material or energetic standards of length and time, and their deformation leads to a violation of the postulates of Euclidean geometry. Therefore the gravitational force is closely linked with geometry.

The first geometrical theory of gravitational force is due to Riemann [Rie 80], [Riem 97], cf. also Treder [Tre 78]. The presently accepted geometrical standard theory of gravitation is due to Einstein [Ein 15] and Hilbert [Hil 15], see also Mehra [Meh 73], who started from different premises but ended up with the same conclusions. With the discovery of the quantum formalism this geometrical theory of gravitation ran into difficulties, as the Hilbert–Einstein Lagrangian leads to a nonrenormalizable quantum field theory. Thus, for this purely formal reason, in the last decades numerous approaches were made to replace the Hilbert–Einstein Lagrangian by a Lagrangian of a renormalizable quantum field theory and to explain conventional gravity as a limiting case of such new theories. The most prominent representatives of such theories are supergravity, strings, superstrings, etc. Apart from the fact that none of these theories has ultimately succeeded, the most important problem is not the formal nonrenormalizability but rather the question: What does quantized geometry mean?

If geometrical gravitational theories are taken seriously, their extension into the quantum domain means that any microscopic process immediately changes the geometrical frame apart from the statistical fluctuations of the metric. This all penetrating flexibility of geometry is in sharp contrast to basic assumptions of quantum theory. In order to exist, quantum theory needs conservation laws providing the basis for quantum numbers. If by a totally flexible geometry all conservation laws are destroyed there are no quantum numbers and thus no quantum theory. On the other hand, quantum theories of microscopic phenomena are the best established theories that ever existed. Thus one has to conclude that geometrical theories of gravity are only compatible with quantum theory if they are referred to fixed frames of reference allowing the formation of conservation laws and quantum numbers. This does not mean total geometrical rigidity, because related with the material or energetic distributions one has to define "effective" metrics, but it means the primacy of quantum theory over flexible geometry.

This point of view was already emphasized by Eddington [Edd 49], and thoroughly discussed by Dehnen, Hönl and Westpfahl [Dehn 60]. Recently, essential contributions were made by Borszeszkowski and Treder [Borz 88] and by Audretsch, Hehl and Lämmerzahl [Aud 92], cf. also the comments of v. Weizsäcker [Weiz 85]. Apart from these basic discussions numerous proposals were made about gravitational theories with fixed (unobservable) background metric and observable effective metric. The drawback of such approaches is in general their ad hoc character. But by fixing the background metric, gravitation is formally described in terms of conventional (quantum) field theory and one can expect a structural incorporation of gravity into the family of the other elementary forces.

The first attempt in this direction was made by the bimetric theory of gravity, see Rosen [Ros 40a,b]. Thirring [Thir 59] discussed the introduction of effective metric in Lorentz covariant gravity theories, see also Sexl and Urbantke [Sexl 87], and from the numerous papers in this field we cite only some recent publications of Logunov [Log 93], Dehnen and Ghaboussi [Dehn 85] and Dehnen and Hitzer [Dehn 94], who formulated gravitational theory as a gauge theory in Minkowski space, and of Rodrigues and Souza [Rod 91].

An alternative systematic approach in this direction is fusion theory: All elementary forces including gravity are to be explained by the formation and interaction of elementary spin–1/2 fields. For relativistic particles this hypothesis was inaugurated by Eddington, see Jordan [Jor 28], who assumed the photon, i.e. the carrier of the electromagnetic forces, to be built up from a "positive" and a "negative" electron, i.e. from spin–1/2 particles. Subsequently de Broglie [Bro 34], [Bro 54] put forth the proposal that the photon is composed of a neutrino and an antineutrino and generalized this ansatz to include fusions of spin–1/2 particles to higher spin states. For instance, the graviton — the carrier of the gravitational force — was assumed to be a spin–2 state composed of four spin–1/2 fermions.

The numerous investigations of such states were performed by equations which described local fusions of spin–1/2 fields and led to free local composite one–particle states. Such equations are for instance the Bargmann–Wigner equations. Recently, there has been a renewed interest in the group theoretical analysis of graviton states resulting from Bargmann–Wigner equations. Rodriguez and Lorente [Rod 84] showed that such gravitons have properties which lead to vanishing curvature or trivial gravity. However, this difficulty can easily be removed if the graviton states are assumed to be multilocal. In addition, one needs nontrivial interactions. Both these requirements can be satisfied if one defines graviton states in the framework of a nonlinear spinorfield quantum theory.

The evaluation of such nonlinear spinorfield quantum theories depends on the ability to formulate an effective composite particle dynamics. A systematic treatment of composite particle dynamics for quantum fields was developed in the preceding chapters. We apply this theory to the problem of the interaction of gravitons with spin–1/2 fermions, i.e. to the derivation of the gravitational force exerted on elementary spin–1/2 fermions following the paper of Stumpf, Borne and Kaus [Stu 93b]. This problem is the most simple one that can be treated within the realm of spin fusion theory of gravity. A first outline of graviton selfinteraction was given in a preceding paper by Stumpf [Stu 88], where also the development of gravitational theories with respect to the above mentioned problems was reviewed. Compared with this previous paper, the discussion performed in this section is based on a more detailed evaluation of the wave functions being involved in this problem. The comparison with phenomenological theory is done by means of a lucid derivation of tetrad gravity and fermions given by Hehl [Hehl 70].

## 7.2 Anholonomic Spinor Connections

According to the preceding section in this chapter we discuss gravitation and its relation to composite particles or, more precisely, the coupling of fermions to gravitation which we interpret as a composite particle effect arising from nonlinear spinorfield dynamics. To prepare this derivation we sketch the classical theory of fermions and gravitation and refer for details to Hehl [Hehl 70]. We expect the reader to be familiar with the basic concepts of the conventional theory of general relativity. Furthermore we mention that, in contrast to the usage in the preceding chapters, in this chapter the indices $i, j, k = 0, 1, 2, 3$ are referred to holonomic coordinates.

The conventional theory of general relativity considers matter only as carrier of energy–momentum and can successfully describe the gravitational behaviour of macroscopic matter. As we are dealing with primarily microscopic phenomena we need an extension of general relativity which includes the spin as the second characteristic property of matter. We therefore assume space–time to be described by a Riemann–Cartan space $U_4$, which is a four–dimensional differentiable manifold with an affine connection

$$\Gamma^k_{ij} = \left\{ \begin{matrix} k \\ ij \end{matrix} \right\} - K_{ij}{}^k . \tag{7.1}$$

The contortion tensor $K_{ij}{}^k$ is given by

$$K_{ij}{}^k := -S_{ij}{}^k + S_j{}^k{}_i - S^k{}_{ij} \quad , \tag{7.2}$$

where Cartan's torsion tensor $S_{ij}{}^k$ is defined by the antisymmetric part of the connection:

$$S_{ij}{}^k := \Gamma^k_{\{ij\}} \, . \tag{7.3}$$

$\left\{ \begin{matrix} k \\ ij \end{matrix} \right\}$ is the Christoffel symbol of the second kind and it is given as a functional of the metrical tensor $g_{ij}(x)$.

The connection $\Gamma^k_{ij}$ describes the change of contravariant vector components $V^k$ under parallel shift along $dx^i$ by

$$dV^k = -\Gamma^k_{ij} V^j dx^i \, . \tag{7.4}$$

Furthermore the connection allows to define a covariant derivative $\overset{\Gamma}{\nabla}_k$ which will be specified later; here for physical reasons we postulate

$$\overset{\Gamma}{\nabla}_k \, g_{ij} = 0 \, . \tag{7.5}$$

Equation (7.5) means that the metric is invariant under parallel shift, i.e. the units of length and time do not change under parallel displacements and thus the local Minkowskian structure of Riemann–Cartan space is established.

A nonvanishing contortion tensor $K_{ij}{}^k$ equips the $U_4$ with rotational degrees of freedom which allow a geometrization of spin analogously to that of energy–momentum. The independent gravitational field quantities are the metric $g_{ij}(x)$ and the torsion $S_{ij}{}^k(x)$ which couple to the energy–momentum tensor or the spin tensor, respectively. For a more detailed discussion we again refer to Hehl [Hehl 70].

The conventional theory of general relativity is obtained by postulating the symmetry of the connection $\Gamma^k_{ij}$ in its lower indices; in this case torsion and contortion tensors vanish and the connection (7.1) describes a Riemannian geometry.

The symmetry group connected with the coordinates of the $U_4$ is that of general coordinate transformations. With respect to this group one can define tensors of any degree in a wellknown manner. Spinors, on the contrary, transform with a representation of the homogeneous Lorentz group which is referred to a pseudo–orthogonal coordinate system. Thus, for formulating spinors in a curved space we have to introduce anholonomic coordinates.

This is done by setting up a pseudo–orthogonal coordinate system at each point of the manifold. The four basis vectors $e_\mu{}^i(x)$, $\mu = 0, 1, 2, 3$, are called tetrads.

With inverse tetrads $e^\nu{}_j(x)$ we have the relations

$$e_\mu{}^i(x) e^\mu{}_j(x) = \delta^i_j \,, \qquad e_\mu{}^i(x) e^\nu{}_i(x) = \delta^\mu_\nu \tag{7.6}$$

and

$$g_{ij}(x) = \eta_{\mu\nu} e_i{}^\mu(x) e_j{}^\nu(x) \,. \tag{7.7}$$

With the help of these relations all tensor expressions can be formulated in anholonomic (tetrad) coordinates.

By means of the transformation properties of Dirac spinors the anholonomic covariant derivative of spinors $\Psi$ can be deduced:

$$\overset{\Gamma}{\nabla}_\mu \Psi(x) = \left[ \partial_\mu + \frac{i}{4} \Gamma_{\mu\rho\sigma}(x) \Sigma^{\rho\sigma} \right] \Psi(x) \tag{7.8}$$

where $\Sigma^{\rho\sigma} = \frac{i}{4}[\gamma^\mu, \gamma^\nu]_-$. The anholonomic connection $\Gamma_{\mu\rho\sigma}$ is given by

$$\Gamma_{\mu\rho\sigma} = -\Omega_{\mu\rho\sigma} + \Omega_{\rho\sigma\mu} - \Omega_{\sigma\mu\rho} - K_{\mu\rho\sigma} \tag{7.9}$$

with the anholonomic contortion tensor $K_{\mu\rho\sigma}$ and the object of anholonomity

$$\Omega_{\mu\rho\sigma} = \eta_{\sigma\lambda} e_\mu{}^i e_\rho{}^j \partial_{\{i} e^\lambda{}_{j\}} \,. \tag{7.10}$$

The connection is antisymmetric in the last two indices, i.e. $\Gamma_{\mu\rho\sigma} = \Gamma_{\mu\{\rho\sigma\}}$.

The Euler–Lagrange equations of the corresponding variational principle yield for the spinorfield coupled to the gravitational field via covariant derivative

$$\left[ i\gamma^\mu \partial_\mu - m - \frac{1}{4} \Gamma_{\mu\rho\sigma}(x) \gamma^\mu \Sigma^{\rho\sigma} \right] \Psi(x) = 0 \,. \tag{7.11}$$

Furthermore the parity transformed equation (7.11) can be shown to read

$$\left[ i\gamma^\mu \partial_\mu(x') - m - \frac{1}{4} \Gamma_{\mu\rho\sigma}(x') \Sigma^{\rho\sigma} \gamma^\mu \right] \Psi'(x') = 0 \,, \tag{7.12}$$

where $\Psi$ is the parity transformed spinor $\Psi'(x') = \eta_p \gamma^0 \Psi(x)$ with the phase $\eta_p$ and $x' = (x^0, -x^1, -x^2, -x^3)$.

Extending (7.12) to isospin and auxiliary field indices, observing that the charge conjugated spinor $\Psi^C_{A\alpha i}(x) := C_{\alpha\beta} \bar{\Psi}_{A\beta i}(x)$ also fulfills equation (7.12) and omitting the labels of the parity transformed quantities, one obtains for the superspinors $\Psi_Z(x)$, $Z = (A, \alpha, i, \Lambda)$, the equation of motion

$$\left[ D^\mu_{Z_1 Z_2} \partial_\mu - m_{Z_1 Z_2} - \frac{1}{4} \Gamma_{\mu\{\rho\sigma\}}(x) (\Sigma^{\rho\sigma} \gamma^\mu)_{Z_1 Z_2} \right] \Psi_{Z_2}(x) = 0 \tag{7.13}$$

with the definitions (1.59), (1.60) and

$$(\Sigma^{\rho\sigma}\gamma^{\mu})_{Z_1 Z_2} := (\Sigma^{\rho\sigma}\gamma^{\mu})_{\alpha_1\alpha_2}\delta_{A_1 A_2}\delta_{\Lambda_1\Lambda_2}\delta_{i_1 i_2} \,. \tag{7.14}$$

Equation (7.13) is the equation describing fermions in a gravitational field which we assert to arise from a weak mapping of spinorfields.

The appearance of torsion in Riemann–Cartan spaces has promoted attempts to formulate a dynamical theory of metric and torsion. These attempts led to the socalled Poincaré gauge theories of gravitation, which were started by Utiyama [Uti 56], Sciama [Sci 62] and Kibble [Kib 61]. As the tetrads and the connections can be considered as a set of ten independent Poincaré gauge potentials, Poincaré gauge theory is directly related to the quantum theory of fermions and therefore opens a way for experimental verification. A discussion of this topic would exceed the scope of this book, in particular as in this chapter only the fermion–boson coupling, i.e. the influence of gravitation on fermions, is studied.

Nevertheless one may hope that a microscopic theory of gravitation which is based on the model of this chapter may give some hints with respect to the special choice of such a gravitational theory including torsion. For a study of such theories we refer to the numerous contributions of Hehl, see for instance Hehl [Hehl 85].

## 7.3 Weak Mapping with Gravitons

According to the discussion in Section 7.1 we consider the appearance of the gravitational force as a composite particle effect. In this picture the gravitational force is excerted and mediated by spin 2 bosons which are assumed to be generated by four subfermion hard core bound states of the spinorfield. As a consequence the dynamical equations which characterize gravitational forces have to be derived as effective theories resulting from the spinorfield equations (1.1), (1.50) or (1.57), respectively. The means for the derivation of effective theories is weak mapping.

Exact weak mapping theorems were derived for the two–fermion hard core bound state dynamics in Chapter 4 and for the combined two–fermion and three–fermion hard core bound state dynamics by Kerschner [Ker 94]. These theorems lead to rather complicated effective theories. If exchange effects between the bound states can be approximately neglected the effective dynamics becomes essentially simplified and can be derived by a short–cut functional calculation technique, see Section 4.7. By this technique one can treat any kind of composite particle configurations. Here we apply it to gravitational boson mapping.

As we will see the weak mapping with four–(sub–)fermion bound states simultaneously enforces the introduction of dressed fermions instead of bare fermions. Hence in the simplest case we have at least to discuss a combined graviton–dressed fermion system. In this section we treat the formal aspects of its effective dynamics.

According to the short–cut calculation technique we define weak mapping by a transformation of the sources $j_I$ of the generating functional state (3.104) or (4.19), respectively. In full generality this transformation should deal with dressed particle states, the formal theory of which was developed in Section 4.8. Dressed particles are assumed to be solutions of the complete functional equation (4.1) and thus contain an infinite number of polarization cloud terms induced by a hard core wave function. To circumvent the difficulties connected with this infinite number of polarization cloud terms one is forced to consider only the lowest order in this expansion. But it should be emphasized that none of those terms is lost, which are omitted in the dressed particle ansatz. Rather, if exact weak mapping theorems are applied, such terms are shifted into effective selfinteractions and mutual interactions of the incompletely dressed particles. So an incomplete dressing does not automatically include approximation.

According to (4.138), (4.139) in the case under consideration the lowest order of the weak mapping transformation reads for gravitons

$$b_{4,k} := C_{4,k}^{I_1I_2I_3I_4} j_{I_1} j_{I_2} j_{I_3} j_{I_4} \tag{7.15}$$

and for fermions

$$f_{1,l} := C_{1,l}^{I_1} j_{I_1} + C_{1,l}^{I_1I_2I_3} j_{I_1} j_{I_2} j_{I_3} \tag{7.16}$$

and leads to source operators $b_{4,k}$ for hard core gravitons and $f_{1,l}$ for first order dressed fermions.

Except for a center of mass amplitude which will be specified in Section 7.4 the set of $C_{4,k}^{I_1I_2I_3I_4}$ is assumed to be a set of solutions of the corresponding four–fermion hard core equation which is defined as the diagonal part of equation (4.1), while the set of $C_{1,l}^{I_1}, C_{1,l}^{I_1I_2I_3}$ is assumed to be a complete set of first order solutions of the full equation (4.1) in the spin–1/2 and fermion number–1 sector of the states (3.104).

The indices $k, l$ denote the quantum numbers of the corresponding states. Furthermore all functions are taken on the hyperplane $t$ = const, i.e. according to our discussion in Section 2.7 we again work in the one–time formalism.

Apart from bound state solutions the set of states $C_{4,k}^{I_1I_2I_3I_4}$ also contains scattering solutions. Due to large fermion masses and the decoupling theorem (cf. Section 4.6) the scattering solutions will be suppressed in the final evaluation of weak mapping results. But for general considerations we refer to the complete set of $C_{4,k}^{I_1I_2I_3I_4}$.

To carry out weak mapping, in addition to the primary functions $C_{4,k}^{I_1I_2I_3I_4}$,$C_{1,l}^{I_1}$ and $C_{1,l}^{I_1I_2I_3}$ their duals $R_{I_1I_2I_3I_4}^{4,k}$, $R_{I_1}^{1,l}$ and $R_{I_1I_2I_3}^{1,l}$ are needed. These dual sets are defined by the orthonormality relations (4.127) for dressed particle states and lead in the lowest order approximation for graviton and fermion states to the relations

$$R_{I_1}^{1,l} C_{1,l'}^{I_1} + R_{I_1I_2I_3}^{1,l} C_{1,l'}^{I_1I_2I_3} = \delta_{ll'} \tag{7.17}$$

and

$$R_{I_1I_2I_3I_4}^{4,k} C_{4,k'}^{I_1I_2I_3I_4} = \delta_{kk'} \quad . \tag{7.18}$$

Furthermore the inverse relations to (7.15) and (7.16) read

$$j_{I_1} = \sum_l R_{I_1}^{1,l} f_{1,l} \quad , \tag{7.19}$$

$$j_{I_1} j_{I_2} j_{I_3} = \sum_l R_{I_1I_2I_3}^{1,l} f_{1,l} \quad , \tag{7.20}$$

$$j_{I_1} j_{I_2} j_{I_3} j_{I_4} = \sum_k R_{I_1I_2I_3I_4}^{4,k} b_{4,k} \quad . \tag{7.21}$$

In specializing the general formulae (4.127), (4.130) and (4.131) to (7.17)–(7.21) we have omitted all other kinds of particles which do not contribute to the graviton–fermion interaction. Furthermore we note that in the present case we do not need the graviton duals $R^{4,k}$ or the completeness of the graviton states $C_{4,k}$, respectively.

Next we assume that the generating functional state of the combined graviton–fermion effective theory is defined by

$$|\mathcal{G}(b,f;a)\rangle := \sum_{m,n} \frac{i^n}{n!} \frac{1}{m!} \Theta^{(m,n)}(k_1 \dots k_m; l_1 \dots l_n|a) b_{4,k_1} \dots b_{4,k_m} f_{1,l_1} \cdots f_{1,l_n} |v\rangle \tag{7.22}$$

with $|v\rangle := |0\rangle_B \otimes |0\rangle_F$. For the construction of functional spaces we refer to Sections 2.2, 3.7 and 6.2. Here and in the following we omit the index $t$, which indicates the hyperplane $t = \text{const}$. The transformation (7.15), (7.16) now induces a rearrangement of the original spinorfield functional state (3.104). This may be expressed by the invariance relation

$$|\mathcal{F}(j;a)\rangle = |\mathcal{G}(b,f;a)\rangle \, . \tag{7.23}$$

If we write the spinorfield energy equation (4.1) in the compact form

$$E_{po}|\mathcal{F}(j;p)\rangle = \mathcal{H}_{\rm F}(j,\partial)|\mathcal{F}(j;p)\rangle \quad , \tag{7.24}$$

then the transformations (7.15), (7.16) induce a transformation of the functional energy operator $\mathcal{H}_{\rm F}(j,\partial)$ and equation (7.24) is transformed into

$$E_{po}|\mathcal{G}(b,f;p)\rangle = \mathcal{H}_{\rm BF}\Big(b,f,\frac{\delta}{\delta b},\frac{\delta}{\delta f}\Big)|\mathcal{G}(b,f;p)\rangle \quad . \tag{7.25}$$

The transformation of $\mathcal{H}_{\rm F}$ is generated by the functional chain rule and by (7.19)–(7.21), apart from higher order polarization cloud terms. Analogously to (4.98) or (6.12) this chain rule reads

$$\partial_I|\mathcal{F}(j;a)\rangle = \left[(\partial_I b_{4,k})\frac{\delta}{\delta b_{4,k}} + (\partial_I f_{1,l})\frac{\delta}{\delta f_{1,l}}\right]|\mathcal{G}(b,f;a)\rangle \quad , \tag{7.26}$$

and repeated application yields $\partial_{I_2}\partial_{I_1}|\mathcal{F}(j;a)\rangle$ and $\partial_{I_3}\partial_{I_2}\partial_{I_1}|\mathcal{F}(j;a)\rangle$ in terms of boson and dressed fermion derivatives acting on $|\mathcal{G}(b,f;a)\rangle$. In lowest order the functional derivatives of $b_{4,k}$ and $f_{1,l}$ are

$$\partial_I b_{4,k} = 4C_{4,k}^{II_2I_3I_4} j_{I_2} j_{I_3} j_{I_4} \, , \tag{7.27}$$

$$\partial_I f_{1,l} = C_{1,l}^I \, , \tag{7.28}$$

and by repeated application of $\partial_I$ higher derivatives of $b_{4,k}$ and $f_{1,l}$ can be derived. In transforming $\mathcal{H}_{\rm F}(j,\partial)$ by means of (7.26)–(7.28) and corresponding higher derivatives, we omit all monomials $j_{I_1} \dots j_{I_n}$ for $n > 4$, as these monomials are connected with higher order polarization terms which are of importance for the boson selfinteraction, but not for the boson–fermion coupling under consideration.

For brevity we introduce the following symbols:

$$b_{4,k} \equiv b_k \, , \quad f_{1,l} \equiv f_l \, , \quad \frac{\delta}{\delta b_{4,k}} \equiv \partial_k^b \, , \quad \frac{\delta}{\delta f_{1,l}} \equiv \partial_l^f \, . \tag{7.29}$$

Using this notation we then obtain with (7.19)–(7.21)

$$
\begin{aligned}
E_{p0}|\mathcal{G}(f;p)\rangle \;=\; & \Big\{\hat{K}_{I_1I_2}\Big[4C^{I_2I_3I_4I_5}_{4,k'}R^{4,k}_{I_1I_3I_4I_5}b_k\partial^b_{k'} + C^{I_2}_{1,\ell'}R^{1,\ell}_{I_1}f_\ell\partial^f_{\ell'}\Big] \\
& +\hat{W}_{I_1I_2I_3I_4}\Big[36C^{I_3I_4K_1K_2}_{4,k}C^{I_2}_{1,\ell'}R^{1,\ell}_{I_1K_1K_2}f_\ell\partial^f_{\ell'}\partial^b_k \\
& -12C^{I_4K_1K_2K_3}_{4,k'}C^{I_3}_{1,\ell_1}C^{I_2}_{1,\ell_2}R^{4,k'}_{I_1K_1K_2K_3}\partial^f_{\ell_2}\partial^f_{\ell_1}b_k\partial^b_{k'} \\
& -C^{I_4}_{1,\ell_1}C^{I_3}_{1,\ell_2}C^{I_2}_{1,\ell_3}R^{1,\ell'}_{I_1}f_{\ell'}\partial^f_{\ell_2}\partial^f_{\ell_1}\partial^f_{\ell_1} \\
& -3F_{I_4K_1}\Big(12C^{I_2I_3K_2K_3}_{4,k}R^{4,k'}_{I_1K_1K_2K_3}b_{k'}\partial^b_k \\
& \qquad +C^{I_2}_{1,\ell_1}C^{I_3}_{1,\ell_2}R^{1,\ell'_1}_{I_1}R^{1,\ell'_2}_{K_1}f_{\ell'_1}f_{\ell'_2}\partial^f_{\ell_2}\partial^f_{\ell_1}\Big) \\
& +(3F_{I_4K_1}F_{I_3K_2}+\frac{1}{4}A_{I_4K_1}A_{I_3K_2})C^{I_2}_{1,\ell'}R^{1,\ell_1}_{I_1}R^{1,\ell_2}_{K_1}R^{1,\ell_3}_{K_2}f_{\ell_1}f_{\ell_2}f_{\ell_3}\partial^f_{\ell'} \\
& -(F_{I_4K_1}F_{I_3K_2}+\frac{1}{4}A_{I_4K_1}A_{I_3K_2})F_{I_2K_3}R^{1,\ell_1}_{I_1}R^{1,\ell_2}_{K_1}R^{1,\ell_3}_{K_2}R^{1,\ell_4}_{K_3}\times \\
& \qquad f_{\ell_1}f_{\ell_2}f_{\ell_3}f_{\ell_4}\Big]\Big\}|\mathcal{G}(b,f;p)\rangle \quad .
\end{aligned}
\tag{7.30}
$$

$\hat{K}$ and $\hat{W}$ are defined by (1.63) and (1.64), $F$ and $A$ are given by (6.41) and (1.56). We again mention that we have omitted the one–time index $t$.

In analogy to equation (6.54) and the arguments given there we apply the equation

$$
E_{po}|\mathcal{G}(b,f;p)\rangle = i\Big[b_k\frac{\partial}{\partial t}\partial^b_k + f_\ell\frac{\partial}{\partial t}\partial^f_\ell\Big]|\mathcal{G}(b,f;p)\rangle \tag{7.31}
$$

and go back to the original functional equations by collecting all terms linear in $b_k$ or $f_\ell$, respectively, and by fulfilling the corresponding expressions separately. This gives the boson equation

$$
\begin{aligned}
i\frac{\partial}{\partial t}\partial^b_k|\mathcal{G}\rangle \;=\; & 4\hat{K}_{I_1I_2}C^{I_2I_3I_4I_5}_{4,k'}R^{4,k}_{I_1I_3I_4I_5}\partial^b_{k'}|\mathcal{G}\rangle \\
& +\hat{W}_{I_1I_2I_3I_4}\Big\{-12C^{I_4K_1K_2K_3}_{4,k'}C^{I_3}_{1,\ell_1}C^{I_2}_{1,\ell_2}R^{4,k}_{I_1K_1K_2K_3}\partial^f_{\ell_2}\partial^f_{\ell_1}\partial^b_{k'} \\
& -36F_{I_4K_1}C^{I_2I_3K_2K_3}_{4,k'}R^{4,k}_{I_1K_1K_2K_3}\partial^b_{k'}\Big\}|\mathcal{G}\rangle \text{ ( + selfinteraction terms)}
\end{aligned}
\tag{7.32}
$$

and the fermion equation

$$
\begin{aligned}
i\frac{\partial}{\partial t}\partial^f_\ell|\mathcal{G}\rangle \;=\; & \hat{K}_{I_1I_2}C^{I_2}_{1,\ell'}R^{1,\ell}_{I_1}\partial^f_{\ell'}|\mathcal{G}\rangle \\
& +W_{I_1I_2I_3I_4}\Big\{36C^{I_3I_4K_1K_2}_{4,k}C^{I_2}_{1,\ell'}R^{1,\ell}_{I_1K_1K_2}\partial^f_{\ell'}\partial^b_k
\end{aligned}
\tag{7.33}
$$

$$-C^{I_4}_{1,\ell'_1}C^{I_3}_{1,\ell'_1}C^{I_2}_{1,\ell'_3}R^{1,\ell}_{I_1}\partial^f_{\ell'_2}\partial^f_{\ell'_1}\partial^f_{\ell'_1}$$

$$-3F_{I_4I_5}C^{I_2}_{1,\ell_1}C^{I_3}_{1,\ell_2}R^{1,\ell}_{I_1}R^{1,\ell'}_{I_5}f_{\ell'}\partial^f_{\ell_2}\partial^f_{\ell_1}$$

$$+(3F_{I_4K_1}F_{I_3K_2}+\frac{1}{4}A_{I_4K_1}A_{I_3K_2})R^{1,\ell}_{I_1}R^{1,\ell'_1}_{K_1}R^{1,\ell'_2}_{K_2}C^{I_2}_{1,\ell'_3}f_{\ell'_1}f_{\ell'_2}\partial^f_{\ell'_3}$$

$$-(F_{I_4K_1}F_{I_3K_2}+\frac{1}{4}A_{I_4K_1}A_{I_3K_2})F_{I_2K_3}R^{1,\ell}_{I_1}R^{1,\ell'_1}_{K_1}R^{1,\ell'_2}_{K_2}R^{1,\ell'_3}_{K_3}f_{\ell'_1}f_{\ell'_2}f_{\ell'_3}\Big\}|\mathcal{G}\rangle$$

We are not interested in the selfinteractions of dressed fermions. Rather we are exclusively interested in the covariant coupling of gravitons to fermions. Thus we study only the linear parts of equation (7.33)

$$\begin{aligned} i\frac{\partial}{\partial t}\partial^f_\ell|\mathcal{G}\rangle &= \hat{K}_{I_1I_2}C^{I_2}_{1,\ell'}R^{1,\ell}_{I_1}\partial^f_{\ell'}|\mathcal{G}\rangle \\ &\quad +36\hat{W}_{I_1I_2I_3I_4}C^{I_3I_4K_1K_2}_{4,k}C^{I_2}_{1,\ell'}R^{1,\ell}_{I_1K_1K_2}\partial^f_{\ell'}\partial^b_k|\mathcal{G}\rangle \end{aligned} \tag{7.34}$$

and evaluate them in the following sections.

## 7.4 Graviton States

To evaluate the graviton–fermion coupling term in equation (7.34) the graviton "wavefunctions" $C^{I_1I_2I_3I_4}_{4,k}$ and the dressed fermion "wavefunctions" $C^{I_1}_{1,\ell}$ and $C^{I_1I_2I_3}_{1,\ell}$ together with their duals $R^{1,\ell}_{I_1}$, $R^{1,\ell}_{I_1I_2I_3}$ are needed. With respect to the graviton states the calculations are easier because in contrast to fermion states their duals are not required. Hence we start with graviton state calculations.

The derivation of graviton states was performed in Section 5.4. We first recapitulate the calculations given there and then further evaluate the result.

Due to our suppression of all higher order polarization cloud terms for graviton states it is sufficient to take into account only their hard core parts. These hard core wavefunctions $C^{I_1I_2I_3I_4}_{4,k}$ are solutions of the diagonal part equation (5.1) in the four–fermion sector. Equation (5.1) reads

$$E^d_{p0}|\mathcal{F}(j^t;p)\rangle^d = [j^t_{I_1}\hat{K}_{I_1I_2}\partial^t_{I_2} - 3\hat{W}_{I_1I_2I_3I_4}F^t_{I_4I_5}j^t_{I_1}j^t_{I_5}\partial^t_{I_3}\partial^t_{I_2}]|\mathcal{F}(j^t;p)\rangle^d \tag{7.35}$$

with a corresponding "diagonal", i.e. hard core functional state $|\mathcal{F}(j^t;p)\rangle^d$.

Due to the algebraic degrees of freedom which are contained in the symbolic indices $I$ of equation (7.35), even for the low fermion numbers $n = 2,3,4$ equation (7.35) is rather complicated. However, these complications can be reduced if equation (7.35) is related to its covariant counterpart equation

$$K_{I_1I_2}\partial_{I_2}|\mathcal{F}(j;a)\rangle^d = 3W_{I_1I_2I_3I_4}F_{I_4K}j_K\partial_{I_3}\partial_{I_2}|\mathcal{F}(j;a)\rangle^d \tag{7.36}$$

with

$$
\begin{aligned}
K_{I_1 I_2} &= (D^{\mu}_{Z_1 Z_2}\partial_\mu(x_1) - m_{Z_1 Z_2})\delta(x_1 - x_2) \\
F_{I_1 I_2} &= i\lambda_{i_1}\delta_{i_1 i_2}(\gamma_5)_{\kappa_1\kappa_2}\left[(i\gamma^\rho\partial_\rho(x_1) + m_{i_1})C\right]_{\alpha_1\alpha_2} \times \\
&\qquad \Delta(x_1 - x_2, m_{i_1}) \\
W_{I_1 I_2 I_3 I_4} &= U_{Z_1\{Z_2 Z_3 Z_4\}_{as}}\delta(x_1 - x_2)\delta(x_1 - x_3)\delta(x_1 - x_4)
\end{aligned}
\tag{7.37}
$$

and covariant source operators $j_I \equiv j_Z(x), \partial_I \equiv \partial_Z(x)$, cf. Section 2.2.

The hard core functional states for $n = 4$ are defined by

$$
|\mathcal{F}(j^t; k, 4)\rangle^d := \frac{1}{4!}\varphi_t^{(4)}(I_1, I_2, I_3, I_4|k) j^t_{I_1} j^t_{I_2} j^t_{I_3} j^t_{I_4}|0\rangle_F \tag{7.38}
$$

or

$$
|\mathcal{F}(j; k, 4)\rangle^d := \frac{1}{4!}\varphi^{(4)}(I_1, I_2, I_3, I_4|k) j_{I_1} j_{I_2} j_{I_3} j_{I_4}|0\rangle_F \tag{7.39}
$$

respectively, where in (7.38) $I \equiv (Z, \mathbf{r})$, while in (7.39) $I \equiv (Z, x)$.

The relation between (7.36) and (7.37) and between (7.38) and (7.39) was established for $n = 2$ by Propositions 5.2 and 5.3. Their generalization to $n = 4$ is straightforward, so we do not explicitly discuss it.

The explicit solution of equation (7.36) can be achieved by application of a reduction technique developed by Schuler and Stumpf [Schu 67] and was performed in Proposition 5.5. We consider the four–dimensional Fourier transform $\tilde{\varphi}^{(4)}(q_1 \dots q_4|k)$ of $\varphi^{(4)}$ in (7.39). Then by means of this reduction technique, from (7.36) an integral equation for a reduced function $\tilde{\varphi}^{(4)}(q_1, q_3|)$ can be derived, and knowing this function the full $\tilde{\varphi}^{(4)}(q_1, q_2, q_3, q_4|k)$ can be reconstructed. However, even this contracted integral equation (5.65) is not exactly soluble.

To properly approximate the wavefunctions of (7.36) we only consider a submanifold of solutions. As we will investigate gravitational effects we are interested in particular in four–fermion spin–2 states which can be related to the tensor expressions of relativistic gravitational theory.

The restriction to spin–2 states suggests itself to imagine these states as being composed of two spin–1 vector boson states., i.e., we consider the four–fermion spin–2 state as a bound state of two vector bosons. In order to utilize this idea for an approximate calculation of the four–fermion states we further assume that in spite of their fusion the internal wave functions of the vector bosons are approximately conserved. In this case the remaining task is the calculation of the wavefunction for the center of mass motion of both bosons, i.e., the solution of the integral equation (7.36) or its contracted version, respectively, can be reduced to the calculation of this wavefunction.

In leading term approximation, i.e. with omission of internal $k$–dependence and internal angular momenta the wavefunctions of the single vector bosons are given in Section 5.3 by (5.42), and with $F_{\mu\nu} := A_\mu k_\nu$ the Fouriertransform can be written as

$$
\begin{aligned}
\tilde{\varphi}^{(2)}_{\substack{i_1 i_2 \\ \kappa_1 \kappa_2 \\ \alpha_1 \alpha_2}}(q_1, q_2) &= \frac{2ig_V}{(2\pi)^4} T^a_{\kappa_1\kappa_2}\delta(k - q_1 - q_2) \quad \times \\
&\left[A_\nu \hat{u}^{\nu}_{\substack{\alpha_1\alpha_2 \\ i_1 i_2}}(q_1 - q_2) + F_{\mu\nu}\hat{v}^{\mu\nu}_{\substack{\alpha_1\alpha_2 \\ i_1 i_2}}(q_1 - q_2)\right]
\end{aligned}
\tag{7.40}
$$

with

$$\hat{u}^{\nu}(p)_{\substack{\alpha_1\alpha_2\\i_1i_2}} = -\lambda_{i_1}\lambda_{i_2}(\frac{p^2}{4} - m_{i_1}m_{i_2})(\frac{p^2}{4} - m_{i_1}^2)^{-1}(\frac{p^2}{4} - m_{i_2}^2)^{-1}(\gamma^{\nu}C)_{\alpha_1\alpha_2}$$
$$=: A_{i_1i_2}(p)(\gamma^{\nu}C)_{\alpha_1\alpha_2} \quad , \tag{7.41}$$
$$\hat{v}^{\mu\nu}(p)_{\substack{\alpha_1\alpha_2\\i_1i_2}} = \frac{i}{2}\lambda_{i_1}\lambda_{i_2}(m_{i_1} + m_{i_2})(\frac{p^2}{4} - m_{i_1}^2)^{-1}(\frac{p^2}{4} - m_{i_2}^2)^{-1}(\Sigma^{\mu\nu}C)_{\alpha_1\alpha_2}$$
$$=: D_{i_1i_2}(p)(\Sigma^{\mu\nu}C)_{\alpha_1\alpha_2} \quad . \tag{7.42}$$

If two of these wavefunctions are combined to form a four–fermion bound state, we assume that the internal wavefunctions (7.41), (7.42) remain unchanged. Then we apply the following ansatz for the four–fermion spin–2 hard core states:

$$\begin{aligned}\tilde{\varphi}^{(4)}_{Z_1Z_2Z_3Z_4}(q_1,q_2,q_3,q_4|k) = N_4 g_V^2 T^a_{\kappa_1\kappa_2} T^b_{\kappa_3\kappa_4}\delta(k - q_1 - q_2 - q_3 - q_4) \times \\ \Big\{\hat{u}^{\nu}_{\substack{\alpha_1\alpha_2\\i_1i_2}}(q_1 - q_2)\hat{u}^{\mu}_{\substack{\alpha_3\alpha_4\\i_3i_4}}(q_3 - q_4)g_{\mu\nu}\Big[(q_1+q_2) - (q_3+q_4)\Big] \\ +\hat{u}^{\nu}_{\substack{\alpha_1\alpha_2\\i_1i_2}}(q_1 - q_2)\hat{v}^{\sigma\rho}_{\substack{\alpha_3\alpha_4\\i_3i_4}}(q_3 - q_4)\Gamma_{\nu\{\sigma\rho\}}\Big[(q_1+q_2) - (q_3+q_4)\Big] \\ +\hat{v}^{\nu\kappa}_{\substack{\alpha_1\alpha_2\\i_1i_2}}(q_1 - q_2)\hat{u}^{\mu}_{\substack{\alpha_3\alpha_4\\i_3i_4}}(q_3 - q_4)\bar{\Gamma}_{\{\nu\kappa\}\mu}\Big[(q_1+q_2) - (q_3+q_4)\Big] \\ +\hat{v}^{\nu\kappa}_{\substack{\alpha_1\alpha_2\\i_1i_2}}(q_1 - q_2)\hat{v}^{\sigma\rho}_{\substack{\alpha_3\alpha_4\\i_3i_4}}(q_3 - q_4)R_{\{\nu\kappa\}\{\sigma\rho\}}\Big[(q_1+q_2) - (q_3+q_4)\Big]\Big\} \; ,\end{aligned} \tag{7.43}$$

where the quantum numbers $k$ on the left–hand side of (7.43) include the four–momentum $k$ and the spin quantum number $s = 2$; for brevity we suppress $s = 2$ and the other quantum numbers in the following.

Substitution of (7.43) into the contracted integral equation leads to a set of integral equations for the calculation of the unknown boson–boson bonding tensor functions $g_{\mu\nu}(z)$, $\Gamma_{\nu\{\sigma\rho\}}(z)$, $\bar{\Gamma}_{\{\sigma\rho\}\nu}(z)$ and $R_{\{\nu\kappa\}\{\sigma\rho\}}(z)$, see Section 5.4, (5.77). In further reducing these integral equations we eliminate the spinor algebra by projection.

As for weak mapping we are mainly interested in this spinor algebra, we can use a further approximation for the calculation of the tensor functions. By applying the mean value theorem on the right–hand side of the integral equations (5.77) we obtain the set of equations (5.82) which after spin projection yield the following equations:

$$\begin{aligned} g_{\mu\nu}(p) &= -\frac{2ig_V}{(2\pi)^4}\int d^4z\, f(-\frac{p}{2} - z, -z)g_{\mu\nu}(p) \\ &+\frac{ig_V}{(2\pi)^4}\frac{A^{-1}}{2^{10}}\int d^4z\, A(\frac{p}{2} - z)A(\frac{p}{2} + z)Q^{\mu'\nu'}_{\nu\mu}\,\tilde{g}_{\mu'\nu'}(0|k) \quad ,\end{aligned} \tag{7.44}$$

$$\Gamma_{\nu\{\sigma\rho\}}(p) = \frac{ig_V}{(2\pi)^4 2^9}A^{-1}\int d^4z\, A(\frac{p}{2} - z)A(\frac{p}{2} + z)Q^{\nu'\sigma'\rho'}_{\nu\sigma\rho}\,\bar{\bar{\Gamma}}_{\{\nu'\sigma'\}\rho'}(0|k) \; , \tag{7.45}$$

$$\begin{aligned} \bar{\Gamma}_{\{\sigma\rho\}\nu}(p) &= -\frac{2ig_V}{(2\pi)^4}\int d^4z\, f(-\frac{p}{2} - z, -z)\bar{\Gamma}_{\{\sigma\rho\}\nu}(p) \\ &+\frac{ig_V}{(2\pi)^4}\frac{A^{-1}}{2^9}\int d^4z\, A(\frac{p}{2} - z)A(\frac{p}{2} + z)Q^{\nu'\sigma'\rho'}_{\nu\sigma\rho}\,\bar{\bar{\Gamma}}_{\{\nu'\sigma'\}\rho'}(0|k) \quad ,\end{aligned} \tag{7.46}$$

$$R_{\{\mu\nu\}\{\sigma\rho\}}(p) = \frac{ig_V}{(2\pi)^4}\frac{AD^{-2}}{2^{10}}\int d^4z\, A(\frac{p}{2}-z)A(\frac{p}{2}+z)Q^{\mu'\nu'}_{\mu\nu,\sigma\rho}\tilde{g}_{\mu'\nu'}(0|k) \tag{7.47}$$

with

$$\begin{aligned} A &:= \int d^4pA(2p)\,, &(7.48)\\ A(p) &:= \sum_{i_1,i_2} A_{i_1i_2}(p)\,, &(7.49)\\ D &:= \sum_{i_1,i_2}\int d^4pD_{i_1i_2}(2p)\,, &(7.50)\\ f(q,p) &:= \sum_{i_1,i_2}\lambda_{i_1}\lambda_{i_2}\frac{-pq+m_{i_1}m_{i_2}}{(p^2-m_{i_1}^2)(q^2-m_{i_2}^2)} &(7.51)\end{aligned}$$

and

$$\begin{aligned} Q^{\nu'\mu'}_{\nu\mu} &:= \eta^{\mu'}_{\nu}\eta^{\nu'}_{\mu} - \eta_{\nu\mu}\eta^{\mu'\nu'} + \eta^{\nu'}_{\nu}\eta^{\mu'}_{\mu}\,, &(7.52)\\ Q^{\nu'\sigma'\rho'}_{\nu\sigma\rho} &:= \eta^{\rho'}_{\sigma}\eta^{\sigma'}_{\rho}\eta^{\nu'}_{\nu}\Big|_{\substack{\mathrm{as}\{\sigma,\rho\}\\ \mathrm{as}\{\sigma',\nu'\}}}\,, &\\ Q^{\mu'\nu'}_{\mu\nu\sigma\rho} &:= \eta^{\nu'}_{\nu}\eta^{\mu'}_{\sigma}\eta_{\mu\rho}\Big|_{\substack{\mathrm{as}\{\sigma\rho\}\\ \mathrm{as}\{\mu\nu\}}}\,, &\end{aligned}$$

while the quantities

$$\tilde{X}(0|k) \in \left\{\tilde{g}_{\mu\nu}(0|k), \tilde{\Gamma}_{\nu\{\sigma\rho\}}(0|k), \tilde{\bar{\Gamma}}_{\{\sigma\rho\}\nu}(0|k), \tilde{R}_{\{\nu\mu\}\{\rho\sigma\}}(0|k)\right\} \tag{7.53}$$

are the bonding tensors at the origin in four–dimensional coordinate space given by (5.81). Due to the four–dimensional translational invariance of equation (7.36) the four–momentum $k$ is a good quantum number of the states (7.39) and therefore the set $\tilde{X}(0|k)$ has to depend on $k$. In Section 5.4 we argued that the $\tilde{X}(0|k)$ even depend only on the three–momentum $\mathbf{k}$, i.e. $\tilde{X}(0|k) \equiv \tilde{X}(0|\mathbf{k})$.

The bonding tensors $\tilde{X}(0|\mathbf{k})$ can be considered as the center of mass amplitudes of the graviton with momentum $\mathbf{k}$. If we resolve equations (7.44)–(7.47) with respect to $X(p|\mathbf{k})$ and substitute the resulting expressions into equation (7.43), then $\varphi^{(4)}(I_1, I_2, I_3, I_4|k)$ is represented by explicitly given functions and the center of mass amplitudes.

Similar to the effective boson theory in the last chapter we now have to determine those quantities which we regard as the dynamical fields of the effective theory. In our case it is obvious to consider the center of mass amplitudes (7.53) as these dynamical quantities. Thus we do not determine the bonding tensors $\tilde{X}(0|\mathbf{k})$ by their eigenvalue equations resulting from (7.36) or approximately from (7.44)–(7.47), but allow their unrestricted variability in order to fulfill the dynamical equations of the corresponding effective theory.

The fact that the center of mass amplitudes are considered as the dynamical fields of the effective theory enforces a discussion of their physical meaning. In the present case the $\tilde{X}(0|\mathbf{k})$ arise from spinorfield products. Therefore it is obvious to assume that $\tilde{X}(0|\mathbf{k})$ have to be identified with anholonomic quantities. For anholonomic coordinates the metrical tensor is $\eta_{\mu\nu}$. Hence, it in accordance with the background field formulation of Section 7.1 we consider $\tilde{g}_{\mu\nu}(0|\mathbf{k})$ as the effective field correction to the background metric $\eta_{\mu\nu}$, in anholonomic coordinates $\tilde{g}_{\mu\nu}(0|\mathbf{k})$ has to vanish, i.e. $\tilde{g}_{\mu\nu}(0|\mathbf{k}) \equiv 0$. This requirement is compatible with equation (7.44). But due to (7.47) it also enforces $R_{\mu\nu\sigma\rho} \equiv 0$. The latter consequence does not mean that $R_{\mu\nu\sigma\rho}$ vanishes at all, because it was only constructed in leading term approximation. Taking into account the contributions of orbital angular momenta, in equation (7.47) $R_{\mu\nu\sigma\rho}$ would couple to $\Gamma_{\nu\{\sigma\rho\}}$ and $\bar{\Gamma}_{\{\sigma\rho\}\nu}$ and thus would not vanish. However, compared with the leading terms the orbital angular momentum terms are much smaller. Therefore, from the beginning we do not take into account these $R$–terms in the weak mapping procedure.

So from all terms of $\tilde{X}(0|\mathbf{k})$ we have to take into account only the $\Gamma_{\nu\{\sigma\rho\}}$– and $\bar{\Gamma}_{\{\sigma\rho\}\nu}$–terms. But even these terms can still be reduced. So far we have done our calculation without antisymmetrization. If we antisymmetrize the ansatz (7.43) with the subsidiary conditions $g_{\mu\nu} = R_{\{\mu\nu\}\{\sigma\rho\}} = 0$, the remaining terms for $\Gamma_{\nu\{\sigma\rho\}}$ and $\bar{\Gamma}_{\{\sigma\rho\}\nu}$ are equal if $\tilde{\Gamma}_{\nu\{\sigma\rho\}} \equiv \tilde{\bar{\Gamma}}_{\{\sigma\rho\}\nu}$. Although there is a slight difference between equations (7.45) and (7.46), this difference does not alter the spin algebra. Hence for a first exploration we restrict ourselves to $\Gamma_{\nu\{\sigma\rho\}}$.

Under these assumptions we Fouriertransform (7.43) into coordinate space and obtain for equal times $t_1 = t_2 = t_3 = t_4 = 0$ the function

$$\begin{aligned}\varphi^{(4)}_{Z_1Z_2Z_3Z_4}(\mathbf{r}_1,\mathbf{r}_2,\mathbf{r}_3,\mathbf{r}_4|\mathbf{k}) = N_4 g_V^2 T^a_{\kappa_1\kappa_2} T^b_{\kappa_3\kappa_4} e^{-i\frac{\mathbf{k}}{4}(\mathbf{r}_1+\mathbf{r}_2+\mathbf{r}_3+\mathbf{r}_4)} \quad \times \\ \tilde{u}^{\nu}_{\substack{\alpha_1\alpha_2\\ i_1i_2}}\left(\frac{\mathbf{r}_1-\mathbf{r}_2}{2}\right)\tilde{v}^{\sigma\rho}_{\substack{\alpha_3\alpha_4\\ i_3i_4}}\left(\frac{\mathbf{r}_3-\mathbf{r}_4}{2}\right)\tilde{\Gamma}_{\nu\{\sigma\rho\}}\left(\frac{\mathbf{r}_1+\mathbf{r}_2}{4}-\frac{\mathbf{r}_3+\mathbf{r}_4}{4}\right) \quad .\end{aligned} \tag{7.54}$$

By means of (7.45), (7.50), (7.51) direct calculation yields

$$\tilde{\Gamma}_{\nu\{\sigma\rho\}}(\mathbf{r}) = w(\mathbf{r})\tilde{\Gamma}_{\nu\{\sigma\rho\}}(0|\mathbf{k}) \tag{7.55}$$

with the regular function $w(\mathbf{r})$ which for $mr \gg 1$ is given by

$$w(\mathbf{r}) = iz_0 g m^{-1} r e^{-4mr} \ , \ z_o \in \mathbb{R} \ , \tag{7.56}$$

and

$$\tilde{\Gamma}_{\nu\{\sigma\rho\}}(0|\mathbf{k}) = Q^{\nu'\sigma'\rho'}_{\nu\sigma\rho}\tilde{\bar{\Gamma}}_{\{\nu'\sigma'\}\rho'}(0|\mathbf{k}) \quad . \tag{7.57}$$

Furthermore, we assume $T^a = T^b = C$. This assumption follows from an investigation of the local limit $\mathbf{r}_1,\mathbf{r}_2,\mathbf{r}_3,\mathbf{r}_4 \to \mathbf{r}$ of the amplitudes (7.54) in Stumpf [Stu 88]. In this limit antisymmetrization enforces that choice of $T^a$ and $T^b$. From a physical point of view by this choice a universal coupling of the gravitation to all kinds of subfermions is attained. With the definitions

$$\begin{aligned}\tilde{u}^{\nu}(\mathbf{r})_{\substack{\alpha_1\alpha_2\\ i_1i_2}} &=: \tilde{A}_{i_1i_2}(\mathbf{r})(\gamma^{\nu}C)_{\alpha_1\alpha_2} \\ \tilde{v}^{\sigma\rho}(\mathbf{r})_{\substack{\alpha_3\alpha_4\\ i_3i_4}} &=: \tilde{D}_{i_3i_4}(\mathbf{r})(\Sigma^{\sigma\rho}C)_{\alpha_3\alpha_4}\end{aligned} \tag{7.58}$$

we then obtain for (7.54)

$$\varphi^{(4)}_{Z_1Z_2Z_3Z_4}(\mathbf{r}_1,\mathbf{r}_2,\mathbf{r}_3,\mathbf{r}_4|\mathbf{k}) = C^{\nu,\rho\sigma}_{Z_1Z_2Z_3Z_4}(\mathbf{r}_1,\mathbf{r}_2,\mathbf{r}_3,\mathbf{r}_4|\mathbf{k})\tilde{\Gamma}_{\nu\{\sigma\rho\}}(0|\mathbf{k}) \tag{7.59}$$

with

$$C^{\nu,\rho\sigma}_{Z_1Z_2Z_3Z_4}(\mathbf{r}_1,\mathbf{r}_2,\mathbf{r}_3,\mathbf{r}_4|\mathbf{k}) = N_4 g_V^2 C_{\kappa_1\kappa_2}C_{\kappa_3\kappa_4}(\gamma^\nu C)_{\alpha_1\alpha_2}(\Sigma^{\sigma\rho}C)_{\alpha_3\alpha_4}\times \quad (7.60)$$
$$e^{-\frac{i}{4}\mathbf{k}(\mathbf{r}_1+\mathbf{r}_2+\mathbf{r}_3+\mathbf{r}_4)}\tilde{A}_{i_1i_2}\left(\frac{\mathbf{r}_1-\mathbf{r}_2}{2}\right)\tilde{D}_{i_3i_4}\left(\frac{\mathbf{r}_3-\mathbf{r}_4}{2}\right)w\left(\frac{\mathbf{r}_1+\mathbf{r}_2}{4}-\frac{\mathbf{r}_3+\mathbf{r}_4}{4}\right).$$

Analogously to the vector boson case in Chapter 5 we choose $\tilde{\Gamma}_{\nu\{\rho\sigma\}}(0|\mathbf{k})$ to be the dynamical field of the effective gravitation theory. This determines the transformation functions of (7.15) to be given by the functions (7.60). Consequently the boson source operators of (7.15) are given by $b_{\nu,\rho\sigma}(\mathbf{k})$, and following the general rules of weak mapping we have to interpret the coefficient functions $\Theta$ of the functional (7.22) as matrix elements of the phenomenological field operators.

To show this we consider a special case of equation (7.23), namely

$$\frac{1}{4!}\varphi^{(4)}(I_1\dots I_4|\mathbf{k})j_{I_1}\dots j_{I_4}|0\rangle_F = \Theta^{(1,0)}(\mathbf{k}'|\mathbf{k})_{\nu,\sigma\rho}b^{4,\mathbf{k}'}_{\nu,\sigma\rho}|v\rangle \quad (7.61)$$

with

$$\Theta^{(1,0)}(\mathbf{k}'|\mathbf{k})_{\nu,\sigma\rho} \equiv \langle 0|\Gamma^{\mathrm{op}}_{\nu\{\sigma\rho\}}(\mathbf{k}')|\mathbf{k}\rangle\,,$$

where we suppressed the quantum numbers beside the momentum $\mathbf{k}$.

On the other hand we obtain with (7.59) and (7.15)

$$\begin{aligned}\frac{1}{4!}\varphi^{(4)}(I_1\dots I_4|\mathbf{k})j_{I_1}\dots j_{I_4}|0\rangle_F & \quad (7.62)\\ =\frac{1}{4!}C^{\nu,\rho\sigma}(I_1\dots I_4|\mathbf{k})\tilde{\Gamma}_{\nu\{\sigma\rho\}}(0|\mathbf{k})j_{I_1}\dots j_{I_4}|0\rangle_F &\\ =\frac{1}{4!}b^{4,\mathbf{k}}_{\nu,\sigma\rho}\tilde{\Gamma}_{\nu\{\sigma\rho\}}(0|\mathbf{k})|v\rangle \quad .&\end{aligned}$$

Comparison of (7.61) and (7.62) leads to

$$\Theta^{(1,0)}(\mathbf{k}'|\mathbf{k})_{\nu,\sigma\rho} = \frac{1}{4!}\tilde{\Gamma}_{\nu\{\sigma\rho\}}(0|\mathbf{k})\delta(\mathbf{k}'-\mathbf{k})\quad,$$

i.e. it is demonstrated that if the expansion functions of (7.15) are given by the functions (7.60), the matrix elements of the effective theory are represented by $\tilde{\Gamma}_{\nu\{\sigma\rho\}}(0|\mathbf{k})$. Of course this result does not mean that all higher effective correlation functions $\Theta^{(n,0)}(\mathbf{k}'_1\dots\mathbf{k}'_n|k)$ have to be represented by products of $\tilde{\Gamma}_{\nu\{\sigma\rho\}}(0|\mathbf{k})$. It simply expresses the fact that (7.60) is the appropriate transformation function and not (7.59).

Finally we want to emphasize that (7.60) has to be antisymmetrized. Due to the various subsymmetries and subantisymmetries the number of independent terms is reduced to six, as will be shown in Section 7.6.

## 7.5 Dressed Fermion State Calculations

The calculation of the (first order) dressed fermion states can be performed along the same lines as that for the boson states. Again we consider the covariant multi–time functional equation and derive the energy equation as well as its solutions by a limit to equal times. However, for dressed particles we need the full covariant functional equation (2.92). This reads in compact notation

$$K_{I_1 I_2}\partial_{I_2}|\mathcal{F}(j;l)\rangle = -W_{I_1 I_2 I_3 I_4} : d_{I_4} d_{I_3} d_{I_2} : |\mathcal{F}(j;l)\rangle \tag{7.63}$$

with

$$d_{I_1} = \partial_{I_1} - F_{I_1 I_2} j_{I_2} \tag{7.64}$$

and definitions (7.37).

In the lowest order of the covariant polarization cloud we represent the dressed fermion state by the functional

$$|\mathcal{F}(j;l)\rangle = i\varphi^{(1)}(I_1|l) j_{I_1}|0\rangle_F - \frac{i}{3!}\varphi^{(3)}(I_1, I_2, I_3|l) j_{I_1} j_{I_2} j_{I_3}|0\rangle_F \quad . \tag{7.65}$$

If we substitute (7.65) into (7.63) and project (7.63) into configuration space we obtain

$$(D^{\mu}_{I_1 I_2}\partial_\mu - m_{I_1 I_2})\varphi^{(1)}(I_2|l) = W_{I_1 I_2 I_3 I_4}\varphi^{(3)}(I_2, I_3, I_4|l) \tag{7.66}$$

and

$$(D^{\mu}_{I_1 I_2}\partial_\mu - m_{I_1 I_2})\varphi^{(3)}(K_1, K_2, I_2|l) = \tag{7.67}$$

$$3W_{I_1 I_2 I_3 I_4}\Big[F_{I_4 K_1} F_{I_3 K_2}\varphi^{(1)}(I_2|l) - F_{I_4 K_2}\varphi^{(3)}(K_1, I_2, I_3|l)\Big]_{\mathrm{as}\{K_1, K_2\}} \, , \tag{7.68}$$

where in (7.67) we suppressed the $\varphi^{(5)}$–term due to our three–source approximation in (7.65). In the same way, due to this approximation we also suppress all higher order equations for $\varphi^{(r)}$ with $r > 3$ which result from equation (7.63).

For a straightforward integration even these remaining equations (7.66) and (7.67) are too complicated. For a first inspection we apply an iteration procedure which we break off in the lowest order. We denote the lowest order functions by $\varphi_0^{(1)}$ and $\varphi_0^{(3)}$ and define the lowest order equations by

$$(D^{\mu}_{I_1 I_2}\partial_\mu - m_{I_1 I_2})\varphi_0^{(1)}(I_2|l) = 0 \tag{7.69}$$

and

$$(D^{\mu}_{I_1 I_2}\partial_\mu - m_{I_1 I_2})\varphi_0^{(3)}(K_1, K_2, I_2|l) = \tag{7.70}$$

$$3W_{I_1 I_2 I_3 I_4}\Big(F_{I_4 K_1} F_{I_3 K_2} - F_{I_4 K_2} F_{I_3 K_1}\Big)\varphi_0^{(1)}(I_2|l) \quad .$$

The other terms in equations (7.66) and (7.67) which do not appear in equations (7.69) and (7.70) are taken into account in the first iteration step.

Equations (7.69) and (7.70) can be directly integrated. While $\varphi_0^{(1)}$ is a Dirac superspinor, $\varphi_0^{(3)}$ is given by

$$\varphi_0^{(3)}(K_1, K_2, K_3) = N_3\Big[3G_{K_3 I_1} W_{I_1 I_2 I_3 I_4}(F_{I_4 K_1} F_{I_3 K_2} - F_{I_4 K_2} F_{I_3 K_1})\varphi_0^{(1)}(I_2)\Big]_{\mathrm{as}\{K_1 K_2 K_3\}} \tag{7.71}$$

with a normalization factor $N_3$.

So far the dressed fermions have been treated in Section 5.5. For the further evaluation we consider the Green function $G_{I_1 I_2}$ in (7.71), which reads

$$G_{I_1 I_2} = \frac{1}{(2\pi)^4}\delta_{\kappa_1\kappa_2}\delta_{i_1 i_2}\int (q_\mu\gamma^\mu - m_i)^{-1}_{\alpha_1\alpha_2} e^{iq(x_1-x_2)}dq \quad . \tag{7.72}$$

Due to the necessity of decoupling, the masses in (7.72) have to be very large. In this limit of large $m_i$ the integral in (7.72) can be approximately represented by

$$\frac{1}{(2\pi)^4}\lim_{m\to\infty}\int (q_\mu\gamma^\mu - m)^{-1}_{\alpha_1\alpha_2} e^{iq(x_1-x_2)}dq \approx -\lim_{m\to\infty}\frac{1}{m}\delta_{\alpha_1\alpha_2}\delta(x_1-x_2) \quad . \tag{7.73}$$

Hence we obtain

$$\lim_{m_i\to\infty} G_{I_1 I_2} \approx -\lim_{m_i\to\infty}\frac{1}{m_i}\delta_{\kappa_1\kappa_2}\delta_{\alpha_1\alpha_2}\delta_{i_1 i_2}\delta(x_1-x_2) \quad . \tag{7.74}$$

If we substitute (7.74) into (7.71) we can perform the integration and afterwards proceed to equal times $t_1 = t_2 = t_3 = 0$. Furthermore, we can represent the propagators in the equal time limit by their leading terms, cf. Pfister [Pfi 90]:

$$F^t_{Z_1 Z_2}(\mathbf{r}_1 - \mathbf{r}_2) \approx -(2\pi)^{-2}\lambda_{i_1} m_{i_1}\delta_{i_1 i_2}\gamma^5_{\kappa_1\kappa_2}C_{\alpha_1\alpha_2}\frac{1}{|\mathbf{r}_1 - \mathbf{r}_2|^2} \quad . \tag{7.75}$$

In general a subsequent auxiliary field summation produces a new order in leading terms. In the case of functions (7.75) we will, however, demonstrate that with respect to auxiliary field summation the leading term property of (7.75) is not destroyed. Hence this approximation is justified. If in addition the algebra is evaluated we obtain from (7.71) the expression

$$\begin{aligned} C^{I_1 I_2 I_3}_{1,l} = {} & N_3 g\lambda_{i_1}\lambda_{i_2}\lambda_{i_3} m_{i_1} m_{i_2} m_{i_3}^{-1}\frac{1}{|\mathbf{r}_3 - \mathbf{r}_1|^2}\frac{1}{|\mathbf{r}_3 - \mathbf{r}_2|^2} \times \\ & \Big\{\Big[-\delta_{\alpha_3\beta}C_{\alpha_1\alpha_2} + \gamma^5_{\alpha_3\beta}(\gamma^5 C)_{\alpha_1\alpha_2}\Big]\delta_{\kappa_3\epsilon}\gamma^5_{\kappa_1\kappa_2} \\ & +\Big[\delta_{\alpha_1\beta}C_{\alpha_3\alpha_2} - \gamma^5_{\alpha_1\beta}(\gamma^5 C)_{\alpha_3\alpha_2}\Big]\delta_{\kappa_1\epsilon}\gamma^5_{\kappa_3\kappa_2} \\ & +\Big[-\delta_{\alpha_2\beta}C_{\alpha_3\alpha_1} + \gamma^5_{\alpha_2\beta}(\gamma^5 C)_{\alpha_3\alpha_1}\Big]\gamma^5_{\kappa_3\kappa_1}\delta_{\kappa_2\epsilon}\Big\}\sum_{i'} C_{\beta\epsilon i'}(\mathbf{r}_3|s,a,j,\mathbf{k}) , \end{aligned} \tag{7.76}$$

where we have already identified the functions $\varphi_0^{(3)}(I_1, I_2, I_3|l)$ and $\varphi_0^{(1)}(I|l)$ with the expansion functions $C^{I_1 I_2 I_3}_{1,l}$ and $C^I_{1,l}$. Later on we will give a comment upon this identification. According to Section 5.6 the quantum numbers $l$ of the function $C^I_{1,l}$ can be explicitly written as the momentum $\mathbf{k}$, the spin quantum number $s$, the isospin–superspin $a$ and the auxiliary field index $j$. From the above iteration procedure, in particular from (7.76) it follows that the quantum numbers $l$ of the polarization cloud indeed are those of the corresponding basic hard core function, in our case of a Dirac super–isospinor.

In the form of (7.76) $C^{I_1 I_2 I_3}_{1,l}$ is not properly antisymmetrized. If one tries to antisymmetrize (7.76), the last two terms of (7.76) lead to wrong results because the arrangement $(\mathbf{r}_3, \alpha_3, \kappa_3, i_3)$ for the arguments of $C^I_{1,l}$ cannot be maintained. Hence we have to omit those terms. Then $C^{I_1 I_2 I_3}_{1,l}$ reads

$$\begin{aligned} C^{I_1 I_2 I_3}_{1,l} = {} & N_3 g\Big\{\left[C_{\alpha_1\alpha_2}\delta_{\alpha_3\alpha_4} - (\gamma^5 C)_{\alpha_1\alpha_2}\gamma^5_{\alpha_3\alpha_4}\right]\gamma^5_{\kappa_1\kappa_2}\delta_{\kappa_3\kappa_4} \times \\ & \lambda_{i_1}\lambda_{i_2}\lambda_{i_3} J\begin{pmatrix}\mathbf{r}_1\ \mathbf{r}_2\ \mathbf{r}_3 \\ i_1\ i_2\ i_3\end{pmatrix}\!\Bigg| l\Bigg)_{\alpha_4\kappa_4}\Big\}_{\mathrm{as}\{\mathbf{r}_1 Z_1, \mathbf{r}_2 Z_2, \mathbf{r}_3 Z_3\}} \end{aligned} \tag{7.77}$$

with

$$J\begin{pmatrix} \mathbf{r}_1\,\mathbf{r}_2\,\mathbf{r}_3 \\ i_1\; i_2\; i_3 \end{pmatrix}\!\!\left|l\right)_{\alpha\kappa} := -m_{i_1} m_{i_2} m_{i_3}^{-1} \frac{1}{|\mathbf{r}_3-\mathbf{r}_1|^2}\frac{1}{|\mathbf{r}_3-\mathbf{r}_2|^2}\sum_{i'} C_{\alpha\kappa i'}(\mathbf{r}_3|s,a,j,\mathbf{k})\ . \quad (7.78)$$

From (7.34) if follows that for the calculation of fermion–graviton coupling we need the dual state to (7.77). In Section 4.8 we have shown the general way to obtain the latter. Dual states can be calculated by means of left–hand side solutions of equation (4.1). We restrict ourselves to functional states (7.65) for the lowest order dressed one–fermion wavefunctions which in one–time limit read $C_{1,\ell}^{I_1}$ and $C_{1,\ell}^{I_1 I_2 I_3}$. Let $L_{I_1}^{1,\ell}, L_{I_1 I_2 I_3}^{1,\ell}$ be a set of corresponding "contravariant" or left–hand side wavefunctions, see Section 3.4, then we have the scalar product (4.135):

$$g_{1,\ell,1,\ell'} = C_{1,\ell}^{I_1} L_{I_1}^{*1,\ell'} + C_{1,\ell}^{I_1 I_2 I_3} L_{I_1 I_2 I_3}^{*1,\ell'} \quad , \qquad (7.79)$$

and by comparison with (7.17) we get

$$R_{I_1}^{1,\ell} = (g_{1,\ell,1,\ell'})^{-1} L_{I_1}^{*1,\ell'} \ , \qquad (7.80)$$

$$R_{I_1 I_2 I_3}^{1,\ell} = (g_{1,\ell,1,\ell'})^{-1} L_{I_1 I_2 I_3}^{*1,\ell'} \qquad (7.81)$$

if $g_{1,\ell,1,\ell'}$ is diagonal.

We do not directly calculate $L_{I_1}^{1,\ell}$ and $L_{I_1 I_2 I_3}^{1,\ell}$ but use the general formula for the scalar product (3.24)

$$g_{m,k,m',k'} = \sum_{n,\ell} C_{m,k}^{*I_1\ldots I_n} G^{-1}_{I_1\ldots I_n, I'_1\ldots I'_\ell} C_{m',k'}^{I'_1\ldots I'_\ell} \qquad (7.82)$$

in order to evaluate (7.80) and (7.81), i.e. to calculate $g_{1,\ell,1,\ell'}$ as well as $L_{I_1}^{1,\ell}$ and $L_{I_1 I_2 I_3}^{1,\ell}$.

The scalar product (7.82) is bilinear in the amplitudes $C_{m,k}$. In evaluating a bilinear expression we have to observe that in general for limiting processes $\lim(CGC) \neq (\lim C)G(\lim C)$ holds and that in addition the right–hand side may not exist at all. Hence we have to do our calculations ab initio, i.e. without performing limits like (7.74).

Specializing to the states $C_{1,l}^{I_1}, C_{1,l}^{I_1 I_2 I_3}$ we obtain by comparison of (7.82) with (7.79)

$$L_{K_1}^{*1,k} = C_{1,k}^{*I_1} G^{-1}_{I_1 K_1} + C_{1,k}^{*I_1 I_2 I_3} G^{-1}_{I_1 I_2 I_3, K_1} \qquad (7.83)$$

$$L_{K_1 K_2 K_3}^{*1,k} = C_{1,k}^{*I_1} G^{-1}_{I_1, K_1 K_2 K_3} + C_{1,k}^{*I_1 I_2 I_3} G^{-1}_{I_1 I_2 I_3, K_1 K_2 K_3}\ . \qquad (7.84)$$

The metrical tensor $G^{-1}$ can be calculated if one observes the iteration procedure (7.69)–(7.71). This procedure implies that for dressed fermions a definite representation is introduced, which is defined by the solutions of (7.69). In this representation the metrical tensor for normalordered basis states reads

$$G_{I_1,K_1} = \langle 0|\psi_{I_1}\psi_{K_1}^{+}|0\rangle \qquad (7.85)$$

$$G_{I_1,K_1 K_2 K_3} = \langle 0|\psi_{I_1}\mathcal{N}(\psi_{K_1}\psi_{K_2}\psi_{K_3})^{+}|0\rangle \qquad (7.86)$$

$$G_{I_1 I_2 I_3,K_1} = \langle 0|\mathcal{N}(\psi_{I_1}\psi_{I_2}\psi_{I_3})\psi_{K_1}^{+}|0\rangle \qquad (7.87)$$

$$G_{I_1 I_2 I_3,K_1 K_2 K_3} = \langle 0|\mathcal{N}(\psi_{I_1}\psi_{I_2}\psi_{I_3})\mathcal{N}(\psi_{K_1}\psi_{K_2}\psi_{K_3})^{+}|0\rangle \qquad (7.88)$$

and its evaluation yields according to (5.94), (5.117):

$$G_{I_1,K_1} = \frac{1}{2}\lambda_i\delta_{I_1K_1} \tag{7.89}$$

$$G_{I_1I_2I_3,K_1K_2K_3} = \frac{1}{8}\lambda_{i_1}\lambda_{i_2}\lambda_{i_3}\delta_{I_1K_1}\delta_{I_2K_2}\delta_{I_3K_3}|_{\mathrm{as}\{I_1,I_2,I_3\}}\ , \tag{7.90}$$

while the nondiagonal elements (7.86), (7.87) vanish. Thus, with (7.83), (7.84), formulae (7.80), (7.81) go over into

$$R^{1,\ell}_{I_1} = (g_{1,\ell,1,\ell'})^{-1}2\lambda_{i_1}^{-1}C^{*I_1}_{1,\ell'} \tag{7.91}$$

$$R^{1,\ell}_{I_1I_2I_3} = 8(g_{1,\ell,1,\ell'})^{-1}\lambda_{i_1}^{-1}\lambda_{i_2}^{-1}\lambda_{i_3}^{-1}C^{*I_1I_2I_3}_{1,\ell'}\ . \tag{7.92}$$

To derive $(g_{1,\ell,1,\ell'})^{-1}$ we substitute the explicit expression (7.77) into (7.82) which in this order reads

$$g_{1,\ell,1,\ell'} = C^{*I_1}_{1,\ell}G^{-1}_{I_1K_1}C^{K_1}_{1,\ell'} + C^{*I_1I_2I_3}_{1,\ell}G^{-1}_{I_1I_2I_3,K_1K_2K_3}C^{K_1K_2K_3}_{1,\ell'}\ . \tag{7.93}$$

Direct evaluation yields

$$g_{1,\ell,1,\ell'} = 2\lambda_j^{-1}\delta_{I_1K_1}C^{*I_1}_{1,\ell}C^{K_1}_{1,\ell'}\left(1+\frac{g^2}{m^4}\right) = 2\lambda_j\delta_{\ell\ell'}(1+\frac{g^2}{m^4}) \tag{7.94}$$

with $\ell = (\mathbf{k}, s, a, j)$. The contribution of the polarization cloud to the one–particle norm is therefore of the order $g^2/m^4$. We will neglect it according to Proposition 5.6.

Collecting the results we obtain for the dual function $R^{1,\ell}_{I_1I_2I_3}$ with (7.92), (7.94) and (7.77)

$$R^{1,\ell}_{I_1I_2I_3} = 4N_3g\lambda_j^{-1}\Big\{\left[C_{\alpha_1\alpha_2}\delta_{\alpha_3\alpha_4} - (\gamma^5C)_{\alpha_1\alpha_2}\gamma^5_{\alpha_3\alpha_4}\right]\gamma^5_{\kappa_1\kappa_2}\delta_{\kappa_3\kappa_4} \times \tag{7.95}$$
$$J\begin{pmatrix}\mathbf{r}_1 & \mathbf{r}_2 & \mathbf{r}_3 \\ i_1 & i_2 & i_3\end{pmatrix}\ell\Bigg)_{\alpha_4\kappa_4}\Big\}_{\mathrm{as}\{\mathbf{r}_1Z_1,\mathbf{r}_2Z_2,\mathbf{r}_3Z_3\}}\ ,$$

where $J\left(\begin{smallmatrix}\mathbf{r}_1 & \mathbf{r}_2 & \mathbf{r}_3 \\ i_1 & i_2 & i_3\end{smallmatrix}\middle|\ell\right)_{\alpha\kappa}$ is given by (7.78).

Finally it should be mentioned that in contrast to Section 7.4 in $\varphi_0^{(3)}$ no center of mass amplitude is separated. This corresponds to the choice of the dynamical fermionic quantities in our effective theory. As we already pointed out one of the rules of weak mapping is to identify the coefficient functions of the corresponding effective functionals, in the present case of (7.22), with expectation values of the dynamical quantities of the theory. In the fermionic sector these dynamical quantities are the Dirac super–isospinors $\varphi(I|l)$ or their projections onto another spinor $\varphi(I|l')$, respectively.

We illustrate the situation in analogy to the boson case (7.61), (7.62) by considering a special case in the pure fermion sector of (7.23), namely

$$i\varphi_t^{(1)}(I|k)j_I - \frac{i}{3!}\varphi_t^{(3)}(I_1,I_2,I_3|k)j_{I_1}j_{I_2}j_{I_3} = i\Theta^{(0,1)}(l|k)f_{1,l}\ . \tag{7.96}$$

We introduce a complete set of one–particle states $C^I_{1,l}$ and obtain

$$\begin{aligned} i\varphi_t^{(1)}(I|k)j_I - \frac{i}{3!}\varphi_t^{(3)}(I_1,I_2,I_3|k)j_{I_1}j_{I_2}j_{I_3} &= iC^I_{1,l}\Theta^{(0,1)}(l|k)R^{1,l'}_I f_{1,l'} \\ &=: i\tilde{\Theta}^{(0,1)}(I|k)\tilde{f}_{1,I} \end{aligned} \tag{7.97}$$

with the "Fourier transforms" $\tilde{\Theta}$ and $\tilde{f}$, which are now related to the coordinate space, and we interpret $\tilde{\Theta}$ as expectation values of Dirac super–isospinors in coordinate space, i.e.

$$\tilde{\Theta}^{(0,1)}(I|k) \equiv \langle 0|\psi_I|k\rangle \,. \tag{7.98}$$

On the other hand according to (7.16) we have in lowest order

$$\begin{aligned} & i\varphi_t^{(1)}(I|k)j_I - \frac{i}{3!}\varphi_t^{(3)}(I_1, I_2, I_3|k)j_{I_1}j_{I_2}j_{I_3} && (7.99)\\ = \; & iC_k^I j_I + iC_k^{I_1,I_2,I_3} j_{I_1}j_{I_2}j_{I_3} && (7.100)\\ = \; & if_{1,k} \\ = \; & iC_{1,k}^I \tilde{f}_{1,I} \quad . \end{aligned}$$

Comparing this with (7.97) and (7.98) we have

$$\tilde{\Theta}^{(0,1)}(I|k) \equiv C_{1,k}^I \equiv \langle 0|\psi_I|k\rangle \quad . \tag{7.101}$$

Thus the choice of $\varphi_t^{(3)}$ for the weak mapping functions $C$ is consistent with the interpretation (7.98).

As among the superspinor quantum numbers also the superspinor mass $m_i$ occurs, the above argumentation is only correct if it is accompanied with a mass renormalization. However, in the present case mass renormalization can be neglected, as the superspinor field is superrenormalizable with finite mass corrections which are absorbed into the practically infinite original masses in the limit of very large $m_i$.

## 7.6 Fermion–Graviton Coupling

The general expression for fermion–graviton coupling is given by the last term of equation (7.34). By means of the wavefunctions of the last two Sections we can evaluate this term in detail. In particular we use the graviton wavefunctions (7.60) and the dual of the dressed fermion wavefunction (7.95). Both of them have to be completely antisymmetrized. We write the mapping term in the following form:

$$\begin{aligned} & \hat{W}_{I_5 I_1 I_2 I_6} C_k^{I_1 I_2 I_3 I_4} R^{1,l}_{I_5 I_3 I_4} C^{I_6}_{1,l'} \partial^f_{1,l'} \partial^b_k = \int d^3r_1 \ldots d^3r_6 d^3K d^3k' \; \times && (7.102)\\ & C^{\mu,\rho\lambda}\left(\begin{array}{cccc|c} \mathbf{r}_1 & \mathbf{r}_2 & \mathbf{r}_3 & \mathbf{r}_4 & \\ \alpha_1 & \alpha_2 & \alpha_3 & \alpha_4 & \\ \kappa_1 & \kappa_2 & \kappa_3 & \kappa_4 & \mathbf{K} \\ i_1 & i_2 & i_3 & i_4 & \end{array}\right) R\left(\begin{array}{ccc|c} \mathbf{r}_5 & \mathbf{r}_3 & \mathbf{r}_4 & \mathbf{k} \\ \alpha_5 & \alpha_3 & \alpha_4 & s \\ \kappa_5 & \kappa_3 & \kappa_4 & a \\ i_5 & i_3 & i_4 & j \end{array}\right) C\left(\begin{array}{c|c} \mathbf{r}_6 & \mathbf{k}' \\ \alpha_6 & s' \\ \kappa_6 & a' \\ i_6 & j' \end{array}\right) \times \\ & \partial^f_{s'a'j'}(\mathbf{k}')\partial^b_{\mu,\rho\lambda}(\mathbf{K}) i^2 g \lambda_{i_5} B^{i_1 i_2 i_6} \gamma^0_{\alpha_5\alpha} V^{\alpha\alpha_1\alpha_2\alpha_6}_{\kappa_5\kappa_1\kappa_2\kappa_6} \delta(\mathbf{r}_5 - \mathbf{r}_1)\delta(\mathbf{r}_5 - \mathbf{r}_2)\delta(\mathbf{r}_5 - \mathbf{r}_6) \quad , \end{aligned}$$

where we keep in mind the abbreviations $I = (\alpha, \kappa, i, \mathbf{r})$ and $l = (s, a, j, \mathbf{k})$.

Substitution of the antisymmetrized wavefunctions (7.60) and (7.95) into (7.102) yields the expression

$$(7.102) \equiv \sum_{i_1 \dots i_6} \int d^3r_3 d^3r_4 d^3r_5 d^3K d^3k' \, e^{\frac{i}{4}\mathbf{K}(2\mathbf{r}_5+\mathbf{r}_3+\mathbf{r}_4)} \frac{N_4 g_V^2}{3!} \quad \times \tag{7.103}$$

$$\Big\{ C_{\kappa_1\kappa_2} C_{\kappa_3\kappa_4} (\gamma^\mu C)_{\alpha_1\alpha_2} (\Sigma^{\sigma\lambda} C)_{\alpha_3\alpha_4} A_{i_1 i_2}(0) D_{i_3 i_4} \left(\frac{\mathbf{r}_3 - \mathbf{r}_4}{2}\right) w \left(\frac{2\mathbf{r}_5 - \mathbf{r}_3 - \mathbf{r}_4}{4}\right)$$

$$+ C_{\kappa_3\kappa_1} C_{\kappa_2\kappa_4} (\gamma^\mu C)_{\alpha_3\alpha_1} (\Sigma^{\sigma\lambda} C)_{\alpha_2\alpha_4} A_{i_3 i_1} \left(\frac{\mathbf{r}_5 - \mathbf{r}_3}{2}\right) D_{i_2 i_4} \left(\frac{\mathbf{r}_5 - \mathbf{r}_4}{2}\right) w \left(\frac{\mathbf{r}_3 - \mathbf{r}_4}{4}\right)$$

$$+ C_{\kappa_1\kappa_4} C_{\kappa_2\kappa_3} (\gamma^\mu C)_{\alpha_1\alpha_4} (\Sigma^{\sigma\lambda} C)_{\alpha_2\alpha_3} A_{i_1 i_4} \left(\frac{\mathbf{r}_5 - \mathbf{r}_4}{2}\right) D_{i_2 i_3} \left(\frac{\mathbf{r}_5 - \mathbf{r}_3}{2}\right) w \left(\frac{\mathbf{r}_4 - \mathbf{r}_3}{4}\right)$$

$$+ C_{\kappa_3\kappa_4} C_{\kappa_1\kappa_2} (\gamma^\mu C)_{\alpha_3\alpha_4} (\Sigma^{\sigma\lambda} C)_{\alpha_1\alpha_2} A_{i_3 i_4} \left(\frac{\mathbf{r}_3 - \mathbf{r}_4}{2}\right) D_{i_1 i_2}(0) w \left(\frac{\mathbf{r}_3 + \mathbf{r}_4 - 2\mathbf{r}_5}{4}\right)$$

$$+ C_{\kappa_3\kappa_1} C_{\kappa_2\kappa_4} (\gamma^\mu C)_{\alpha_2\alpha_4} (\Sigma^{\sigma\lambda} C)_{\alpha_3\alpha_1} A_{i_2 i_4} \left(\frac{\mathbf{r}_5 - \mathbf{r}_4}{2}\right) D_{i_1 i_3} \left(\frac{\mathbf{r}_5 - \mathbf{r}_3}{2}\right) w \left(\frac{\mathbf{r}_4 - \mathbf{r}_3}{4}\right)$$

$$+ C_{\kappa_2\kappa_3} C_{\kappa_1\kappa_4} (\gamma^\mu C)_{\alpha_2\alpha_3} (\Sigma^{\sigma\lambda} C)_{\alpha_1\alpha_4} A_{i_2 i_3} \left(\frac{\mathbf{r}_5 - \mathbf{r}_3}{2}\right) D_{i_1 i_4} \left(\frac{\mathbf{r}_5 - \mathbf{r}_4}{2}\right) w \left(\frac{\mathbf{r}_3 - \mathbf{r}_4}{4}\right) \Big\}$$

$$\times \frac{1}{3!} \sum_h g\lambda_{i_5} \gamma^0_{\alpha_5\alpha} \Big\{ v^h_{\alpha\alpha_1} (v^h C)_{\alpha_2\alpha_6} \delta_{\kappa_5\kappa_1} \gamma^5_{\kappa_2\kappa_6}$$

$$+ v^h_{\alpha\alpha_2} (v^h C)_{\alpha_1\alpha_6} \delta_{\kappa_5\kappa_2} \gamma^5_{\kappa_1\kappa_6}$$

$$+ v^h_{\alpha\alpha_6} (v^h C)_{\alpha_1\alpha_2} \delta_{\kappa_5\kappa_6} \gamma^5_{\kappa_1\kappa_2} \Big\} \times \quad (-1) N_3 g \lambda_{j'}^{-1} \frac{4}{3!} \sum_{h=1}^{2} \quad \times$$

$$\Big\{ \gamma^5_{\kappa_5\kappa_3} \delta_{\kappa_4\kappa_7} (v^h C)_{\alpha_5\alpha_3} (v^h)_{\alpha_7\alpha_4} \left[ J^* \begin{pmatrix} \mathbf{r}_5 & \mathbf{r}_3 & \mathbf{r}_4 \\ i_5 & i_3 & i_4 \end{pmatrix} l \Big)_{\alpha_7\kappa_7} + J^* \begin{pmatrix} \mathbf{r}_3 & \mathbf{r}_5 & \mathbf{r}_4 \\ i_3 & i_5 & i_4 \end{pmatrix} l \Big)_{\alpha_7\kappa_7} \right]$$

$$+ \gamma^5_{\kappa_3\kappa_4} \delta_{\kappa_5\kappa_7} (v^h C)_{\alpha_3\alpha_4} (v^h)_{\alpha_7\alpha_5} \left[ J^* \begin{pmatrix} \mathbf{r}_3 & \mathbf{r}_4 & \mathbf{r}_5 \\ i_3 & i_4 & i_5 \end{pmatrix} l \Big)_{\alpha_7\kappa_7} + J^* \begin{pmatrix} \mathbf{r}_4 & \mathbf{r}_3 & \mathbf{r}_5 \\ i_4 & i_3 & i_5 \end{pmatrix} l \Big)_{\alpha_7\kappa_7} \right]$$

$$+ \gamma^5_{\kappa_4\kappa_5} \delta_{\kappa_3\kappa_7} (v^h C)_{\alpha_4\alpha_5} (v^h)_{\alpha_7\alpha_3} \left[ J^* \begin{pmatrix} \mathbf{r}_4 & \mathbf{r}_5 & \mathbf{r}_3 \\ i_4 & i_5 & i_3 \end{pmatrix} l. \Big)_{\alpha_7\kappa_7} + J^* \begin{pmatrix} \mathbf{r}_5 & \mathbf{r}_4 & \mathbf{r}_3 \\ i_5 & i_4 & i_3 \end{pmatrix} l \Big)_{\alpha_7\kappa_7} \right] \Big\} \times$$

$$C_{\alpha_6\kappa_6 i_6}(\mathbf{r}_5|\mathbf{k}', s', a', j') \partial^f_{s'a'j'}(\mathbf{k}') \partial^b_{\mu,\sigma\lambda}(\mathbf{K})$$

where we have to observe $l = (\mathbf{k}, s, a, j)$ according to definition (7.78). In particular $\lambda_{j'}^{-1}$ in (7.103) is referred to these quantum numbers.

Performing the multiplications in (7.103) gives $6 \cdot 3 \cdot 3 = 72$ terms, where the summations over $m$ are in each case collected in one term. These terms can be characterized by the combinations $(x, y, z)$, $1 \le x \le 6$, $1 \le y \le 3$, $1 \le z \le 3$. Due to the special form of (7.78) the sum over $\lambda_{i_5}$ in $z = 1, 3$ yields the expression $\sum_{i_5} m_{i_5} \lambda_{i_5} = 0$. Hence due to regularization it is $(x, y, 1) = 0, (x, y, 3) = 0$ for all $x, y$. Thus we have to calculate only the terms $(x, y, 2)$. By exact evaluation we get for the algebraic parts of $(2,2,2)$, $(3,2,2)$, $(5,1,2)$ and $(6,1,2)$: $-4\delta_{\kappa_6\kappa_7}(\gamma^0\Sigma^{\sigma\lambda}\gamma^\mu)_{\alpha_7\alpha_5}$, while all other terms $(x, y, 2)$ vanish. Inserting these results into (7.103) and observing the symmetry properties of the coefficient functions we obtain

$$(7.102) \quad = \quad N g_V^2 g^2 \sum_{i_1 \dots i_6} \int d^3r_3 d^3r_4 d^3r_5 d^3K d^3k' \, e^{\frac{i}{4}\mathbf{K}(2\mathbf{r}_5+\mathbf{r}_3+\mathbf{r}_4)} \quad \times \tag{7.104}$$

$$\lambda_{j'}^{-1} \lambda_{i_5} A_{i_3 i_1} \left(\frac{\mathbf{r}_5 - \mathbf{r}_3}{2}\right) D_{i_2 i_4} \left(\frac{\mathbf{r}_5 - \mathbf{r}_4}{2}\right) w \left(\frac{\mathbf{r}_3 - \mathbf{r}_4}{4}\right) \quad \times$$

$$\left[J^*\begin{pmatrix}\mathbf{r}_3\,\mathbf{r}_4\,\mathbf{r}_5 \\ i_3\ i_4\ i_5\end{pmatrix}l\Bigg)_{\alpha_7\kappa_6} + J^*\begin{pmatrix}\mathbf{r}_4\,\mathbf{r}_3\,\mathbf{r}_5 \\ i_4\ i_3\ i_5\end{pmatrix}l\Bigg)_{\alpha_7\kappa_6}\right] \times$$
$$C_{\alpha_6\kappa_6 i_6}(\mathbf{r}_5|\mathbf{k}',s',a',j')\left(\gamma^0\Sigma^{\sigma\lambda}\gamma^\mu\right)_{\alpha_7\alpha_6}\partial^f_{s'a'j'}(\mathbf{k}')\partial^b_{\mu,\sigma\lambda}(\mathbf{K}) \quad .$$

Furthermore substitution of the definition (7.78) for $J$ yields

$$(7.104) = N' g_V^2 g^2 \lambda_{j'}^{-1} \int \left[\sum_{i_1,i_3} A_{i_1 i_3}\left(\frac{\mathbf{r}_5-\mathbf{r}_3}{2}\right) m_{i_3}\right]\left[\sum_{i_2,i_4} D_{i_2 i_4}\left(\frac{\mathbf{r}_5-\mathbf{r}_4}{2}\right) m_{i_4}\right] \tag{7.105}$$
$$w\left(\frac{\mathbf{r}_3-\mathbf{r}_4}{4}\right)\left(\sum_{i_5}\lambda_{i_5} m_{i_5}^{-1}\right)\frac{1}{|\mathbf{r}_5-\mathbf{r}_3|^2}\frac{1}{|\mathbf{r}_5-\mathbf{r}_4|^2} e^{\frac{i}{4}\mathbf{K}(2\mathbf{r}_5+\mathbf{r}_3+\mathbf{r}_4)} \times$$
$$\left(\gamma^0\Sigma^{\sigma\lambda}\gamma^\mu\right)_{\alpha_7\alpha_6}\sum_{i'} C^*_{\alpha_7\kappa_6 i'}(\mathbf{r}_5|\mathbf{k},s,a,j)\sum_{i_6}\tilde{\partial}^f_{\alpha_6\kappa_6 i_6}(\mathbf{r}_5)\partial^b_{\mu,\sigma\lambda}(\mathbf{K})d^3r_3 d^3r_4 d^3r_5 d^3K$$

with the "Fourier transformed" $\tilde{\partial}^f_I = C^I_{1,l}\partial^f_l$, see (7.97), which is a consequence of (7.19) and the commutation relation between the functional sources.

Now we have to calculate the expressions $\sum_{i_1,i_3} A_{i_1 i_3} m_{i_3}$ and $\sum_{i_2,i_4} D_{i_2 i_4} m_{i_4}$. This can be done by setting $m_i = m + \Delta_i$ and by assuming $\Delta_i$ very small, $\Delta_i \ll m$ in comparison with $m$. In the one–time limit the leading terms of a power series expansion of $A_{i_1 i_3}$ and $D_{i_2 i_4}$ in $\Delta_i$ can then be calculated and yield together with the regularization conditions (1.32)

$$\sum_{i_1,i_3} A_{i_1 i_3}(\mathbf{r}) m_{i_3} = i c_1 m^3 \int e^{-i\mathbf{p}\cdot\mathbf{r}}(\mathbf{p}^2+m^2)^{-\frac{7}{2}} d^3p\,, \tag{7.106}$$
$$\sum_{i_2,i_4} D_{i_2 i_4}(\mathbf{r}) m_{i_4} = c_2 m \int e^{-i\mathbf{p}\cdot\mathbf{r}}(\mathbf{p}^2+m^2)^{-\frac{5}{2}} d^3p\,. \tag{7.107}$$

These Fouriertransforms can be directly calculated. They read

$$\sum_{i_1,i_3} A_{i_1 i_3}(\mathbf{r}) m_{i_3} = i c_1' m^{\frac{1}{2}} r^{\frac{3}{2}} e^{-mr} \tag{7.108}$$
$$\sum_{i_2,i_4} D_{i_2 i_4}(\mathbf{r}) m_{i_4} = c_2' m^{-\frac{3}{2}} r^{\frac{1}{2}} e^{-mr} \tag{7.109}$$

for $mr \gg 1$ with real constants $c_1', c_2'$. For small $r$ in particular at $r = 0$ the integrals (7.106) and (7.107) are finite. Hence the functions (7.108), (7.109) are suitable testfunctions which are regular at the origin and have a rapid decrease at infinity. Therefore the integrals in (7.104) are finite. Introducing $\mathbf{r}_5 - \mathbf{r}_3 = \mathbf{u}$, $\mathbf{r}_5 - \mathbf{r}_4 = \mathbf{v}$ and polar coordinates in $\mathbf{u}$ and $\mathbf{v}$, we can evaluate the integral over $\mathbf{r}_3$ and $\mathbf{r}_4$ and due to high concentration of the testfunctions at the origin $\mathbf{u} = 0$, $\mathbf{v} = 0$ we can replace $\exp[\frac{i}{4}\mathbf{K}(2\mathbf{r}_5+\mathbf{r}_3+\mathbf{r}_4)]$ by $\exp(i\mathbf{K}\mathbf{r}_5)$.

For further evaluation we write the one–particle states explicitly as

$$C_{\alpha\kappa i}(\mathbf{r}|\mathbf{k},s,a,j) = \lambda_j \delta_{ij} e^{-i\mathbf{r}\mathbf{k}}\chi_{\alpha\kappa}(\mathbf{k},s,a)\,, \tag{7.110}$$
$$\chi_{\alpha\kappa}(\mathbf{k},s,a) = \delta_{\kappa a}\chi^s_\alpha(\mathbf{k})$$

with ordinary Dirac spinors $\chi^s_\alpha(\mathbf{k})$.

We now retransform the open index $l$ of equation (7.34) into coordinate space. This is done by multiplying (7.105) from the left with $C_{\alpha\kappa i}(\mathbf{r}'|\mathbf{k},s,a,j)$ and summing over $\mathbf{k},s,a,j$. We finally obtain the result

$$\sum_{s,a,j}\int d^3k\, C_{\alpha\kappa i}(\mathbf{r}'|\mathbf{k},s,a,j)\;\times\;(7.105) = \tag{7.111}$$
$$\kappa\lambda_i\left(\gamma^0\Sigma^{\sigma\lambda}\gamma^\mu\right)_{\alpha\beta}\sum_{i'}\tilde{\partial}^f_{\beta\kappa i'}(\mathbf{r})\partial^b_{\mu,\sigma\lambda}(\mathbf{r})$$

with the real coupling constant $\kappa$.

In the same way the other terms of equation (7.34) can be treated. So we eventually obtain

$$i\frac{\partial}{\partial t}\tilde{\partial}^f_{\alpha\kappa i}(\mathbf{r},t)|\mathcal{G}\rangle = \left[-i(\gamma^0\gamma^k)_{\alpha\beta}\partial_k(\mathbf{r})+\gamma^0_{\alpha\beta}m_i\right]\tilde{\partial}^f_{\alpha\kappa i}(\mathbf{r},t)|\mathcal{G}\rangle \tag{7.112}$$
$$+\kappa\lambda_i\left(\gamma^0\Sigma^{\sigma\lambda}\gamma^\mu\right)_{\alpha\beta}\sum_{i'}\tilde{\partial}^f_{\beta\kappa i'}(\mathbf{r},t)\partial^b_{\mu,\sigma\lambda}(\mathbf{r},t)|\mathcal{G}\rangle \quad .$$

Equation (7.112) is the functional formulation of a Dirac field coupled to anholonomic connections. The physical content of (7.112) can be illustrated if one retranslates the quantized equation (7.112) into its classical precursor. This can be achieved by the ansatz, see Stumpf [Stu 88]:

$$|\mathcal{G}(\tilde{f},b;a)\rangle = e^{Z_0[b,\tilde{f};a]}|v\rangle \tag{7.113}$$

with

$$Z_0[b,\tilde{f};a] := i\int \tilde{f}_{\alpha\kappa i}(\mathbf{r},t)\phi_{\alpha\kappa i}(\mathbf{r},t|a)d^3r + \int b_{\mu,\sigma\lambda}(\mathbf{r},t)\Gamma_{\mu,\sigma\lambda}(\mathbf{r},t|a)d^3r \tag{7.114}$$

which corresponds to the classical approximation in statistical mechanics and where in quantum field theory the classical amplitudes $\phi_{\alpha\kappa i}(\mathbf{r},t),\Gamma_{\mu,\sigma\rho}(\mathbf{r},t)$ are interpreted as one–particle transition matrix elements of the corresponding quantized fields $\psi^{\text{op}}$ and $\Gamma^{\text{op}}$:

$$\phi_{\alpha\kappa i}(\mathbf{r},t|a) \;:=\; \langle 0|\psi_{\alpha\kappa i}(\mathbf{r},t)^{\text{op}}|a\rangle\,, \tag{7.115}$$
$$\Gamma_{\mu,\sigma\rho}(\mathbf{r},t|a) \;:=\; \langle 0|\Gamma_{\mu,\sigma\rho}(\mathbf{r},t)^{\text{op}}|a\rangle\,. \tag{7.116}$$

Substitution of (7.113) in (7.112) leads to the equation

$$\frac{\partial}{\partial t}\phi_{\alpha\kappa i}(\mathbf{r},t) \;=\; -\left[(\gamma^0\gamma^k)_{\alpha\beta}\partial_k(\mathbf{r})+i\gamma^0_{\alpha\beta}m_i\right]\phi_{\beta\kappa i}(\mathbf{r},t) \tag{7.117}$$
$$+i\kappa\lambda_i\left(\gamma^0\Sigma^{\sigma\lambda}\gamma^\mu\right)_{\alpha\beta}\Gamma_{\mu,\sigma\lambda}(\mathbf{r},t)\sum_{i'}\phi_{\beta\kappa i'}(\mathbf{r},t) \quad .$$

Introducing superindices $Z=(\kappa,\alpha,i)$ one immediately shows that (7.117) is identical with the parity transformed equation (7.13) except for two features concerning the auxiliary fields: The summation over $i'$ and the constants $\lambda_i$ are a relic of our nonperturbative Pauli–Villars regularization. This relic disappears if the transition from nonobservable subfermions to observable fermions is performed. Furthermore, the transformation from anholonomic to holonomic coordinates can be performed as soon as not only the effective fermion equation, but also the effective graviton equation (7.32) is thoroughly investigated and its differential geometric meaning is clarified. This has not been elaborated so far in detail.

Finally a conclusive comparison with the "phenomenological" equations requires an investigation of the coupling constant $\kappa$.

# 8 Weak Mapping and Gauge Fields

## 8.1 Introduction

Atoms and molecules in quantum electrodynamics, and nuclei and mesons in quantum chromodynamics are considered as composite particles built up of elementary leptons, baryons or quarks. Hence the study of the formation of composite particles and the derivation of their corresponding effective dynamics have to be some of the basic aims of abelian and nonabelian gauge theories with respect to their application to atomic and nuclear physics. In the course of the evaluation of these theories numerous investigations were performed concerning the formation of bound states by means of Bethe–Salpeter equations and Schrödinger equations, see Itzykson and Zuber [Itz 80], and the effective dynamics were studied by Fock space methods, e.g. by Meyer [Mey 92], Buchmüller and Dietz [Buch 80] or by path integral evaluation, see e.g. Cahill and Roberts [Cah 85]. Summarizing the results of these investigations one observes the following drawbacks: The bound state calculations are in general not related to a corresponding effective dynamics, the Fock space methods are in contradiction to algebraic representation theory of quantum fields, cf., e.g., Haag [Haag 92], Honegger and Rieckers [Hon 90], and path integrals imply conceptional as well as technical difficulties, in particular with respect to effective dynamics for composite particles, see Rivers [Riv 88]. For a more detailed criticism we refer to Chapters 10 and 11.

To avoid these drawbacks, in Chapter 4 a composite particle theory was developed which is in accordance with algebraic representation theory. One of the main ingredients of this theory is the weak mapping method which can be rigorously formulated by weak mapping theorems.

In Chapter 4 weak mapping was confined to the treatment of a fully regularized quantum field theory, namely to the super–renormalizable spinorfield theory of Chapter 1. In the present chapter we extend the weak mapping formalism to the case of renormalizable quantum field theories. We exemplify this treatment for the most prominent models of gauge theories, namely for abelian spinor quantum electrodynamics and, at least partially, for non–abelian quantum chromodynamics.

In the algebraic formulation of the quantum field dynamics the time evolution of the quantum fields is assumed to be a one–parameter automorphism group on the field algebra. This kind of description of quantum fields is closely related to the Hamilton formalism which is not manifestly relativistically covariant. In order to adapt spinor quantum electrodynamics to this formalism we cannot use a covariant gauge with superfluous field variables. Rather we have to apply a noncovariant gauge with a minimal number of physical field variables, and for spinor quantum electrodynamics the Coulomb gauge seems to be the most appropriate one. Although the Coulomb gauge is a noncovariant gauge, it was recently demonstrated that it is inherently relativistic invariant even in nonabelian theories, cf. Besting and Schütte [Best 90], Kim, Namgung, Soh and Yee [Kim 91]. We therefore assume that effective dynamics derived in Coulomb gauge and Hamilton formalism will be inherently relativistic invariant, too, but we will not try to explicitly prove this property here. The treatment of effective dynamics given in this chapter follows the paper of Stumpf, Fauser and Pfister [Stu 93a].

Finally we mention: In QED the Coulomb gauge is especially adapted to the formation of composite particles. On the other hand, in the preceding chapters with respect to the derivation of effective theories we preferred another gauge, namely the temporal gauge. This is, however, no contradiction. While in the preceding chapters the basic model was a spinorfield model and the gauge theories were derived as effective theories, in this chapter we consider the gauge theories as fundamental and look for effective theories "above" them. So what was useful for the derivation of effective gauge theories need not be useful for the treatment of basic gauge theories.

Nevertheless the Coulomb gauge of QED is an exception. Even in QCD we will have to return to the temporal gauge, because for nonabelian theories due to the nonlinearity of the Gauß law an exact classical elimination of the longitudinal and scalar fields is impossible, apart from dubious perturbative resolutions.

But even the Coulomb gauge of QED is not as exceptional as it seems. Although gauge transformations at the quantum level do not work, it was the content of the classical paper of Bleuler [Bleu 50] to show the equivalence of Lorentz gauge and Coulomb gauge in QED. Without explicit calculation we thus assume that by a similar procedure also the equivalence of temporal gauge and Coulomb gauge in QED can be verified, cf. also Christ and Lee [Chri 80], what eventually justifies the treatment of QED and QCD in different gauges.

## 8.2 Spinor Electrodynamics in Coulomb Gauge

The bare Lagrangian density of spinor electrodynamics reads

$$\mathcal{L} := -\frac{1}{4}F_{\mu\nu}F^{\mu\nu} + \bar{\psi}(i\gamma^\mu\partial_\mu - m_0)\psi + e_0 A_\mu : \bar{\psi}\gamma^\mu\psi : \tag{8.1}$$

with

$$F_{\mu\nu} := \partial_\mu A_\nu - \partial_\nu A_\mu \ , \tag{8.2}$$

$\psi \equiv \psi_\alpha(x)$ and the bare fermion mass $m_0$ and charge $e_0$. We do not take (8.1) as the starting point for evaluating a classical theory of spinors and electromagnetic fields. Rather we want to derive a quantized theory. As an anticipation of the quantum version we assume the $\psi$–fields to be formal anticommuting or Grassmann–variables. This provides a formulation of the quantized theory which e.g. is explicitly invariant under charge conjugation, c.f. Roman [Rom 69].

According to Section 8.1 we treat spinor electrodynamics in a non–covariant gauge. With respect to composite particle dynamics the Coulomb gauge is definitely distinguished under all non–covariant gauges, as in this gauge the basic binding force between charged fermions, i.e., the Coulomb force or its potential, respectively, explicitly appears in the Hamiltonian. Furthermore, we can work in this gauge with a minimal number of physical field variables which enables us to define consistent field quantization rules and to prevent the admixture of unphysical states to physical bound states.

For further evaluation we need the corresponding field equations. If we combine $\psi$ and its charge conjugated field $\psi^c$ into a superspinor field $\Psi$ defined by

$$\Psi_{\alpha\Lambda}(x) := \begin{cases} \psi_\alpha(x) & \text{if} \quad \Lambda = 1 \\ \psi^c_\alpha(x) = C_{\alpha\beta}\bar{\psi}_\beta(x) & \text{if} \quad \Lambda = 2 \end{cases} \tag{8.3}$$

with $C := i\gamma^2\gamma^0$, we get from (8.1) in a compact notation

$$\partial_\nu F^{\mu\nu} + \frac{ie_0}{2}\,\Psi C\gamma^\mu\sigma^2\Psi = 0 \tag{8.4}$$

$$(i\gamma^\mu\partial_\mu - m_0)\Psi + e_0\,A_\mu\gamma^\mu\sigma^3\Psi = 0 \;, \tag{8.5}$$

where the Pauli matrices $\sigma^\mu$ act on the index $\Lambda$.

In addition the Hamilton formalism requires a formulation of the field equations in terms of canonically conjugate variables, in particular with respect to the electromagnetic field. We therefore introduce $E^k := F^{0k}$ in the above field equations. Equations (8.4) and (8.5) can then be rewritten as

$$\begin{aligned} i\partial_0\Psi &= -(i\alpha^k\partial_k - \beta m_0)\Psi - e_0\,\gamma^0\gamma^\mu A_\mu\sigma^3\Psi \\ i\partial_0 E^k &= -\frac{1}{2}e_0\,\Psi C\gamma^k\sigma^2\Psi + i(\partial_j\partial^k A^j - \partial_j\partial^j A^k) \\ i\partial_0 A^k &= iE^k - i\partial_k A^0 \\ \partial_k E^k &= -i\frac{e_0}{2}\,\Psi C\gamma^0\sigma^2\Psi \end{aligned} \tag{8.6}$$

with $\alpha^k = \gamma^0\gamma^k$ and $\beta = \gamma^0$.

We now impose the gauge condition

$$\partial_k A^k = 0 \tag{8.7}$$

and decompose $E^k$ and $A^k$ into transversal and longitudinal parts:

$$E^k = E^k_{\text{tr}} + E^k_\ell\,, \quad A^k = A^k_{\text{tr}} + A^k_\ell\;. \tag{8.8}$$

The decomposition is performed by the application of projection operators

$$P_{\text{tr}} := 1 - \Delta^{-1}\nabla\otimes\nabla\cdot\;,\; P_\ell := \Delta^{-1}\nabla\otimes\nabla\cdot \tag{8.9}$$

with $P_{\text{tr}}\mathbf{V} = \mathbf{V}_{\text{tr}}$ and $P_\ell\mathbf{V} = \mathbf{V}_\ell$. Furthermore it is

$$\nabla\cdot\mathbf{V}_{\text{tr}} = 0\,, \quad \nabla\times\mathbf{V}_\ell = 0 \tag{8.10}$$

for any vector $\mathbf{V}$. For suitable boundary conditions which we assume to be fulfilled the decomposition (8.8) is unique; in this case from the gauge condition (8.7) it follows that $\mathbf{A}_\ell = P_\ell\mathbf{A} = 0$, i.e. $\mathbf{A} = \mathbf{A}_{\text{tr}}$.

Applying the projection operators $P_{\text{tr}}$ and $P_\ell$ to (8.6), we obtain the following set of equations

$$i\partial_0\Psi = -(i\alpha^k\partial_k - \beta m_0)\Psi - e_0\,A^0\sigma^3\Psi + e_0\,\alpha^k A^k_{\text{tr}}\sigma^3\Psi \tag{8.11}$$

$$i\partial_0 E^k_{\text{tr}} = -P_{\text{tr}}\,\frac{e_0}{2}\,\Psi C\gamma^k\sigma^2\Psi + i\Delta A^k_{\text{tr}} \tag{8.12}$$

$$i\partial_0 A^k_{\text{tr}} = iE^k_{\text{tr}} \tag{8.13}$$

$$i\partial_0 E^k_\ell = -P_\ell\,\frac{e_0}{2}\,\Psi C\gamma^k\sigma^2\Psi \tag{8.14}$$

$$E^k_\ell = \partial_k A^0 \tag{8.15}$$

$$\partial_k E^k_\ell = -i\frac{e_0}{2}\,\Psi C\gamma^0\sigma^2\Psi \;. \tag{8.16}$$

If we observe current conservation $\partial_\mu \bar{\Psi}\gamma^\mu\Psi = 0$ it turns out that for unique decomposition (8.8) equation (8.14) is a consequence of equation (8.16). Hence in the following we can omit (8.14). Substitution of equation (8.15) into (8.16) yields with appropriate boundary conditions

$$A^0 = -\frac{i}{2}e_0\Delta^{-1}\Psi C\gamma^0\sigma^2\Psi\,. \tag{8.17}$$

By this equation (8.15) and (8.16) are satisfied and we can replace (8.15) and (8.16) by (8.17). Finally we substitute (8.17) into the spinorial equations (8.11). Therefore, from (8.6) the following set of independent dynamical equations remains:

$$\begin{aligned} i\partial_0\Psi &= -(i\alpha^k\partial_k - \beta m_0)\Psi + \frac{i}{2}e_0^2\,(\Delta^{-1}\,\Psi C\gamma^0\sigma^2\Psi)\sigma^3\Psi + e_0\,\alpha^k A^k_{\text{tr}}\sigma^3\Psi \\ i\partial_0 E^k_{\text{tr}} &= -P_{\text{tr}}\,\frac{e_0}{2}\,\Psi C\gamma^k\sigma^2\Psi + i\Delta A^k_{\text{tr}} \\ i\partial_0 A^k_{\text{tr}} &= iE^k_{\text{tr}} \quad . \end{aligned} \tag{8.18}$$

In these equations all superfluous field variables are eliminated in favour of a minimal set of independent variables, namely $\Psi$, $E^k_{\text{tr}}$, $A^k_{\text{tr}}$.

According to Chapter 3 we now apply the perturbation independent scheme of Algebraic Schrödinger Representation to spinor QED.

For convenience we introduce some abbreviations. The first is an additional index $\eta$ to combine the fields $E^k_{\text{tr}}$, $A^k_{\text{tr}}$ into a superfield

$$B^k_\eta := \begin{cases} A^k_{\text{tr}} & \text{if} \quad \eta = 1 \\ E^k_{\text{tr}} & \text{if} \quad \eta = 2 \end{cases} \quad . \tag{8.19}$$

The vector fields are then characterized by the arguments $(\mathbf{r}, k, \eta)$, while the spinorfield has the arguments $(\mathbf{r}, \alpha, \Lambda)$. For both kinds of arguments we symbolically use only one index $K := (\mathbf{z}, k, \eta)$ or $I := (\mathbf{r}, \alpha, \Lambda)$, respectively. With these abbreviations equations (8.18) can be rewritten as

$$\begin{aligned} i\partial_0\Psi_{I_1} &= D_{I_1I_2}\Psi_{I_2} + W^K_{I_1I_2}B_K\Psi_{I_2} + U^{J_1J_2J_3}_{I_1}\Psi_{J_1}\Psi_{J_2}\Psi_{J_3}\,, \\ i\partial_0 B_{K_1} &= L_{K_1K_2}B_{K_2} + J^{I_1I_2}_{K_1}\Psi_{I_1}\Psi_{I_2} \end{aligned} \tag{8.20}$$

with

$$\begin{aligned} D_{I_1I_2} &:= -(i\alpha^k\partial_k - \beta m_0)_{\alpha_1\alpha_2}\delta(\mathbf{r}_1 - \mathbf{r}_2)\delta_{\Lambda_1\Lambda_2} \\ W^K_{I_1I_2} &:= e_0\,\alpha^k_{\alpha_1\alpha_2}\delta(\mathbf{r}_1 - \mathbf{r}_2)\delta(\mathbf{r}_1 - \mathbf{z})\delta_{1\eta}\sigma^3_{\Lambda_1\Lambda_2} \\ U^{J_1J_2J_3}_I &:= -\frac{i}{8\pi}e_0^2\left[(C\gamma^0)_{\alpha_1\alpha_2}\delta_{\alpha\alpha_3}\sigma^2_{\Lambda_1\Lambda_2}\sigma^3_{\Lambda\Lambda_3}\frac{\delta(\mathbf{r}_1 - \mathbf{r}_2)\delta(\mathbf{r} - \mathbf{r}_3)}{|\mathbf{r} - \mathbf{r}_1|}\right]_{\text{as}\{J_1,J_2,J_3\}} \\ L_{K_1K_2} &:= i\delta(\mathbf{z}_1 - \mathbf{z}_2)\delta_{k_1k_2}\delta_{\eta_11}\delta_{\eta_22} + i\Delta(\mathbf{z}_1)\delta(\mathbf{z}_1 - \mathbf{z}_2)\delta_{k_1k_2}\delta_{\eta_12}\delta_{\eta_21} \\ J^{I_1I_2}_K &:= -\frac{1}{2}e_0\,P_{\text{tr}}(\mathbf{z} - \mathbf{r}_1)\delta(\mathbf{r}_1 - \mathbf{r}_2)(C\gamma^k)_{\alpha_1\alpha_2}\delta_{2\eta}\sigma^2_{\Lambda_1\Lambda_2} \quad . \end{aligned} \tag{8.21}$$

We note that we use the symbols $I$, $D$, $W$, $U$, etc. in a different meaning compared with that in the preceding chapters and hope that there will arise no confusion about this fact.

As it stands, the field theory defined by (8.1), (8.2) or by (8.20), (8.21) is not renormalized, i.e. the perturbative evaluation of the corresponding quantum theory leads to divergent physical quantities. It can, however, be shown that with respect to perturbation theory the theory can be renormalized if the classical Lagrangian is supplemented by corresponding counterterms and therefore the dynamical equations (8.20) are corrected, too.

Usually, renormalization is performed by means of multiplicative $Z$–factors for fields and charge. It can, however, be proven that this multiplicative renormalization can be transformed into a mass and charge renormalization, see, e.g., Bogoliubov and Shirkov [Bogo 59], Gupta [Gup 77], Nakanishi and Ojima [Naka 90]. Following this line we introduce

$$m_0 = m - \delta m \ , \quad e_0 = e - \delta e \ , \tag{8.22}$$

where $m$ is the observed mass and $e$ the observed charge of the fermion, and $\delta m$ and $\delta e$ lead to the only counterterms of the theory. This renormalization can be performed in any gauge; thus we can substitute (8.22) into (8.20), (8.21) without any difficulty.

In general, renormalization is properly defined only in perturbation theory. On the other hand we rejected perturbation theory as an inadequate means to treat composite particle dynamics. Thus if we perform the renormalization (8.22) the question arises whether we can give renormalization a definite meaning beyond perturbative treatment. How this difficulty can be resolved will be discussed in the following section.

## 8.3 Quantization of Spinor Electrodynamics

We turn to the quantization of (8.20). With respect to the fermion field we impose the canonical equal time anticommutation relations

$$[\Psi_{I_1}, \Psi_{I_2}]_+^t = A_{I_1 I_2}\mathbf{1} \ , \quad A_{I_1 I_2} = (C\gamma^0)_{\alpha_1\alpha_2}\sigma^1_{\Lambda_1\Lambda_2}\delta(\mathbf{r}_1 - \mathbf{r}_2) \ . \tag{8.23}$$

These relations correspond to the relations (1.53), (1.54) without the introduction of auxiliary fields. Furthermore we assume for equal times $t$

$$[B_K, \Psi_I]_-^t \ = \ 0 \tag{8.24}$$

$$[B_{K_1}, B_{K_2}]_-^t \ = \ C_{K_1 K_2}\mathbf{1} \ , \tag{8.25}$$

where $C_{K_1K_2}$ has to be a purely imaginary antisymmetric generalized c–number function, i.e. $C_{K_1K_2} = -C_{K_2K_1}$. Usually, the special form of $C_{K_1K_2}$ is postulated by means of the canonical quantization procedure. However, as will be seen below, by the requirement of a consistent quantum theory $C_{K_1K_2}$ is uniquely determined for the field equations (8.20) and the fermion anticommutator (8.23) . Hence, we do not specify the form of the commutator $C_{K_1K_2}$ from the beginning, rather it will come out as a result of our calculation.

In the algebraic formulation of quantum field theory the quantum states $|a\rangle$ of our system can be characterized by the set of matrix elements for equal times $t$. This was extensively discussed in Chapter 3 for spinorfields. In the case of quantum electrodynamics we have to extend this formalism to simultaneously include boson fields. This leads to the set of matrix elements

$$\langle 0|\,\mathcal{A}(\Psi_{I_1}\dots\Psi_{I_n})_t\,\mathcal{S}(B_{K_1}\dots B_{K_m})_t\,|a\rangle \ , \quad n, m = 1\dots\infty \ . \tag{8.26}$$

The operator $\mathcal{A}$ antisymmetrizes the product of field operators $\Psi_{I_1}\dots\Psi_{I_n}$ with respect to the indices $I_1, \dots, I_n$, whereas the operator $\mathcal{S}$ symmetrizes the boson operators with respect to the indices $K_1, \dots K_m$.

Now we have to establish an equation which allows, at least in principle, the determination of the matrix elements (8.26). If we assume the existence of a Hamilton operator $\mathbf{H}$, we have the Heisenberg equation

$$i\partial_t\,[\mathcal{A}(\Psi_{I_1}\dots\Psi_{I_n})_t\,\mathcal{S}(B_{K_1}\dots B_{K_m})_t] = [\mathcal{A}(\Psi_{I_1}\dots\Psi_{I_n})_t\,\mathcal{S}(B_{K_1}\dots B_{K_m})_t\,,\mathbf{H}]_-\,. \tag{8.27}$$

We require the states $|0\rangle$ and $|p\rangle$, which are characterized by a complete set of quantum numbers, to be eigenstates of $\mathbf{H}$. Then we obtain from (8.27)

$$\begin{aligned} E_{p0}\,\langle 0|\,\mathcal{A}(\Psi_{I_1}\dots\Psi_{I_n})_t\,\mathcal{S}(B_{K_1}\dots B_{K_m})_t\,|p\rangle = \\ i\partial_t\,\langle 0|\,\mathcal{A}(\Psi_{I_1}\dots\Psi_{I_n})_t\,\mathcal{S}(B_{K_1}\dots B_{K_m})_t\,|p\rangle \end{aligned} \tag{8.28}$$

with $E_{p0} := E_p - E_0$. For brevity we suppress the index $t$ in the following.

With the help of (8.20), we can evaluate the right–hand side of (8.28) to give

$$\begin{aligned} & i\partial_t\,\langle 0|\,\mathcal{A}(\Psi_{I_1}\dots\Psi_{I_n})\mathcal{S}(B_{K_1}\dots B_{K_m})\,|p\rangle = \\ & \quad \sum_{l=1}^{n}\langle 0|\,\mathcal{A}(\Psi_{I_1}\dots(D_{I_lJ}\Psi_J)\dots\Psi_{I_n})\,\mathcal{S}(B_{K_1}\dots B_{K_m})\,|p\rangle \\ & \quad +\sum_{l=1}^{n}\langle 0|\,\mathcal{A}(\Psi_{I_1}\dots(W^L_{I_lJ}\Psi_J)\dots\Psi_{I_n})\,B_L\,\mathcal{S}(B_{K_1}\dots B_{K_m})\,|p\rangle \\ & \quad +\sum_{l=1}^{n}\langle 0|\,\mathcal{A}(\Psi_{I_1}\dots(U_{I_l}^{J_1J_2J_3}\Psi_{J_1}\Psi_{J_2}\Psi_{J_3})\dots\Psi_{I_n})\,\mathcal{S}(B_{K_1}\dots B_{K_m})\,|p\rangle \\ & \quad +\sum_{l=1}^{m}\langle 0|\,\mathcal{A}(\Psi_{I_1}\dots\Psi_{I_n})\,\mathcal{S}(B_{K_1}\dots(L_{K_lL_l}B_{L_l})\dots B_{K_m})\,|p\rangle \\ & \quad +\sum_{l=1}^{m}\langle 0|\,\mathcal{A}(\Psi_{I_1}\dots\Psi_{I_n})\Psi_{J_1}\Psi_{J_2}\,\mathcal{S}(B_{K_1}\dots B_{K_l-1}J_{K_l}^{J_1J_2}B_{K_l+1}\dots B_{K_m})\,|p\rangle \\ & =:\; T_1+T_2+T_3+T_4+T_5 \quad . \end{aligned} \tag{8.29}$$

In order to get an equation for the matrix elements (8.26) we have to represent the right–hand side of (8.29) as a linear combination of the matrix elements (8.26). As a first step in this direction, we transform the expressions $T_1,\dots,T_5$ into a standard representation, cf. Proposition 3.10. As this proposition was derived in detail, we only briefly sketch the necessary rearrangements for the present case of a boson–fermion coupling theory.

This standard form is achieved by shifting the operators which are not (anti–) symmetrized with respect to the surrounding operators to the left. We begin with the evaluation of $T_1$. By repeated anticommutations we shift $\Psi_J$ to the left, and after relabeling we get

$$\begin{aligned} T_1 \;=\; & \sum_{p\in S_n}\frac{\mathrm{sgn}(p)}{n!}\sum_{l=1}^{n}D_{I_{p1}J}\Big[\langle 0|\,\Psi_J\Psi_{I_{p2}}\dots\Psi_{I_{pn}}\,\mathcal{S}(B_{K_1}\dots B_{K_m})\,|p\rangle \\ & -(l-1)A_{I_{p2}J}\,\langle 0|\,\Psi_{I_{p3}}\dots\Psi_{I_{pn}}\,\mathcal{S}(B_{K_1}\dots B_{K_m})\,|p\rangle\Big] \quad . \end{aligned} \tag{8.30}$$

Making use of the formula

$$\sum_{l=1}^{n}(l-1)\dots(l-k) = \frac{1}{k+1}n(n-1)\dots(n-k) \tag{8.31}$$

we obtain from (8.30)

$$T_1 = \sum_{p \in S_n} \frac{\operatorname{sgn}(p)}{n!} D_{I_{p1} J} \Big[ n \langle 0 | \Psi_J \Psi_{I_{p2}} \dots \Psi_{I_{pn}} \mathcal{S}(B_{K_1} \dots B_{K_m}) | p \rangle \\ - \frac{1}{2} n(n-1) A_{I_{p2} J} \langle 0 | \Psi_{I_{p3}} \dots \Psi_{I_{pn}} \mathcal{S}(B_{K_1} \dots B_{K_m}) | p \rangle \Big] . \quad (8.32)$$

The evaluation of the terms $T_2 \dots T_5$ runs along the same lines. The fermion fields $\Psi_J$ and the boson fields $B_L$ are shifted to the left; then after relabeling and using (8.31) we obtain

$$T_2 = \sum_{p \in S_n} \frac{\operatorname{sgn}(p)}{n!} W^L_{I_{p1} J} \Big[ n \langle 0 | \Psi_J \Psi_{I_{p2}} \dots \Psi_{I_{pn}} B_L \mathcal{S}(B_{K_1} \dots B_{K_m}) | p \rangle \\ - \frac{1}{2} n(n-1) A_{I_{p2} J} \langle 0 | \Psi_{I_{p3}} \dots \Psi_{I_{pn}} B_K \mathcal{S}(B_{K_1} \dots B_{K_m}) | p \rangle \Big] , \quad (8.33)$$

$$T_3 = \sum_{p \in S_n} \frac{\operatorname{sgn}(p)}{n!} U^{J_1 J_2 J_3}_{I_{p1}} \Big[ n \langle 0 | \Psi_{J_1} \Psi_{J_2} \Psi_{J_3} \Psi_{I_{p2}} \dots \Psi_{I_{pn}} \mathcal{S}(B_{K_1} \dots B_{K_m}) | p \rangle \\ - \frac{3}{2} n(n-1) A_{I_{p2} J_1} \langle 0 | \Psi_{J_2} \Psi_{J_3} \Psi_{I_{p3}} \dots \Psi_{I_{pn}} \mathcal{S}(B_{K_1} \dots B_{K_m}) | p \rangle \\ + n(n-1)(n-2) A_{I_{p2} J_1} A_{I_{p3} J_3} \langle 0 | \Psi_{J_2} \Psi_{I_{p4}} \dots \Psi_{I_{pn}} \mathcal{S}(B_{K_1} \dots B_{K_m}) | p \rangle \\ + \frac{1}{4} n(n-1)(n-2)(n-3) A_{I_{p2} J_1} A_{I_{p3} J_2} A_{I_{p4} J_3} \quad \times \\ \langle 0 | \Psi_{I_{p5}} \dots \Psi_{I_{pn}} \mathcal{S}(B_{K_1} \dots B_{K_m}) | p \rangle \Big] , \quad (8.34)$$

$$T_4 = \sum_{p \in S_m} \frac{1}{m!} L_{K_{p1} L} \Big[ m \langle 0 | \mathcal{A}(\Psi_{I_1} \dots \Psi_{I_n}) B_L B_{K_{p2}} \dots B_{K_{pm}} | p \rangle \\ + \frac{1}{2} m(m-1) C_{K_{p2} L} \langle 0 | \mathcal{A}(\Psi_{I_1} \dots \Psi_{I_n}) B_{K_{p3}} \dots B_{K_{pm}} | p \rangle \Big] \quad (8.35)$$

and finally

$$T_5 = \sum_{p \in S_n} \frac{\operatorname{sgn}(p)}{n!} \sum_{p' \in S_m} \frac{1}{m!} J^{J_1 J_2}_{K_{p'_1}} \quad \times \quad (8.36) \\ \Big[ m \langle 0 | \Psi_{J_1} \Psi_{J_2} \Psi_{I_{p1}} \dots \Psi_{I_{pn}} B_{K_{p'_2}} \dots B_{K_{p'_m}} | p \rangle \\ + 2nm A_{I_{p1} J_1} \langle 0 | \Psi_{J_2} \Psi_{I_{p2}} \dots \Psi_{I_{pn}} B_{K_{p'_2}} \dots B_{K_{p'_m}} | p \rangle \\ - n(n-1) m A_{I_{p1} J_1} A_{I_{p2} J_2} \langle 0 | \Psi_{I_{p3}} \dots \Psi_{I_{pn}} B_{K_{p'_2}} \dots B_{K_{p'_m}} | p \rangle \Big] .$$

With (8.32)–(8.36) we have brought the right–hand side of (8.29) into the above mentioned standard form. The next step is to decompose this standard representation into a sum of matrix elements of the form (8.26). This decomposition can be described in a compact and concise manner by means of functional equations. In accordance with Chapter 3 we introduce the functional sources $j_I$, $b_K$ and their duals $\partial_I$, $\partial^b_K$ which are defined by their anticommutation or commutation relations, respectively,

$$[j_I \, , j_{I'}]_+ = 0 \, , \qquad [j_I \, , \partial_{I'}]_+ = \delta_{II'} \, , \\ [b_K \, , b_{K'}]_- = 0 \, , \qquad [\partial^b_K \, , b_{K'}]_- = \delta_{KK'} \, , \quad (8.37)$$

while all other commutators vanish. Furthermore we have to introduce a functional space. This can be done by defining a Fock space structure with respect to the sources $j_I$, $b_K$ with a corresponding source vacuum $|v\rangle$:

$$|v\rangle := |0\rangle_F \otimes |0\rangle_B \tag{8.38}$$

with $\partial_I|v\rangle = \partial^b_K|v\rangle = \langle v|j_I = \langle v|b_K = 0$, where $|0\rangle_F$ and $|0\rangle_B$ are functional Fock vacua for fermions and bosons. For brevity we again suppressed the one–time index $t$. It should be noted that this construction does not fix the algebraic representation of the original fields and thus is not in conflict with our general intentions of formulating a consistent non–perturbative quantum theory. Rather this functional construction is only a means for a compact formulation.

We substitute (8.32)–(8.36) into (8.29) and (8.29) into (8.28), multiply from the right with

$$\frac{i^n}{n!\,m!} j_{I_1} \dots j_{I_n} b_{K_1} \dots b_{K_m}|v\rangle \tag{8.39}$$

and perform the sumation over $n, m$ from 0 to $\infty$. Then we get after relabeling

$$\begin{aligned}
E_{p0}\,|\mathcal{G}(j,b;p)\rangle\rangle &= iD_{I_1I_2}j_{I_1}\,|\mathcal{G}_{I_2}(j,b;p)\rangle\rangle + \frac{1}{2}D_{I_1I_2}A_{I_3I_2}j_{I_1}j_{I_3}|\mathcal{G}(j,b;p)\rangle\rangle \\
&+W^K_{I_1I_2}j_{I_1}\left[i\,|\mathcal{G}^K_{I_2}(j,b;p)\rangle\rangle + \frac{1}{2}A_{I_3I_2}j_{I_3}\,|\mathcal{G}^K(j,b;p)\rangle\rangle\right] \\
&+iU^{J_1J_2J_3}_{I_1}j_{I_1}\Big[|\mathcal{G}_{J_1J_2J_3}(j,b;p)\rangle\rangle - \frac{3i}{2}A_{I_2J_1}j_{I_2}\,|\mathcal{G}_{J_2J_3}(j,b;p)\rangle\rangle \\
&-A_{I_1J_1}A_{I_2J_3}j_{I_1}j_{I_2}\,|\mathcal{G}_{J_2}(j,b;p)\rangle\rangle - \frac{i}{4}A_{I_1J_1}A_{I_2J_2}A_{I_3J_3}j_{I_1}j_{I_2}j_{I_3}\,|\mathcal{G}(j,b;p)\rangle\rangle\Big] \\
&+L_{K_1K_2}\left[b_{K_1}\,|\mathcal{G}^{K_2}(j,b;p)\rangle\rangle + \frac{1}{2}C_{K_3K_2}b_{K_1}b_{K_3}|\mathcal{G}(j,b;p)\rangle\rangle\right] \\
&+J^{I_1I_2}_K b_K\Big[|\mathcal{G}_{I_1I_2}(j,b;p)\rangle\rangle + 2iA_{I_3I_1}j_{I_3}\,|\mathcal{G}_{I_2}(j,b;p)\rangle\rangle \\
&\qquad +A_{I_3I_1}A_{I_4J_2}j_{I_3}j_{I_4}|\mathcal{G}(j,b;p)\rangle\rangle\Big] \quad ,
\end{aligned} \tag{8.40}$$

where we used the definition

$$\begin{aligned}
&\left|\mathcal{G}^{K_1\dots K_s}_{J_1\dots J_r}(j,b;a)\right\rangle := \\
&\sum_{n,m=0}^{\infty}\frac{i^n}{n!m!}\langle 0|\,\Psi_{J_1}\dots\Psi_{J_r}\Psi_{I_1}\dots\Psi_{I_n}B_{K_1}\dots B_{K_s}B_{L_1}\dots B_{L_m}\,|a\rangle \quad\times \\
&\qquad j_{I_1}\dots j_{I_n}\,b_{L_1}\dots b_{L_m}|v\rangle \quad .
\end{aligned} \tag{8.41}$$

The bosonic and fermionic sources in (8.41) enforce a partial symmetrization or antisymmetrization, respectively, of the coefficient functions in their bosonic or fermionic indices. This means that (8.41) defines a generating functional for matrix elements like (8.26).

For the further evaluation we prove the following proposition, which is an extension of Proposition 3.11:

- **Proposition 8.1**: The recursion formula

$$\langle 0|\,\Psi_J\,\mathcal{A}(\Psi_{I_1}\ldots\Psi_{I_n})\mathcal{S}(B_{K_1}\ldots B_{K_m})\,|a\rangle = \sum_{p\in S_n}\frac{\mathrm{sgn}(p)}{n!}\frac{1}{n+1}\times \tag{8.42}$$

$$\Bigg[\sum_{l=1}^{n+1}(-1)^{l+1}\,\langle 0|\,\Psi_{I_{p_1}}\ldots\Psi_L\Psi_{I_{p_l}}\ldots\Psi_{I_{p_n}}\mathcal{S}(B_{K_1}\ldots B_{K_m})\,|a\rangle$$

$$+\sum_{l=1}^{n} lA_{JI_{p_1}}\,\langle 0|\,\Psi_{I_{p_2}}\ldots\Psi_{I_{p_n}}\,\mathcal{S}(B_{K_1}\ldots B_{K_m})\,|a\rangle\Bigg]$$

which generates the relation between non–antisymmetric standard products and antisymmetric products of field operators can be expressed by the functional relation

$$|\mathcal{G}_J(j,b;a)\rangle \;=\; (\frac{1}{i}\partial_J + \frac{i}{2}A_{JI}j_I)|\mathcal{G}(j,b;a)\rangle \quad . \tag{8.43}$$

**Proof:** We use the identity

$$\sum_{p\in S_n}\frac{\mathrm{sgn}(p)}{n!}\,\langle 0|\,\Psi_J\Psi_{I_{p_1}}\ldots\Psi_{I_{p_n}}\,\mathcal{S}(B_{K_1}\ldots B_{K_m})\,|a\rangle = \tag{8.44}$$

$$\sum_{p\in S_n}\frac{\mathrm{sgn}(p)}{n!}\frac{1}{n+1}\sum_{l=1}^{n+1}\langle 0|\,\Psi_J\Psi_{I_{p_1}}\ldots\Psi_{I_{p_n}}\,\mathcal{S}(B_{K_1}\ldots B_{K_m})\,|a\rangle$$

and shift in the summand $l'$ on the right–hand side of (8.44) the operator $\Psi_J$ by repeated anticommutations in the position $l'$, i.e. to

$$\langle 0|\,\Psi_{I_{p_1}}\ldots\Psi_{I_{p(l'-1)}}\Psi_J\Psi_{I_{p_{l'}}}\ldots\Psi_{I_{p_n}}\,\mathcal{S}(B_{K_1}\ldots B_{K_m})\,|a\rangle \quad .$$

After relabeling we get (8.42).

If we multiply (8.42) from the right with

$$\frac{i^n}{n!\,m!}j_{I_1}\ldots j_{I_n}\,b_{K_1}\ldots b_{K_m}|0\rangle_F \quad ,$$

we obtain, with the use of (8.31) and after relabeling, equation (8.43). ◇

The relation between non–symmetric standard products and symmetric products of boson field operators is given by

- **Proposition 8.2:**

$$|\mathcal{G}^L(j,b;a)\rangle = (\partial^b_L + \frac{1}{2}C_{LK}b_K)|\mathcal{G}(j,b;a)\rangle \quad . \tag{8.45}$$

**Proof:** The proof is analogous to the proof of Proposition 8.1 and will be omitted for brevity. ◇

- **Proposition 8.3:** For the general state (8.41) the relation

$$\left|\mathcal{G}_{I_1\dots I_r}^{K_1\dots K_s}(j,b;a)\right\rangle = \tag{8.46}$$
$$(\frac{1}{i}\partial_{I_r} + \frac{i}{2}A_{I_r I'_r} j_{I'_r}) \dots (\frac{1}{i}\partial_{I_1} + \frac{i}{2}A_{I_1 I'_1} j_{I'_1}) \times$$
$$(\partial^b_{K_s} + \frac{1}{2}C_{K_s K'_s} b_{K'_s}) \dots (\partial^b_{K_1} + \frac{1}{2}C_{K_1 K'_1} b_{K'_1})|\mathcal{G}(j,b;a)\rangle$$

holds.

**Proof:** Repeated application of Propositions 8.1 and 8.2. ◊

With the help of (8.46) we can rearrange equation (8.40) and obtain

$$E_{p0}|\mathcal{G}(j,b;p)\rangle = \tag{8.47}$$
$$\Big[j_{I_1} D_{I_1 I_2}\partial_{I_2} + b_{K_1} L_{K_1 K_2}\partial^b_{K_2}$$
$$+ j_{I_1} W_{I_1 I_2}^{K_1}\left(\partial_{I_2}\partial^b_{K_1} + \frac{1}{2}C_{K_1 K_2} b_{K_2}\partial_{I_2}\right)$$
$$+ j_{I_1} U_{I_1}^{J_1 J_2 J_3}\left(\partial_{J_1}\partial_{J_2}\partial_{J_3} - \frac{1}{4}A_{J_2 I_2} A_{J_1 I_3} j_{I_2} j_{I_3}\partial_{J_3}\right)$$
$$+ J_K^{J_1 J_2} b_K\left(\partial_{J_1}\partial_{J_2} + A_{I_1 J_1} j_{I_1}\partial_{J_2} + \frac{1}{4}A_{I_1 J_1} A_{I_2 J_2} j_{I_1} j_{I_2}\right)\Big]|\mathcal{G}(j,b;p)\rangle \ .$$

We now turn to the discussion of the bosonic commutator $C_{kk'}$ and consider the time derivation of the commutator

$$[\Psi_I, B_K]_- = 0 \quad .$$

Evaluating it by means of the dynamical equations (8.20) and the commutation relations (8.23)–(8.25) yields the consistency condition

$$C_{K_1 K_2} W_{I_1 I_2}^{K_2} = 2A_{I_1 I_3} J_{K_1}^{I_3 I_2} \quad . \tag{8.48}$$

Equation (8.48) relates the anticommutator $A_{I_1 I_2}$ and the dynamics with the commutator $C_{K_1 K_2}$ of the boson operators. If we substitute (8.48) into (8.47) we finally obtain

$$E_{p0}|\mathcal{G}(j,b;p)\rangle = \tag{8.49}$$
$$\Big[j_{I_1} D_{I_1 I_2}\partial_{I_2} + j_{I_1} W_{I_1 I_2}^{K}\partial_{I_2}\partial^b_K + b_{K_1} L_{K_1 K_2}\partial^b_{K_2}$$
$$+ j_{I_1} U_{I_1}^{J_1 J_2 J_3}(\partial_{J_1}\partial_{J_2}\partial_{J_3} - \frac{1}{4}A_{J_2 I_2} A_{J_1 I_3} j_{I_2} j_{I_3}\partial_{J_3})$$
$$+ J_K^{J_1 J_2} b_K(\partial_{J_1}\partial_{J_2} + \frac{1}{4}A_{I_1 J_1} A_{I_2 J_2} j_{I_1} j_{I_2})\Big]|\mathcal{G}(j,b;p)\rangle \quad .$$

By projection with

$$\frac{1}{i^n}\langle v|\partial^b_{K_m}\dots\partial^b_{K_1}\partial_{I_n}\dots\partial_{I_1}$$

we immediately obtain from (8.49) the equations for the matrix elements (8.26).

To start calculations which include renormalization we have to substitute equations (8.22) into equation (8.49). Equation (8.49) is valid beyond perturbation theory and is thus suitable for the discussion of general properties of spinor quantum electrodynamics.

With respect to the boson commutator we formulate the following proposition:

- **Proposition 8.4:** The concistency equation (8.48) together with the field dynamics (8.18) uniquely determines the commutator $C_{K_1K_2}$ to be the usual canonical commutator for the $A$– and $E$–fields in Coulomb gauge.

**Proof:** If we substitute the expressions for $A_{I_1I_2}$, $W^K_{I_1I_2}$ and $J^K_{I_1I_2}$ into (8.48), we get

$$e_0\int d^3z_2 C_{\substack{k_1k_2\\ \eta_1\eta_2}}(\mathbf{z}_1,\mathbf{z}_2)\alpha^{k_2}_{\alpha_1\alpha_2}\delta(\mathbf{r}_1-\mathbf{r}_2)\delta(\mathbf{r}_1-\mathbf{z}_2)\sigma^3_{\Lambda_1\Lambda_2}\delta_{\eta_2 1} = \tag{8.50}$$
$$-e_0\int d^3r_3(C\gamma^0)_{\alpha_1\alpha_3}\sigma^1_{\Lambda_1\Lambda_3}\delta(\mathbf{r}_1-\mathbf{r}_3)\,P_{\mathrm{tr}}(\mathbf{z}_1-\mathbf{r}_3)\delta(\mathbf{r}_3-\mathbf{r}_2)(C\gamma^k)_{\alpha_3\alpha_2}\sigma^2_{\Lambda_3\Lambda_2}\delta_{\eta_1 2}\,.$$

After performing the summations and integrations, we obtain

$$C_{\substack{k_1k_2\\ \eta_1 2}}(\mathbf{z}_1,\mathbf{r}_1) = -iP_{\mathrm{tr}}(\mathbf{z}_1-\mathbf{r}_1)\delta_{k_1k_2}\delta_{\eta_1 2}\quad . \tag{8.51}$$

If we choose $\eta_1 = 1$, we infer from (8.51)

$$\left[A^{k_1}_{\mathrm{tr}}(\mathbf{z}_1,t)\,,\,A^{k_2}_{\mathrm{tr}}(\mathbf{r}_1,t)\right]_- = 0 \tag{8.52}$$

and for $\eta_1 = 2$ we get

$$\left[E^{k_1}_{\mathrm{tr}}(\mathbf{z}_1,t)\,,\,A^{k_2}_{\mathrm{tr}}(\mathbf{r}_1,t)\right]_- = -iP_{\mathrm{tr}}(\mathbf{z}_1-\mathbf{r}_1)\delta_{k_1k_2}\quad . \tag{8.53}$$

The commutator relation between the transversal parts of the electric field cannot be read off directly from (8.48). However, we can deduce it with the help of (8.52), (8.53) and the dynamics. From (8.53) we obtain

$$\begin{aligned}\partial_0&\left[E^{k_1}_{\mathrm{tr}}(\mathbf{z}_1,t)\,,\,A^{k_2}_{\mathrm{tr}}(\mathbf{r}_1,t)\right]_- \\ &= \left[\partial_0E^{k_1}_{\mathrm{tr}}(\mathbf{z}_1,t)\,,\,A^{k_2}_{\mathrm{tr}}(\mathbf{r}_1,t)\right]_- + \left[E^{k_1}_{\mathrm{tr}}(\mathbf{z}_1,t)\,,\,\partial_0A^{k_2}_{\mathrm{tr}}(\mathbf{r}_1,t)\right]_- \\ &= 0\quad .\end{aligned} \tag{8.54}$$

Substitution of the field equations (8.18) into (8.54) gives

$$\left[\frac{ie_0}{2}P^{\mathbf{z}}_{\mathrm{tr}}\Psi(\mathbf{z})C\gamma^k\sigma^2\Psi(\mathbf{z}) + \Delta(\mathbf{z})A^k_{\mathrm{tr}}(\mathbf{z},t)\,,\,A^j_{\mathrm{tr}}(\mathbf{r},t)\right]_- + \left[E^k_{\mathrm{tr}}(\mathbf{z},t)\,,\,E^j_{\mathrm{tr}}(\mathbf{r},t)\right]_- = 0\,. \tag{8.55}$$

Hence, with (8.55), (8.24) and (8.52), we obtain

$$\left[E^k_{\mathrm{tr}}(\mathbf{z},t)\,,\,E^j_{\mathrm{tr}}(\mathbf{r},t)\right]_- = 0\quad . \tag{8.56}$$

Equations (8.52), (8.53) and (8.56) are the well known canonical commutation relations for the boson operators in Coulomb gauge. ◇

## 8.4 Composite Particle Dynamics

In spinor QED the fermions are commonly identified with electrons and positrons. Due to decay channels into gamma quanta no stable bound states between electrons and positrons exist, while bound states between electrons do not occur due to Coulomb repulsion. However, these "unstable" bound states have to be solutions of the full dynamical equations and thus are referred to dressed particle states.

As we pointed out in Chapter 4 there is some arbitrariness in defining those bound states to which weak mapping is referred to. In our case we again consider hard core bound states as eigenstates of the fermionic diagonalpart of the functional energy equation. As will be demonstrated for this part the gamma decay channels are closed and there appear truly bound states between electrons and positrons. If afterwards with these states the weak mapping procedure is performed, this stability is destroyed by interaction terms which appear as a consequence of this weak mapping. Thus the stability is an intermediate step which is afterwards removed by the theory. Therefore, in principle the weak mapping theorems can also be used for the description of decay processes.

In this section we apply the procedure of weak mapping to the mapping of spinor QED onto a theory of electron–positron bound states in the electromagnetic field. The treatment runs along the same lines as was exemplified in Chapter 4.

Bound states are described by corresponding matrix elements in terms of generating functionals. These matrix elements may be decomposed into correlated and uncorrelated parts. While the former describe genuine fermion processes, the latter represent contributions of the vacuum to the field dynamics. In order to have a "pure" composite particle theory such vacuum contributions have to be removed from the beginning. In particular for the treatment of two–fermion bound states the uncorrelated fermion propagators have to be eliminated from the generating functional; this is done by the nonperturbative normalordering of Section 2.5.

The propagators which appear in the generating functional are referred to the physical vacuum, i.e., they are the renormalized propagators. But the renormalized propagator which corresponds to the exact solution of the vacuum state is just (by renormalization definition) the ordinary *free* electron–positron propagator with the observable parameters $e$ and $m$. Thus we apply the transformation

$$|\mathcal{G}(j,b;a)\rangle = \exp\left(-\frac{1}{2}j_{I_1}F_{I_1I_2}j_{I_2}\right)|\mathcal{F}(j,b;a)\rangle \quad , \tag{8.57}$$

where $F_{I_1I_2}$ is the fermion propagator for equal times. According to Sections 3.8 and 2.5 the normaltransformation (8.57) induces the replacement of $\partial_I$ in equation (8.49) by

$$d_{I_1} := \partial_{I_1} - F_{I_1I_2}j_{I_2} \quad . \tag{8.58}$$

The normaltransformed equation (8.49) reads

$$E_{p0}\,|\mathcal{F}(j,b;p)\rangle = \tag{8.59}$$

$$\Big\{D_{I_1I_2}j_{I_1}\partial_{I_2} - Z_{I_1I_3}F_{I_3I_2}j_{I_1}j_{I_2} + L_{K_1K_2}b_{K_1}\partial^b_{K_2}$$

$$+W^K_{I_1I_2}\left[j_{I_1}\partial_{I_2}\partial^b_K - F_{I_2I_3}j_{I_1}j_{I_3}\partial^b_K\right]$$

$$-U_J^{I_1 I_2 I_3} j_J \Big[ \partial_{I_3} \partial_{I_2} \partial_{I_1} - 3F_{I_3 I_4} j_{I_4} \partial_{I_2} \partial_{I_1}$$

$$+ \Big( \frac{1}{4} A_{I_3 I_4} A_{I_2 I_5} + 3F_{I_3 I_4} F_{I_2 I_5} \Big) j_{I_4} j_{I_5} \partial_{I_1}$$

$$- \Big( \frac{1}{4} A_{I_3 I_4} A_{I_2 I_5} F_{I_1 I_6} + F_{I_3 I_4} F_{I_2 I_5} F_{I_1 I_6} \Big) j_{I_4} j_{I_5} j_{I_6} \Big]$$

$$+ J_K^{I_1 I_2} b_K \Big[ \partial_{I_1} \partial_{I_2} - 2F_{I_1 I_3} j_{I_3} \partial_{I_2}$$

$$+ \Big( \frac{1}{4} A_{I_1 I_3} A_{I_2 I_4} + F_{I_1 I_3} F_{I_2 I_4} \Big) j_{I_3} j_{I_4} \Big] \Big\} |\mathcal{F}(j, b; p)\rangle \quad ,$$

where we observed $j_{I_1} D_{I_1 I_2} F_{I_2 I_3} j_{I_3} = 0$, which is valid due to symmetry arguments. $D_{I_1 I_2}$ is referred to $m$ instead of $m_0$ and the counterterm $Z_{I_1 I_2}$ is given by

$$Z_{I_1 I_2} := -\,\delta m\, \gamma^0_{\alpha_1 \alpha_2} \delta_{\Lambda_1 \Lambda_2} \delta(\mathbf{r}_1 - \mathbf{r}_2) \quad . \tag{8.60}$$

Furthermore the normaltransformed functional state is defined by

$$|\mathcal{F}(j, b; a)\rangle := \tag{8.61}$$

$$\sum_{n,m=0}^{\infty} \frac{i^n}{m!n!} \varphi^{(n+m)} (I_1 \dots I_n; K_1 \dots K_m | a)\, j_{I_1} \dots j_{I_n}\, b_{K_1} \dots b_{K_m} |v\rangle \quad .$$

Analogously to the vertex renormalization in Section 2.6 in (8.59) we omitted terms which contain divergent integrals involving propagators at the origin.

According to weak mapping theory we define bound states to be solutions of the normal-transformed equation (8.59) with respect to their fermionic diagonal part. This diagonal part equation reads

$$E^d_{p0} |\mathcal{F}(j; p)\rangle^d = \Big( D_{I_1 I_2} j_{I_1} \partial_{I_2} + 3U_{I_1}^{I_2 I_3 I_4} F_{I_4 I_5} j_{I_1} j_{I_5} \partial_{I_3} \partial_{I_2} \Big) |\mathcal{F}(j; p)\rangle^d \quad , \tag{8.62}$$

where $m_0$ and $e_0$ are replaced by $m$ and $e$ and all counterterm expressions with $\delta m$ and $\delta e$ are excluded from (8.62); i.e. the bound states are referred to the observable quantities $m$ and $e$.

The set of solutions (8.62) can be assumed to be a complete set in $\mathbb{R}^{3n}$, $n = 1 \dots$, i.e., they describe all possible $n$–fermion states. In particular we denote the two–particle solutions by

$$|\mathcal{F}(j; 2, k)\rangle^d = -\frac{1}{2} C_k^{I_1 I_2} j_{I_1} j_{I_2} |0\rangle_F \, . \tag{8.63}$$

This set contains bound states as well as scattering states. In spite of the completeness, it is, however, in general *not* an orthogonal set since the eigenvalue equation (8.62) is *not* a Schrödinger equation and the solutions $C_k^{I_1 I_2}$ in general have *not* the meaning of quantum mechanical probability amplitudes. The non–Schrödinger properties of the $C_k^{I_1 I_2}$, in particular the non–orthogonality, were treated in detail in Chapter 3 and are of no relevance for weak mapping. It suffices to introduce dual functions $R^k_{I_1 I_2}$ which are defined by the orthogonality and completeness relations

$$\sum_{I_1,I_2} R^{k_1}_{I_1 I_2} C^{I_1 I_2}_{k_2} = \delta_{k_1 k_2} \quad , \tag{8.64}$$

$$\sum_{k} R^{k}_{I_1 I_2} C^{I'_1 I'_2}_{k} = \frac{1}{2}\left(\delta_{I_1 I'_1}\delta_{I_2 I'_2} - \delta_{I_1 I'_2}\delta_{I_2 I'_1}\right) \quad .$$

As the initial set of functions $C^{I_1 I_2}_k$ is explicitly known, the duals $R^k_{I_1 I_2}$ can explicitly be constructed, cf. Sections 4.8, 7.5.

We now study the functional mapping of equation (8.59) for the simplest case of a transformation to a two–fermion bound state theory. In order to perform such a mapping we have to define an effective bound state functional

$$|\mathcal{B}(b,c;a)\rangle := \sum_{m,n} \frac{1}{m!n!} \rho^{(m+n)}(K_1 \dots K_m; k_1 \dots k_n|a) b_{K_1} \dots b_{K_m} c_{k_1} \dots c_{k_n} |0\rangle_B \otimes |0\rangle_B \tag{8.65}$$

where the $c_k$ are independent bosonic sources which carry the quantum numbers of the two–particle solutions $C^{I_1 I_2}_k$. The exact relation between (8.61) and (8.65) is given by

$$\varphi^{(m+2n)}\left(K_1 \dots K_m; I_1 \dots I_{2n} \,|a\right) := C^{\{I_1 I_2}_{k_1} \dots C^{I_{2n-1} I_{2n}\}}_{k_n} \rho^{(m+n)}(K_1 \dots K_m; k_1 \dots k_n|a) \tag{8.66}$$

where $\{I_1 \dots I_{2n}\}$ means antisymmetrization in all included indices (cf. Section 4.4). If the $C^{I_1 I_2}_k$ build up a complete set, there is a one–to–one correspondence between the $\varphi$– and the $\rho$–coefficients of (8.61) and (8.65). However, in the case of a pure bound state dynamics we consider by definition only those states $|a\rangle$ for which the expansion (8.66) contains only these bound states and nothing else, i.e. no two–fermion scattering states which also appear in the complete set of $C^{I_1 I_2}_k$.

By (8.66) a mapping between the elementary QED state functionals (8.61) and QED fermionic bound state functionals (8.65) is established. The consequences of such a mapping with respect to the functional equation (8.59) were formulated by the weak mapping theorem, cf. Proposition 4.2. This theorem was proven for a pure spinor theory, but the proof can be taken over to QED without any modifications as far as fermionic bound states are considered.

In its full length the weak mapping theorem contains all exchange effects of the bound states which are due to their fermionic substructure. In order to reduce these rather lengthy expressions in Chapter 4 a short–cut calculation technique was given which leads to a better manageable version of the theory. It was demonstrated that this short–cut calculation can be justified if the bound state wavefunctions are highly concentrated. However, the corresponding estimates are rather complicated and should be completed by other criteria. For instance, one should expect that for low densities of the bound states, i.e. in ideal gas approximation for the composite particles, exchange effects can be neglected. Although this criterion has not yet been investigated, we assume that in quantum electrodynamics of bound states there exist experimental situations where it can be realized. Hence for a first draft of composite particle theory in quantum electrodynamics, for brevity we ignore exchange effects in the following and apply the short–cut calculation technique of Chapter 4.

In analogy to (4.94) or (6.6) we postulate the invariance relation

$$|\mathcal{B}(b,c;a)\rangle = |\mathcal{F}(j,b;a)\rangle \quad , \tag{8.67}$$

which implies

$$c_k = \frac{1}{2} C_k^{I_1 I_2} j_{I_1} j_{I_2} \ , \tag{8.68}$$

i.e., in the short–cut approach the sources $j_I, c_k$ are not considered to be independent of each other. With (8.67) and (8.68) we can work out the functional derivative

$$\partial_I \,|\mathcal{F}(j,b;a)\rangle \equiv \partial_I \,|\mathcal{B}(b,c[j];a)\rangle \quad , \tag{8.69}$$

which yields by direct calculation the functional chain rule

$$\partial_{I_1} \,|\mathcal{B}(b,c;a)\rangle = C_k^{I_1 I_2} j_{I_2} \partial_k^c \,|\mathcal{B}(b,c;a)\rangle \quad , \tag{8.70}$$

where $\partial_k^c \equiv \delta/\delta c_k$ is the "dual" of $c_k$. This dual is defined independendly of (8.68), i.e. it is not obtained by taking the dual of the right–hand side of (8.68).

Successive application of (8.70) yields

$$\begin{aligned} \partial_{I_2}\partial_{I_1} \,|\mathcal{B}(b,c;p)\rangle &= -C_{k_1}^{I_1 I_3} C_{k_2}^{I_2 I_4} j_{I_3} j_{I_4} \partial_{k_1}^c \partial_{k_2}^c \,|\mathcal{B}(b,c;a)\rangle \\ &\quad + C_k^{I_1 I_2} \partial_k^c \,|\mathcal{B}(b,c;a)\rangle \quad , \end{aligned} \tag{8.71}$$

$$\begin{aligned} \partial_{I_3}\partial_{I_2}\partial_{I_1} \,|\mathcal{B}(b,c;a)\rangle &= -C_{k_1}^{I_1 I_4} C_{k_2}^{I_2 I_5} C_{k_3}^{I_3 I_6} j_{I_4} j_{I_5} j_{I_6} \partial_{k_1}^c \partial_{k_2}^c \partial_{k_3}^c \,|\mathcal{B}(b,c;a)\rangle \\ &\quad -3 \left( C_{k_1}^{I_1 I_3} C_{k_2}^{I_2 I_4} j_{I_4} \partial_{k_1}^c \partial_{k_2}^c \right)_{\mathrm{as}\{I_1,I_2,I_3\}} |\mathcal{B}(b,c;a)\rangle \quad . \end{aligned} \tag{8.72}$$

With (8.67), (8.70)–(8.72) and the relations

$$\begin{aligned} j_{I_1} j_{I_2} &= 2R_{I_1 I_2}^k c_k \ , \\ j_{I_1} j_{I_2} j_{I_3} j_{I_4} &= 4R_{\{I_1 I_2}^{k_1} R_{I_3 I_4\}}^{k_2} c_{k_1} c_{k_2} \end{aligned} \tag{8.73}$$

we can map the functional equation (8.59) onto a functional equation for the functional state $|\mathcal{B}(b,c;a)\rangle$. The result is

$$\begin{aligned} E_{p0} \,|\mathcal{B}(b,c;p)\rangle = \Big\{ & R_{I_1 I_2}^{k'} D_{I_1 I_3} C_k^{I_3 I_2} c_{k'} \partial_k^c - R_{I_1 I_2}^k Z_{I_1 I_3} F_{I_3 I_2} c_k \\ + \ & W_{I_1 I_2}^K \Big[ R_{I_1 I_3}^k C_{k'}^{I_2 I_3} c_k \partial_{k'}^c \partial_K^b - R_{I_1 I_3}^k F_{I_2 I_3} c_k \partial_K^b \Big] + b_{K_1} L_{K_1 K_2} \partial_{K_2}^b \\ - \ & U_J^{I_1 I_2 I_3} \Big[ 4R_{\{J I_4}^{k_1} R_{I_5 I_6\}}^{k_2} C_{k_1'}^{I_1 I_4} C_{k_2'}^{I_2 I_5} C_{k_3'}^{I_3 I_6} c_{k_1} c_{k_2} \partial_{k_1'}^c \partial_{k_1'}^c \partial_{k_3'}^c \\ - \ & 3R_{J I_4}^k C_{k_1'}^{I_1 I_3} C_{k_2'}^{I_2 I_4} c_k \partial_{k_1'}^c \partial_{k_2'}^c + 12 R_{\{J I_4}^{k_1} R_{I_5 I_6\}}^{k_2} F_{I_3 I_4} C_{k_1'}^{I_1 I_5} C_{k_2'}^{I_2 I_6} c_{k_1} c_{k_2} \partial_{k_1'}^c \partial_{k_2'}^c \\ - \ & 3R_{J I_4}^k F_{I_3 I_4} C_{k'}^{I_1 I_2} c_k \partial_{k'}^c + \Big( 3F_{I_3 I_4} F_{I_2 I_5} + \frac{1}{4} A_{I_3 I_4} A_{I_2 I_5} \Big) R_{\{J I_4}^{k_1} R_{I_5 I_6\}}^{k_2} C_{k'}^{I_1 I_6} c_{k_1} c_{k_2} \partial_{k'}^c \\ - \ & 4 \Big( F_{I_3 I_4} F_{I_2 I_5} + \frac{1}{4} A_{I_3 I_4} A_{I_2 I_5} \Big) F_{I_1 I_6} R_{\{J I_4}^{k_1} R_{I_5 I_6\}}^{k_2} c_{k_1} c_{k_2} \Big] \\ + \ & J_K^{I_1 I_2} b_K \Big[ \Big( \frac{1}{4} A_{I_1 I_3} A_{I_2 I_4} + F_{I_1 I_3} F_{I_2 I_4} \Big) R_{I_3 I_4}^k c_k - 2R_{I_3 I_4}^k F_{I_1 I_3} C_{k'}^{I_2 I_4} c_k \partial_{k'}^c \\ - \ & R_{I_3 I_4}^k C_{k_1'}^{I_2 I_3} C_{k_2'}^{I_1 I_4} c_k \partial_{k_1'}^c \partial_{k_2'}^c + C_{k'}^{I_2 I_1} \partial_{k'}^c \Big] \Big\} \,|\mathcal{B}(b,c;p)\rangle \quad . \end{aligned} \tag{8.74}$$

As we asserted above there are terms in (8.74) which describe the electron–positron decay, e.g. the terms $c_k \partial^c_{k'} \partial^b_K$.

Apart from exchange effects equation (8.74) is strictly valid without any reference to perturbation theory. In order to properly renormalize the bound state theory we bring perturbation theory into play by treating all bound state–bound state interactions nonperturbatively, but use perturbation theory for the interaction of the transversal electromagnetic field with the bound states. In this way the contradiction between nonperturbative functional equations and perturbative renormalization is removed. It remains to give a comment with respect to renormalization for bound states.

This question is concerned with the determination of the values of the counterterms $\delta m$ and $\delta e$ if we apply the weak mapping procedure, i.e. transform quantum electrodynamics into a bound state dynamics. Primarily, counterterms arise from the renormalization of free electron (positron) mass and charge. Additionally, they are destined to remove all divergencies from the theory under consideration. As in the renormalized equations the empirical mass $m$ and charge $e$ are fixed, it follows that the counterterms are uniquely determined by the condition that the calculation of the selfenergy and electromagnetic coupling constant of the dressed electron (and positron) may not alter the original empirical values of $m$ and $e$; i.e. $\delta m$ and $\delta e$ have to compensate all expressions which would change these empirical values. It is part of the renormalization theory of quantum electrodynamics that simultaneously by this condition also all infinities from the interaction picture with free–field representation (shortly: interaction representation) are removed.

Now we consider weak mapping. For instance, we assume a mapping by the two–fermion states, given by (8.68):

$$c_k := \frac{1}{2} \int C^k_{Z_1 Z_2}(\mathbf{r}_1, \mathbf{r}_2) j_{Z_1}(\mathbf{r}_1) j_{Z_2}(\mathbf{r}_2) d^3 r_1 d^3 r_2 \quad , \tag{8.75}$$

where $\{C^k_{Z_1 Z_2}(\mathbf{r}_1, \mathbf{r}_2)\}$ is a complete set of two–fermion wave functions. Among this set we have (Coulomb problem!) bound state and scattering state solutions. The scattering state solutions refer to ingoing or outgoing electrons (positrons). By dressing these particles with vacuum– and radiative–corrections we have the same situation as in the case of the interaction representation with perturbation expansion: The counterterms have to compensate all terms which would alter the already used empirical values $m$ and $e$. Although the Coulomb forces in the Coulomb gauge prevent an asymptotic behaviour in terms of free electrons (positrons), the conclusion is that in the bound state representation the counterterms $\delta m$ and $\delta e$ are already fixed by the interaction representation. This means that there is no freedom with respect to bound states for a renewed adaption of the counterterms, as bound states and scattering states occur in one map. On the other hand it must be guaranteed that by this choice for $\delta m$ and $\delta e$ also the radiative correction singularities for bound states are compensated. This is *not* selfevident. However, if this should not be the case, there is no escape of the conclusion that renormalization cannot be applied to bound state problems.

In the literature it is demonstrated for hydrogen bound states that this procedure indeed works, see, e.g., Schweber, Bethe and Hoffmann [Schwe 55], and for one–electron bound states the Furry picture [Fur 51] is used for renormalization. A general proof of renormalizability for bound states, however, seems not to be available.

Let us give an argument that renormalizability may also work in the general case. The argument depends crucially on the possibility of a Hamiltonian, i.e. a Schrödinger representation of quantum electrodynamics. For this representation it suffices to consider all states in a one–time representation which is used as the basis for the weak mapping procedure. In this representation for instance the two–particle solutions $C_k^{I_1 I_2}$ can be resolved into a Fourier series

$$C^k_{Z_1 Z_2}(\mathbf{r}_1, \mathbf{r}_2) = \int \tilde{C}^k_{Z_1 Z_2}(\mathbf{p}_1, \mathbf{p}_2) e^{-i(\mathbf{p}_1 \mathbf{r}_1 + \mathbf{p}_2 \mathbf{r}_2)} d^3 p_1 d^3 p_2 \,, \tag{8.76}$$

i.e. a plane wave representation which is simultaneously the basis of the pure interaction representation. Thus, in the Hamilton formalism the bound state representation and the interaction representation can be referred to the same basis system. Thus, if renormalization works in this basis system it must necessarily work in both representations.

Finally we want to mention that the attempt to introduce dressed electrons and positrons by weak mapping already from the beginning, and only afterwards study the formation of bound states will not solve the above mentioned renormalization problem, but is much more difficult to perform. In this way the weak mapping with hard core states seems to be the most promising approach to the discussion of bound state dynamics in renormalizable spinor quantum electrodynamics.

## 8.5 Nonabelian Quantum Fields in Temporal Gauge

Due to their applications in quantum chromodynamics, electroweak theory and in numerous more general elementary particle theories, nonabelian gauge fields are at present one of the centers of interest in theoretical physical research. But in contrast to the abelian gauge group of quantum electrodynamics, the nonabelian gauge groups lead to nonlinear relations between vector potentials and field strength tensors causing severe difficulties with respect to quantization and renormalization. As one way to circumvent these difficulties the use of path integrals has been proposed. But this again offers basic as well as technical difficulties which were gradually recognized and which we will discuss in the summary of Chapter 10. Hence to escape this situation, with increasing intensity canonical quantization of nonabelian gauge fields is studied, see, e.g., Gaigg, Kummer and Schweda [Gai 90]. As quantization is connected with gauge fixing, numerous versions of quantization related to different gauges have been studied.

In a rough classification one can distinguish the quantization prescriptions by the use of covariant or noncovariant gauges. With respect to the treatment of covariant gauges the most prominent one seems to be the BRS–formalism which leads to new global symmetries. While such covariant gauges are of basic theoretical interest, the use of noncovariant gauges has turned out to be crucial for an improved understanding of gauge field phenomena. "The basic reason for this ... can be traced back to the fact that such gauges lead to the reduction of redundant degrees of freedom. However, the price to be paid for these more physical features in formal approaches has been serious complications at the mathematical level. These problems are almost exclusively tied to the appearance of spurious poles in the propagator of the gauge field and may even lead to a nonstandard renormalization..."

"On the other hand, new original ideas have emerged recently which seem to pave a way around these obstacles", see Gaigg, Kummer and Schweda [Gai 90]. It would exceed the scope of this book to describe the present situation in this field in detail. Nevertheless, it is remarkable that by several authors the temporal gauge is considered as the "physical" gauge which sets the standard for the treatment of nonabelian gauge theories, see Cheng [Che 90], Tsai [Tsai 90], Haller [Hall 90], Huang [Hua 92], Jackiw [Jack 80], Goldstone and Jackiw [Gold 78] and Leibbrand [Leib 87]. This opinion is in accordance with our results which distinguish the temporal gauge, because it is related to the energy representation of quantum fields. Thus, in order to compare the results of weak mapping with phenomenological theories we derive the energy representation of nonabelian gauge fields in the temporal gauge.

The quantization in this gauge is equivalent to the BRS–formalism, cf. Tsai [Tsai 90], and unitarity and relativistic covariance can be established, cf. Cheng [Che 90]. However, on account of nonstandard renormalization, the nontrivial Gauß law as a weak constraint and confinement, we restrict ourselves to the derivation of the energy representation without attempting to develop a complete composite particle theory for nonabelian quantum fields in analogy to quantum electrodynamics. In particular in contrast to quantum electrodynamics the Coulomb gauge cannot be exactly evaluated at the classical level because of the nonlinearity of the Gauß law for nonabelian fields, and thus this case is ruled out as a candidate of gauge fixing. Hence at any rate we must deviate from the treatment of quantum electrodynamics, and this may serve as a further justification for starting with the temporal gauge, see also the remarks in Section 8.1.

In spite of the fact that the full treatment of composite particle theory for nonabelian fields remains for future investigations, we adjust our discussion with regard to this aim. This in particular means that we include the coupling of the bosons to nonabelian fermions from the beginning. In this case the Lagrangian density has the general form

$$\mathcal{L} := -\frac{1}{4} F^a_{\mu\nu} F^{\mu\nu}_a + \bar{\chi}(i\gamma^\mu D_\mu - m)\chi \tag{8.77}$$

where

$$F^a_{\mu\nu} := \partial_\mu A^a_\nu - \partial_\nu A^a_\mu + g c_{abc} A^b_\mu A^c_\nu \, , \tag{8.78}$$

$$D_\mu := \partial_\mu - i g A^a_\mu L_a \tag{8.79}$$

with $L_a$, $a = 1 \dots N^2 - 1$ the matrix representation of the generators of the nonabelian $SU(N)$ group with the Lie algebra $[L_a, L_b]_- = i c_{abc} L_c$ and $\chi \equiv \chi_{\alpha A}(x)$ the $N$–dimensional fermion field representation of the group, while $A^a_\nu$ is the nonabelian vector potential. The corresponding field equations read

$$\nabla^{ab}_\mu F^{\nu\mu}_b = j^\nu_a \tag{8.80}$$

and

$$\begin{aligned} (i\gamma^\mu \partial_\mu - m)\chi + g\gamma^\mu A^a_\mu L_a \chi &= 0 \, , \\ (i\gamma^\mu \partial_\mu - m)\chi^c - g\gamma^\mu A^a_\mu L^T_a \chi^c &= 0 \end{aligned} \tag{8.81}$$

with

$$\nabla_\mu^{ab} := \delta_{ab}\partial_\mu + g c_{cba} A_\mu^c \tag{8.82}$$

$$\begin{aligned} j_a^\mu &:= g(:\bar{\chi}\gamma^\mu L_a \chi :) \\ &=: \frac{g}{2}\left[\bar{\chi}_{\alpha A}\gamma^\mu_{\alpha_1\beta}(L_a)_{AB}\chi_{\beta B} - \chi_{\beta B}\gamma^\mu_{\alpha\beta}(L_a^T)_{AB}\bar{\chi}_{\alpha A}\right] \end{aligned} \tag{8.83}$$

and $\chi^c_{\alpha A} = C_{\alpha\beta}\bar{\chi}_{\beta A}$ as the "charge" conjugated spinorfield. As in quantum electrodynamics we also introduce an abbreviated notation for nonabelian fields in order to accomplish the rearrangements of our "algebraic" treatment. As a first step we combine $\chi$ and $\chi^c$ into one superspinor $\Psi$ in accordance with (8.3). This gives

$$\Psi_{\alpha A\Lambda} := \begin{cases} \chi_{\alpha A} & \text{for} \quad \Lambda = 1 \\ \chi^c_{\alpha A} & \text{for} \quad \Lambda = 2\,, \end{cases} \tag{8.84}$$

and in addition we need the definition of the nonabelian superspinor generators

$$(\hat{L}_a)_{\substack{\Lambda\Lambda' \\ AA'}} := \delta_{\Lambda 1}\delta_{\Lambda' 1}(L_a)_{AA'} + \delta_{\Lambda 2}\delta_{\Lambda' 2}(L_a^t)_{AA'}\,. \tag{8.85}$$

With these definitions the field equations (8.80), (8.81) go over into the set of equations

$$\nabla_\mu^{ab}\, F_b^{\nu\mu} = -i\frac{g}{2}\Psi C\gamma^\nu\sigma^2\hat{L}_a\Psi \tag{8.86}$$

and

$$(i\gamma^\mu\partial_\mu - m)\Psi + gA_\mu^a\gamma^\mu\sigma^3\hat{L}_a\Psi = 0 \quad , \tag{8.87}$$

where $C \equiv C_{\alpha\beta}$ is the charge conjugation matrix in spinor space, while $\sigma^i \equiv \sigma^i_{\Lambda\Lambda'}$, $i = 1,2,3$, are the Pauli matrices in superspinor space. If we introduce the nonabelian "electric" field strength $E_a^k := F_a^{k0}$ we obtain from (8.86) for $\nu = 1,2,3$

$$\partial_0 E_a^\ell = -g c_{cba} A_0^c E_b^\ell - \nabla_k^{ab} F_b^{\ell k} - i\frac{g}{2}\Psi C\gamma^\ell\sigma^2\hat{L}_a\Psi \tag{8.88}$$

and for $\nu = 0$

$$\nabla_k^{ab}\, E_b^k = i\frac{g}{2}\Psi C\gamma^0\sigma^2\hat{L}_a\Psi \quad , \tag{8.89}$$

while with the definition of $E_a^k$ it follows from (8.78)

$$\partial_0 A_a^k = -E_a^k - \partial_k A_a^0 - g c_{abc} A_b^0 A_c^k \quad . \tag{8.90}$$

We now pass from the general form of equations (8.86), (8.87) into a special gauge, namely the temporal gauge, by the gauge fixing condition $A_a^0 \equiv 0$, which we assume to apply to classical as well as to quantum theory. If we substitute definitions (8.78), (8.82) into (8.88), (8.89), these equations and (8.87), (8.90) read in temporal gauge

$$i\partial_0\Psi = -(i\alpha^k\partial_k - \beta m)\Psi + gA_a^k\alpha^k\sigma^3\hat{L}_a\Psi \tag{8.91}$$

$$i\partial_0 A_a^k = -iE_a^k \tag{8.92}$$

$$\begin{aligned} i\partial_0 E_a^k = i\Big[&-\Delta A_a^k + \partial_k\partial_j A_a^j + g c_{abc}(\partial_j A_b^j)A_c^k + 2g c_{abc} A_b^j(\partial_j A_c^k) \\ &- g c_{abc} A_b^j(\partial_k A_c^j) - g^2 c_{abc}c_{cde} A_b^j A_d^j A_e^k\Big] + \frac{g}{2}\Psi C\gamma^k\sigma^2\hat{L}_a\Psi \end{aligned} \tag{8.93}$$

$$\partial_k E_a^k = g c_{abc} A_b^k E_c^k - i\frac{g}{2}\Psi C\gamma^0\sigma^2\hat{L}_a\Psi \tag{8.94}$$

with $\alpha^k = \gamma^0\gamma^k$, $\beta = \gamma^0$. In analogy to (8.19) we introduce the superindex $K := (\mathbf{z}, j, a, \eta)$ and define

$$B_K := \begin{cases} A_a^j(\mathbf{z}) & \text{for} \quad \eta = 1 \\ E_a^j(\mathbf{z}) & \text{for} \quad \eta = 2 \,, \end{cases} \tag{8.95}$$

and $\Psi_I := \Psi_{\alpha A \Lambda}(\mathbf{r})$ with $I := (\mathbf{r}, \alpha, A, \Lambda)$. Then the dynamical equations (8.91), (8.92), (8.93) go over into

$$i\partial_0 \Psi_{I_1} = D_{I_1 I_2}\Psi_{I_2} + W^K_{I_1 I_2} B_K \Psi_{I_2} \tag{8.96}$$

$$\begin{aligned} i\partial_0 B_{K_0} = {} & L_{K_0 K_1} B_{K_1} + M^{K_1 K_2}_{K_0} B_{K_1} B_{K_2} \\ & + V^{K_1 K_2 K_3}_{K_0} B_{K_1} B_{K_2} B_{K_3} + J^{I_1 I_2}_{K_0} \Psi_{I_1} \Psi_{I_2} \,, \end{aligned} \tag{8.97}$$

while the constraint equation (8.94) can be written

$$N^L_K B_K + S^L_{K_1 K_2} B_{K_1} B_{K_2} + R^L_{I_1 I_2} \Psi_{I_1} \Psi_{I_2} = 0 \,. \tag{8.98}$$

For the formulation of these equations the following definitions are used:

$$D_{I_1 I_2} := -(i\alpha^k \partial_k - \beta m)_{\alpha_1 \alpha_2} \delta_{\Lambda_1 \Lambda_2} \delta_{A_1 A_2} \delta(\mathbf{r}_1 - \mathbf{r}_2) \tag{8.99}$$

$$W^K_{I_1 I_2} := g\alpha^j_{\alpha_1 \alpha_2} \sigma^3_{\Lambda_1 \Lambda_3} (\hat{L}_a)_{\substack{\Lambda_3 \Lambda_2 \\ A_1 A_2}} \delta_{1\eta} \delta(\mathbf{r}_1 - \mathbf{r}_2) \delta(\mathbf{r}_1 - \mathbf{z}) \tag{8.100}$$

$$\begin{aligned} L_{K_1 K_2} := {} & -i\delta_{j_1 j_2} \delta_{a_1 a_2} \Big[ \delta_{\eta_1 1} \delta_{\eta_2 2} \delta(\mathbf{z}_1 - \mathbf{z}_2) - \delta_{\eta_1 2} \delta_{\eta_2 1} \partial_k^2(\mathbf{z}_1) \delta(\mathbf{z}_1 - \mathbf{z}_2) \Big] \\ & + i\delta_{\eta_1 2} \delta_{\eta_2 1} \delta_{a_1 a_2} \partial_{j_1}(\mathbf{z}_1) \partial_{j_2}(\mathbf{z}_1) \delta(\mathbf{z}_1 - \mathbf{z}_2) \end{aligned} \tag{8.101}$$

$$\begin{aligned} M^{K_1 K_2}_{K_0} := {} & ig \Big\{ \delta_{\eta_0 2} \delta_{\eta_1 1} \delta_{\eta_2 1} c_{a_0 a_1 a_2} \Big[ \delta_{j_0 j_2} \delta(\mathbf{z}_0 - \mathbf{z}_2) \partial_{j_1}(\mathbf{z}_0) \delta(\mathbf{z} - \mathbf{z}_1) \\ & + 2\delta_{j_0 j_2} \delta(\mathbf{z}_0 - \mathbf{z}_1) \partial_{j_1}(\mathbf{z}_0) \delta(\mathbf{z} - \mathbf{z}_2) \\ & - \delta_{j_1 j_2} \delta(\mathbf{z}_0 - \mathbf{z}_1) \partial_{j_0} \delta(\mathbf{z}_0 - \mathbf{z}_2) \Big] \Big\}_{\mathrm{sym}[\mathrm{K}_1, \mathrm{K}_2]} \end{aligned} \tag{8.102}$$

$$\begin{aligned} V^{K_1 K_2 K_3}_{K_0} := {} & -\Big[ ig^2 \delta_{\eta_0 2} \delta_{\eta_1 1} \delta_{\eta_2 1} \delta_{\eta_3 1} c_{a_0 a_1 b} c_{b a_2 a_3} \times \\ & \delta_{j_1 j_2} \delta_{j_0 j_3} \delta(\mathbf{z}_0 - \mathbf{z}_1) \delta(\mathbf{z}_0 - \mathbf{z}_2) \delta(\mathbf{z}_0 - \mathbf{z}_3) \Big]_{\mathrm{sym}[\mathrm{K}_1, \mathrm{K}_2, \mathrm{K}_3]} \end{aligned} \tag{8.103}$$

$$J^{I_1 I_2}_K := \frac{g}{2} \delta_{\eta 2} (C\gamma^j)_{\alpha_1 \alpha_2} \sigma^2_{\Lambda_1 \Lambda_3} (\hat{L}_a)_{\substack{\Lambda_3 \Lambda_2 \\ A_1 A_2}} \delta(\mathbf{z} - \mathbf{r}_1) \delta(\mathbf{z} - \mathbf{r}_2) \tag{8.104}$$

$$N^L_{K_1} := \delta_{a a_1} \delta_{\eta_1 2} \partial_{j_1}(\mathbf{z}) \delta(\mathbf{z} - \mathbf{z}_1) \tag{8.105}$$

$$S^L_{K_1 K_2} := g \Big[ c_{a a_1 a_2} \delta_{j_1 j_2} \delta(\mathbf{z} - \mathbf{z}_1) \delta(\mathbf{z} - \mathbf{z}_2) \delta_{\eta_1 1} \delta_{\eta_2 2} \Big]_{\mathrm{sym}[\mathrm{K}_1, \mathrm{K}_2]} \tag{8.106}$$

$$R^L_{I_1 I_2} := -i\frac{g}{2} (C\gamma^0)_{\alpha_1 \alpha_2} \sigma^2_{\Lambda_1 \Lambda_3} (\hat{L}_a)_{\substack{\Lambda_3 \Lambda_2 \\ A_1 A_2}} \delta(\mathbf{z} - \mathbf{r}_1) \delta(\mathbf{z} - \mathbf{r}_2) \tag{8.107}$$

with the reduced index $L = (\mathbf{z}, a)$.

Next we have to quantize the dynamical variables of our theory. As in quantum electrodynamics, also for nonabelian fields it can be shown that the fermion field quantization fixes the boson field quantization. Analogously to Proposition 8.4 from the fermion field anticommutator

$$\left[\Psi_{I_1}, \Psi_{I_2}\right]_+ = A_{I_1 I_2} \tag{8.108}$$

with

$$A_{I_1 I_2} := (C\gamma^0)_{\alpha_1\alpha_2}\delta_{A_1 A_2}\sigma^1_{\Lambda_1\Lambda_2}\delta(\mathbf{r}_1 - \mathbf{r}_2)\mathbb{I} \tag{8.109}$$

and from vanishing commutators between $B$– and $\Psi$–fields the $c$–number boson field commutator

$$\left[B_{K_1}, B_{K_2}\right]_- = C_{K_1 K_2}\mathbb{I} \tag{8.110}$$

is determined to have the form

$$C_{K_1 K_2} = -\delta_{j_1 j_2}\sigma^2_{\eta_1\eta_2}\delta_{a_1 a_2}\delta(\mathbf{z}_1 - \mathbf{z}_2) \quad . \tag{8.111}$$

With the relations (8.108)–(8.111) the fermion and boson field quantization is defined and we can apply it to the system under consideration. This in particular means that the dynamical equations (8.96) and (8.97) are considered as operator equations in the Heisenberg picture, while in order to avoid contradictions the constraint equations (8.98) are assumed to be valid only by acting on physical states $|a\rangle$ of the corresponding state space. This weak implementation is in accordance with our general algebraic treatment of quantum fields, i.e., we can proceed along the same lines as we did for theories without constraint. First we define the state functionals for all physical states $|a\rangle$ by

$$|\mathcal{G}(f,b;a)\rangle \;:=\; \sum_{n,m=0}^{\infty} \frac{i^n}{n!m!}\langle 0|\mathcal{S}\left(B_{K_1}\ldots B_{K_m}\right)\mathcal{A}\left(\Psi_{I_1}\ldots\Psi_{I_n}\right)|a\rangle \times \tag{8.112}$$
$$f_{I_1}\ldots f_{I_1} b_{K_1}\ldots b_{K_m}|v\rangle \, .$$

Next we derive the dynamical equation for (8.112), and then the constraint equation. The dynamical equation follows by analogous rearrangements to that which we performed in Sections 3.7 and 8.3 at great length. For short we do not explicitly repeat such calculations for the nonabelian case. Rather we give only the results of these extensive calculations which were carried out by Pfister [Pfi 93]. The corresponding energy functional equations read

$$E_{p0}|\mathcal{G}(f,b;p)\rangle = \tag{8.113}$$

$$\Big[f_{I_1} D_{I_1 I_2}\partial^f_{I_2} + f_{I_1} W^K_{I_1 I_2}\partial^f_{I_2}\partial^b_K + b_{K_1} L_{K_1 K_2}\partial^b_{K_2}$$

$$+J^K_{I_1 I_2} b_K\left(\partial^f_{I_1}\partial^f_{I_2} + \frac{1}{4}A_{I_1 I_3}A_{I_2 I_4} f_{I_3} f_{I_4}\right)$$

$$+M^{K_1 K_2}_{K_0} b_{K_0}\left(\partial^b_{K_1}\partial^b_{K_2} + \frac{1}{12}C_{K_1 K_3}C_{K_2 K_4} b_{K_3} b_{K_4}\right)$$

$$+V^{K_1 K_2 K_3}_{K_0} b_{K_0}\left(\partial^b_{K_1}\partial^b_{K_2}\partial^b_{K_3} + \frac{1}{4}C_{K_0 K_4}C_{K_2 K_5} b_{K_4} b_{K_5}\partial^b_{K_3}\right)\Big]|\mathcal{G}(f,b;p)\rangle \quad .$$

With respect to the constraint equation (8.98) we apply this equation to a physical state $|a\rangle$ and operate from the left with the GNS–basis vector $\langle 0|\mathcal{S}(B_{K_1}\dots B_{K_m})\times\mathcal{A}(\Psi_{I_1}\dots\Psi_{I_n})$ on this equation. Then we can use the rearrangement techniques of Sections 3.7 and 8.3 and obtain the functional constraint equation

$$\Big[N^L_{K_1}\Big(\partial^b_{K_1}-\frac{1}{2}C_{K_1K_2}b_{K_2}\Big) \tag{8.114}$$

$$+S^L_{K_1K_2}\Big(\partial^b_{K_1}\partial^b_{K_2}-C_{K_1K_3}b_{K_3}\partial^b_{K_2}+\frac{1}{4}C_{K_2K_3}C_{K_1K_4}b_{K_3}b_{K_4}\Big)$$

$$+R^L_{I_1I_2}\Big(\partial^f_{I_2}\partial^f_{I_1}-A_{I_2I_3}f_{I_3}\partial^f_{I_1}+\frac{1}{4}A_{I_2I_3}A_{I_1I_4}f_{I_3}f_{I_4}\Big)\Big]|\mathcal{G}(f,b;a)\rangle=0$$

with $L=(\mathbf{z},a)$.

From these considerations it follows that functional quantum theory is ideally suited for the treatment of constraints, as the formulation of the weak implementation is in accordance with the general lines of Algebraic Schrödinger Representation.

For comparison with the results of weak mapping performed in Chapter 6 it is convenient to normaltransform equation (8.113); for a further discussion of normaltransformation we refer to Sections 2.5 and 6.5. In the combined boson–fermion case we introduce normalordered functional states by

$$|\mathcal{N}(f,b;a)\rangle:=\exp\Big[-\frac{1}{2}b_{K_1}F_{K_1K_2}b_{K_2}\Big]\exp\Big[-\frac{1}{2}f_{I_1}F_{I_1I_2}f_{I_2}\Big]|\mathcal{G}(f,b;a)\rangle \tag{8.115}$$

with the propagators $F_{K_1K_2}:=\langle 0|\mathcal{S}(B_{K_1}B_{K_2})|0\rangle$ and the fermion propagator $F_{I_1I_2}$.

Observing (2.89) and the corresponding bosonic relation

$$\partial^b_{K_1}\exp\Big[-\frac{1}{2}b_{K_2}F_{K_2K_3}b_{K_3}\Big]=\exp\Big[-\frac{1}{2}b_{K_2}F_{K_2K_3}b_{K_3}\Big]d^b_{K_1} \tag{8.116}$$

with

$$d^b_{K_1}:=\partial^b_{K_1}+F_{K_1K_2}b_{K_2} \tag{8.117}$$

we obtain from (8.113) the normalordered equation

$$E_{p0}|\mathcal{N}(f,b;p)\rangle= \tag{8.118}$$

$$\Big[f_{I_1}D_{I_1I_2}d^f_{I_2}+f_{I_1}W^K_{I_1I_2}d^f_{I_2}d^b_K+b_{K_1}L_{K_1K_2}d^b_{K_2}$$

$$+J^K_{I_1I_2}b_K\Big(d^f_{I_1}d^f_{I_2}+\frac{1}{4}A_{I_1I_3}A_{I_2I_4}f_{I_3}f_{I_4}\Big)$$

$$+M^{K_1K_2}_{K_0}b_{K_0}\Big(d^b_{K_1}d^b_{K_2}+\frac{1}{12}C_{K_1K_3}C_{K_2K_4}b_{K_3}b_{K_4}\Big)$$

$$+V^{K_1K_2K_3}_{K_0}b_{K_0}\Big(d^b_{K_1}d^b_{K_2}d^b_{K_3}+\frac{1}{4}C_{K_0K_4}C_{K_2K_5}b_{K_4}b_{K_5}d^b_{K_3}\Big)\Big]|\mathcal{N}(f,b;p)\rangle$$

with definition (8.117) and the corresponding fermion operator

$$d^f_{I_1} := \partial^f_{I_1} + F_{I_1 I_2} f_{I_2} \quad . \tag{8.119}$$

Finally the resolution of the superindices in (8.118) by means of (8.99)–(8.104) yields

$$\begin{aligned}
&E_{p0}|\mathcal{N}(f,b;p)\rangle \qquad (8.120)\\
&= \Big\{ -f_{\alpha_1 A_1 \Lambda_1}(\mathbf{r}) \Big[ i\alpha^k \partial_k(\mathbf{r}) - \beta m \Big]_{\alpha_1\alpha_2} d^f_{\alpha_2 A_1 \Lambda_1}(\mathbf{r}) \\
&\quad + g f_{\alpha_1 A_1 \Lambda_1}(\mathbf{r}) \alpha^j_{\alpha_1\alpha_2} \sigma^3_{\Lambda_1\Lambda_2} (\hat{L}^a)_{\substack{\Lambda_2\Lambda_3\\A_1A_2}} d^f_{\alpha_2 A_2 \Lambda_3}(\mathbf{r}) d^A_{j,a}(\mathbf{r}) \\
&\quad - i b^A_{j,a}(\mathbf{r}) \partial^E_{j,a}(\mathbf{r}) + i b^E_{j,a}(\mathbf{r}) \partial_k \partial_k d^A_{j,a}(\mathbf{r}) - i b^E_{j,a}(\mathbf{r}) \partial_j \partial_k d^A_{k,a}(\mathbf{r}) \\
&\quad - \frac{1}{2} g b^E_{j,a}(\mathbf{r}) \big( C\gamma^j \big)_{\alpha_1\alpha_2} \sigma^2_{\Lambda_1\Lambda_2} (\hat{L}^a)_{\substack{\Lambda_2\Lambda_3\\A_1A_2}} \Big[ d^f_{\alpha_1 A_1 \Lambda_1}(\mathbf{r}) d^f_{\alpha_2 A_2 \Lambda_3}(\mathbf{r}) \\
&\qquad + \frac{1}{4} A_{\substack{\alpha_1\alpha_3\\A_1A_3\\\Lambda_1\Lambda_3}} A_{\substack{\alpha_2\alpha_4\\A_2A_4\\\Lambda_2\Lambda_4}} f_{\alpha_3 A_3 \Lambda_3}(\mathbf{r}) f_{\alpha_4 A_4 \Lambda_4}(\mathbf{r}) \Big] \\
&\quad + i g b^E_{j,a}(\mathbf{r}) c_{abc} \Big[ \big( \partial_k d^A_{b,k}(\mathbf{r}) \big) d^A_{c,j}(\mathbf{r}) + 2 d^A_{b,k}(\mathbf{r}) \partial_k d^A_{c,j}(\mathbf{r}) - d^A_{b,k}(\mathbf{r}) \partial_j d^A_{c,k}(\mathbf{r}) \Big] \\
&\quad - \frac{1}{12} i g b^E_{j,a}(\mathbf{r}) c_{abc} \Big[ \big( \partial_k b^E_{b,k}(\mathbf{r}) \big) b^E_{c,j}(\mathbf{r}) + 2 b^E_{b,k}(\mathbf{r}) \partial_k b^E_{c,j}(\mathbf{r}) - b^E_{b,k}(\mathbf{r}) \partial_j b^E_{c,k}(\mathbf{r}) \Big] \\
&\quad - i g^2 b^E_{j,a}(\mathbf{r}) c_{abc} c_{cde} \Big[ d^A_{b,k}(\mathbf{r}) d^A_{d,k}(\mathbf{r}) d^A_{e,j}(\mathbf{r}) \\
&\qquad - \frac{1}{4} b^E_{b,k}(\mathbf{r}) b^E_{d,k}(\mathbf{r}) b^A_{e,j}(\mathbf{r}) \Big] \Big\} |\mathcal{N}(f,b;p)\rangle \quad .
\end{aligned}$$

In (8.120) the following definitions have to be observed:

$$d^f_{\alpha_1 A_1 \Lambda_1}(\mathbf{r}_1) := \partial^f_{\alpha_1 A_1 \Lambda_1}(\mathbf{r}_1) + \int d^3 r_2 F_{\substack{\alpha_1\alpha_2\\A_1A_2\\\Lambda_1\Lambda_2}} (\mathbf{r}_1 - \mathbf{r}_2) f_{\alpha_2 A_2 \Lambda_2} \tag{8.121}$$

$$d^X_{a,j}(\mathbf{r}) := \partial^X_{a,j}(\mathbf{r}) + \int d^3 r' F^{XX'}_{\substack{aa'\\jj'}} (\mathbf{r} - \mathbf{r}') b^{X'}_{a',j'} \tag{8.122}$$

with $X = A, E$, the boson propagator $F^{XX'}_{12} = \langle 0|S(X_1 X'_2)|0\rangle$ and the fermion propagator $F_{I_1 I_2}$. Furthermore the anticommutator $A_{II'}$ is given by (8.109).

For a comparison with the weak mapping equations from Chapter 6 we have to specialize the general $SU(N)$–equations (8.120) to $N = 2$. This is easily done by choosing the structure constants $c_{abc} = \varepsilon_{abc}$. Comparing the bosonic sector of (8.120) for $N = 2$ with the normaltransformed weak mapping equation (6.90) it is easy to show that the two equations are identical apart from a propagator modification in (6.93) and apart from additional quantization terms contained in (8.120), namely the terms $b^E b^E b^E$ and $b^E b^E b^E b^A$. Analogously to the discussion given in Section 6.5 we can redefine the propagators which are modified at the origin of the light cone just by these terms and expect these modifications to be high energy effects. Thus the effective bosonic $SU(2)$–equations resulting from weak mapping are identical with the phenomenological equations in a low energy limit.

Concerning the boson–fermion coupling part we have not explicitly given the corresponding normalordered equations in Chapter 6. Nevertheless we can compare the coupling terms from (8.113) with those which were derived by weak mapping in (6.96). Resolution of the superindices and comparing these terms shows that they are also identical apart from an additional quantization term $b^E ff$ in (8.120). We expect that a thorough analysis of this term will show that it is also negligible in a low energy limit.

# 9 Superconductivity and Higgs Fields

## 9.1 Introduction

Higgs fields play an essential role in the formulation of the Standard model. But from a theoretical point of view they are not very well understood, apart from the fact that so far they have not been discovered in experiments.

Higgs fields originate from an analogue to Ginzburg–Landau equations in superconductivity, cf. Rickayzen [Rick 65] and Buckel [Buck 77]. But even these equations lack a thorough analysis with respect to their microscopic justification and especially their connection with inequivalent representations is of interest. So we assume the Ginzburg–Landau equations to be a suitable model to examine the effects of the groundstate and the corresponding representations in the derivation of these equations. In accordance with our preceding investigations it is reasonable to try to better understand these phenomena by applying the method of weak mapping to an appropriate superconductivity model. Naturally in doing so our intention is to clarify the just mentioned representation problem and not to give a complete discussion of superconductivity phenomena.

On the other hand, superconductivity theory is closely related to statistical mechanics. This branch of many body physics requires additional efforts to incorporate temperature and other characteristic thermodynamic quantities in the framework of algebraic theory. Concerning weak mapping we have to consider this as a challenge to enlarge our formalism into this field.

The application of quantum field theory to problems of statistical mechanics is an important branch of modern quantum theory. It requires an extension of the concept of quantum fields to the concept of quantum thermofields or, algebraically, of thermostates. The primary ingredients of statistical mechanics are statistical weighting and statistical averages for distributions of states. Thus, in order to adjust our "algebraic" method to such problems we have to include statistical distributions of states. In statistical mechanics these distributions are defined for thermal equilibrium states or for nonequilibrium processes. In a first step one should confine oneself to the treatment of thermal equilibrium states, i.e. to statistical weighting and averaging by means of thermal equilibrium density operators, or more general, to an extension of a theory for pure states to a theory for mixed states.

However, also the extension of the theory for pure states to a theory for mixed states will not be treated in full generality; rather we will restrict ourselves to sketch the general lines of this extension in Section 9.6. As a first step towards this extension it is convenient and instructive to discuss a definite model of statistical mechanics for zero temperature in order to become familiar with the problems of statistical mechanics and to improve the insight into the formalism of weak mapping. Especially we are interested to learn in which way the functional equations leading to Higgs field equations reflect the possibility of various inequivalent representations. With respect to the content of the preceding chapters, the most suitable model for such a discussion is the BCS–model of superconductivity, as the BCS–model is a special case of a spinor quantum electrodynamics with an additional nonlinear four–fermion interaction, cf., e.g., Fetter and Walecka [Fett 71].

Since the discovery of quantum mechanics the theory of superconductivity has been one of the most fascinating problems of modern physics and an enormous number of papers has been published in this field. The general content of these papers can be characterized in a twofold way: either the models themselves are derived from physical considerations about the mechanisms of superconductivity or (and) given models are exactly or approximately evaluated. It is not the aim of this book to discuss the foundations of superconductivity. Rather we are interested in the evaluation of specific models and to relate this evaluation to the construction and derivation of an effective composite particle dynamics.

In the case of the BCS–model the composite particles are represented by Cooper pairs, and in evaluating this model we will describe the formation and physical reactions of Cooper pairs in terms of composite particle dynamics. In spite of the fact that the BCS–model has been treated so extensively we will demonstrate that by the evaluation of this model according to our methods in connection with composite particles a new treatment of the BCS–model can be performed which is different from the commonly accepted symmetry breaking model and which allows a very transparent and elegant derivation of essential properties of this model.

The essential feature of this treatment is the combination of the BCS–Hamiltonian with an evaluation according to the ideas of Schafroth [Schaf 54a,b], Blatt and Butler [Bla 54], Matsubara and Blatt [Mat 60]. Although in the BCS–theory the notion of a Cooper pair was introduced, the mathematical evaluation deals with anomalous Green functions and not with composite particles. On the other hand, Schafroth, Butler, Blatt and Matsubara based their model on the idea of a charged boson gas consisting of two–electron bound states, i.e. of genuine composite particles. The obvious thing to do is to combine the successfull BCS–Hamiltonian with this boson gas idea in order to achieve an evaluation in accordançe with the composite particle formalism developed in the preceding chapters. We treat the model for zero temperature in the following sections, and sketch the treatment of the thermofields in the last section.

Finally it should be mentioned that at present the genuine boson model of superconductivity has got a renewed application in high temperature superconductivity where the bosons are assumed to be bipolarons, see Mott [Mott 93], i.e., in this case the bound states are dressed particles. In treating bosonic bipolarons, another Hamiltonian would be required as a starting point and all complications connected with phonon clouds, etc., would have to be mastered, which is beyond the scope of this book.

## 9.2 Selfconsistent Propagators and States

In Section 4.3 and in connection with normalordering we discussed the need for a selfconsistent calculation of propagators with respect to corresponding hard core states what so far we have not seriously set about. In the case of the BCS–model (and, in general, of superconductivity quantum field theory) this task is unavoidable as the selfconsistent calculation of propagators is considered as one of the essentials of the model. In contrast to spinorfield theory we now want to give a selfconsistent solution of the corresponding full functional equations exhibiting the mutual influence of states and propagators, i.e. the influence of the physical vacuum on the state calculations.

In the literature about the BCS–model either to the model itself or to its evaluation crude meanfield approximations etc. have been applied, the justification of which is doubtful. Although we have to apply approximations, too, it is our aim to justify them in a more plausible way and thus to provide for a better understanding of the theory.

The corrections already begin with the choice of a modified Hamiltonian. In contrast to the common use of non–antisymmetric four–fermion interaction terms in the BCS–model, due to subsequent fermion quantization and due to the fully antisymmetrized matrix elements occuring in our formalism we use antisymmetrized interaction terms. It is convenient to formulate this antisymmetrization by the introduction of superspinors.

Let $\psi_\alpha(\mathbf{r},t)$, $\alpha = 1,2$, be a two–component nonrelativistic spinorfield and $\psi^+_\alpha(\mathbf{r},t)$ its Hermitean conjugate. Then we define by $\psi_{\alpha\Lambda}(\mathbf{r},t)$, $\alpha = 1,2$, $\Lambda = 1,2$, a nonrelativistic superspinor

$$\begin{aligned} \psi_{\alpha 1}(\mathbf{r},t) &:= \psi_\alpha(\mathbf{r},t) \\ \psi_{\alpha 2}(\mathbf{r},t) &:= \psi^+_\alpha(\mathbf{r},t) . \end{aligned} \tag{9.1}$$

It is convenient to combine the two indices $\alpha, \Lambda$ into only one index $Z$ by means of the identification

$$\begin{array}{ccccccc} (\alpha,\Lambda) & = & (1,1), & (2,1), & (1,2), & (2,2) \\ Z & = & 1, & 2, & 3, & 4, \end{array} \tag{9.2}$$

i.e., the superspinor has four components.

In this formulation we replace the usual local electronic BCS–Hamiltonian, see for instance Fetter and Walecka [Fett 71], by the following antisymmetrized Hamiltonian:

$$\begin{aligned} \mathbf{H}_{\mathrm{BCS}} := \int d^3r \Big[ & \psi_{Z_1}(\mathbf{r})(\gamma^0\gamma^5)_{Z_1 Z_2} \frac{1}{2m} \Delta_r \psi_{Z_2}(\mathbf{r}) \\ & - g\varepsilon_{Z_1 Z_2 Z_3 Z_4} \psi_{Z_1}(\mathbf{r})\psi_{Z_2}(\mathbf{r})\psi_{Z_3}(\mathbf{r})\psi_{Z_4}(\mathbf{r}) \Big] \end{aligned} \tag{9.3}$$

with $\psi_Z(\mathbf{r}) \equiv \psi_Z(\mathbf{r},0)$. In particular no chemical potential is introduced, as on the microscopic level no thermodynamic quantity is allowed to appear. The electromagnetic field will be taken into account at a later stage of the theory. We observe that the Hamiltonian (9.3) is invariant against global gauge transformations $\psi'_\alpha(\mathbf{r}) = \exp(i\lambda)\psi_\alpha(\mathbf{r})$ and in the following we will make no attempt to break this invariance. Rather this invariance is essential for the derivation of the correct Ginzburg–Landau equations.

The spinorfield is quantized by the anticommutation relations

$$\left[\psi_\alpha(\mathbf{r}), \psi^+_\beta(\mathbf{r}')\right]_+ = \delta_{\alpha\beta}\delta(\mathbf{r}' - \mathbf{r}) \tag{9.4}$$

or

$$[\psi_Z(\mathbf{r}), \psi_{Z'}(\mathbf{r}')]_+ = \gamma^5_{ZZ'}\delta(\mathbf{r}' - \mathbf{r})$$

where $\gamma^5$ is an element of the Dirac algebra and where all other anticommutators vanish.

In accordance with our treatment of quantum fields in Chapter 3 we define for the fermion fields $\psi_Z(\mathbf{r})$ the generating one–time functional states

$$|\mathcal{A}(j;a)\rangle := \sum_{n=0}^{\infty} \frac{i^n}{n!} \tau_t^{(n)}(I_1 \dots I_n|a) j_{I_1} \dots j_{I_n}|0\rangle_F \tag{9.5}$$

with

$$\tau_t^{(n)}(I_1 \dots I_n|a) := \langle 0|\mathcal{A}(\psi_{I_1} \dots \psi_{I_n})|a\rangle \tag{9.6}$$

and $I = (Z, \mathbf{r})$, where for brevity in contrast to (3.72), (3.73) we omit the time parameter $t$. The fermionic sources $j_I$ generate a Fock–like functional space, cf. Sections 2.2 and 3.7. In order to calculate the functional states (9.5) we have to map the Heisenberg equation (3.77) for $\mathbf{H} \equiv \mathbf{H}_{\mathrm{BCS}}$ into functional space. Application of Proposition 3.12 to the modified BCS–Hamiltonian (9.3) yields the functional energy eigenvalue equation

$$\begin{aligned} E_{p0}|\mathcal{A}(j;p)\rangle &= K_{I_1 I_2} j_{I_1} \partial_{I_2} |\mathcal{A}(j;p)\rangle \\ &+ W_{I_1 I_2 I_3 I_4}\left[ j_{I_1} \partial_{I_4} \partial_{I_3} \partial_{I_2} + \frac{1}{4} A_{I_4 I_5} A_{I_3 I_6} j_{I_1} j_{I_5} j_{I_6} \partial_{I_2} \right] |\mathcal{A}(j;p)\rangle \end{aligned} \tag{9.7}$$

with

$$K_{I_1 I_2} := \frac{1}{m} \Delta_{r_1} \delta_{Z_1 Z_2} \delta(\mathbf{r}_1 - \mathbf{r}_2) \quad , \tag{9.8}$$

$$W_{I_1 I_2 I_3 I_4} := 4g(\gamma^0\gamma^5)_{Z_1 Z_5} \varepsilon_{Z_5 Z_2 Z_3 Z_4} \delta(\mathbf{r}_1 - \mathbf{r}_2)\delta(\mathbf{r}_1 - \mathbf{r}_3)\delta(\mathbf{r}_1 - \mathbf{r}_4) \,. \tag{9.9}$$

According to the "algebraic" method developed in Chapter 3, states are characterized by a set of corresponding matrix elements $a^{(n)}$, $n = 1, \dots \infty$. Those matrix elements which correspond to energy eigenstates are solutions of equation (9.7). In this formalism propagators, i.e. vacuum expectation values of antisymmetrized products of two field operators, are contained in the set of states and are also solutions of (9.7). It is our aim to calculate propagators and states simultaneously and consistently. Thus in our formalism equation (9.7) is the suitable starting point for this task.

The simultaneous calculation of various matrix elements is the basic feature of Heisenberg's matrix mechanics. Therefore we shall at least partially transfer the method of matrix mechanics to quantum field theory, and under certain restrictions we will demonstrate that this also includes our former hard core equations.

As already mentioned we observe that the propagators are contained in the set of states $|\mathcal{A}(j;p)\rangle$ for $|p\rangle \equiv |0\rangle$ and that $|\mathcal{A}(j;0)\rangle$ has to satisfy (9.7) for $E_{p0} = 0$. For $|p\rangle \equiv |0\rangle$ the functional for generalized propagators is given by

$$|\mathcal{A}(j;0)\rangle = \sum_{n=0}^{\infty} \frac{(-1)^n}{(2n)!} G^{(2n)}(I_1 \dots I_{2n}) j_{I_1} \dots j_{I_{2n}}|0\rangle_F \tag{9.10}$$

with $G^{(2n)}(I_1 \dots I_{2n}) := \langle 0|\mathcal{A}[\psi_{I_1} \dots \psi_{I_{2n}}]|0\rangle$, and for states with an even electron number we write

$$|\mathcal{A}(j;a)\rangle = \sum_{n=0}^{\infty} \frac{(-1)^n}{(2n)!} \phi^{(2n)}(I_1 \dots I_{2n}|a) j_{I_1} \dots j_{I_{2n}}|0\rangle_F \tag{9.11}$$

with $\phi^{(2n)}(I_1 \dots I_{2n}|a) := \langle 0|\mathcal{A}(\psi_{I_1} \dots \psi_{I_{2n}})|a\rangle$.

Alternative substitution of (9.10) or (9.11) into (9.7) and projection from the left by $\langle 0|(\partial_{K_1}\partial_{K_2})$ yields the exact equations

$$\left(\frac{\Delta_{r1}}{m}+\frac{\Delta_{r2}}{m}\right)G^{(2)}_{X_1X_2}(\mathbf{r}_1,\mathbf{r}_2) \tag{9.12}$$
$$= 4g\sum_{p\in S_2}\mathrm{sgn}(p)(\gamma^0\gamma^5)_{X_{p_1}Z_1}\varepsilon_{Z_1Z_2Z_3Z_4}G^{(4)}_{X_{p_2}Z_4Z_3Z_2}(\mathbf{r}_{p_2},\mathbf{r}_{p_1},\mathbf{r}_{p_1},\mathbf{r}_{p_1})$$

and

$$\left(\frac{\Delta_{r1}}{m}+\frac{\Delta_{r2}}{m}-E_{p0}\right)\phi^{(2)}_{X_1X_2}(\mathbf{r}_1,\mathbf{r}_2|p) \tag{9.13}$$
$$= 4g\sum_{p\in S_2}\mathrm{sgn}(p)(\gamma^0\gamma^5)_{X_{p_1}Z_1}\varepsilon_{Z_1Z_2Z_3Z_4}\phi^{(4)}_{X_{p_2}Z_4Z_3Z_2}(\mathbf{r}_{p_2},\mathbf{r}_{p_1},\mathbf{r}_{p_1},\mathbf{r}_{p_1}|p) \quad .$$

The transition to matrix mechanics is performed by the introduction of intermediate states on the right–hand sides of (9.12) and (9.13). By replacing $\psi_Z(\mathbf{r})$ simply by $\binom{\mathbf{r}}{Z}$ we obtain

$$\left(\frac{\Delta_{r1}}{m}+\frac{\Delta_{r2}}{m}\right)\langle 0|\mathcal{A}\begin{pmatrix}\mathbf{r}_1\mathbf{r}_2\\X_1X_2\end{pmatrix}|0\rangle \tag{9.14}$$
$$\begin{aligned}&= 4g(\gamma^0\gamma^5)_{X_1Z_1}\varepsilon_{Z_1Z_2Z_3Z_4}\mathcal{A}\sum_a\left[\langle 0|\begin{pmatrix}\mathbf{r}_2\mathbf{r}_1\\X_2Z_4\end{pmatrix}|a\rangle\langle a|\begin{pmatrix}\mathbf{r}_1\mathbf{r}_1\\Z_3Z_2\end{pmatrix}|0\rangle\right]\\
&\quad -4g(\gamma^0\gamma^5)_{X_2Z_1}\varepsilon_{Z_1Z_2Z_3Z_4}\mathcal{A}\sum_a\left[\langle 0|\begin{pmatrix}\mathbf{r}_1\mathbf{r}_2\\X_1Z_4\end{pmatrix}|a\rangle\langle a|\begin{pmatrix}\mathbf{r}_2\mathbf{r}_2\\Z_3Z_2\end{pmatrix}|0\rangle\right]\end{aligned}$$

and

$$\left(\frac{\Delta_{r1}}{m}+\frac{\Delta_{r2}}{m}-E_{p0}\right)\langle 0|\mathcal{A}\begin{pmatrix}\mathbf{r}_1\mathbf{r}_2\\X_1X_2\end{pmatrix}|p\rangle \tag{9.15}$$
$$\begin{aligned}&= 4g(\gamma^0\gamma^5)_{X_1Z_1}\varepsilon_{Z_1Z_2Z_3Z_4}\mathcal{A}\sum_a\left[\langle 0|\begin{pmatrix}\mathbf{r}_2\mathbf{r}_1\\X_2Z_4\end{pmatrix}|a\rangle\langle a|\begin{pmatrix}\mathbf{r}_1\mathbf{r}_1\\Z_3Z_2\end{pmatrix}|p\rangle\right]\\
&\quad -4g(\gamma^0\gamma^5)_{X_2Z_1}\varepsilon_{Z_1Z_2Z_3Z_4}\mathcal{A}\sum_a\left[\langle 0|\begin{pmatrix}\mathbf{r}_1\mathbf{r}_2\\X_1Z_4\end{pmatrix}|a\rangle\langle a|\begin{pmatrix}\mathbf{r}_2\mathbf{r}_2\\Z_3Z_2\end{pmatrix}|p\rangle\right] \quad .\end{aligned}$$

With respect to the calculation of the propagators $G^{(2)}$ and the state amplitudes $\phi^{(2)}$ the system of equations (9.14) and (9.15) is not closed, because additional matrix elements are involved in these equations. To prevent a further extension of the system (9.14), (9.15) by additional equations and to obtain a closed system of equations we have to take into account some specific properties of the model under consideration. We discuss these properties and formulate their consequences as propositions or assumptions.

First of all we assume that the model is characterized by a state space $\mathbb{H}_S$ which is spanned by a basis of pure quantum states $\mathbb{H}_B := \{|p\rangle\}$ where any element of $\mathbb{H}_B$ is an eigenstate of the Hamiltonian (9.3) as well as of the particle number operator **N**. This assumption is in conformity with the global gauge invariance of (9.3) and can be formulated as follows:

- **Proposition 9.1:** Any basis state $|p\rangle \in \mathbb{H}_B$ with inclusion of the ground state $|0\rangle$ can be characterized by quantum numbers $p := (N, k, l)$, where $N$ is the particle number, $k$ is the four–momentum and $l$ expresses further properties like spin, orbital angular momentum and internal excitations.

Due to gauge invariance the number of nonvanishing matrix elements is reduced; we express this fact in the following way:

- **Proposition 9.2:** Let $|N\rangle \equiv |N, k, l\rangle$ be an $N$–particle state. Then there exist only matrix elements of the following kind: $\langle N|\mathcal{A}\psi_{\alpha_1}\psi^+_{\alpha_2}|N\rangle$, $\langle N|\mathcal{A}\psi^+_{\alpha_1}\psi_{\alpha_2}|N\rangle$, $\langle N|\mathcal{A}\psi_{\alpha_1}\psi_{\alpha_2}|N+2\rangle$ and $\langle N|\mathcal{A}\psi^+_{\alpha_1}\psi^+_{\alpha_2}|N-2\rangle$ which are unequal zero.

This proposition follows directly from the orthogonality of eigenstates of the particle number operator $\mathbf{N}$.

Furthermore we assume that because of the conservation of the spin of the propagators $\langle 0|\mathcal{A}(\psi^+_\alpha\psi_\beta)|0\rangle$ and their hermitean conjugates these quantities vanish for $\alpha \neq \beta$; this assumption is compatible with the evaluation of the propagator equations which we will perform in the following. Thus we have to perform the intended selfconsistent calculation with respect to the propagators $\langle 0|\mathcal{A}\begin{pmatrix}\mathbf{r}_1\mathbf{r}_2\\ 3\,1\end{pmatrix}|0\rangle$ and $\langle 0|\mathcal{A}\begin{pmatrix}\mathbf{r}_1\mathbf{r}_2\\ 4\,2\end{pmatrix}|0\rangle$. With respect to the transition matrix elements we restrict ourselves to states where the electron pairs have opposite spins. This leads to matrix elements $\langle 0|\mathcal{A}\begin{pmatrix}\mathbf{r}_1\mathbf{r}_2\\ 1\,2\end{pmatrix}|N_0+2, k, l\rangle$ and $\langle 0|\mathcal{A}\begin{pmatrix}\mathbf{r}_1\mathbf{r}_2\\ 3\,4\end{pmatrix}|N_0-2, k, l\rangle$ with $N_0$ as the particle number of the vacuum state $|0\rangle$ and allows a selfconsistent calculation.

The model Hamiltonian (9.3) is derived by elimination of the phonon–electron interaction. The phonons describe the deviation of the lattice ion cores from the ideal lattice positions. If the phonon modes are eliminated, the electrons of the effective Hamiltonian are only subjected to the residual interaction with the ion cores at the ideal lattice positions, i.e., to a periodic potential. From this it follows the conclusion

- **Proposition 9.3:** The states $|p\rangle \in \mathbb{H}_B$ can be characterized by a wave vector $\mathbf{k}$ with $\mathbf{k} \in \Omega_B$, where $\Omega_B$ is the first Brillouin zone of the crystal.

If the antisymmetrizations in equations (9.14) and (9.15) are explicitly evaluated, if the intermediate states $|a\rangle$ are taken from the basis $\mathbb{H}_B$ and if Propositions 9.2 and 9.3 are taken into account, this yields the following system of equations:

$$\left(\frac{\Delta_{r_1}}{m}+\frac{\Delta_{r_2}}{m}\right)\langle 0|\mathcal{A}\begin{pmatrix}\mathbf{r}_1\mathbf{r}_2\\ 3\,1\end{pmatrix}|0\rangle \tag{9.16}$$

$$\begin{aligned}= 4g\sum_{\mathbf{k}\in\Omega_B,l}\Big[&\langle 0|\mathcal{A}\begin{pmatrix}\mathbf{r}_1\mathbf{r}_2\\ 3\,4\end{pmatrix}|N_0-2,k,l\rangle\langle N_0-2,k,l|\mathcal{A}\begin{pmatrix}\mathbf{r}_2\mathbf{r}_2\\ 2\,1\end{pmatrix}|0\rangle\\
&+\langle 0|\mathcal{A}\begin{pmatrix}\mathbf{r}_2\mathbf{r}_2\\ 2\,1\end{pmatrix}|N_0+2,k,l\rangle\langle N_0+2,k,l|\mathcal{A}\begin{pmatrix}\mathbf{r}_1\mathbf{r}_2\\ 3\,4\end{pmatrix}|0\rangle\\
&+\langle 0|\mathcal{A}\begin{pmatrix}\mathbf{r}_1\mathbf{r}_2\\ 3\,1\end{pmatrix}|N_0,k,l\rangle\langle N_0,k,l|\mathcal{A}\begin{pmatrix}\mathbf{r}_2\mathbf{r}_2\\ 4\,2\end{pmatrix}|0\rangle\\
&+\langle 0|\mathcal{A}\begin{pmatrix}\mathbf{r}_2\mathbf{r}_2\\ 4\,2\end{pmatrix}|N_0,k,l\rangle\langle N_0,k,l|\mathcal{A}\begin{pmatrix}\mathbf{r}_1\mathbf{r}_2\\ 3\,1\end{pmatrix}|0\rangle\end{aligned}$$

$$+\langle 0|\mathcal{A}\binom{\mathbf{r}_1\mathbf{r}_2}{3\,2}|N_0,k,l\rangle\langle N_0,k,l|\mathcal{A}\binom{\mathbf{r}_2\mathbf{r}_2}{1\,4}|0\rangle$$
$$+\langle 0|\mathcal{A}\binom{\mathbf{r}_2\mathbf{r}_2}{4\,1}|N_0,k,l\rangle\langle N_0,k,l|\mathcal{A}\binom{\mathbf{r}_2\mathbf{r}_1}{2\,3}|0\rangle$$
$$+\langle 0|\mathcal{A}\binom{\mathbf{r}_2\mathbf{r}_1}{1\,2}|N_0+2,k,l\rangle\langle N_0+2,k,l|\mathcal{A}\binom{\mathbf{r}_1\mathbf{r}_1}{4\,3}|0\rangle$$
$$+\langle 0|\mathcal{A}\binom{\mathbf{r}_1\mathbf{r}_1}{4\,3}|N_0-2,k,l\rangle\langle N_0-2,k,l|\mathcal{A}\binom{\mathbf{r}_2\mathbf{r}_1}{1\,2}|0\rangle$$
$$+\langle 0|\mathcal{A}\binom{\mathbf{r}_2\mathbf{r}_1}{1\,3}|N_0,k,l\rangle\langle N_0,k,l|\mathcal{A}\binom{\mathbf{r}_1\mathbf{r}_1}{2\,4}|0\rangle$$
$$+\langle 0|\mathcal{A}\binom{\mathbf{r}_1\mathbf{r}_1}{4\,2}|N_0,k,l\rangle\langle N_0,k,l|\mathcal{A}\binom{\mathbf{r}_1\mathbf{r}_2}{3\,1}|0\rangle$$
$$+\langle 0|\mathcal{A}\binom{\mathbf{r}_2\mathbf{r}_1}{1\,4}|N_0,k,l\rangle\langle N_0,k,l|\mathcal{A}\binom{\mathbf{r}_1\mathbf{r}_1}{3\,2}|0\rangle$$
$$+\langle 0|\mathcal{A}\binom{\mathbf{r}_1\mathbf{r}_1}{3\,2}|N_0,k,l\rangle\langle N_0,k,l|\mathcal{A}\binom{\mathbf{r}_2\mathbf{r}_1}{1\,4}|0\rangle\Bigg] \quad ,$$

$$\left(\frac{\Delta_{r_1}}{m}+\frac{\Delta_{r_2}}{m}\right)\langle 0|\mathcal{A}\binom{\mathbf{r}_1\mathbf{r}_2}{4\,2}|0\rangle \tag{9.17}$$
$$= 4g\sum_{\mathbf{k}\in\Omega_B,l}\Bigg[\langle 0|\mathcal{A}\binom{\mathbf{r}_1\mathbf{r}_2}{4\,3}|N_0-2,k,l\rangle\langle N_0-2,k,l|\mathcal{A}\binom{\mathbf{r}_2\mathbf{r}_2}{1\,2}|0\rangle$$
$$+\langle 0|\mathcal{A}\binom{\mathbf{r}_2\mathbf{r}_2}{2\,1}|N_0+2,k,l\rangle\langle N_0+2,k,l|\mathcal{A}\binom{\mathbf{r}_2\mathbf{r}_1}{3\,4}|0\rangle$$
$$+\langle 0|\mathcal{A}\binom{\mathbf{r}_1\mathbf{r}_2}{4\,2}|N_0,k,l\rangle\langle N_0,k,l|\mathcal{A}\binom{\mathbf{r}_2\mathbf{r}_2}{3\,1}|0\rangle$$
$$+\langle 0|\mathcal{A}\binom{\mathbf{r}_2\mathbf{r}_2}{3\,1}|N_0,k,l\rangle\langle N_0,k,l|\mathcal{A}\binom{\mathbf{r}_1\mathbf{r}_2}{4\,2}|0\rangle$$
$$+\langle 0|\mathcal{A}\binom{\mathbf{r}_1\mathbf{r}_2}{4\,1}|N_0,k,l\rangle\langle N_0,k,l|\mathcal{A}\binom{\mathbf{r}_2\mathbf{r}_2}{2\,3}|0\rangle$$
$$+\langle 0|\mathcal{A}\binom{\mathbf{r}_2\mathbf{r}_2}{2\,3}|N_0,k,l\rangle\langle N_0,k,l|\mathcal{A}\binom{\mathbf{r}_1\mathbf{r}_2}{4\,1}|0\rangle$$
$$+\langle 0|\mathcal{A}\binom{\mathbf{r}_2\mathbf{r}_1}{2\,1}|N_0+2,k,l\rangle\langle N_0+2,k,l|\mathcal{A}\binom{\mathbf{r}_1\mathbf{r}_1}{3\,4}|0\rangle$$
$$+\langle 0|\mathcal{A}\binom{\mathbf{r}_1\mathbf{r}_1}{3\,4}|N_0-2,k,l\rangle\langle N_0-2,k,l|\mathcal{A}\binom{\mathbf{r}_2\mathbf{r}_1}{2\,1}|0\rangle$$
$$+\langle 0|\mathcal{A}\binom{\mathbf{r}_2\mathbf{r}_1}{2\,4}|N_0,k,l\rangle\langle N_0,k,l|\mathcal{A}\binom{\mathbf{r}_1\mathbf{r}_1}{1\,3}|0\rangle$$
$$+\langle 0|\mathcal{A}\binom{\mathbf{r}_1\mathbf{r}_1}{3\,1}|N_0,k,l\rangle\langle N_0,k,l|\mathcal{A}\binom{\mathbf{r}_1\mathbf{r}_2}{4\,2}|0\rangle$$

$$+\langle 0|\mathcal{A}\begin{pmatrix}\mathbf{r}_2\mathbf{r}_1\\ 2\,3\end{pmatrix}|N_0,k,l\rangle\langle N_0,k,l|\mathcal{A}\begin{pmatrix}\mathbf{r}_1\mathbf{r}_1\\ 4\,1\end{pmatrix}|0\rangle$$

$$+\langle 0|\mathcal{A}\begin{pmatrix}\mathbf{r}_1\mathbf{r}_1\\ 4\,1\end{pmatrix}|N_0,k,l\rangle\langle N_0,k,l|\mathcal{A}\begin{pmatrix}\mathbf{r}_2\mathbf{r}_1\\ 2\,3\end{pmatrix}|0\rangle\Big]$$

and

$$\left(\frac{\Delta_{r_1}}{m}+\frac{\Delta_{r_2}}{m}-E_{p0}\right)\langle 0|\mathcal{A}\begin{pmatrix}\mathbf{r}_1\mathbf{r}_2\\ 2\,1\end{pmatrix}|p\rangle \tag{9.18}$$

$$\begin{aligned}
= \;& 4g\sum_{\mathbf{k}\in\Omega_B,l}\Big[\langle 0|\mathcal{A}\begin{pmatrix}\mathbf{r}_1\mathbf{r}_2\\ 2\,4\end{pmatrix}|N_0,k,l\rangle\langle N_0,k,l|\mathcal{A}\begin{pmatrix}\mathbf{r}_2\mathbf{r}_2\\ 2\,1\end{pmatrix}|p\rangle\\
&+\langle 0|\mathcal{A}\begin{pmatrix}\mathbf{r}_1\mathbf{r}_2\\ 2\,1\end{pmatrix}|N_0+2,k,l\rangle\langle N_0+2,k,l|\mathcal{A}\begin{pmatrix}\mathbf{r}_2\mathbf{r}_2\\ 4\,2\end{pmatrix}|p\rangle\\
&+\langle 0|\mathcal{A}\begin{pmatrix}\mathbf{r}_2\mathbf{r}_2\\ 4\,2\end{pmatrix}|N_0,k,l\rangle\langle N_0,k,l|\mathcal{A}\begin{pmatrix}\mathbf{r}_2\mathbf{r}_1\\ 2\,1\end{pmatrix}|p\rangle\\
&+\langle 0|\mathcal{A}\begin{pmatrix}\mathbf{r}_2\mathbf{r}_2\\ 2\,1\end{pmatrix}|N_0+2,k,l\rangle\langle N_0+2,k,l|\mathcal{A}\begin{pmatrix}\mathbf{r}_1\mathbf{r}_2\\ 2\,4\end{pmatrix}|p\rangle\\
&-\langle 0|\mathcal{A}\begin{pmatrix}\mathbf{r}_2\mathbf{r}_1\\ 1\,2\end{pmatrix}|N_0+2,k,l\rangle\langle N_0+2,k,l|\mathcal{A}\begin{pmatrix}\mathbf{r}_1\mathbf{r}_1\\ 3\,1\end{pmatrix}|p\rangle\\
&-\langle 0|\mathcal{A}\begin{pmatrix}\mathbf{r}_2\mathbf{r}_1\\ 1\,3\end{pmatrix}|N_0,k,l\rangle\langle N_0,k,l|\mathcal{A}\begin{pmatrix}\mathbf{r}_1\mathbf{r}_1\\ 1\,2\end{pmatrix}|p\rangle\\
&-\langle 0|\mathcal{A}\begin{pmatrix}\mathbf{r}_1\mathbf{r}_1\\ 2\,1\end{pmatrix}|N_0+2,k,l\rangle\langle N_0+2,k,l|\mathcal{A}\begin{pmatrix}\mathbf{r}_1\mathbf{r}_2\\ 3\,1\end{pmatrix}|p\rangle\\
&-\langle 0|\mathcal{A}\begin{pmatrix}\mathbf{r}_1\mathbf{r}_1\\ 3\,1\end{pmatrix}|N_0,k,l\rangle\langle N_0,k,l|\mathcal{A}\begin{pmatrix}\mathbf{r}_2\mathbf{r}_1\\ 1\,2\end{pmatrix}|p\rangle\Big]
\end{aligned}$$

and a corresponding equation for $\langle 0|\mathcal{A}\left(\begin{smallmatrix}\mathbf{r}_1\mathbf{r}_2\\ 4\,3\end{smallmatrix}\right)|p\rangle$.

For the further evaluation of (9.16)–(9.18) we have to concretize our conception of the physical vacuum and of the various matrix elements.

First we assume the vacuum state $|0\rangle$ to describe an $N_0$–particle condensate of Cooper pairs with vanishing momentum and other quantum numbers; i.e. we can write $|0\rangle \equiv |N_0,0,0\rangle$. Second we interpret the matrix elements containing the products $\psi_\alpha\psi_\beta$ and $\psi_\alpha^+\psi_\beta^+$ as describing Cooper pairs. In particular we are interested in the matrix element $\phi_{21}^{(2)}(\mathbf{r}_1,\mathbf{r}_2|p) \equiv \langle 0|\mathcal{A}[\psi_2(\mathbf{r}_1)\psi_1(\mathbf{r}_2)]|N_0+2,k,l\rangle$ which we interpret as a Cooper pair with quantum numbers $k,l$ moving "above" the filled groundstate, while the pairs described by $\langle 0|\mathcal{A}[\psi_1^+(\mathbf{r}_1)\psi_2^+(\mathbf{r}_2)]|N_0-2,0,0\rangle$ are the subunits of the condensed groundstate.

According to Proposition 9.2 there are matrix elements $\langle 0|\mathcal{A}(\psi_\alpha^+\psi_\beta)|N_0,k,l\rangle$. For $k\neq 0$ we interpret the latter as describing electron–hole excitations which conserve the particle number of the vacuum but move with a nonvanishing momentum $\mathbf{k}$. These types of states are not of primary interest and for a first exploration of the theory we suppress them. This is expressed by the following assumption:

- **Assumption 9.1:** Contributions of electron–hole excitations in the equations (9.14), (9.15) will be omitted; i.e. we assume the set of states $\langle 0|\mathcal{A}\left(\begin{smallmatrix}\mathbf{r}_1\mathbf{r}_2\\ Z_1Z_2\end{smallmatrix}\right)|N_0,k,l\rangle$ to be reduced to the set of $\langle 0|\mathcal{A}\left(\begin{smallmatrix}\mathbf{r}_1\mathbf{r}_2\\ Z_1Z_2\end{smallmatrix}\right)|0\rangle$, where $\langle 0| \equiv \langle N_0,0,0|$ and where the allowed $Z$–indices are given by $(Z_1,Z_2) \in \{(1,3),(1,4),(2,3),(2,4)\}$.

With Assumption 9.1 equations (9.16)–(9.18) are reduced to the equations

$$\begin{aligned}
&\left(\frac{\Delta_{r1}}{m}+\frac{\Delta_{r2}}{m}\right)\langle 0|\mathcal{A}\begin{pmatrix}\mathbf{r}_1\mathbf{r}_2\\ 3\,1\end{pmatrix}|0\rangle \qquad (9.19)\\
&= 4g\sum_{\mathbf{k}\in\Omega_B,l}\Bigg[\langle 0|\mathcal{A}\begin{pmatrix}\mathbf{r}_1\mathbf{r}_2\\ 3\,4\end{pmatrix}|N_0-2,k,l\rangle\langle N_0-2,k,l|\mathcal{A}\begin{pmatrix}\mathbf{r}_2\mathbf{r}_2\\ 2\,1\end{pmatrix}|0\rangle\\
&\qquad+\langle 0|\mathcal{A}\begin{pmatrix}\mathbf{r}_2\mathbf{r}_2\\ 2\,1\end{pmatrix}|N_0+2,k,l\rangle\langle N_0+2,k,l|\mathcal{A}\begin{pmatrix}\mathbf{r}_1\mathbf{r}_2\\ 3\,4\end{pmatrix}|0\rangle\Bigg]\\
&\quad+8g\langle 0|\mathcal{A}\begin{pmatrix}\mathbf{r}_1\mathbf{r}_2\\ 3\,1\end{pmatrix}|0\rangle\langle 0|\mathcal{A}\begin{pmatrix}\mathbf{r}_2\mathbf{r}_2\\ 4\,2\end{pmatrix}|0\rangle\\
&\quad+4g\sum_{\mathbf{k}\in\Omega_B,l}\Bigg[\langle 0|\mathcal{A}\begin{pmatrix}\mathbf{r}_2\mathbf{r}_1\\ 1\,2\end{pmatrix}|N_0+2,k,l\rangle\langle N_0+2,k,l|\mathcal{A}\begin{pmatrix}\mathbf{r}_1\mathbf{r}_1\\ 4\,3\end{pmatrix}|0\rangle\\
&\qquad+\langle 0|\mathcal{A}\begin{pmatrix}\mathbf{r}_1\mathbf{r}_1\\ 4\,3\end{pmatrix}|N_0-2,k,l\rangle\langle N_0-2,k,l|\mathcal{A}\begin{pmatrix}\mathbf{r}_2\mathbf{r}_1\\ 1\,2\end{pmatrix}|0\rangle\Bigg]\\
&\quad+8g\langle 0|\mathcal{A}\begin{pmatrix}\mathbf{r}_2\mathbf{r}_1\\ 1\,3\end{pmatrix}|0\rangle\langle 0|\mathcal{A}\begin{pmatrix}\mathbf{r}_1\mathbf{r}_1\\ 2\,4\end{pmatrix}|0\rangle \quad ,
\end{aligned}$$

$$\begin{aligned}
&\left(\frac{\Delta_{r1}}{m}+\frac{\Delta_{r2}}{m}\right)\langle 0|\mathcal{A}\begin{pmatrix}\mathbf{r}_1\mathbf{r}_2\\ 4\,2\end{pmatrix}|0\rangle \qquad (9.20)\\
&= 4g\sum_{\mathbf{k}\in\Omega_B,l}\Bigg[\langle 0|\mathcal{A}\begin{pmatrix}\mathbf{r}_1\mathbf{r}_2\\ 4\,3\end{pmatrix}|N_0-2,k,l\rangle\langle N_0-2,k,l|\mathcal{A}\begin{pmatrix}\mathbf{r}_2\mathbf{r}_2\\ 1\,2\end{pmatrix}|0\rangle\\
&\qquad+\langle 0|\mathcal{A}\begin{pmatrix}\mathbf{r}_2\mathbf{r}_2\\ 2\,1\end{pmatrix}|N_0+2,k,l\rangle\langle N_0+2,k,l|\mathcal{A}\begin{pmatrix}\mathbf{r}_2\mathbf{r}_1\\ 3\,4\end{pmatrix}|0\rangle\Bigg]\\
&\quad+8g\langle 0|\mathcal{A}\begin{pmatrix}\mathbf{r}_1\mathbf{r}_2\\ 4\,2\end{pmatrix}|0\rangle\langle 0|\mathcal{A}\begin{pmatrix}\mathbf{r}_2\mathbf{r}_2\\ 3\,1\end{pmatrix}|0\rangle\\
&\quad+4g\sum_{\mathbf{k}\in\Omega_B,l}\Bigg[\langle 0|\mathcal{A}\begin{pmatrix}\mathbf{r}_2\mathbf{r}_1\\ 2\,1\end{pmatrix}|N_0+2,k,l\rangle\langle N_0+2,k,l|\mathcal{A}\begin{pmatrix}\mathbf{r}_1\mathbf{r}_1\\ 3\,4\end{pmatrix}|0\rangle\\
&\qquad+\langle 0|\mathcal{A}\begin{pmatrix}\mathbf{r}_1\mathbf{r}_1\\ 3\,4\end{pmatrix}|N_0-2,k,l\rangle\langle N_0-2,k,l|\mathcal{A}\begin{pmatrix}\mathbf{r}_2\mathbf{r}_1\\ 2\,1\end{pmatrix}|0\rangle\Bigg]\\
&\quad+8g\langle 0|\mathcal{A}\begin{pmatrix}\mathbf{r}_2\mathbf{r}_1\\ 2\,4\end{pmatrix}|0\rangle\langle 0|\mathcal{A}\begin{pmatrix}\mathbf{r}_1\mathbf{r}_1\\ 1\,3\end{pmatrix}|0\rangle
\end{aligned}$$

and

$$\left(\frac{\Delta_{r_1}}{m} + \frac{\Delta_{r_2}}{m} - E_{p0}\right) \langle 0|\mathcal{A}\begin{pmatrix}\mathbf{r}_1\mathbf{r}_2\\ 2\,1\end{pmatrix}|p\rangle \tag{9.21}$$

$$\begin{aligned}
= \;& 4g \sum_{\mathbf{k}\in\Omega_B, l} \Bigg[\langle 0|\mathcal{A}\begin{pmatrix}\mathbf{r}_1\mathbf{r}_2\\ 2\,1\end{pmatrix}|N_0+2,k,l\rangle\langle N_0+2,k,l|\mathcal{A}\begin{pmatrix}\mathbf{r}_2\mathbf{r}_2\\ 4\,2\end{pmatrix}|p\rangle \\
& + \langle 0|\mathcal{A}\begin{pmatrix}\mathbf{r}_2\mathbf{r}_2\\ 2\,1\end{pmatrix}|N_0+2,k,l\rangle\langle N_0+2,k,l|\mathcal{A}\begin{pmatrix}\mathbf{r}_1\mathbf{r}_2\\ 2\,4\end{pmatrix}|p\rangle \\
& - \langle 0|\mathcal{A}\begin{pmatrix}\mathbf{r}_2\mathbf{r}_1\\ 1\,2\end{pmatrix}|N_0+2,k,l\rangle\langle N_0+2,k,l|\mathcal{A}\begin{pmatrix}\mathbf{r}_1\mathbf{r}_1\\ 3\,1\end{pmatrix}|p\rangle \\
& - \langle 0|\mathcal{A}\begin{pmatrix}\mathbf{r}_1\mathbf{r}_1\\ 2\,1\end{pmatrix}|N_0+2,k,l\rangle\langle N_0+2,k,l|\mathcal{A}\begin{pmatrix}\mathbf{r}_1\mathbf{r}_2\\ 3\,1\end{pmatrix}|p\rangle\Bigg] \\
& + 4g\Bigg[\langle 0|\mathcal{A}\begin{pmatrix}\mathbf{r}_1\mathbf{r}_2\\ 2\,4\end{pmatrix}|0\rangle\langle 0|\mathcal{A}\begin{pmatrix}\mathbf{r}_2\mathbf{r}_2\\ 2\,1\end{pmatrix}|p\rangle \\
& + \langle 0|\mathcal{A}\begin{pmatrix}\mathbf{r}_2\mathbf{r}_2\\ 4\,2\end{pmatrix}|0\rangle\langle 0|\mathcal{A}\begin{pmatrix}\mathbf{r}_2\mathbf{r}_1\\ 2\,1\end{pmatrix}|p\rangle \\
& - \langle 0|\mathcal{A}\begin{pmatrix}\mathbf{r}_2\mathbf{r}_1\\ 1\,3\end{pmatrix}|0\rangle\langle 0|\mathcal{A}\begin{pmatrix}\mathbf{r}_1\mathbf{r}_1\\ 1\,2\end{pmatrix}|p\rangle \\
& - \langle 0|\mathcal{A}\begin{pmatrix}\mathbf{r}_1\mathbf{r}_1\\ 3\,1\end{pmatrix}|0\rangle\langle 0|\mathcal{A}\begin{pmatrix}\mathbf{r}_2\mathbf{r}_1\\ 1\,2\end{pmatrix}|p\rangle\Bigg]
\end{aligned}$$

where $|p\rangle$ contains the particle number $N_0 + 2$.

Next we bring into play the fact that the groundstate is a condensate of the quasibosonic Cooper pairs. In accordance with the idea of condensation this means that all bosons of the groundstate are in the same one–particle state, and if the bosons are two–fermion bound states then all fermion pairs also belong to the same bound state. In particular if the condensate is at rest, the groundstate contains no contributions of two–fermion bound states with non-vanishing wave vector $\mathbf{k}$. We observe that the cooper pairs in the groundstate are described by the matrixelements $\langle 0|\mathcal{A}\left(\begin{smallmatrix}\mathbf{r}_1\mathbf{r}_2\\ 3\,4\end{smallmatrix}\right)|N_0-2\rangle \equiv \langle 0|\mathcal{A}[\psi_1^+(\mathbf{r}_1)\psi_2^+(\mathbf{r}_2)]|N_0-2\rangle$, extend these arguments to the quantum numbers $l$ and express this property by

- **Assumption 9.2:** For groundstates with condensate structure the set of states $\langle 0|\mathcal{A}\left(\begin{smallmatrix}\mathbf{r}_1\mathbf{r}_2\\ 3\,4\end{smallmatrix}\right)|N_0-2,k,l\rangle$ is reduced to the set of $\langle 0|\mathcal{A}\left(\begin{smallmatrix}\mathbf{r}_1\mathbf{r}_2\\ 3\,4\end{smallmatrix}\right)|N_0-2,0,0\rangle$, where $|N_0-2,0,0\rangle$ may be considered as the groundstate for $N_0 - 2$ electrons.

The content of Assumption 9.2 becomes clear, if one represents the states $\langle 0|$ and $|N_0-2,k,l\rangle$ in Fock space. Naturally condensation of fermion pairs is confronted with antisymmetry. In Section 9.3 it will be discussed how the condensation is affected by antisymmetrization.

Furthermore we observe that in equations (9.19)–(9.21) the terms like $\langle 0|\mathcal{A}\left(\begin{smallmatrix}\mathbf{r}\,\mathbf{r}\\ 3\,1\end{smallmatrix}\right)|0\rangle$ and $\langle 0|\mathcal{A}\left(\begin{smallmatrix}\mathbf{r}\,\mathbf{r}\\ 4\,2\end{smallmatrix}\right)|0\rangle$ are singular. In accordance with Section 2.6 these singular terms can be removed by vertex renormalization. Without mentioning this as a particular assumption we introduce vertex renormalization in (9.3) from the beginning.

Taking into account this vertex renormalization and Assumption 9.2 we obtain from (9.19)–(9.21) the equations

$$\left(\frac{\Delta_{r_1}}{m}+\frac{\Delta_{r_2}}{m}\right)\langle 0|\mathcal{A}\begin{pmatrix}\mathbf{r}_1\mathbf{r}_2\\ 3\,1\end{pmatrix}|0\rangle \tag{9.22}$$
$$= 4g\Big[\langle 0|\mathcal{A}\begin{pmatrix}\mathbf{r}_1\mathbf{r}_2\\ 3\,4\end{pmatrix}|N_0-2,0,0\rangle\langle N_0-2,0,0|\mathcal{A}\begin{pmatrix}\mathbf{r}_2\mathbf{r}_2\\ 2\,1\end{pmatrix}|0\rangle$$
$$+\langle 0|\mathcal{A}\begin{pmatrix}\mathbf{r}_1\mathbf{r}_1\\ 4\,3\end{pmatrix}|N_0-2,0,0\rangle\langle N_0-2,0,0|\mathcal{A}\begin{pmatrix}\mathbf{r}_2\mathbf{r}_1\\ 1\,2\end{pmatrix}|0\rangle\Big]$$
$$+4g\sum_{\mathbf{k}\in\Omega_B,l}\Big[\langle 0|\mathcal{A}\begin{pmatrix}\mathbf{r}_2\mathbf{r}_2\\ 2\,1\end{pmatrix}|N_0+2,k,l\rangle\langle N_0+2,k,l|\mathcal{A}\begin{pmatrix}\mathbf{r}_1\mathbf{r}_2\\ 3\,4\end{pmatrix}|0\rangle$$
$$+\langle 0|\mathcal{A}\begin{pmatrix}\mathbf{r}_2\mathbf{r}_1\\ 1\,2\end{pmatrix}|N_0+2,k,l\rangle\langle N_0+2,k,l|\mathcal{A}\begin{pmatrix}\mathbf{r}_1\mathbf{r}_1\\ 4\,3\end{pmatrix}|0\rangle\Big]\quad ,$$

$$\left(\frac{\Delta_{r_1}}{m}+\frac{\Delta_{r_2}}{m}\right)\langle 0|\mathcal{A}\begin{pmatrix}\mathbf{r}_1\mathbf{r}_2\\ 4\,2\end{pmatrix}|0\rangle \tag{9.23}$$
$$= 4g\Big[\langle 0|\mathcal{A}\begin{pmatrix}\mathbf{r}_1\mathbf{r}_2\\ 4\,3\end{pmatrix}|N_0-2,0,0\rangle\langle N_0-2,0,0|\mathcal{A}\begin{pmatrix}\mathbf{r}_2\mathbf{r}_2\\ 1\,2\end{pmatrix}|0\rangle$$
$$+\langle 0|\mathcal{A}\begin{pmatrix}\mathbf{r}_1\mathbf{r}_1\\ 3\,4\end{pmatrix}|N_0-2,0,0\rangle\langle N_0-2,0,0|\mathcal{A}\begin{pmatrix}\mathbf{r}_2\mathbf{r}_1\\ 2\,1\end{pmatrix}|0\rangle\Big]$$
$$+4g\sum_{\mathbf{k}\in\Omega_B,l}\Big[\langle 0|\mathcal{A}\begin{pmatrix}\mathbf{r}_2\mathbf{r}_2\\ 2\,1\end{pmatrix}|N_0+2,k,l\rangle\langle N_0+2,k,l|\mathcal{A}\begin{pmatrix}\mathbf{r}_2\mathbf{r}_1\\ 3\,4\end{pmatrix}|0\rangle$$
$$+\langle 0|\mathcal{A}\begin{pmatrix}\mathbf{r}_2\mathbf{r}_1\\ 2\,1\end{pmatrix}|N_0+2,k,l\rangle\langle N_0+2,k,l|\mathcal{A}\begin{pmatrix}\mathbf{r}_1\mathbf{r}_1\\ 3\,4\end{pmatrix}|0\rangle\Big]$$

and

$$\left(\frac{\Delta_{r_1}}{m}+\frac{\Delta_{r_2}}{m}-E_{p0}\right)\langle 0|\mathcal{A}\begin{pmatrix}\mathbf{r}_1\mathbf{r}_2\\ 2\,1\end{pmatrix}|p\rangle \tag{9.24}$$
$$= 4g\sum_{\mathbf{k}\in\Omega_B,l}\Big[\langle 0|\mathcal{A}\begin{pmatrix}\mathbf{r}_1\mathbf{r}_2\\ 2\,1\end{pmatrix}|N_0+2,k,l\rangle\langle N_0+2,k,l|\mathcal{A}\begin{pmatrix}\mathbf{r}_2\mathbf{r}_2\\ 4\,2\end{pmatrix}|p\rangle$$
$$+\langle 0|\mathcal{A}\begin{pmatrix}\mathbf{r}_2\mathbf{r}_2\\ 2\,1\end{pmatrix}|N_0+2,k,l\rangle\langle N_0+2,k,l|\mathcal{A}\begin{pmatrix}\mathbf{r}_1\mathbf{r}_2\\ 2\,4\end{pmatrix}|p\rangle$$
$$-\langle 0|\mathcal{A}\begin{pmatrix}\mathbf{r}_2\mathbf{r}_1\\ 1\,2\end{pmatrix}|N_0+2,k,l\rangle\langle N_0+2,k,l|\mathcal{A}\begin{pmatrix}\mathbf{r}_1\mathbf{r}_1\\ 3\,1\end{pmatrix}|p\rangle$$
$$-\langle 0|\mathcal{A}\begin{pmatrix}\mathbf{r}_1\mathbf{r}_1\\ 2\,1\end{pmatrix}|N_0+2,k,l\rangle\langle N_0+2,k,l|\mathcal{A}\begin{pmatrix}\mathbf{r}_1\mathbf{r}_2\\ 3\,1\end{pmatrix}|p\rangle\Big]$$

$$+4g\left[\langle 0|\mathcal{A}\begin{pmatrix}\mathbf{r}_1\mathbf{r}_2\\24\end{pmatrix}|0\rangle\langle 0|\mathcal{A}\begin{pmatrix}\mathbf{r}_2\mathbf{r}_2\\21\end{pmatrix}|p\rangle\right.$$

$$\left.-\langle 0|\mathcal{A}\begin{pmatrix}\mathbf{r}_2\mathbf{r}_1\\13\end{pmatrix}|0\rangle\langle 0|\mathcal{A}\begin{pmatrix}\mathbf{r}_1\mathbf{r}_1\\12\end{pmatrix}|p\rangle\right] .$$

For the further evaluation of equations (9.22) and (9.23) we assume the electron number $N_0$ of the groundstate $|0\rangle \equiv |N_0, k = 0, l = 0\rangle$ to be very large and we compare matrix elements between states for different $N$. For very large $N$ it is plausible that the difference between such matrix elements is negligible. Thus we formulate

- **Assumption 9.3:** For very large $N$ it holds

$$\lim_{N\to\infty}\left[\langle N-2,0,0|\mathcal{A}\begin{pmatrix}\mathbf{r}_1\mathbf{r}_2\\Z_1Z_2\end{pmatrix}|N,0,0\rangle - \langle N,0,0|\mathcal{A}\begin{pmatrix}\mathbf{r}_1\mathbf{r}_2\\Z_1Z_2\end{pmatrix}|N+2,0,0\rangle\right] = 0 \tag{9.25}$$

where $N = N_0, N_0 - 2, N_0 + 2$ are the particle numbers of the various groundstates.

Furthermore for the evaluation of the sum over $\mathbf{k}$ and $l$ in (9.22) and (9.23) we now turn to the discussion of the Cooper pair states. Due to translational invariance these states must have the form

$$\begin{aligned}\phi^{(2)}_{Z_1Z_2}(\mathbf{r}_1,\mathbf{r}_2|k,l) :&= \langle 0|\mathcal{A}\begin{pmatrix}\mathbf{r}_1\mathbf{r}_2\\Z_1Z_2\end{pmatrix}|N,k,l\rangle \\ &= e^{-\frac{i}{2}\mathbf{k}(\mathbf{r}_1+\mathbf{r}_2)}\chi_{Z_1Z_2}(\mathbf{r}_1-\mathbf{r}_2|k,l)\end{aligned} \tag{9.26}$$

for $(Z_1, Z_2) \in \{(1,2),(3,4)\}$ and $N = N_0 + 2$.

In addition by translational invariance it follows that

$$G^{(2)}_{Z_1Z_2}(\mathbf{r}_1,\mathbf{r}_2) \equiv G^{(2)}_{Z_1Z_2}(\mathbf{r}_1-\mathbf{r}_2) \tag{9.27}$$

for $(Z_1, Z_2) \in \{(1,3),(2,4)\}$.

Equations (9.22) and (9.23) can be considerably simplified if we suppress the $k$– and $l$–dependence of $\chi$; the latter essentially means that we omit spin–1 states in some cases.

- **Assumption 9.4:** As an approximation (strong internal coupling limit) the $k$– and $l$–dependence of $\chi$ is omitted, i.e. $\chi_{Z_1Z_2}(\mathbf{r}_1-\mathbf{r}_2|k,l) \approx \chi_{Z_1Z_2}(\mathbf{r}_1-\mathbf{r}_2)$.

Applying appropriate Hermitean conjugations and taking into account (9.25) and (9.26) we obtain for (9.22) and (9.23):

$$\begin{aligned}&\left(\frac{\Delta_{r_1}}{m}+\frac{\Delta_{r_1}}{m}\right)\langle 0|\mathcal{A}\begin{pmatrix}\mathbf{r}_1\mathbf{r}_2\\31\end{pmatrix}|0\rangle \\ &= 4g\left(1+\sum_{k\in\Omega_B}e^{\frac{i}{2}\mathbf{k}(\mathbf{r}_1-\mathbf{r}_2)}\right)\left[\chi^*_{21}(\mathbf{r}_2-\mathbf{r}_1)\chi_{21}(0)+\chi_{21}(\mathbf{r}_1-\mathbf{r}_2)\chi^*_{21}(0)\right]\end{aligned} \tag{9.28}$$

and

$$\left(\frac{1}{m}\Delta_{r_1} + \frac{1}{m}\Delta_{r_1}\right) \langle 0|\mathcal{A}\begin{pmatrix}\mathbf{r}_1\mathbf{r}_2\\ 4\,2\end{pmatrix}|0\rangle \tag{9.29}$$

$$= \quad 4g\left(1 + \sum_{k\in\Omega_B} e^{\frac{i}{2}\mathbf{k}(\mathbf{r}_1-\mathbf{r}_2)}\right)\left[\chi_{21}^*(\mathbf{r}_1-\mathbf{r}_2)\chi_{21}(0) + \chi_{21}(\mathbf{r}_2-\mathbf{r}_1)\chi_{21}^*(0)\right] \quad .$$

If we introduce center of mass coordinates $\mathbf{z} = \frac{1}{2}(\mathbf{r}_1 + \mathbf{r}_2)$ and $\mathbf{u} = \mathbf{r}_1 - \mathbf{r}_2$ and observe (9.27), equations (9.28) and (9.29) can be reduced to the following set of equations:

$$\frac{2}{m}\Delta_u G_{31}^{(2)}(\mathbf{u}) = 4gV(\mathbf{u})\left[\chi_{21}^*(-\mathbf{u})\,\chi_{21}(0) + \chi_{12}(-\mathbf{u})\,\chi_{12}^*(0)\right] \tag{9.30}$$

and

$$G_{42}^{(2)}(\mathbf{u}) = G_{31}^{(2)}(-\mathbf{u}) \tag{9.31}$$

with

$$V(\mathbf{u}) := \left(1 + \sum_{k\in\Omega_B} e^{\frac{i}{2}\mathbf{k}\mathbf{u}}\right) \quad . \tag{9.32}$$

We now turn to the evaluation of (9.24). However, before going into details we have to explore the physical meaning of (9.24). Ignoring for the moment our preceding approximations, equation (9.24) is equivalent to (9.13). In order to calculate $\phi^{(2)}$ from (9.13) we have to supplement (9.13) by an equation for $\phi^{(4)}$ which for its part contains $\phi^{(6)}$ and so on; i.e. we have to calculate a solution of the full functional equation (9.7) in order to get the matrix element $\phi^{(2)}$. This means that $\phi^{(2)}$ is calculated as the hard core of a dressed particle state with two electrons added to the groundstate. Therefore equation (9.24) expresses the condition for the calculation of dressed particle states, modified by the application of matrix mechanics.

Although in the following we shall need a first order polarization cloud wave function for weak mapping, we emphasize that the concept of the Cooper pair is primarily that of a hard core state and not that of a dressed particle state. Thus in order to catch up with conventional superconductivity theory and simultaneously to simplify our calculations, we introduce the following definition:

- **Definition 9.1:** Cooper pair states are defined by hard core bound states of two electrons above the groundstate or within the groundstate.

We are justified to use the notion "definition" because this kind of definition does not imply any kind of approximation. It is rather a way to decompose the whole electronic system of the superconductor into physically plausible subunits. Any part of electronic interactions that is excluded by the formation of these subunits is later on taken into account in the procedure of weak mapping, so in principle apart from suitable but not necessary approximations by this definition nothing is lost.

By comparison of (9.24) with the relativistic two–fermion hard core equation (5.33) one easily verifies that the hard core part of equation (9.24) is given by

$$\left(\frac{\Delta_{r_1}}{m} + \frac{\Delta_{r_2}}{m} - E_{p0}\right) \langle 0|\mathcal{A}\begin{pmatrix}\mathbf{r}_1\mathbf{r}_2\\ 2\,1\end{pmatrix}|p \equiv (N_0+2,k,l)\rangle \tag{9.33}$$

$$= 4g\Big[\langle 0|\mathcal{A}\begin{pmatrix}\mathbf{r}_1\mathbf{r}_2\\ 2\,4\end{pmatrix}|0\rangle\langle 0|\mathcal{A}\begin{pmatrix}\mathbf{r}_2\mathbf{r}_2\\ 2\,1\end{pmatrix}|N_0+2,k,l\rangle$$

$$-\langle 0|\mathcal{A}\begin{pmatrix}\mathbf{r}_2\mathbf{r}_1\\ 1\,3\end{pmatrix}|0\rangle\langle 0|\mathcal{A}\begin{pmatrix}\mathbf{r}_1\mathbf{r}_1\\ 1\,2\end{pmatrix}|N_0+2,k,l\rangle\Big] \quad .$$

In addition we need the matrix elements $\langle 0|\mathcal{A}\left(\begin{smallmatrix}\mathbf{r}_1\mathbf{r}_2\\ 3\,4\end{smallmatrix}\right)|N_0-2,0,0\rangle$, which describe bound states below the Fermi level. By means of Assumption 9.3 we can relate them to the hard core states of equation (9.33) by the relation

$$\langle 0|\mathcal{A}\begin{pmatrix}\mathbf{r}_1\mathbf{r}_2\\ 3\,4\end{pmatrix}|N_0-2,0,0\rangle = \langle 0|\mathcal{A}\begin{pmatrix}\mathbf{r}_2\mathbf{r}_1\\ 2\,1\end{pmatrix}|N_0+2,0,0\rangle^* \quad . \tag{9.34}$$

Introduction of center of mass coordinates and observation of (9.26) yields then for (9.33)

$$\left(\frac{2}{m}\Delta_u - E_{p0}\right)\chi_{21}(\mathbf{u}) = 4g\left[G_{24}^{(2)}(\mathbf{u})\,\chi_{21}(0) - G_{31}^{(2)}(\mathbf{u})\,\chi_{21}(0)\right] \quad . \tag{9.35}$$

According to (9.2) we can resolve the indices of $G^{(2)}$ and $\chi$ and obtain

$$\begin{aligned} G_{31}^{(2)}(\mathbf{u}) &\equiv \langle 0|\mathcal{A}[\psi_1^+(\mathbf{r}_1)\psi_1(\mathbf{r}_2)]|0\rangle\,, \\ G_{42}^{(2)}(\mathbf{u}) &\equiv \langle 0|\mathcal{A}[\psi_2^+(\mathbf{r}_1)\psi_2(\mathbf{r}_2)]|0\rangle\,, \\ \chi_{21}(\mathbf{u}) &\equiv \langle 0|\mathcal{A}[\psi_2(\mathbf{r}_1)\psi_1(\mathbf{r}_2)]|N_0+2,0,0\rangle\,. \end{aligned} \tag{9.36}$$

The formulas (9.36) suggest to introduce the ansatz

$$\chi_{\alpha_1\alpha_2}(\mathbf{u}) = \sigma^2_{\alpha_1\alpha_2}\chi(\mathbf{u}) \quad . \tag{9.37}$$

Then we have due to (9.37)

$$\begin{aligned} \chi_{21}(\mathbf{u}) &= 2i\chi(\mathbf{u}) \\ \chi_{12}(\mathbf{u}) &= -2i\chi(\mathbf{u}) \quad . \end{aligned} \tag{9.38}$$

If $\chi(\mathbf{u})$ is assumed to be real and with (9.31) we obtain for (9.30) and (9.35)

$$\frac{1}{m}\Delta_u G_{31}^{(2)}(\mathbf{u}) = 16gV(\mathbf{u})\chi(\mathbf{u})\chi(0) \tag{9.39}$$

and

$$\left(\frac{2}{m}\Delta_u - E_{p0}\right)\chi(\mathbf{u}) = 8gG_{31}^{(2)}(\mathbf{u})\,\chi(0) \quad . \tag{9.40}$$

Finally we substitute (9.39) into (9.40) and obtain

$$\chi(\mathbf{u}) = 2^7 mg^2\left(\frac{2}{m}\Delta_u - E_{p0}\right)^{-1}\Delta_u{}^{-1}\left[V(\mathbf{u})\chi(\mathbf{u})\right]\chi(0)^2\,. \tag{9.41}$$

The latter equation is an eigenvalue equation for $E$ where in addition $\chi(0)$ has to be determined. The constant $\chi(0)$ is not arbitrary. Due to the fact that the states $|0\rangle$ and $|N_0+2\rangle$ are assumed to be normalized, the amplitude $\chi(0) \equiv \langle 0|\mathcal{A}\psi(\mathbf{r})\psi(\mathbf{r})|N_0+2\rangle$ is completely fixed. This is expressed by (3.40) and (3.46) if (3.46) is applied to $\langle N_0+2|N_0+2\rangle = 1$. Hence by requiring the Fredholm determinant of equation (9.41) to vanish and by applying (3.46) to $\langle N_0+2|N_0+2\rangle$ and thus to $\chi$ we have two algebraic equations which allow to calculate the eigenvalue $E$ as well as $\chi(0)$.

For the evaluation of (3.46) the metrical tensor for two fermion states is needed. This metrical tensor is in a close connection with the propagators, which can be written with (9.4) in the form

$$\begin{aligned} G^{(2)}_{Z_1 Z_2}(\mathbf{r}_1-\mathbf{r}_2) &= \langle 0|\psi^+_{\alpha_1}(\mathbf{r}_1)\psi_{\alpha_2}(\mathbf{r}_2)|0\rangle - \frac{1}{2}\delta_{\alpha_1\alpha_2}\delta(\mathbf{r}_1-\mathbf{r}_2) \\ &=: g_{\alpha_1\alpha_2}(\mathbf{r}_1-\mathbf{r}_2) - \frac{1}{2}\delta_{\alpha_1\alpha_2}\delta(\mathbf{r}_1-\mathbf{r}_2) \end{aligned} \tag{9.42}$$

for $(Z_1, Z_1) \in \{(4,2),(3,1)\}$. Substitution of (9.42) into (9.39) and the corresponding equation for $G^{(2)}_{42}(\mathbf{u})$ gives

$$g_{\alpha_1\alpha_2}(\mathbf{u}) = g(\mathbf{u}) = \frac{1}{2}\delta(\mathbf{u}) + 4mg\Delta_u{}^{-1}V(\mathbf{u})\chi(\mathbf{u})\chi(0) \tag{9.43}$$

for $\alpha_1 = \alpha_2$. According to our previous assumptions the propagators $G^{(2)}_{32}(\mathbf{u})$ and $G^{(2)}_{41}(\mathbf{u})$ vanish; thus the metric tensor has the form

$$g_{\alpha_1\alpha_2}(\mathbf{u}) = \delta_{\alpha_1\alpha_2} g(\mathbf{u})\,. \tag{9.44}$$

So far, by (9.43) we have derived the metric tensor for one–fermion states. Although as long as we do not perform the thermodynamic limit we use a Fock space for finite particle numbers, formula (9.43) shows that the metric tensor is different from the Fock space tensor which is solely given by $\delta_{\alpha_1\alpha_2}\delta(\mathbf{u})$. This is due to the fact that we have shifted the usual Fock space vacuum to the new groundstate of condensed Cooper pairs. Hence (9.43) reflects the nontrivial vacuum structure on account of the Cooper pair condensation. By means of equations (9.7) the four–point metric tensor can be calculated by inserting (9.43). But for a first draft we assume the four–point metric tensor to be the direct product of $g_{\alpha\beta}$ with itself. As $g_{\alpha\beta}$ is represented by $\chi$, even in this approximation equations (9.41) and (3.46) become a highly nonlinear system in $E$ as well as in $\chi(0)$.

## 9.3 Spectrum of Bound States

In the commonly accepted theory of superconductivity the condensation phenomenon is related to symmetry breaking of the global fermionic gauge invariance by means of anomalous Green functions. In our approach we do not make use of this mechanism, but try to explain superconductivity by the formation of bound states. While the lowest bound states are the objects of boson condensation, excited bound states should be responsible for the thermal breakdown of superconductivity. All that depends on the bound state spectrum which we will investigate in this section.

In contrast to anomalous vacuum condensation the bound state model is much more evident. But this evidence has its prize. If the condensation is not due to an abstract vacuum phenomenon, but is considered as an effect of the $N$–electron groundstate of the superconductor, it is a genuine many–particle problem and the question arises: How can one make statements about the groundstate if one knows only two–electron bound states? The answer is indeed that one cannot ignore the many–particle structure of the groundstate, and this includes that one cannot simply postulate $|0\rangle$ to be the groundstate and all other things follow automatically. In this way the decoupling of symmetry breaking and condensation phenomena enforces a detailed resolution of the groundstate configuration.

One of the most far–reaching effects of many fermion systems is the influence of exchange forces on the formation of bound states. These forces are caused by the antisymmetry condition of the wavefunctions, and antisymmetrization is the minimal requirement which has to be imposed on many–fermion wavefunctions. Therefore we have to analyze the influence of antisymmetry on the condensate. Let (9.26) be the two–electron condensate wavefunction. According to Assumption 9.2 and the ansatz (9.37) we assume that the groundstate $|0\rangle$ may be explicitly constructed in position space by

$$|0\rangle = \prod_{i=1}^{N} \int d\tau \sigma^2_{\alpha_{2i-1}\alpha_{2i}} \chi(\mathbf{r}_{2i-1} - \mathbf{r}_{2i}) \psi^+_{\alpha_{2i-1}}(\mathbf{r}_{2i-1}) \psi^+_{\alpha_{2i}}(\mathbf{r}_{2i}) |f\rangle \,, \tag{9.45}$$

where $|f\rangle$ is the Fock space vacuum and $d\tau$ the position space volume element for all coordinates. If we represent $\psi^+_\alpha(\mathbf{r})$ in momentum space

$$\psi^+_\alpha(\mathbf{r}) = \int a^+_\alpha(\mathbf{k}) e^{i\mathbf{k}\mathbf{r}} d^3k \tag{9.46}$$

equation (9.45) goes over into

$$|0\rangle = \prod_{i=1}^{N} \int d^3k_i c(\mathbf{k}_i) a^+_\uparrow(\mathbf{k}_i) a^+_\downarrow(-\mathbf{k}_i) |f\rangle \tag{9.47}$$

with

$$c(\mathbf{k}) := \int \chi(\mathbf{u}) e^{2i\mathbf{k}\mathbf{u}} d^3u \tag{9.48}$$

and $a_\downarrow := a_1$, $a_\uparrow := a_2$.

In order to get a clear idea what (9.47) means we consider the system under the assumption of periodicity with respect to a macroscopic volume $V$. Then in connection with the elementary lattice periodicity the integral over $\mathbf{k}$ in (9.48) goes over into a discrete sum over $\mathbf{k}$ vectors in the Brillouin zone. As the superconductor is a metal, its Fermi surface is located in the interior of the Brillouin zone. We now postulate: There is a biunique relation between the k–vectors in the Fermi sphere and the electrons, i.e., we have as many electrons as there are k–vectors in the Fermi sphere (for the groundstate!). This postulate is in accordance with the usual BCS–ansatz; we formulate it as

- **Assumption 9.5:** For the groundstate of the superconductor all $\mathbf{k}$–states within the Fermi sphere are completely filled up with electrons and only the electrons of the Fermi surface form bound states.

For brevity we shift in $|0\rangle$ the electrons in the interior of the Fermi sphere into a new groundstate and concentrate on the Fermi surface by writing (9.47) in the form

$$|0\rangle = \prod_{i=1}^{N} \sum_{k_i \in FS} c(\mathbf{k}_i) a_\uparrow^+(\mathbf{k}_i) a_\downarrow^+(-\mathbf{k}_i)|f'\rangle \tag{9.49}$$

where $|f'\rangle$ contains the valence electron states in the interior of the Fermi surface. Now any $\mathbf{k}_i$ runs through all $\mathbf{k}$–vectors of the Fermi sphere $FS := \{\mathbf{K}_1 \ldots \mathbf{K}_N\}$. However, due to the fermionic anticommutation relations the state (9.49) is antisymmetric in spin– and momentum–variables. Thus it cannot contain any product of $a_\uparrow^+(\mathbf{k}_i)$ and $a_\uparrow^+(\mathbf{k}_j)$ with $\mathbf{k}_i \equiv \mathbf{K}_m$ and $\mathbf{k}_j \equiv \mathbf{K}_m$. Together with the complete occupation of all surface levels by electrons this leads to the following representation of the groundstate (9.49):

$$|0\rangle = N \prod_{\alpha=1}^{N} c(\mathbf{K}_\alpha) a_\uparrow^+(\mathbf{K}_\alpha) a_\downarrow^+(-\mathbf{K}_\alpha)|f'\rangle \quad . \tag{9.50}$$

Hence the antisymmetry of the groundstate enforces the two–electron condensate function to be a "pure" Cooper pair. The corresponding wavefunction in position space is then given by

$$\chi_{\alpha_1\alpha_2}(\mathbf{r}_1 - \mathbf{r}_2|\mathbf{K}_m) = \sigma^2_{\alpha_1\alpha_2} e^{\frac{i}{2}\mathbf{K}_m(\mathbf{r}_1-\mathbf{r}_2)} \quad . \tag{9.51}$$

Therefore, if we want to study the lowering of energy by bound states we have to substitute (9.51) into (9.41). For a first inspection we set $V(\mathbf{u}) \equiv 1$. Omitting the index of $\mathbf{K}_m$ we then obtain

$$1 = 2^7 m g^2 \left(\frac{\mathbf{K}^2}{2m} - E\right)^{-1} \left(\frac{\mathbf{K}^2}{2m}\right)^{-1} \chi(0)^2 \quad . \tag{9.52}$$

As $\mathbf{K}$ is an element of the Fermi surface, any $\mathbf{K}$–vector leads to the same energy $E_{FS}$. Hence (9.52) goes over into

$$1 = 2^7 g^2 \left(E_{\mathrm{FS}} - E\right)^{-1} E_{\mathrm{FS}}^{-1} \chi(0)^2 \quad . \tag{9.53}$$

From (9.53) we obtain

$$(E_{\mathrm{FS}} - E) = 2^7 g^2 E_{\mathrm{FS}}^{-1} \chi(0)^2 > 0 \, , \tag{9.54}$$

i.e., we obtain a lowering of the energy in comparison with the energy of ordinary band electrons.

We now turn to excited states given by (9.26). In the excited state $|N_0 + 2, k, l\rangle$ the groundstate electrons occupy the whole Fermi sphere, while two additional electrons have to be outside of the Fermi sphere. This means that if we represent (9.26) in Fourier space, the domains of $\mathbf{k}_1$ and $\mathbf{k}_2$ are two $\mathbf{k}$–values above the Fermi sphere.

## 9.4 Ginzburg–Landau Equation

The physical behaviour of superconductors is governed by the reactions of Cooper pairs on external electromagnetic fields. In phenomenological and in microscopic theory this behaviour is described by Ginzburg–Landau equations coupled to the electromagnetic field equations, where the latter are supplied by Ohm's law for infinite conductivity, i.e. by Landau's first equation. The derivation of Ohm's law is connected with the elimination of phonons from the theory and thus, if one works with the pure electronic BCS–model, Ohm's law has to be considered as a prerequisite and not as an object of derivation. On the other hand Ginzburg–Landau equations describe the behaviour of the electronic condensate and therefore must be derived within the BCS–model.

In general the equations for the electromagnetic vector potential coupled to the Cooper–pair current are added to the genuine dynamical Cooper pair equations. For brevity we omit these equations by considering the vector potential as an external field. It is, however, no difficulty to derive also these equations by application of quantum electrodynamics to the superconductor model and performing the calculations along the lines of Chapter 8.

The derivation of the Ginzburg–Landau equations for the BCS–model is due to Gorkov [Gor 59], who solved this problem by means of perturbation theory using the argument of a small gap function. But in a consequent treatment of Cooper pairs this argument cannot be correct because boson condensation occurs in theories with continuous momentum spectrum, and as we saw in the preceding section this holds true also for Cooper pairs. Hence we have to understand Ginzburg–Landau equations from another point of view, and if the Cooper pair picture is correct this can only be the effective, nonperturbative dynamics of these Cooper pairs.

As we indicated in the introduction we first treat the effective Cooper pair dynamics for pure states, i.e. for $T = 0$, and in this section we introduce the electromagnetic field into the BCS–model from the beginning (as an external field!).

The modified Hamiltonian of the coupled system of electrons and fields reads, see Fetter and Walecka [Fett 71] and (9.3):

$$\mathbf{H}_{\mathrm{A}} := \int d^3 r \psi_{Z_1}(\mathbf{r})(\gamma^0\gamma^5)_{Z_1Z_2}\frac{1}{2m}\left[-i\nabla_r + e\mathbf{A}(\mathbf{r})\right]^2\psi_{Z_2}(\mathbf{r}) \tag{9.55}$$

$$-g\varepsilon_{Z_1Z_2Z_3Z_4}\int d^3 r \psi_{Z_1}(\mathbf{r})\psi_{Z_2}(\mathbf{r})\psi_{Z_3}(\mathbf{r})\psi_{Z_4}(\mathbf{r}) \quad .$$

In this Hamiltonian the electric scalar potential as well as the electromagnetic Hamiltonian itself has been omitted, i.e., one considers a reduced theory without electric and electromagnetic effects and interactions, solely adapted to the treatment of magnetic fields which are indeed responsible for the most striking phenomena of superconductivity. We follow this general usage as in particular already at this stage Ginzburg–Landau equations can be derived.

The vector potential $\mathbf{A}(\mathbf{r})$ in (9.55) is considered as a classical field and equation (9.7) applies to (9.55), too, provided the definition (9.8) is changed into

$$\dot{K}_{I_1I_2} := \delta_{Z_1Z_2}\frac{1}{2m}\left[-i\nabla(\mathbf{r}_1) + e\mathbf{A}(\mathbf{r}_1)\right]^2\delta(\mathbf{r}_1 - \mathbf{r}_2) \quad . \tag{9.56}$$

To derive the effective dynamics for Cooper pairs, the uncorrelated parts have to be removed from the state functional (9.5). This leads to the normalordered spinorfield equations (4.28), (4.29), where for the Hamiltonian (9.55) the operator $K_{I_1 I_2}$ in (4.29) has to be replaced by the operator (9.56), the operator $W_{I_1 I_2 I_3 I_4}$ in (4.29) by (9.9), $A_{I_1 I_2}$ by (9.5) and the propagator $F_{I_1 I_2}$ in (4.29) by $G^{(2)}_{I_1 I_2}$.

The map of (4.28) onto the effective dynamics of Cooper pairs is performed in the chain rule approximation by means of the Cooper pair wavefunctions (9.26). In contrast to Assumption 9.2, in the case of non–vanishing vector potential $\mathbf{A}(\mathbf{r})$ we now also admit quantum numbers $k, l \neq 0$ for the Cooper pair functions. This reflects the idea of possible excitations of the condensate groundstate by means of the magnetic field.

We define the functional Cooper pair operators by

$$c_k := \int \langle 0|\mathcal{A}\begin{pmatrix}\mathbf{r}_1\mathbf{r}_2\\ 3\,4\end{pmatrix}|N_0 - 2, k, l\rangle j_{Z_3}(\mathbf{r}_1) j_{Z_4}(\mathbf{r}_2) d^3r_1 d^3r_2 + \dots, \tag{9.57}$$

$$c_k^+ := \int \langle 0|\mathcal{A}\begin{pmatrix}\mathbf{r}_1\mathbf{r}_2\\ 1\,2\end{pmatrix}|N_0 + 2, k, l\rangle j_{Z_1}(\mathbf{r}_1) j_{Z_2}(\mathbf{r}_2) d^3r_1 d^3r_2 + \dots$$

where we explicitly write out only the first order terms of the dressed particle sources, and introduce the superoperators $c_{k,\Lambda}$ with $c_{k,1} := c_k$ and $c_{k,2} := c_k^+$ which we represent in the general form

$$c_{k,\Lambda} := \int \phi^{(2)}_{Z_1 Z_2}(\mathbf{r}_1, \mathbf{r}_2|k, l, \Lambda)\, j_{Z_1}(\mathbf{r}_1) j_{Z_2}(\mathbf{r}_2) d^3r_1 d^3r_2 + \dots \tag{9.58}$$

Analogously to Section 4.8 the functional sources $c_{k,\Lambda}$ fulfill bosonic commutation relations. Finally we combine $k, l, \Lambda$ into one index $\ell$, i.e. $\ell = (k, l, \Lambda)$.

In Chapter 4 we saw that a physically meaningful interpretation of effective functional equations in general enforces the use of dressed particle states for weak mapping. This requirement is in contrast to the model of the electronic state of a superconductor which we hitherto used, where the subunits of this state were defined as hard core states.

However, without any contradiction we can pass from the hard core picture to the dressed state picture because by this transition nothing is lost. Rather only a partial shift of energies, interactions etc. is induced by redefining the subunits of the system. Hence without conceptional difficulties we can enlarge our subunits to be dressed fermion Cooper pairs in order to derive effective Cooper pair dynamics. The corresponding calculations were performed in Section 4.9 for a combined system of dressed bosons and dressed fermions, and we can take over the corresponding formulas to the model under consideration by assuming the fermion sources $f$ to be zero and by applying definitions (9.9), (9.56), etc.

Let $|\mathcal{G}(c;p)\rangle$ be the bosonic image functional state of the normalordered functional state $|\mathcal{F}(j;p)\rangle$. Then according to (4.158) the transformed energy equation is given by

$$E_{p0}|\mathcal{G}(c;p)\rangle = \mathcal{H}_B(c, \partial^c)|\mathcal{G}(c;p)\rangle \tag{9.59}$$

with

$$\mathcal{H}_B := c_\ell \left[K^{\ell k}\partial^c_k + M^{\ell k}\partial^c_k + N^{\ell k_1 k_2}\partial^c_{k_1}\partial^c_{k_2} + H^{\ell k_1 k_2 k_3}\partial^c_{k_1}\partial^c_{k_2}\partial^c_{k_3}\right] \tag{9.60}$$

and

$$K^{\ell k} := 2\hat{K}_{I_1 I_2}\phi^k_{I_2 I_3} R^\ell_{I_1 I_3} \tag{9.61}$$

$$M^{\ell k} := W_{I_4 I_1 I_2 I_3}\Big[-6G^{(2)}_{I_3 K}\phi^k_{I_1 I_2} R^\ell_{I_4 K} \tag{9.62}$$

$$+2\left(3G^{(2)}_{I_1 K_1} G^{(2)}_{I_4 K_2} + \frac{1}{4} A_{I_1 K_1} A_{I_4 K_2}\right)\phi^k_{I_3 K_3} R^\ell_{I_2 K_1 K_2 K_3}\Big]$$

$$N^{\ell k_1 k_2} := W_{I_4 I_1 I_2 I_3}\Big[12 G^{(2)}_{I_3 K_1}\phi^{k_1}_{I_1 K_2}\phi^{k_2}_{I_2 K_3} R^\ell_{I_4 K_1 K_2 K_3} \tag{9.63}$$

$$+12\phi^{k_1}_{I_1 I_2}\phi^{k_2}_{I_3 K} R^\ell_{I_4 K}\Big]$$

$$H^{\ell k_1 k_2 k_3} := -8 W_{I_4 I_1 I_2 I_3}\phi^{k_1}_{I_1 K_1}\phi^{k_2}_{I_2 K_2}\phi^{k_3}_{I_3 K_3} R^\ell_{I_4 K_1 K_2 K_3} \tag{9.64}$$

with the corresponding duals $R$, where the $c_\ell$ are considered as dressed particle operators.

Without any calculation one can show that (9.63) has to vanish.

- **Proposition 9.4:** If the fermion theory is invariant under global gauge transformations $\psi' = \exp(i\lambda)\psi$, then $N^{\ell k_1 k_2}$ must be zero.

**Proof:** We assume to have a complete set of two–fermion functions $\phi^{(2)}$ for the map. In this case the map is faithful. This means in particular that if fermion theory is gauge invariant the boson image theory must be gauge invariant, too. The elements of states in functional space are the matrix elements. According to our algebraic representation the theory is formulated by the functional equations (9.59) or its projections into coordinate space, respectively.

These projected equations may be written in a condensed formulation as matrix equations, cf. Section 3.6:

$$A_{mn}\phi_n = E\phi_m \ . \tag{9.65}$$

The gauge transformed equations read

$$UA_{mn}U^{-1}U\phi_n = EU\phi_m \ , \tag{9.66}$$

where $U$ is a unitary operator in function space.

According to (1.78) the wavefunctions $\phi_n$ transform as classical functions, and as $\phi_n$ for $n \neq m$ in (9.65) describes a polarization cloud with respect to $\phi_m$, it has the same quantum numbers like $\phi_m$ and thus $\phi_m$ and $\phi_n$ transform identical for all $m$ and $n$. So in our case we obtain

$$UA_{mn}U^{-1}\phi_n = E\phi_m \tag{9.67}$$

and thus $A_m = UA_mU^{-1}$, i.e. the invariance of the "matrices" $K^{\ell k}, M^{\ell k}, N^{\ell k_1 k_2}$ and $H^{\ell k_1 k_2 k_3}$. We conclude that the image theory is gauge invariant if all matrix element combinations in the coefficient functions (9.61)–(9.64) are invariant.

The states $\phi^{(2)}_{21}(\mathbf{r}_1, \mathbf{r}_2)$ transform with $\exp(2i\lambda)$ while the states $\phi^{(2)}_{34}(\mathbf{r}_1, \mathbf{r}_2)$ transform with $\exp(-2i\lambda)$. Due to their definition all dual states and the corresponding polarization cloud terms transform reciprocally to their original states. The function $G^{(2)}$ is invariant. Therefore for boson states (9.57) and their duals, the term (9.63) cannot be gauge invariant. Hence (9.63) must vanish. ◊

According to Proposition 9.4, (9.60) is reduced to

$$\mathcal{H}_{\mathrm{B}} = c_\ell \left[ K^{\ell k} \partial_k^c + M^{\ell k} \partial_k^c + H^{\ell k_1 k_2 k_3} \partial_{k_1}^c \partial_{k_2}^c \partial_{k_3}^c \right] \quad . \tag{9.68}$$

For the evaluation of the remaining terms we use the following form of our boson functions, referring to (9.26), (9.38) and the results for $G^{(2)}$ in Section 9.2:

$$\phi^{(2)}_{Z_1 Z_2}(\mathbf{r}_1, \mathbf{r}_2 | k, l, \Lambda = 1) = \begin{pmatrix} 0 & 0 \\ 0 & \sigma^2 \end{pmatrix}_{Z_1 Z_2} e^{-\frac{i}{2}\mathbf{k}(\mathbf{r}_1 + \mathbf{r}_2)} \chi(\mathbf{r}_1 - \mathbf{r}_2 | l) \tag{9.69}$$

$$\phi^{(2)}_{Z_1 Z_2}(\mathbf{r}_1, \mathbf{r}_2 | k, l, \Lambda = 2) = \begin{pmatrix} \sigma^2 & 0 \\ 0 & 0 \end{pmatrix}_{Z_1 Z_2} e^{-\frac{i}{2}\mathbf{k}(\mathbf{r}_1 + \mathbf{r}_2)} \chi(\mathbf{r}_1 - \mathbf{r}_2 | l) \tag{9.70}$$

$$G^{(2)}_{Z_1 Z_2}(\mathbf{r}_1, \mathbf{r}_2) = (\gamma^0 \gamma^5)_{Z_1 Z_2} g(\mathbf{r}_1 - \mathbf{r}_2) \quad , \tag{9.71}$$

while the dual functions are given by

$$R_{Z_1 Z_2}(\mathbf{r}_1, \mathbf{r}_2 | k, l, \Lambda = 1) = c \begin{pmatrix} 0 & 0 \\ 0 & \sigma^{2T} \end{pmatrix} e^{\frac{i}{2}\mathbf{k}(\mathbf{r}_1 + \mathbf{r}_2)} \chi(\mathbf{r}_1 - \mathbf{r}_2 | l) \tag{9.72}$$

$$R_{Z_1 Z_2}(\mathbf{r}_1, \mathbf{r}_2 | k, l, \Lambda = 2) = c \begin{pmatrix} \sigma^{2T} & 0 \\ 0 & 0 \end{pmatrix} e^{\frac{i}{2}\mathbf{k}(\mathbf{r}_1 + \mathbf{r}_2)} \chi(\mathbf{r}_1 - \mathbf{r}_2 | l) \tag{9.73}$$

with a normalization constant $c$. For a first inspection we do not take into account the continous part of the duals very carefully which yields only a numerical constant. Finally, due to fermion number conservation the polarization cloud $R_{I_1 I_2 I_3 I_4}$ must have the algebraic part $\sim (\gamma^0 \gamma^5) \otimes \phi^{(2)}$. By direct calculation it can be shown that the terms $K$ and $M$ are connected with $c\partial^c + c^+ \partial^{c^+}$, while $H$ is connected with $c^+ \partial^{c^+} \partial^c \partial^c + c \partial^c \partial^{c^+} \partial^{c^+}$.

The corresponding classical equations to the functional equation (9.59) are obtained by the ansatz $|\mathcal{G}\rangle = \exp \int [\Psi(\mathbf{r}) c(\mathbf{r}) + \Psi^*(\mathbf{r}) c^+(\mathbf{r})] \, d^3\mathbf{r} |0\rangle_B$, cf. (7.113), where $c(\mathbf{r})$ is the Fourier transformed of $c_k$. If this ansatz is substituted and if the above mentioned structure of the terms (9.61), (9.62), (9.64) is taken into account, then with an equation analogous to (6.54) one obtains after projection with $\langle 0 | \partial^c$:

$$\frac{\partial}{\partial t} \Psi(\mathbf{r}, t) = \left[ -i\nabla + e\mathbf{A}(\mathbf{r}) \right]^2 \Psi + m\Psi + a|\Psi|^2 \Psi \tag{9.74}$$

with real constants $m$ and $a$, and the complex conjugate equation. For stationary states this equation goes over into the Ginzburg–Landau equation

$$\left[ -i\nabla + e\mathbf{A} \right]^2 \Psi + m\Psi + a(\Psi^* \Psi)\Psi = 0 \tag{9.75}$$

etc. Usually $\Psi$ is interpreted as the vacuum condensate density. In our derivation $\Psi$ is a transition matrix element. This, however, is no contradiction: According to (9.57) these transition matrix elements just describe a part of the condensate wavefunction.

In high energy physics equations (9.74) and (9.75) are commonly denoted as Higgs field equations. They have to be considered as rough approximations of the functional equations (9.59) with the functional Hamiltonian (9.68). It is this Hamiltonian where the influence of the specific representation defined by the groundstate condensate structure can be explicitly studied, because the evaluation of (9.68) depends on the specific choice of the dressed particle operators (9.59). A detailed study of the various possibilities of this choice has not been performed so far. So we confine ourselves to this formal analysis of the problem.

## 9.5 Electrical Resistance

In superconductors the electrical resistance vanishes in a certain range of temperature. It is of interest to explain why this has to happen. The theoretical treatment of superconductors is different from that of ordinary metals. While in the latter the electron–phonon interaction is directly transferred to irreversible statistical mechanics, for superconductors the phonon modes are eliminated from the (reversible) electron–phonon dynamics resulting in an attractive electron–electron interaction. If this proceeding could be strictly performed it would not allow the definition of electrical resistance at all, if one disregards other sources of resistance. Fortunately the elimination of phonon modes is not exactly possible and thus a residual electron–phonon interaction remains in the theory which can be considered as the cause of resistance and therefore has to be transferred into irreversible dynamics.

Nevertheless, this proceeding brings up the question: Where is the borderline between reversible and irreversible dynamics? We neither can answer this question, nor can we give the exact expression of the remaining residual electron–phonon interaction. Rather we are forced to give a more qualitative argument for the vanishing of resistance in superconductors. We assume that in spite of the phonon mode elimination the residual electron–phonon interaction is similar to that of electron–phonon coupling in ordinary metals, and we investigate the consequences of a superconducting groundstate with respect to this interaction. Therefore we have to supply the BCS–Hamiltonian (9.3) by an appropriate electron–phonon coupling term while we can omit the phonon selfenergy as it plays no role in our discussion. This yields the modified Hamiltonian

$$H_{\mathrm{EP}} := H_{\mathrm{BCS}} + \kappa \int \psi_{Z_1}(\mathbf{r})(\gamma^0\gamma^5)_{Z_1 Z_2}\psi_{Z_2}(\mathbf{r}) \left[ \sum_{\mathbf{w}} \left( b^+_{\mathbf{w}} e^{-i\mathbf{w}\cdot\mathbf{r}} - b_{\mathbf{w}} e^{i\mathbf{w}\cdot\mathbf{r}} \right) \right] d^3r \quad (9.76)$$

with $\kappa := -4i\pi e/\lambda^2\ (1/2\gamma V\omega_{\mathbf{w}})^{1/2}$, where the phonon modes $b_{\mathbf{w}}$ and $b^+_{\mathbf{w}}$ are acoustical modes. With respect to the definition of constants, etc., see Haken [Hak 73].

From experiments one learns that on the whole only the electrons on the Fermi surface contribute to the electrical conductivity, i.e. in the case of superconductivity the Cooper pairs. Thus, in order to study resistance we have to study the effective dynamics of Cooper pairs under the influence of electron–phonon coupling. Without any further extension we can apply weak mapping to this problem. Instead of (9.8) or (9.56) we define

$$\tilde{K}_{I_1 I_2} := \delta_{Z_1 Z_2} \left[ \frac{\Delta_{r_1}}{2m} + \kappa \sum_{\mathbf{w}} \left( b^+_{\mathbf{w}} e^{-i\mathbf{w}\mathbf{r}_1} - b_{\mathbf{w}} e^{i\mathbf{w}\mathbf{r}_1} \right) \right] \delta(\mathbf{r}_1 - \mathbf{r}_2) \qquad (9.77)$$

and obtain by weak mapping in accordance with (9.60) and (9.61)

$$\mathcal{H}_b = c_\ell \tilde{K}^{\ell k} \partial^c_k + \dots \qquad (9.78)$$

with

$$\tilde{K}^{\ell k} := 2\tilde{K}_{I_1 I_2}\phi^k_{I_2 I_3} R^\ell_{I_1 I_3} \quad . \qquad (9.79)$$

We substitute any wavefunction of a Cooper pair and its dual into (9.79) and calculate the Cooper pair–phonon coupling term. This yields for the coupling term $CT$:

$$CT \propto \int e^{-i\mathbf{w}\mathbf{r}_1}\delta(\mathbf{r}_1-\mathbf{r}_2)e^{-i\mathbf{K}_\mu(\mathbf{r}_2-\mathbf{r}_3)}e^{i\mathbf{K}_\nu(\mathbf{r}_1-\mathbf{r}_3)}d^3r_1d^3r_2d^3r_3 \qquad (9.80)$$
$$\propto \delta(\mathbf{w})\delta(\mathbf{K}_\mu-\mathbf{K}_\nu)=0\,,$$

as $\mathbf{w}$ is assumed to be unequal zero. Hence the electron–phonon coupling vanishes if the condensate is at rest.

Now we consider the case where an electrical pulse was applied to the superconductor. As a reaction to this pulse the condensate is driven into a state of collective motion. This is expressed by a uniform shift of the Fermi surface along the direction of the electrical field pulse, i.e., any surface vector $\mathbf{K}_m$ is shifted by a constant vector $\mathbf{k}_0$ to $\mathbf{K}'_m = \mathbf{K}_m + \mathbf{k}_0$. By this shift the Cooper pair wavefunction $\exp[i\mathbf{K}_m(\mathbf{r}_1-\mathbf{r}_2)]$ is transformed into the wavefunction $\exp[i\mathbf{K}'_m(\mathbf{r}_1-\mathbf{r}_2)-i\mathbf{k}_0(\mathbf{r}_1-\mathbf{r}_2)]$. For the latter wavefunction the same result is obtained by substitution into (9.79) as for the wavefunction of the condensate at rest. Hence the coupling term $CT$ vanishes also for the condensate in motion. This means that there can be no resistance due to coupling to phonon modes if a supercurrent is present in the superconductor.

## 9.6 Thermostates and Weak Mapping

Nonrelativistic quantum field theoretic models of matter and recently even some high energy models of matter are subject to thermal statistics. Thermostatistics is introduced in quantum theory by means of the density operator for thermal equilibrium or nonequilibrium distributions. In the algebraic formulation of quantum theory this is expressed by the introduction of thermostates. Thus, if one intends to apply the formalism of the preceding chapters to models with thermostatistics, the central questions are: How is weak mapping related to quantum statistical theory and are thermostates and weak mapping compatible? Hence we will outline in which direction one has to proceed.

In statistical mechanics of quantum fields mainly two lines of theoretical treatment can be distinguished. One line is concerned with the extension of conventional covariant quantum field theory into the realm of thermodynamic systems. This is achieved by the introduction of thermofields. The other line is the direct algebraic approach to a state representation theory for thermal equilibrium and nonequilibrium situations. While thermofields are treated by means of path integral methods and perturbation theory, in the algebraic approach emphasis is laid on the properties of thermal states which survive the thermodynamic limit. With respect to the first line we refer to the review of Landsmann and van Weert [Land 87], with respect to the algebraic approach to Haag [Haag 92], Bratteli and Robinson [Brat 79] and Sewell [Sew 86].

The thermofield approach suffers from the general difficulties of the covariant quantum field theory. In contrast, in the algebraic approach thermal states can be rigorously treated, cf., e.g., Sewell [Sew 86], Rieckers [Riec 91], [Riec 93], but the treatment has been restricted to the investigation of thermal expectation values.

In our approach the transition matrix elements play a dominant role, while the subset of expectation values appears only in the calculation of the metrical tensor and in the selfconsistency calculation of Section 9.2. On the other hand, the peculiarities of statistical mechanics enforce the use of expectation values as the primary quantities. How to reconcile composite particle dynamics with this requirement is therefore one question we have to deal with. In doing so we remember that the central equation from which our previous investigations started is Heisenberg's time translation equation $\dot{\mathbf{A}} = i[\mathbf{H}, \mathbf{A}]_-$. So the question is: Can one find an analogous equation in quantum statistical mechanics? If this is possible, then in principle we should be able to transfer our previous methods to this quantum statistical equivalent of Heisenberg's equation.

The answer to this question is affirmative. Starting with the quantum Liouville equation in statistical mechanics we can find such an equivalent to Heisenberg's equation. The quantum Liouville equation formally reads

$$\frac{\partial}{\partial t} f(t) = -i\mathcal{L} \cdot f(t) \tag{9.81}$$

where $f(t)$ is the statistical operator; see for instance Rieckers and Stumpf [Rieck 77]. Equation (9.81) can be embedded in representation spaces in a twofold way. Let $\mathbb{H}$ be the Hilbert space spanned by a set of basis vectors $|e_i\rangle$, $i = 1 \ldots \infty$, as defined in Section 3.4 and referred to the original theory with Hamiltonian $\mathbf{H}$. Writing for short $|i\rangle \equiv |e_i\rangle$ one can represent the Liouville–von Neumann operator $\mathcal{L}$ as a fourth–order tensor over $\mathbb{H}$ in the following form:

$$\mathcal{L} = \sum_{i,j,k,\ell} |i\rangle\langle j| \otimes |k\rangle\langle \ell| \left[ H_{i\ell}\delta_{jk} - H^+_{jk}\delta_{i\ell} \right] \tag{9.82}$$

with $H_{ij} := \langle i|\mathbf{H}|j\rangle$ and $\mathcal{L} \in \mathbb{H} \otimes \mathbb{H}^d \otimes \mathbb{H} \otimes \mathbb{H}^d$, while $f$ has to be represented in the form

$$f = \sum_{i,j} |i\rangle f_{ij} \langle j| \tag{9.83}$$

with $f_{ij} := \langle i|f|j\rangle$ and $f \in \mathbb{H} \otimes \mathbb{H}^d$.

On the other hand, a general quantum statistical state has to be described by a statistical operator $f$ which has to be selfadjoint, positive and belongs to the trace class with unity normalization, i.e.

$$f^+ = f\,, \quad f \geq 0\,, \quad \mathrm{tr}\, f = 1 \quad . \tag{9.84}$$

It is convenient to consider an enlarged set of operators which are selfadjoint, positive and belong to the trace class without unity normalization, i.e. $\mathrm{tr}|g| < \infty$ where $|g|$ means the positive square root of $g^+g$. By means of these properties one can introduce a further representation space for $\mathcal{L}$ which is called the Liouville space $\mathbb{L}$. With respect to this construction for finite dimensional systems some essential statements can be derived which we transfer to our case without a further discussion and which we cite without proof; for the proofs we refer to Rieckers and Stumpf [Riec 77].

- **Proposition 9.5:**

  i) The set of operators of the trace class $\mathbf{S} \subset \mathbb{H} \otimes \mathbb{H}^d$ constitutes a $^*$–algebra.

  ii) Defining a scalar product $(f, g) := \mathrm{tr}[f^+g]$, $\mathbf{S}$ is a pre–Hilbert space. The closure of $\mathbf{S}$ with respect to $(f, g)$ represents a Hilbert space which is denoted by $\mathbb{L}$.

  iii) The operators of finite rank are dense in $\mathbf{S}$ and $\mathbb{L}$.

The Hilbert space $\mathbb{L}$ can be spanned by a complete set of basis vectors. For basis vectors $|a)$, $|b)$ we can represent the Liouville–von Neumann operator $\mathcal{L}$ in the alternative form

$$\mathcal{L} = \sum_{a,b} |a)(b| L_{ab} \tag{9.85}$$

with $\mathcal{L} \in \mathbb{L}^d \otimes \mathbb{L}$.

With respect to this alternative representation the following property can be proven:

- **Proposition 9.6:** The Liouville–von Neumann operator $\mathcal{L}$ is a linear, Hermitean operator on $\mathbb{L}$, i.e. $\mathcal{L}^+ = \mathcal{L}$.

Furthermore for all finite time translations in $\mathbb{H}$ it holds

- **Proposition 9.7:** If the Hamilton operator $\mathbf{H} \in \mathbb{H} \otimes \mathbb{H}^d$ generates a continuous, unitary Schrödinger dynamics, then the corresponding Liouville–von Neumann operator $\mathcal{L}$ is the generator of a one–parameter continuous group of unitary transformations in $\mathbb{L}$.

For further comments about properties of $\mathcal{L}$ and their consequences we again refer to Rieckers and Stumpf [Riec 77].

The above propositions allow to represent the system dynamics in $\mathbb{L}$ and to consider the Liouville equation (9.81) as a Schrödinger equation in $\mathbb{L}$. Therefore, applying the usual rules of quantum mechanics we can proceed in $\mathbb{L}$ in the same way as we did for ordinary Schrödinger equations in $\mathbb{H}$. In particular we can define a Heisenberg picture for operators $\mathcal{R}$ on the Liouville space $\mathbb{L}$. Let us assume that $\mathcal{R} \equiv \mathcal{R}(0)$ is an operator in the Schrödinger picture, then with $\mathcal{L} \equiv \mathcal{L}(0)$ we define by

$$\mathcal{R}(t) = e^{-i\mathcal{L}(0)t} \mathcal{R}(0) e^{i\mathcal{L}(0)t} \tag{9.86}$$

the corresponding Heisenberg operator. From (9.86) it immediately follows

$$\dot{\mathcal{R}}(t) = i[\mathcal{L}(0), \mathcal{R}(t)]_- \,, \tag{9.87}$$

i.e. the Heisenberg equation in Liouville space.

The fact that (9.87) is considered as the Heisenberg equation in Liouville space can be used to derive the image operator $\mathcal{O}_i$ on $\mathbb{L}$ of the field operators $\psi_I$ on $\mathbb{H}$. The natural condition for such a construction is the postulate, that the dynamical equation (9.87) is in accordance with the Heisenberg equation $\dot{\psi} = i[\mathbf{H}, \psi]_-$. With this condition one can directly construct $\mathcal{O}_i$ as a direct product of the original field operators $\psi$, and in this way reduce (9.87) to the original Heisenberg equation $\dot{\psi} = i[\mathbf{H}, \psi]_-$. We do not further elaborate this superfield construction, because in the following we will directly relate our functional equations for states in $\mathbb{L}$ to states in $\mathbb{H}$.

For an infinite number of degrees of freedom, i.e. for a quantum field theory, equations (9.81) and (9.87) possess inequivalent representations, i.e., we have to expect a set of different Liouville spaces $\mathbb{L}_\alpha$, $\alpha = 1, \ldots$, which refer to corresponding GNS–constructions. Let $|\Omega_\alpha)$ be a cyclic vector with $|\Omega_\alpha) \in \mathbb{L}_\alpha$ and let $\mathcal{O}_i$ be a (super–) field operator which is used for the formulation of $\mathcal{L}$. Then a set of cyclic basis vectors $|E_i)$ can be defined by

$$|E_i) := \mathcal{O}_1^+ \ldots \mathcal{O}_n^+ |\Omega_\alpha) \tag{9.88}$$

which spans $\mathbb{L}_\alpha$. We then obtain from (9.87)

$$\frac{\partial}{\partial t}\left(\mathcal{O}_1^+ \dots \mathcal{O}_n^+\right) = i\left[\mathcal{L}(0), (\mathcal{O}_1^+ \dots \mathcal{O}_n^+)\right]_- \quad . \tag{9.89}$$

If we furthermore consider the stationary Liouville–Schrödinger equation

$$U_\kappa|\kappa) = \mathcal{L}(0)|\kappa) \tag{9.90}$$

which defines the eigenstates of $\mathcal{L}(0)$ we can project equation (9.89) onto these states and obtain

$$i\frac{\partial}{\partial t}(\kappa'|\mathcal{O}_1^+ \dots \mathcal{O}_n^+|\kappa) = (U_\kappa - U_{\kappa'})(\kappa'|\mathcal{O}_1^+ \dots \mathcal{O}_n^+|\kappa) \quad . \tag{9.91}$$

The results of this discussion can be summarized by the following statement:

- **Proposition 9.8:** For systems with an infinite number of degrees of freedom the field operator algebra and the Heisenberg dynamics can be represented in Liouville spaces $\mathbb{L}_\alpha$, which are generated by corresponding cyclic vectors $|\Omega_\alpha)$ belonging to dynamically invariant states.

Guided by this proposition we now draw some conclusions which eventually allow the explicit construction of the functional equation for state vectors in $\mathbb{L}_\alpha$.

As in Liouville space the Heisenberg equation (9.87) holds, formally the only difference between the theories for pure and mixed states is the replacement of the ordinary Hilbert space $\mathbb{H}$ by the Liouville space $\mathbb{L}$ with corresponding "states". According to the preceding discussion a "state" in $\mathbb{L}$ is simultaneously a statistical operator $f$ on $\mathbb{H}$, and if we consider the "vacuum state" $|\Omega_T) \in \mathbb{L}$, we assume this state to be a statistical operator for thermal equilibrium with temperature $T \neq 0$, i.e. we have $|\Omega_T) = \exp(-\frac{1}{2kT}\mathbf{H})$ for canonical systems. The factor $1/2$ has to be inserted in order to reproduce thermal expectation values by algebraic states. If in contrast we consider excited states $|\kappa)$, we have to imagine these states as being related to additional particles moving "above" this thermal equilibrium state $|\Omega_T)$.

With respect to the GNS–construction, $|\Omega_T)$ can serve as a cyclic vector leading to the basis states (9.88). Therefore according to our "algebraic" formalism all information about the states above and below the thermal equilibrium and this thermal equilibrium itself is contained in their projections onto the basis (9.88) with $|\Omega_\alpha) \equiv |\Omega_T)$. This idea opens the way for a short–cut calculation of the corresponding "thermofield" functional equations. For simplicity we refer our derivation to the superconductor model of the preceding sections.

For this model an equation of the kind (3.98) holds. We write this equation in configuration space and obtain the general form

$$E_{p0}\langle 0|\mathcal{A}(\psi_{I_1}\dots\psi_{I_n})|p\rangle = \sum_m K(I_1\dots I_n, J_1\dots J_m)\langle 0|\mathcal{A}(\psi_{J_1}\dots\psi_{J_m})|p\rangle \,. \tag{9.92}$$

Equation (9.92) can be generalized to arbitrary energy differences $E_{pp'} := E_p - E_{p'}$, and in an abbreviated form it then reads

$$E_{pp'}\langle p'|\mathbf{A}_n|p\rangle = \sum_m K_{nm}\langle p'|\mathbf{A}_m|p\rangle \,. \tag{9.93}$$

With $\dot{\mathbf{A}} = i[\mathbf{H}, \mathbf{A}]_-$ it is

$$E_{pp'}\langle p'|\mathbf{A}_n|p\rangle = i\frac{\partial}{\partial t}\langle p'|\mathbf{A}_n|p\rangle \tag{9.94}$$

and thus

$$i\frac{\partial}{\partial t}\langle p'|\mathbf{A}_n|p\rangle = \sum_m K_{nm}\langle p'|\mathbf{A}_m|p\rangle\,. \tag{9.95}$$

Equation (9.95) is referred to the Hilbert space $\mathbb{H}$. Our aim is to transform this equation into the Liouville space $\mathbb{L}$; toward this end and according to our above assumption we define $|\Omega_T)$ to be represented by

$$|\Omega_T) = \sum_p |p\rangle \exp(-\frac{1}{2}\beta E_p)\langle p| \tag{9.96}$$

with $\beta = \frac{1}{kT}$, where we assume all energy states $|p\rangle$ to be states of an $N$–fermion system, i.e. $|p\rangle \equiv |p, N\rangle$. Then (9.96) is a canonical equilibrium distribution.

In accordance with Section 9.2 we consider a state $|\kappa) \in \mathbb{L}$, which describes a configuration of two fermions above $|\Omega_T)$, i.e. a state with a total number of $N+2$ fermions. This state $|\kappa)$ is a nonequilibrium state, because according to our GNS–construction the formulation of the theory is referred to the cyclic groundstate $|\Omega_T)$, which is assumed to be the only thermal equilibrium state. Any other equilibrium state may serve as a new GNS–vacuum, but not as an "excited" state over $|\Omega_T)$. Thus the statistical operator which characterizes this $(N+2)$–fermion state has the general form

$$|\kappa) \equiv |\kappa, N+2) = \sum_{p,p'} |p\rangle f^{\kappa}_{pp'}\langle p'|\,. \tag{9.97}$$

With the help of the representations (9.96) and (9.97) we can directly relate the "wavefunctions" $(\Omega_T|\mathcal{O}_n|\kappa)$ to expressions in ordinary Hilbert space and also transfer the dynamical equation (9.95) from $\mathbb{H}$ to $\mathbb{L}$. Using definitions (9.96) and (9.97) we multiply equation (9.95) from the left with $|p'\rangle\exp(-\frac{1}{2}\beta E_{p'})$ and from the right with $\sum_{p''} f^{\kappa}_{pp''}\langle p''|$ and sum over $p$ and $p'$. Taking the trace of the resulting expression and observing Proposition 9.5 we obtain for (9.95):

$$i\frac{\partial}{\partial t}(\Omega_T|\mathcal{O}_n|\kappa) = \sum_m K_{nm}(\Omega_T|\mathcal{O}_m|\kappa) \tag{9.98}$$

with

$$(\Omega_T|\mathcal{O}_n|\kappa) := \mathrm{tr}\left[\sum_{p,p',p''} |p\rangle \exp(-\frac{1}{2}\beta\mathbf{H})\langle p|\mathbf{A}_n|p'\rangle f^{\kappa}_{p'p''}\langle p''|\right], \tag{9.99}$$

or with (9.91)

$$(U_\kappa - U_0)(\Omega_T|\mathcal{O}_n|\kappa) = \sum_m K_{nm}(\Omega_T|\mathcal{O}_m|\kappa)\,. \tag{9.100}$$

In this way we have succeeded in transforming equation (9.95) from $\mathbb{H}$ into $\mathbb{L}$. Although the set of hierarchy functions $(\Omega_T|\mathbf{A}_n|\kappa)$ for $n = 1, \ldots \infty$ is now defined by (9.99), i.e. completely different from $\langle 0|\mathbf{A}_n|p\rangle$, it is remarkable that apart from different energy eigenvalues equations (9.92) and (9.100) are functionally equivalent. This fact confirms us in considering the Liouville spaces as representation spaces of the system dynamics.

So again as in Hilbert space also in Liouville space the question remains: How can one characterize such a specific representation for the calculation of (9.99), if the corresponding equations (9.92) or (9.100) are independent of any representation? The answer was already given in preceding chapters and confirmed in the present one by the selfconsistent calculation: The specific representation of the respective operator algebra is fixed by the corresponding propagators and transferred to any matrix element by the application of normaltransformation to the dynamical equations (9.92) and (9.100), respectively.

There is, however, an essential difference between pure and mixed states. While for pure states the selfconsistent calculation of Section 9.2 is an appropriate means for fixing the representation, for mixed states this does no longer hold. It is the peculiarity of mixed states that by coupling to a heat reservoir the system is forced to take a thermal equilibrium state as groundstate and in this way in a certain sense the freedom of the system to choose its appropriate representation is destroyed. This also means that the thermal propagators cannot be selfconsistently calculated, rather they are determined by the thermal groundstate itself. In this case the propagators have to be calculated by direct evaluation of

$$(\Omega_T|\mathcal{O}_n\mathcal{O}_n^+|\Omega_T) = \mathrm{tr}\left[\sum_{p,p'}|p\rangle\exp(-\frac{1}{2}\beta\mathbf{H})\langle p|\mathbf{A}_n\mathbf{A}_n^+|p'\rangle\exp(-\frac{1}{2}\beta\mathbf{H})\langle p'|\right] . \quad (9.101)$$

Such a calculation is usually performed for a finite number of degrees of freedom, where Fock representations can be applied, and afterwards the thermodynamic limit can be explicitly performed, see Rieckers [Riec 91], [Riec 93].

In the exactly evaluated molecular field models, see the following discussion, the thermal expectation values, i.e. thermally weighted traces, are directly calculated by means of exact solutions of the local low dimensional eigenvalue problem of $h_\Lambda$, which are substituted into the thermal traces. In our case due to the use of the fully local gauge invariant Hamiltonian (9.3) such a low dimensional representation does not exist. Hence the only way to master the direct calculation of these thermal traces consists in the determination of the $T=0$–spectrum by means of our functional equation (9.7) and to substitute this energy spectrum and the transition matrix elements into suitably decomposed thermal expectation values.

As we already mentioned, a detailed discussion would exceed the scope of this book. Hence with equation (9.100) we finish our outlook on the possibility of weak mapping in statistical mechanics of quantum fields, because it is now obvious that also for thermal representations we can proceed along the same lines as in the previous chapters.

Finally it is worthwhile to discuss the differences between the model treated above and models which can be treated by means of operator algebraic methods. For a review see Rieckers [Riec 91], [Riec 93].

In the simplest version the latter models are based on a family of Hamilton operators

$$\begin{aligned} h_\Lambda &= \sum_{k\in\Lambda}(\epsilon-\mu)(a_{k\uparrow}^* a_{k\uparrow} + a_{-k\downarrow}^* a_{-k\downarrow}) \\ &\quad - \frac{g}{|\Lambda|}\sum_{k,k'\in\Lambda} a_{k\uparrow}^* a_{-k\downarrow}^* a_{-k'\downarrow} a_{k'\uparrow} \end{aligned} \quad (9.102)$$

being defined for a set of increasing volumes $\Lambda$ in the lattice space of $k$–vectors, where finally the thermodynamic limit $\Lambda\to\infty$ can be explicitly performed. These hamiltonians $h_\Lambda$ can be rewritten as

$$h_\Lambda = 2\sum_{k\in\Lambda}(\epsilon-\mu)n(k) - \frac{g}{|\Lambda|}\sum_{k,k'\in\Lambda} b^*(k)b(k') \tag{9.103}$$

with

$$\begin{aligned} n(k) &:= a^*_{k\uparrow}a_{k\uparrow} + a^*_{-k\downarrow}a_{-k\downarrow} \\ b(k) &:= a_{-k\downarrow}a_{k\uparrow} \quad . \end{aligned} \tag{9.104}$$

Due to the special form of (9.103) one can define

$$m_\Lambda(a) := |\Lambda|^{-1}\sum_{k\in\Lambda} a(k) \tag{9.105}$$

and obtains for (9.103)

$$h_\Lambda = 2|\Lambda|(\epsilon-\mu)m_\Lambda(n) - g|\Lambda|m_\Lambda(b^*)m_\Lambda(b) \quad , \tag{9.106}$$

where $m_\Lambda(a)$ is a so–called molecular field operator.

If the system is quantized by passing from $a^*$ to $a^+$ and by considering $a^+$ and $a$ as creation– and annihilation operators one can derive corresponding commutation relations for the molecular field operators (9.105). This, however, means that essential features of the system can be treated only in terms of molecular field operators without resort to the original field algebra of creation– and annihilation operators. The algebra of molecular field operators is a low dimensional algebra and thus the eigenvalue problem of $h_\Lambda$ can be exactly solved. This in turn allows an exact evaluation of the thermodynamic states referred to a cyclic vacuum state $|\Omega_T)$ in Liouville space in the thermodynamic limit $\Lambda \to \infty$, see Thirring [Thir 67], [Thir 68], Jelinek [Jel 68] and Fleig [Flei 83].

The analysis of these states reveals a wealth of mathematical structures contained in this model. In particular a GNS–state $\omega^\beta$ in $\mathbb{L}$ can be decomposed into a set of pure phase states $\omega^{\beta\vartheta}$ which are characterized by effective mean field Hamiltonians $h_\vartheta$ which for themselves induce an effective dynamics in the macroscopic phase angle $\vartheta$. Thus an insight into the macroscopic dynamics of superconducting phases can be gained. This should be considered as a special advantage of this simple mean field model over the much more complicated short range interacting models. For the latter model class the techniques of algebraic quantum field theory were originally developed by Ruelle [Rue 69] and Robinson [Brat 79]. In general they provided structural insights into the states, dynamics and appropriate representations in the thermodynamical limit. However, for such models the demonstration of a phase transition could not be carried through.

On the other hand in the thermodynamic limit the commutators of the molecular field operators (9.105) vanish, i.e. the molecular fields, which are a superposition of Cooper pairs, become classical expressions, situated in the center of the algebra. Furthermore by the ansatz for $h_\Lambda$ the invariance against local gauge transformations is lost what prevents a correct coupling of the electromagnetic vector potential to $h_\Lambda$. This is a general shortcoming of the BCS model.

In contrast to these results in our method the global as well as the local gauge invariance of the Hamiltonian (9.3) is guaranteed and the Cooper pairs are quantum states which lead to an effective quantum dynamics, where in principle all contributions of the Hamilton operator (9.3) can be taken into account. In this context we note that the ansatz (9.102) corresponds to a bilocal Hamiltonian in coordinate space, in contrast to our strictly local Hamiltonian (9.3).

So the use of the Hamiltonian (9.102) seems to be somewhat complementary to our model: While in the molecular field BCS–approach (9.102) the macroscopic phase dynamics can be correctly interpreted but the electromagnetic effects are difficult to treat, in the enlarged BCS–model with Hamiltonian (9.3) just the reverse holds.

As Section 9.4 shows, it is easy to derive the basic equations which govern the electromagnetic reactions of a superconductor but so far a derivation of the phase structure of a superconductor with corresponding macroscopic phase angle on the microscopic level has not been elaborated. Rather one has to employ the common recipe for the derivation of this phase angle by decomposing the amplitude of the "classical" Higgs field into its absolute value and its phase, see Fetter and Walecka [Fett 71]. In view of the fact that any superconductor is finite, i.e. that the thermodynamic limit cannot play such a decisive role and that the concept of itinerant and localized Cooper pair quantum states is simple and natural and holds also for superconductors with a complicated geometrical structure, the model treated in this chapter deserves further attention because it is applicable to these more realistic situations.

# 10 Path Integrals and Effective Theories

## 10.1 Introduction

To an increasing extent the canonical quantization of fields is replaced by path integral representations . It is assumed that path integrals not only implicitly contain the quantization rules but also offer the complete solution of quantum field dynamics, i.e. they should describe all possible field configurations and processes. In particular with respect to composite particle dynamics they should implicitly contain all effective theories describing composite particle processes. Thus the "only" problem is the evaluation of such path integrals. Although great efforts have been made with respect to their numerical evaluation, for complicated systems, e.g. for interacting composite particles, one cannot circumvent the analytical treatment.

In the following we will discuss this problem by analyzing a representative example in the realm of quantum chromodynamics. With respect to the evaluation of such path integrals it is, however, advisable to study the derivation of path integrals first. In particular we relate the path integrals to the covariant formalism of quantized fields in Chapter 2 in order to obtain a better understanding of their meaning.

The derivation of path integrals for nonrelativistic quantum mechanical systems with a finite number of degrees of freedom is performed by a decomposition of transition amplitudes into infinitesimal steps, which is achieved by introducing intermediate states. As long as one deals only with such finite systems this method is correct. However, serious problems are caused by the extrapolation of these techniques to quantum fields with an infinite number of degrees of freedom, because in this case the choice of intermediate states is not unique. This feature is directly related to the problem of inequivalent representations for quantum fields. Hence in our discussion we will not rely on the method of intermediate states, rather we have to take the problem of inequivalent representations seriously.

## 10.2 Functional Perturbation Theory and Path Integrals

In analogy to quantum mechanics we are interested in the question whether in principle path integrals can correctly describe quantum field dynamics, i.e. whether they can replace canonical quantization. This problem is closely connected with algebraic representation theory. As already pointed out in preceding chapters in contrast to quantum mechanics the algebra of quantum fields with canonical commutation relations allows infinitely many inequivalent representations, and the choice of appropriate representations is part of the problem of solving the dynamical field equations with inclusion of corresponding boundary conditions, etc. Under definite physical conditions one is not free to choose an arbitrary representation at will. Rather the dynamics of the quantum fields enforces the use of adapted representations. So, in order to clarify the physical and mathematical meaning of path integrals the central question is: What kind of representations of the field operator algebra are induced by path integrals?

To answer this question we relate path integrals to the ordinary canonical quantization procedure of quantum fields which is relativistically invariant and independent of the representation of the algebra. To grasp the essence it is sufficient to consider a scalar field with nonlinear dynamics. Spinorfields etc. can be treated along the same lines and with the same result. We give a streamlined version of the arguments of Lurie [Luri 68] and Rivers [Riv 88].

We start with the Lagrangian density

$$\mathcal{L}[\varphi] := \frac{1}{2}\partial_\mu\varphi\partial^\mu\varphi - \frac{1}{2}m^2\varphi^2 - U[\varphi] = \mathcal{L}_0[\varphi] - U[\varphi] \tag{10.1}$$

for a scalar field $\varphi \equiv \varphi(x)$, where $U$ is a nonlinear potential in $\varphi$. The corresponding field equation reads

$$\left(\Box + m^2\right)\varphi(x) + U'\Big[\varphi(x)\Big] = 0 \tag{10.2}$$

with $\Box := \partial_\mu\partial^\mu$. The generating covariant functional for the canonically quantized field $\varphi(x)$ is defined by

$$Z[j] := \langle 0|T\exp\left[i\int d^4x j(x)\varphi(x)\right]|0\rangle \tag{10.3}$$

with the time–ordering operator $T$. Applying the methods of Chapter 2 we see that this functional has to be a solution of the functional equation

$$\left\{\left(\Box + m^2\right)\Big[-i\partial(x)\Big] + U'\Big[-i\partial(x)\Big] - j(x)\right\}Z[j] = 0 \tag{10.4}$$

with $\partial(x) := \delta/\delta j(x)$ in contrast to $\partial_\mu := \partial/\partial x^\mu$, which is the functional version of corresponding Schwinger–Dyson–Freese equations. If (10.3) is expanded into a power series in $j(x)$, the corresponding coefficient functions $\tau^{(n)}(x_1\ldots x_n)$ are given by

$$\tau^{(n)}(x_1\ldots x_n) := \langle 0|T\Big[\varphi(x_1)\ldots\varphi(x_n)\Big]|0\rangle, \qquad n = 1,\ldots,\infty, \tag{10.5}$$

and in principle equation (10.4) should allow to explicitly calculate these functions for the model under consideration. As the $\tau^{(n)}$–functions are assumed to completely characterize the quantum field system, equation (10.4) is a formulation of a canonically quantized field theory which additionally is relativistically invariant (cf. Chapter 2) and independent of representations of the field algebra, i.e., the algebraic representation is not fixed. Rather a specific representation is determined by the kind of solution of equation (10.4), i.e. any specific solution fixes a specific representation.

We mention that equation (10.4) is the analogue to the functional equations (2.77) etc., which were derived in Chapter 2, except for the fact that (10.4) is formulated without the concept of the functional state space and that the functionals (10.3) are confined to VEV's of field operators.

A general theory of solutions of equation (10.4) which takes into account the relation of these solutions to specific algebraic representations does not exist at present; see also Section 2.7. So one is forced to study exact solutions which are accidentally known.

In order to obtain such a solution we first consider the functional equation (10.4) for vanishing interaction $U \equiv 0$:

$$\left\{(\Box + m^2)\left[-i\partial(x)\right] - j(x)\right\} Z_0[j] = 0 \quad . \tag{10.6}$$

Its solution is given by

$$Z_0[j] = \exp\left[-\frac{1}{2}i\int d^4x d^4y j(x)\Delta_F(x-y)j(y)\right] \quad , \tag{10.7}$$

where we used the definition for the Feynman propagator $\Delta_F(x-y) := -K^{-1}(x,y)$ with

$$K(x,y) := (\Box + m^2)\,\delta(x-y) \quad . \tag{10.8}$$

Furthermore it is $\langle 0|T\left[\varphi^f(x)\varphi^f(y)\right]|0\rangle = -i\Delta_F$ which analogously to (10.3) relates $Z_0[j]$ to its power series definition in terms of VEVs of time–ordered products for free fields $\varphi^f(x)$.

It is now assumed that the solution of equation (10.4) has the form

$$Z[j] = \hat{Z}[\partial] Z_0[j] \quad , \tag{10.9}$$

where in particular $\hat{Z}[\partial]$ is given by

$$\hat{Z}[\partial] = \exp\left\{-i\int F\left[-i\partial(x)\right]d^4x\right\} \quad . \tag{10.10}$$

Due to the fact that $Z_0[j]$ is a solution of equation (10.6) we have the identity

$$\begin{aligned} &\left\{(\Box + m^2)\left[-i\partial(x)\right] - j(x)\right\}\hat{Z}[\partial]Z_0[j] \\ &= \left[\left\{(\Box + m^2)\left[-i\partial(x)\right] - j(x)\right\}, \hat{Z}[\partial]\right]_- Z_0[j] \\ &= -\left[j(x), \hat{Z}[\partial]\right]_- Z_0[j] \quad , \end{aligned} \tag{10.11}$$

and with (10.10) we obtain

$$-\left[j(x), \hat{Z}[\partial]\right]_- = -F'\left[-i\partial(x)\right]\hat{Z}[\partial] \quad . \tag{10.12}$$

Therefore by inserting (10.9) into (10.4) and observing (10.11), (10.12) we get

$$\begin{aligned} &\left\{(\Box + m^2)\left[-i\partial(x)\right] + U'\left[-i\partial(x)\right] - j(x)\right\}\hat{Z}[\partial]Z_0[j] \\ &= \left\{-F'\left[-i\partial(x)\right] + U'\left[-i\partial(x)\right]\right\}\hat{Z}[\partial]Z_0[j] \\ &= 0 \quad . \end{aligned} \tag{10.13}$$

From this equation it follows that $F \equiv U$ apart from a constant. Hence we have

$$Z[j] = \exp\left\{-i\int U\left[-i\partial(x)\right]d^4x\right\} Z_0[j] \tag{10.14}$$

as a solution of (10.4). The functional (10.14) is the well–known functional formulation of field theoretic perturbation series in terms of corresponding Feynman propagators.

Applying a functional Fourier transformation to (10.14) we can rewrite (10.14) as a path integral. We first consider the functional Fourier transform of $Z_0[j]$. This transformation is formally expressed by

$$Z_0[j] = \int D[\chi]\tilde{Z}_0[\chi] \exp\left[i \int d^4x j(x)\chi(x)\right] \quad , \tag{10.15}$$

where $D[\chi]$ denotes integration over an infinite dimensional function space of classical fields $\chi(x)$. Such integrals are assumed to arise as limiting values of integrations over finite dimensional function spaces if the number of dimensions is gradually increased. In particular for $Z_0[j]$ one obtains the following Fourier transform, cf. Rivers [Riv 88],

$$\tilde{Z}_0[\chi] = \exp\left\{i \int d^4x \left[-\frac{1}{2}\chi(x)\left(\Box + m^2\right)\chi(x)\right]\right\} \quad , \tag{10.16}$$

where for convenience a constant prefactor has been incorporated into the measure $D[\chi]$. Furthermore, in order to emphasize the relation to the $\varphi$–field dynamics one can relabel the $\chi$–field into the $\varphi$–field.

A proof of the existence of (10.15) is only possible if the field $\varphi(x)$ and $\Delta_F$ are analytically continued into Euclidian space. In this case (10.15) goes over into the Fourier transform of a Gaussian measure, cf. Glimm and Jaffe [Gli 87], provided one replaces $j(x)$ by $ij(x)$. This replacement is enforced by the transition from $d^4x$ in Minkowski space to $(id^4x)_E$ in Euclidian space. But since $j(x)$ is largely a book–keeping artefact this change has no physical significance and is omitted in the subsequent sections. In Euclidian space (10.15) and (10.16) yield

$$Z_0^E[j] = \int D[\varphi] \exp\left\{-\int \mathcal{L}_0^E[\varphi(x)](d^4x)_E + \int j(x)\varphi(x)(d^4x)_E\right\} \quad . \tag{10.17}$$

In the following we simply write $d^4x$ instead of $(d^4x)_E$.

If (10.17) is substituted into (10.14) we finally get the path integral

$$Z^E[j] = \int D[\varphi] \exp\left\{-\int \left[\mathcal{L}^E[\varphi(x)] - j(x)\varphi(x)\right] d^4x\right\} \quad , \tag{10.18}$$

where $\mathcal{L}^E$ is the Euclidean Lagrange functional. Hence, the path integral (10.18) is a Fourier transform of the Euclidean continuation of the perturbation solution (10.14). This implies that the corresponding representation of the field $\varphi(x)$ which is induced by the path integral, is the interaction picture with free field representation, and thus the severe mathematical and physical problems of perturbation theory are still present.

Usually path integrals are derived by an evaluation of transition amplitudes using intermediate states. In this respect the most elaborate version is due to Faddeev and Slavnov [Fadd 80]. In this version the evaluation is directly performed in terms of the interaction picture with free field representation of creation and annihilation operators of a scalar field $\varphi(x)$ for arbitrary potential $U[\varphi]$, and in addition of Fermi fields $\psi_\alpha(x)$ and $\psi_\alpha^+(x)$. It is emphasized by Faddeev and Slavnov that the path integral is a reformulation of perturbation theory. Thus our results coincide with those of these authors. In the same way Rivers [Riv 88] stresses that in all applications the path integral remains an algorithm for generating perturbation series, rather than an entity in its own right.

From a mathematical point of view a more profound insight into path integrals can be gained by measure theory. Defining the measure

$$d\mu[\varphi] := D[\varphi] \exp\left\{-\int \mathcal{L}^E\Big[\varphi(x)\Big] d^4x\right\} \tag{10.19}$$

we can rewrite $Z^E[j]$ in the form

$$Z^E[j] = \int d\mu[\varphi] \exp\left\{\int j(x)\varphi(x) d^4x\right\} \quad . \tag{10.20}$$

In this way the path integral is reformulated as a Fourier transform (10.20) of the specific measure (10.19). For state functionals of classical statistical mechanics the Bochner–Minlos theorem secures the existence of such Fourier representations, cf. Bratteli and Robinson [Brat 79]. The situation is more complicated in nontrivial quantum field theory as the measure (10.19) becomes non–Gaussian. While for Gaussian measures, i.e. for free fields, the theory is simple, the non–Gaussian measures offer considerable difficulties. To be mathematically precise one has to show that $Z^E[j]$ is a continuous positive–definite functional over a test function space $\mathcal{T} \ni j$, which is a nuclear topological space. The measure (10.19) is then defined on $\mathcal{T}'$, the topological dual space of $\mathcal{T}$. This is a large distribution space containing many singular fields $\varphi(x)$ which have no physical meaning. In particular it is not probable that the measure $d\mu[\varphi]$ is supported by smooth fields with a well defined (least) action functional.

A more positive mathematical perspective would be, that, by the Bochner–Minlos theorem, $d\mu[\varphi]$ is in the whole, but not in the part $D[\varphi]$ of the representation (10.19), a well defined measure which may be approximated by measures on finite dimensional subspaces of $\mathcal{T}'$. This opens the possibility that the generating functional $Z^E[j]$ is approximated by finite dimensional integrals which are not necessarily analytic in the coupling constant. However, a thorough discussion of this question would exceed the scope of this book.

Within perturbation theory Glimm and Jaffe [Gli 87] discussed convergence and existence properties of (10.20). They were able to give rigorous results only in $d = 2$ dimensions. A further analysis of this problem was performed by Sardanashvily and Zakharov [Sar 91]. These authors emphasize that (10.19) is not a true measure in general and that it is restricted to perturbation expansion, while the true measures for interacting fields in (10.20) lead to inequivalent representations with respect to the interaction picture with free field representation.

In the following section path integrals are applied to non–abelian gauge theories. This gives rise to additional difficulties which are discussed for instance by Gutzwiller [Gutz 90].

## 10.3 Hadronization of QCD

In this section we treat the composite particle techniques for path integrals and discuss a representative example in QCD. We review this approach according to Cahill and Roberts [Cah 85], Roberts and Cahill [Rob 87], Praschifka, Roberts and Cahill [Pra 87a,b], Roberts, Cahill and Praschifka [Rob 88], Cahill, Praschifka and Burden [Cah 89a], Cahill [Cah 89b], Roberts, Praschifka and Cahill [Rob 89].

The generating functional for QCD in Euclidean metric is based on the fermionic quark fields $q \equiv q_{\mu,f,c}(x)$ and the gluon fields $A^a_\mu(x)$ and reads

$$\begin{aligned} Z\left[J^a_\mu, \bar\eta, \eta\right] &= \int D[\bar q] D[q] \prod_{a,\mu} D[A^a_\mu] \Delta_f[A^a_\mu] \quad \times \\ & \exp\left\{ -S\left[A^a_\mu, \bar q, q\right] + \int d^4x \left(\bar\eta q + \bar q \eta + J^a_\mu A^a_\mu\right)\right\} \end{aligned} \tag{10.21}$$

where

$$S\left[A^a_\mu, \bar q, q\right] = \int d^4x \left( \frac{1}{4} F^a_{\mu\nu} F^a_{\mu\nu} + \frac{1}{2\xi} \left(\partial_\mu A^a_\mu\right)^2 + \bar q \mathcal{D} q \right) \tag{10.22}$$

$$F^a_{\mu\nu} := \partial_\mu A^a_\nu - \partial_\nu A^a_\mu + g f^{abc} A^b_\mu A^c_\nu \tag{10.23}$$

and

$$\mathcal{D} := \gamma_\mu \left( \partial_\mu - ig\frac{1}{2}\lambda^a A^a_\mu \right) + M\ . \tag{10.24}$$

The meaning and interpretation of symbols is standard and can be found elsewhere, see, e.g., Cheng and Li [Che 86], Nachtmann [Nach 86].

The functional (10.21) can be rewritten as

$$\begin{aligned} Z\left[J^a_\mu, \bar\eta, \eta\right] &= \int D[\bar q] D[q] \exp\left\{ - \int d^4x \left[ \bar q (\gamma_\mu \partial_\mu + M) q - \bar\eta q - \bar q \eta \right] \right\} \quad \times \\ & \exp\left[ \int d^4x \left( ig\bar q \frac{\lambda^a}{2} \gamma_\mu q \frac{\delta}{\delta J^a_\mu} \right) \right] \exp\Big( W[J^a_\mu] \Big) \end{aligned} \tag{10.25}$$

with

$$\begin{aligned} \exp\Big( W[J^a_\mu] \Big) &= \int \prod_{a,\mu} D[A^a_\mu] \Delta_f [A^a_\mu] \times \\ & \exp\left\{ - \int d^4x \left[ \frac{1}{4} F^a_{\mu\nu} F^a_{\mu\nu} + \frac{1}{2\xi} \left(\partial_\mu A^a_\mu\right)^2 - J^a_\mu A^a_\mu \right] \right\}\ , \end{aligned} \tag{10.26}$$

where $W[J^a_\mu]$ is the generating functional of connected gluon Green's functions. Using the relation

$$\exp\left[ \int d^4x \left( ig\bar q \frac{1}{2} \lambda^a \gamma_\mu q \frac{\delta}{\delta J^a_\mu} \right) \right] \exp\left( W\left[J^a_\mu\right] \right) = \exp\left( W\left[ J^a_\mu + ig\bar q \frac{1}{2} \lambda^a \gamma_\mu q \right] \right) \tag{10.27}$$

we can rewrite (10.25) for $J^a_\mu \equiv 0$ as

$$\begin{aligned} Z[0, \bar\eta, \eta] &= \int D[\bar q] D[q] \exp\left\{ - \int d^4x \left[ \bar q (\gamma_\mu \partial_\mu + M) q - \bar\eta q - \bar q \eta \right] \right\} \quad \times \\ & \exp\left( W\left[ ig\bar q \frac{1}{2} \lambda^a \gamma_\mu q \right] \right)\ . \end{aligned} \tag{10.28}$$

Putting $J_\mu^a \equiv 0$ is no inadmissible restriction as we are exclusively interested in hadron or quark dynamics, respectively.

Expansion of (10.26) in a power series in $J_\mu^a$ and omission of all higher order terms leads to

$$W\left[J_\mu^a\right] \approx \int d^4x_1 d^4x_2 \frac{1}{2} D_{\mu_1\mu_2}^{a_1a_2}(x_1,x_2)\, J_{\mu_1}^{a_1}(x_1) J_{\mu_2}^{a_2}(x_2)\,. \tag{10.29}$$

In this approximation (10.28) becomes

$$Z[0,\bar{\eta},\eta] = \int D[q]D[\bar{q}]\exp\left[-S[\bar{q},q] + \int d^4x(\bar{\eta}q + \bar{q}\eta)\right] \tag{10.30}$$

with

$$\begin{aligned} S[\bar{q},q] &= \int d^4x d^4y \Big[\bar{q}(x)\gamma_\mu \partial_\mu \delta^4(x-y)q(y) + M\bar{q}(x)\delta^4(x-y)q(y) \\ &\quad + \frac{g^2}{2}\bar{q}(x)\frac{\lambda^a}{2}\gamma_\mu q(x) D(x-y)\bar{q}(y)\frac{\lambda^a}{2}\gamma_\mu q(y)\Big]\,, \end{aligned} \tag{10.31}$$

where the gluon propagator $D_{\mu\nu}^{ab}(x,y)$ is assumed to have the form

$$D_{\mu\nu}^{ab}(x,y) = \delta_{ab}\delta_{\mu\nu}D(x-y)\,.$$

The action (10.31) is referred to as Global Color–symmetry Model (GCM) of QCD.

The physical meaning of (10.30) is not changed if it is multiplied by a constant factor. This factor is chosen to be

$$C := \int \prod_A D[B^A]\exp\left[-\int d^4x d^4y\, \frac{B^A(x,y)B^A(y,x)}{2g^2 D(x-y)}\right]\,, \tag{10.32}$$

where the index $A = (a,\mu)$ corresponds to $\Lambda^A \equiv \lambda^a\gamma_\mu$ and the $B^A(x,y)$–fields are assumed to be hermitean. Then for $\hat{Z} = CZ$ it is

$$\hat{Z}[0,\bar{\eta},\eta] = \int D[q]D[\bar{q}]\prod_A D[B^A]\exp\left[-S\left[\bar{q},q,B^A\right] + \int d^4x\,(\bar{\eta}q + \bar{q}\eta)\right]\,, \tag{10.33}$$

and $S[\bar{q},q,B^A]$ is the sum of $S[\bar{q},q]$ and the quadratic $B^A$–term from (10.32). Provided the $\bar{q}$– and $q$–fields are assumed to be Grassmann–variables and provided furthermore that the measure $D[\hat{B}^A]$ is invariant with respect to the shift

$$\hat{B}^A(x,y) = B^A(x,y) + g^2 D(x-y)\bar{q}(y)\frac{\Lambda^A}{2}q(x) \tag{10.34}$$

equation (10.33) goes over into

$$\hat{Z}[0,\bar{\eta},\eta] = \int D[q]D[\bar{q}]\prod_A D[\hat{B}^A]\exp\left\{-\hat{S}\left[\bar{q},q,\hat{B}^A\right] + \int d^4x\,(\bar{\eta}q + \bar{q}\eta)\right\} \tag{10.35}$$

with

$$\hat{S}\left[\bar{q}, q, \hat{B}^A\right] := \int d^4x d^4y \left[\bar{q}(x)\left(\gamma_\mu \partial_\mu + M\right)\delta^4(x-y)q(y) - \bar{q}(x)\hat{B}^A(x,y)\frac{\Lambda^A}{2}q(y)\right] + \int d^4x d^4y \, \frac{\hat{B}^A(x,y)\hat{B}^A(y,x)}{2g^2 D(x-y)} \quad . \tag{10.36}$$

The exponent in (10.35) is thus quadratic in the fermion fields and so the integration over fermion fields can be performed with the result

$$\hat{Z}[0,\bar{\eta},\eta] = \int \prod_A D[\hat{B}^A] \exp\left\{-s\left[\hat{B}^A\right] + \int d^4x d^4y \bar{\eta}(x) G(x,y,\hat{B}^A)\eta(y)\right\} \ , \tag{10.37}$$

where

$$s\left[\hat{B}^A\right] = -\ln\det|G^{-1}(x,y,\hat{B}^A)| + \int d^4x d^4y \, \frac{\hat{B}^A(x,y)\hat{B}^A(y,x)}{2g^2 D(x-y)} \tag{10.38}$$

and

$$G^{-1}(x,y,\hat{B}^A) := (\gamma_\mu \partial_\mu + M)\delta^4(x-y) - \hat{B}^A(x,y)\frac{\Lambda^A}{2} \tag{10.39}$$

is formulated as an integral kernel.

If the determinant in (10.38) is evaluated by means of $\det|A| = \exp(\mathrm{tr}\ln A)$, (10.38) becomes

$$s[\hat{B}^A] = -\mathrm{tr}\ln\left[G^{-1}(x,y,\hat{B}^A)\right] + \int d^4x d^4y \, \frac{\hat{B}^A(x,y)\hat{B}^A(y,x)}{2g^2 D(x-y)} \ . \tag{10.40}$$

Thus if the fermion integration is correct, equations (10.35) and (10.37) have to be equivalent, i.e., we can study either (10.35) or (10.37). Furthermore, if path integrals correctly describe quantum field theory in the sense that (10.38) is an effective action functional, too, we can equivalently use either (10.36) or (10.38) in order to draw further conclusions.

## 10.4 Composite Particles and Field Operator Products

In Section 3 we formally introduced a bosonization of the (approximated) QCD path integral (10.30) which resulted in the path integral (10.35). Thus the question arises: What is the physical meaning of this procedure? In particular the meaning of the $\hat{B}^A$–field is of interest.

If we apply the action principle to (10.36) we obtain the field equation

$$\hat{B}^A(y,x) = g^2 D(x-y)\bar{q}(x)\frac{\Lambda^A}{2}q(y) \quad , \tag{10.41}$$

i.e. in this version the $\hat{B}^A$–field is a classical composite field. By the authors cited in Section 10.3 the classical character of the $\hat{B}^A$–field is emphasized and only the classical actions (10.36) or (10.38) are studied. However, if one wishes to theoretically describe composite particles, this interpretation is incomplete and unsatisfactory. In general, if the elementary constituents of composite particles are quantum particles (or fields) the composite particle itself is expected to be a quantum particle (or field). This includes that as a consequence of the subcomponent dynamics apart from the classical Lagrangian all correlation functions and the corresponding quantization rules for the composite particle have to be derived directly from the subcomponent equations. In this way the purely classical interpretation has to be rejected.

To overcome these difficulties one could try to quantize the $\hat{B}^A$–field itself via a path integral. To examine the consequences of this procedure we postulate: The measure

$$d\mu\left[q,\bar{q},\hat{B}^A\right] := D[q]D[\bar{q}]D[\hat{B}^A]\exp\left\{-\hat{S}\left[\bar{q},q,\hat{B}^A\right]\right\} \tag{10.42}$$

which results from the bosonized path integral (10.35) is the measure of the generating functional of $q$–, $\bar{q}$– *and* $\hat{B}^A$–fields. Thus we define the generating functional

$$\begin{aligned}\hat{Z}[0,\eta,\bar{\eta},Q] \ := \ & \int d\mu\left[q,\bar{q},\hat{B}^A\right] \ \times \\ & \exp\left\{\int d^4x(\bar{\eta}q+\eta\bar{q}) + \int d^4xd^4yQ^A(x,y)\hat{B}^A(x,y)\right\} ,\end{aligned} \tag{10.43}$$

where $Q^A(x,y)$ are the sources of the $\hat{B}^A$–fields. For $Q^A \equiv 0$ we regain the path integral (10.35).

In the following discussion for brevity we introduce the notation

$$\partial^\eta(x) := \frac{\delta}{\delta\eta(x)}\,, \quad \bar{\partial}^\eta(x) := \frac{\delta}{\delta\bar{\eta}(x)}\,, \quad \partial^A(x,y) := \frac{\delta}{\delta Q^A(x,y)}\,. \tag{10.44}$$

Then the correlation functions for the $\hat{B}^A$–fields are defined by

$$\begin{aligned}&\langle \hat{B}^{A_1}(x_1,y_1)\dots\hat{B}^{A_n}(x_n,y_n)\rangle \\ &:= \ \partial^{A_1}(x_1,y_1)\dots\partial^{A_n}(x_n,y_n)\hat{Z}[0,\eta,\bar{\eta},Q]_{|\eta,\bar{\eta},Q=0} \ .\end{aligned} \tag{10.45}$$

In order to draw further conclusions we pass through formulae (10.14) to (10.18) in the reverse direction. In application to the generating functional (10.43) this yields

$$\begin{aligned}\hat{Z}[0,\eta,\bar{\eta},Q] = \exp&\left[\int d^4xd^4y\partial^\eta(x)\partial^A(x,y)\frac{\Lambda^A}{2}\bar{\partial}^\eta(y)\right] \ \times \\ \exp&\left\{\int d^4xd^4y\left[\bar{\eta}(x)S_E(x-y)\eta(y) + \frac{1}{2}g^2Q^A(x,y)D(x-y)Q^A(y,x)\right]\right\}\end{aligned} \tag{10.46}$$

with the Euclidian propagator $S_E(x-y)$.

By means of this representation we can show that (10.46) satisfies a corresponding set of Schwinger–Dyson–Freese equations. Applying the method of Section 10.2 by direct calculation one can verify that these equations in their functional version read as follows:

$$\left[(\gamma_\mu\partial_\mu + M)\,\bar{\partial}^\eta(x) + \frac{1}{2}\int d^4y\partial^A(x,y)\Lambda^A\bar{\partial}^\eta(y)\right]\hat{Z} \;=\; \eta(x)\hat{Z}\,, \tag{10.47}$$

$$\left[\partial^\eta(x)\,(\gamma_\mu\partial_\mu + M) + \frac{1}{2}\int d^4y\partial^\eta(y)\partial^A(y,x)\Lambda^A\right]\hat{Z} \;=\; -\bar{\eta}(x)\hat{Z}\,, \tag{10.48}$$

$$\left[\frac{1}{g^2}D(x-y)^{-1}\partial^A(x,y) - \frac{1}{2}\partial^\eta(y)\Lambda^A\bar{\partial}^\eta(x)\right]\hat{Z} \;=\; Q^A(y,x)\hat{Z}\,. \tag{10.49}$$

Combined with appropriate boundary conditions we consider the Schwinger–Dyson–Freese equations as a complete characterization of the dynamics of a quantum field or several quantum fields (cf. Chapters 2,3). Thus, starting from the definition of the generating functional (10.43) for the $q$– and $\bar{q}$–fields as well as for the $\hat{B}^A$–fields we have obtained a description of a composite particle theory which is physically and mathematically consistent and complete.

We now discuss the consequences of this description. First of all we learn from (10.46) that $\hat{Z}$ can be expanded in a power series with respect to the $\eta$–, $\bar{\eta}$– and $Q^A$–sources. Thus, without any restriction of generality we may write $\hat{Z}$ in the following form:

$$\begin{aligned}\hat{Z}[0,\eta,\bar{\eta},Q] &= \hat{Z}_0[\eta,\bar{\eta}] \\ &+ \int d^4x_1d^4y_1Q^{A_1}(x_1,y_1)\hat{Z}_1^{A_1}\left[\eta,\bar{\eta}\right](x_1,x_2) \\ &+ \int d^4x_1d^4y_1d^4x_2d^4y_2Q^{A_1}(x_1,y_1)Q^{A_2}(x_2,y_2)\hat{Z}_2^{A_1A_2}\left[\eta,\bar{\eta}\right](x_1,y_1,x_2,y_2) \\ &+ \ldots\end{aligned} \tag{10.50}$$

where the expansion coefficients $\hat{Z}_1^{A_1}\left[\eta,\bar{\eta}\right](x_1,x_2)$ etc. are functionals of the $\eta$– and $\bar{\eta}$–sources. Then by putting $\eta = \bar{\eta} = 0$ we obtain the generating functional for the composite particle field $\hat{B}^A$ alone. This was our intention in the first place. To achieve this, the general form (10.46) of $\hat{Z}$ is not very well suited because it is not disentangled with respect to the $Q$–, $\eta$– and $\bar{\eta}$–sources as has to be required for the expansion (10.50). Thus we need a reordering of (10.46) in order to verify (10.50). This reordering can be done by the use of equations (10.47) to (10.49).

We rewrite equation (10.49) in the form

$$\partial^A(x,y)\hat{Z} = g^2D(x-y)\left[\partial^\eta(y)\frac{\Lambda^A}{2}\bar{\partial}^\eta(x) + Q^A(y,x)\right]\hat{Z} \tag{10.51}$$

and substitute (10.51) into equations (10.47), (10.48). This gives

$$\begin{aligned}&\left[(\gamma_\mu\partial_\mu + M)\bar{\partial}^\eta(x) + \frac{1}{4}g^2\int d^4yD(x-y)\partial^\eta(y)\Lambda^A\bar{\partial}^\eta(x)\Lambda^A\bar{\partial}^\eta(y)\right]\hat{Z} \\ &\quad= \left[\eta(x)\hat{Z} - \frac{1}{2}g^2\int d^4yD(x-y)Q^A(y,x)\Lambda^A\bar{\partial}^\eta(y)\right]\hat{Z}\end{aligned} \tag{10.52}$$

and the corresponding adjoint equation. If we substitute the expansion (10.50) into (10.52) we see that $\hat{Z}_0[\eta,\bar{\eta}]$ has to be a solution of

$$\left[ (\gamma_\mu \partial_\mu + M)\,\bar{\partial}^\eta(x) + \frac{1}{4} g^2 \int d^4y D(x-y) \partial^\eta(y) \Lambda^A \bar{\partial}^\eta(x) \Lambda^A \bar{\partial}^\eta(y) \right] \hat{Z}_0 = \eta(x) \hat{Z}_0 \tag{10.53}$$

and the corresponding adjoint equation.

We are not interested in the calculation of the full $q$–, $\bar{q}$– and $\hat{B}^A$– dynamics, i.e. the calculation of $\hat{Z}[0, \eta, \bar{\eta}, Q]$ itself. Rather as already indicated we want to study only the $\hat{B}^A$– field dynamics, i.e. $\hat{Z}[0, 0, 0, Q^A]$. This can be achieved by means of equation (10.51) which has to hold in any case. Once we know $\hat{Z}_0[\eta, \bar{\eta}]$ we can solve equation (10.51) by an iteration procedure. In the first step we substitute (10.50) into (10.51) and afterwards put $Q^A \equiv 0$. This gives

$$\hat{Z}_1^{A_1}[\eta, \bar{\eta}](x_1, y_1) = \frac{1}{2} g^2 D(x_1 - y_1) \partial^\eta(y_1) \Lambda^{A_1} \bar{\partial}^\eta(x_1) \hat{Z}_0[\eta, \bar{\eta}] \quad . \tag{10.54}$$

In the next step we apply $\partial^{A_2}(x_2, y_2)$ to equation (10.51), substitute (10.50) into this equation and afterwards put $Q^A \equiv 0$. This gives

$$\begin{aligned} \hat{Z}_2^{A_1 A_2}[\eta, \bar{\eta}](x_1, y_1, x_2, y_2) &= \frac{1}{2} g^2 D(x_1 - y_1) \partial^\eta(y_1) \Lambda^{A_1} \bar{\partial}^\eta(x_1) \hat{Z}_1^{A_2}[\eta, \bar{\eta}](x_2, y_2) \\ &\quad + g^2 D(x_1 - y_1) \delta(x_2 - y_1) \delta(y_2 - x_1) \hat{Z}_0[\eta, \bar{\eta}] \\ &= \frac{1}{4} g^4 D(x_1 - y_1) D(x_2 - y_2) \partial^\eta(y_1) \Lambda^{A_1} \bar{\partial}^\eta(x_1) \partial^\eta(y_2) \Lambda^{A_2} \bar{\partial}^\eta(x_2) \hat{Z}_0[\eta, \bar{\eta}] \\ &\quad + g^2 D(x_1 - y_1) \delta(x_2 - y_2) \delta(y_2 - x_1) \hat{Z}_0[\eta, \bar{\eta}] \quad . \end{aligned} \tag{10.55}$$

The quantities $\hat{Z}_0\,[0,0]$ etc. are the expansion coefficients of the functional $\hat{Z}[0,0,0,Q]$ of the $\hat{B}^A$–fields and therefore coincide with the boson correlation functions (10.45). If we define the power series expansion of $\hat{Z}_0[\eta, \bar{\eta}]$ by

$$\hat{Z}_0[\eta, \bar{\eta}] = \sum_n \int \xi^{(n)}(x_1, y_1 \ldots x_n, y_n) \eta(x_1) \bar{\eta}(y_1) \ldots \eta(x_n) \bar{\eta}(y_n) d^4x_1 \ldots d^4y_n \tag{10.56}$$

we have

$$\frac{\partial \hat{Z}[0, \eta, \bar{\eta}, Q]}{\bar{\partial}^\eta(x_1) \partial^\eta(y_1) \ldots \bar{\partial}^\eta(x_n) \partial^\eta(y_n)}_{|\eta, \bar{\eta}, Q=0} = \xi^{(n)}(y_1, x_1, \ldots, y_n, x_n) \quad ,$$

and then from (10.54) and (10.55) for $\eta, \bar{\eta} \equiv 0$ it follows

$$\begin{aligned} \hat{Z}_1^{A_1}[0,0](x_1, y_1) &= \frac{1}{2} g^2 D(x_1 - y_1) \Lambda^{A_1} \xi^{(1)}(y_1, x_1) \qquad (10.57) \\ \hat{Z}_2^{A_1 A_2}[0,0](x_1, y_1, x_2, y_2) &= \frac{1}{4} g^4 D(x_1 - y_1) D(x_2 - y_2) \Lambda^{A_1} \otimes \Lambda^{A_2} \xi^{(2)}(y_1, x_1, y_2, x_2) \end{aligned}$$

and so on. In this way all boson field correlation functions are given by the fermion field correlation functions. Comparing this identity of correlation functions with equation (10.41) we therefore have to conclude that equation (10.41) has to hold as an *operator equation.* So by requiring a complete quantum field theoretic description of the fusioned particle based on the path integral (10.43) we ultimately arrive at the former field theoretic way to describe composite particles by field operator products.

However, by involving the path integral we have to pay for it by even stronger restrictions. As we saw in Section 2, the path integral enforces the interaction picture with the free field representation. This means that also the representation of $q(x)$ and $\bar{q}(x)$ is fixed to be the free field representation. Thus we may represent $q$ and $\bar{q}$ in terms of quark and antiquark creation and annihilation operators by the expansion

$$q(x) = (2\pi)^{-3/2} \int d^3p \left(\frac{m}{E(\mathbf{p})}\right)^{1/2} \times \sum_{r=1}^{2} \left\{a_r(\mathbf{p}) w^r(\mathbf{p}) e^{-ipx} + d_r^+(\mathbf{p}) v^r(\mathbf{p}) e^{ipx}\right\} , \tag{10.58}$$

$$\bar{q}(y) = (2\pi)^{-3/2} \int d^3p \left(\frac{m}{E(\mathbf{p})}\right)^{1/2} \times \sum_{r=1}^{2} \left\{a_r^+(\mathbf{p}) \bar{w}^r(\mathbf{p}) e^{ipy} + d_r(\mathbf{p}) \bar{v}^r(\mathbf{p}) e^{-ipy}\right\} \tag{10.59}$$

with

$$\left[a_r^+(\mathbf{p}), a_{r'}(\mathbf{p}')\right]_+ = \left[d_r^+(\mathbf{p}), d_{r'}(\mathbf{p}')\right]_+ = \delta_{rr'}\delta(\mathbf{p}-\mathbf{p}') \tag{10.60}$$

and $a_r(\mathbf{p})|0\rangle = d_r(\mathbf{p})|0\rangle = 0$, the interpretation of which is standard. (For simplicity we assume flavor–independent quark masses.) With $X = \frac{1}{2}(x+y)$ and $z = (x-y)$ this yields for $\hat{B}^A(x,y)$ in (10.41):

$$\begin{aligned} \hat{B}^A(y,x) \equiv \hat{B}^A(X,z) = (2\pi)^{-3} g^2 \int d^3p d^3p' \, m \left[E(\mathbf{p})E(\mathbf{p}')\right]^{-\frac{1}{2}} \times \\ \sum_{r,r'=1}^{2} \Big[ a_r^+(\mathbf{p}) a_{r'}(\mathbf{p}') \bar{w}_r(\mathbf{p}) \Lambda^A w_{r'}(\mathbf{p}') e^{-iX(p'-p)+\frac{i}{2}z(p+p')} \\ + d_r(\mathbf{p}) d_{r'}^+(\mathbf{p}') \bar{v}_r(\mathbf{p}) \Lambda^A v_{r'}(\mathbf{p}') e^{+iX(p'-p)-\frac{i}{2}z(p+p')} \\ + a_r^+(\mathbf{p}) d_{r'}^+(\mathbf{p}') \bar{w}_r(\mathbf{p}) \Lambda^A v_{r'}(\mathbf{p}') e^{+iX(p+p')+\frac{i}{2}z(p-p')} \\ + d_r(\mathbf{p}) a_{r'}(\mathbf{p}') \bar{v}_r(\mathbf{p}) \Lambda^A w_r(\mathbf{p}') e^{-iX(p+p')-\frac{i}{2}z(p-p')} \Big] D(z) \quad . \end{aligned} \tag{10.61}$$

For convenience we consider $\tilde{B}^A := \hat{B}^A D^{-1}$ and define the following set of functions:

$$\Gamma_k^A(X,z) := \frac{1}{(2\pi)^4} \int e^{iX(q+q')} v_s(q) W_{s'}^A(q') \chi_k \begin{pmatrix} s, s' \\ q, q' \end{pmatrix} e^{i\frac{1}{2}z(q-q')} d^4q d^4q' \tag{10.62}$$

$$\hat{\Gamma}_k^A(X,z) := \frac{1}{(2\pi)^4} \int e^{-iX(q+q')} w_s(q) V_{s'}^A(q') \chi_k \begin{pmatrix} s, s' \\ q, q' \end{pmatrix} e^{-i\frac{1}{2}z(q-q')} d^4q d^4q'$$

$$\Pi_k^A(X,z) := \frac{1}{(2\pi)^4} \int e^{-iX(q-q')} w_s(q) W_{s'}^A(q') \chi_k \begin{pmatrix} s, s' \\ q, q' \end{pmatrix} e^{-i\frac{1}{2}z(q+q')} d^4q d^4q'$$

$$\hat{\Pi}_k^A(X,z) \quad := \quad \frac{1}{(2\pi)^4}\int e^{-iX(q'-q)} v_s(q) V_{s'}^A(q')\chi_k\begin{pmatrix} s, s' \\ q, q' \end{pmatrix} e^{i\frac{1}{2}z(q+q')} d^4q d^4q'$$

where $V_s^A$ and $W_s^A$ are the duals of $\Lambda^A \bar{v}_s$ and $\Lambda^A \bar{w}_s$. The corresponding dual set $\{{}^*\Gamma_k^A, {}^*\hat{\Gamma}_k^A, {}^*\Pi_k^A, {}^*\hat{\Pi}_k^A\}$ is given by complex conjugation of c–numbers and taking the duals of spinor amplitudes. For the proof of duality one has to observe that $v_s(q) \equiv v_s(q)\Theta(q_0)$ and $w_s(q) \equiv w_s(q)\Theta(q_0)$. Furthermore, the expansion functions $\chi_k$ are chosen to be real and antisymmetric and have to fulfill

$$\sum_{s,s'}\int \chi_k\begin{pmatrix} s, s' \\ q, q' \end{pmatrix}\chi_{k'}\begin{pmatrix} s, s' \\ q, q' \end{pmatrix} d^4q d^4q' = \delta_{kk'} \ . \tag{10.63}$$

With these definitions we obtain the expansion

$$\tilde{B}^A = \sum_k \left\{ -b_k {}^*\Gamma_k^A - b_k^{+\,*}\hat{\Gamma}_k^A + \pi_k {}^*\Pi_k^A - \hat{\pi}_k {}^*\hat{\Pi}_k^A \right\} \quad , \tag{10.64}$$

where the second minus sign stems from the combination of antisymmetry and Hermitean conjugation.

Herein, the corresponding operators are defined by projection with $\Gamma_k^A$, $\hat{\Gamma}_k^A$, $\Pi_k^A$ or $\hat{\Pi}_k^A$, respectively, on $\tilde{B}^A$ and read

$$b_k \quad := \quad \sum_{r,r'}\int d^3p d^3p' \hat{\chi}_k\begin{pmatrix} r, r' \\ \mathbf{p}, \mathbf{p}' \end{pmatrix} d_r(\mathbf{p}) a_{r'}(\mathbf{p}') \tag{10.65}$$

$$b_k^+ \quad := \quad \sum_{r,r'}\int d^3p d^3p' \hat{\chi}_k\begin{pmatrix} r, r' \\ \mathbf{p}, \mathbf{p}' \end{pmatrix} a_r^+(\mathbf{p}) d_{r'}^+(\mathbf{p}')$$

$$\pi_k \quad := \quad \sum_{r,r'}\int d^3p d^3p' \hat{\chi}_k\begin{pmatrix} r, r' \\ \mathbf{p}, \mathbf{p}' \end{pmatrix} a_r^+(\mathbf{p}) a_{r'}(\mathbf{p}')$$

$$\hat{\pi}_k \quad := \quad \sum_{r,r'}\int d^3p d^3p' \hat{\chi}_k\begin{pmatrix} r, r' \\ \mathbf{p}, \mathbf{p}' \end{pmatrix} d_r(\mathbf{p}) d_{r'}^+(\mathbf{p}')$$

with

$$\hat{\chi}_k\begin{pmatrix} r, r' \\ \mathbf{p}, \mathbf{p}' \end{pmatrix} := 2\pi g^2 N_c N_f m \left[E(\mathbf{p})E(\mathbf{p}')\right]^{-\frac{1}{2}} \chi_k\begin{pmatrix} r, r' \\ p, p' \end{pmatrix}_{|p_0=E(\mathbf{p}),\ p_0'=E(\mathbf{p}')} \quad , \tag{10.66}$$

where $N_c$ and $N_f$ are the numbers of colors or flavors, respectively.

We mention that the set of functions (10.62) is not complete and that for simplicity in (10.64) we neglected diquark–terms, etc.

The operators $b_k^+$ and $b_k$ have to be interpreted as meson creation and annihilation operators, while $\pi_k$ and $\hat{\pi}_k$ describe plasma oscillations of quarks and antiquarks.

If we write $b_k$ and $b_k^+$ in the abbreviated form

$$b_k = C_{ss'}^k a_s d_{s'} \ ; \ b_{k'}^+ = C_{tt'}^{k'} d_t^+ a_{t'}^+ \ , \tag{10.67}$$

then with $\left[a_k, a_{k'}^+\right]_+ = \delta_{kk'}$ and $\left[d_k^+, d_{k'}\right]_+ = \delta_{kk'}$ we obtain the noncanonical commutation relations

$$\left[b_k, b_{k'}^+\right]_- = -\mathrm{tr}\left[C^k C^{k'}\right] + C^k_{s't} C^{k'}_{tt'}\left[d_{s'} d_{t'}^+ + a_{s'} a_{t'}^+\right] . \tag{10.68}$$

Furthermore, with $N_k := b_k^+ b_k$ for the meson number operators it follows from (10.68)

$$\begin{aligned}[N_k, N_{k'}]_- &= \mathrm{tr}\left[C^k C^{k'}\right]\left(b_{k'}^+ b_k - b_k^+ b_{k'}\right) \\ &\quad + C^{k'}_{s't} C^k_{tt'}\left[b_k^+\left(d_{t'} d_{s'}^+ + a_{t'} a_{s'}^+\right) b_{k'} - b_{k'}^+\left(d_{s'} d_{t'}^+ + a_{s'} a_{t'}^+\right) b_k\right] .\end{aligned} \tag{10.69}$$

Thus, the meson numbers for the modes $k$ and $k'$ are not simultaneously measurable. This clearly contradicts the idea of generating independent meson modes by the expansion (10.64).

But the expansion of (10.41) resulting in the operator products (10.65) not only violates the idea of independent composite particles, but also leads to inconsistencies if applied in functional integrals. This can be seen by considering the intermediate states

$$|e_n\rangle := |k_1 \dots k_n\rangle := (n!)^{-\frac{1}{2}} b_{k_1}^+ \dots b_{k_n}^+ |0\rangle . \tag{10.70}$$

If (10.67) is substituted into (10.70) one immediately sees that the metrical tensor

$$G_{nm} := \langle e_n | e_m \rangle = \langle k_1 \dots k_n | k_1' \dots k_m' \rangle \tag{10.71}$$

is nondiagonal: $G_{nm} \neq \delta_{nm}$ and therefore $G^{nm} \equiv G^{-1}_{nm} \neq \delta_{nm}$, where $\delta_{nm}$ is the symmetric unit operator in $k_1 \dots k_n$–space. On the other hand, as was stated in Section 2, the functional path integrals are evaluated by inserting intermediate states in the interaction picture with free field representation, i.e., under the assumption $G^{nm} \equiv \delta_{nm}$. In the present case this means that the bosonized path integral (10.43) is in contradiction to a path integral for the field operators (10.64) and (10.65) which is evaluated "correctly" referring to the free field representation of the interaction picture. Therefore it is not clear if and how a sensible physical meaning can be given to the bosonized path integral at all.

Besides from application in path integrals we conclude that field operator products of the kind (10.64), (10.65) generally are not an appropriate description of definite composite particle states. It should be noted that this result also holds if no interaction picture with free field representation for the quark field is assumed.

## 10.5 Evaluation of Fermion Determinants

A crucial point in the evaluation of bosonized path integrals like (10.37), (10.38) is the fermion determinant. We first review the treatment of this determinant as performed by the authors cited in Section 10.3. If by definition we consider in (10.38) only that part of the action $s$ which results from the determinant we have

$$s_F\left[\hat{B}^A\right] := -\ln \det\left|G^{-1}(x, y, \hat{B}^A)\right| \tag{10.72}$$

with $G^{-1}$ given by (10.39). To interpret this effective action, $\hat{B}^A$ is transformed to center of mass coordinates $X = \frac{1}{2}(x + y)$ and $z = (x - y)$, and the following decomposition is introduced:

$$\hat{B}^A(X,z) = \sum_k \varphi_k^A(X)\Gamma_k^A(z)\,, \tag{10.73}$$

where $\{\Gamma_k^A(z)\}$ is a complete orthonormal set of functions in the relative coordinate $z$ and the local fields $\varphi_k^A(X)$ are considered as meson fields. We obtain $G^{-1}(x,y,\hat{B}^A) = G^{-1}(x,y,\varphi_k^A)$ with

$$G^{-1}(x,y,\varphi_k^A) := (\gamma_\mu\partial_\mu + M)\,\delta^4(x-y) - \sum_k \varphi_k^A\left(\frac{x+y}{2}\right)\Gamma_k^A(x-y)\frac{\Lambda^A}{2}\,. \tag{10.74}$$

Furthermore, by means of the formula $\det|A| = \exp(\operatorname{tr}\ln A)$, (10.72) yields

$$s_F\left[\varphi_k^A\right] = -\operatorname{tr}\ln\left[G^{-1}(x,y,\varphi_k^A)\right]\,. \tag{10.75}$$

We rewrite (10.74) into the form

$$G^{-1} = (\gamma_\mu\partial_\mu(x) + M)\left[\delta^4(x-y) - \int G_0(x,z)\sum_k \varphi_k^A\left(\frac{y+z}{2}\right)\Gamma_k^A(z-y)\frac{\Lambda^A}{2}d^4z\right] \tag{10.76}$$

with $G_0 := (\gamma_\mu\partial_\mu + M)^{-1}$, and renormalize (10.75) by substraction of $\operatorname{tr}\ln(\gamma_\mu\partial_\mu + M)$. This gives

$$s_F^{\text{ren}}\left[\varphi_k^A\right] = s_F\left[\varphi_k^A\right] + \operatorname{tr}\,\ln(\gamma_\mu\partial_\mu + M) \tag{10.77}$$

or

$$s_F^{\text{ren}}\left[\varphi_k^A\right] = -\operatorname{tr}\,\ln\left[\delta^4(x-y) - \int G_0(x,z)\sum_k \varphi_k^A\left(\frac{y+z}{2}\right)\Gamma_k^A(z-y)\frac{\Lambda^A}{2}d^4z\right]\,. \tag{10.78}$$

Denoting the second term in (10.78) by $V$ with

$$V(x,y) := -\int G_0(x,z)\sum_k \varphi_k^A\left(\frac{y+z}{2}\right)\Gamma_k^A(z-y)\frac{\Lambda^A}{2}d^4z \tag{10.79}$$

we have

$$s_F^{\text{ren}}\left[\varphi_k^A\right] = -\operatorname{tr}\ln(\mathbf{1} + V)\,. \tag{10.80}$$

Expansion of $\ln(\mathbf{1} + V)$ eventually leads to

$$s_F^{\text{ren}}\left[\varphi_k^A\right] = \sum_{n=1}^{\infty}\operatorname{tr}\frac{(-1)^n}{n}V^n\,. \tag{10.81}$$

According to the authors cited above due to the non–locality in (10.79) all trace terms of $V^n$ should give finite results. The full expansion leads to a nonlocal nonpolynomial effective action for the meson fields, where, however, no kinetic meson energies appear.

Returning to the original determinant we obtain with (10.76) and (10.79)

$$\det\left|G^{-1}(x,y,\hat{B}^A)\right| = \det\left|G_0^{-1}\right|\det\left|(\mathbf{1} + V)\right|\,. \tag{10.82}$$

The additive renormalization in (10.77) corresponds in connection with the determinant to a multiplicative renormalization by $\det\left|G_0^{-1}\right|^{-1}$. Hence we get

$$\det\left|G^{-1}(x,y,\hat{B}^A)\right|^{\text{ren}} = \exp\left[\operatorname{tr}\ln(\mathbb{I}+V)\right] \tag{10.83}$$

for the renormalized fermion determinant, provided

$$\det\left|(\mathbb{I}+V)\right| = \exp\left[\operatorname{tr}\ln(\mathbb{I}+V)\right] . \tag{10.84}$$

Relations (10.81) and (10.84) are the essential formulas for the evaluation of the fermion determinant in the papers quoted above; in the following we will critically discuss their application in this context.

As (10.81) or (10.84), respectively, are a part of the measure of the corresponding path integral (10.43), an investigation of (10.81) or (10.84), respectively, has to include this integral. We are exclusively interested in the $\hat{B}^A$–field dynamics and thus assume $\eta = \bar{\eta} = 0$. Furthermore, for the moment we are not interested in discussing fermionic Gaussian path integration. Hence without any comment we integrate out the $q$–, $\bar{q}$–fields, and under these assumptions we obtain for $\hat{Z}$ of (10.43) the reduced expression

$$\hat{Z}[0,0,0,Q^A] = \int D[\hat{B}^A]\exp\left\{-s[\hat{B}^A] + Q^A\hat{B}^A\right\} \tag{10.85}$$

where $s[\hat{B}^A]$ is given by (10.38).

For a general investigation of the existence and convergence properties of (10.85) we do not need the resolution into local fields. Hence $G^{-1}$ of definition (10.38) is given by (10.39). If we perform the renormalization (10.83), equation (10.85) yields

$$\begin{aligned}\hat{Z}[0,0,0,Q^A]^{\text{ren}} &= \int D[\hat{B}^A]\exp\left(Q^A\hat{B}^A\right) \times \\ &\quad \exp\left[\operatorname{tr}\ln(\mathbb{I}+V) - \int d^4x d^4y \frac{\hat{B}^A(x,y)\hat{B}^A(y,x)}{2g^2D(x-y)}\right]\end{aligned} \tag{10.86}$$

with

$$V(x,y) := \frac{1}{2}\int G_0(x,z)\hat{B}^A(z,y)\Lambda^A d^4z \quad . \tag{10.87}$$

As it stands, the integral (10.86) is ill–defined. Neither the left–hand side nor the right–hand side of (10.83) has any meaning in the continuum. They are only defined for operators in finite dimensional linear vector spaces. In the same way an integration over continuous fields in $\mathbb{R}^4$ is not defined a priori. Thus, in order to give these formal expressions a definite meaning we have to start from a finite dimensional lattice and to perform the limit to the continuum. With respect to a rigorous mathematical foundation this limit should be proven to be independent of the kind of lattice approximation. However, for a first draft we restrict ourselves to the study of a specific limiting process.

For simplicity we suppress the dependence on the $A$–index and thus replace $\Lambda^A$ by 1. By means of the continuum operators we then define the following finite dimensional matrix representations

$$\begin{aligned} B_{ik} &:= \hat{B}(x_i, y_k)\varepsilon \\ G_{ik} &:= G_0(x_i, y_k)\varepsilon \\ D_{ik} &:= D(x_i - y_k)\varepsilon \\ Q_{ik} &:= Q(x_i, y_k)\varepsilon \\ V_{ik} &:= \sum_l G_{il} B_{lk} \, , \end{aligned} \tag{10.88}$$

where $\{x_i\}$ and $\{y_k\}$, $i, k = 1, \ldots, N$, are the sets of lattice points, while $\varepsilon$ is the lattice spacing. Furthermore we assume that for a given $\varepsilon$ the lattice is confined to a volume $V = \varepsilon^{-1}$, i.e. with decreasing lattice spacing we have an increasing volume.

By definition we then introduce the following functional $\hat{Z}_\varepsilon$ for a definite lattice spacing $\varepsilon$:

$$\hat{Z}[Q]_\varepsilon := \int \prod_{i,k} dB_{ik} \exp\left\{ \sum_{l,h} Q_{lh} B_{lh} \right\} \quad \times \tag{10.89}$$

$$\exp\left\{ \operatorname{tr} \ln\left(\delta_{i'k'} + V_{i'k'}\right) - \sum_{i',k'} \frac{B_{i'k'} B_{k'i'}}{2g^2 D_{i'k'}} \right\} C_\varepsilon^{-1}$$

with $C_\varepsilon := \int \prod_{i,k} dB_{ik} \exp\left\{ -\sum_{i'k'} B_{i'k'} B_{k'i'} / 2g^2 D_{i'k'} \right\}$, which is assumed to be the lattice version of (10.86).

According to Cahill et al. the matrix $B_{ik}$ is a Hermitean matrix. For brevity we assume it to be real. Furthermore, in accordance with (10.32) we normalized the Gaussian integral on the lattice such that $C = 1$. As in the integral (10.89) the field $B_{ik}$ varies in the range $\{-\infty \le B_{ik} \le \infty\}$, it follows that the expansion (10.81) does not converge. Hence we are forced to replace $\exp[\operatorname{tr} \ln(1+V)]$ by its original definition $\det|(1+V)|$. For this determinant we can apply the expansion, see, e.g., Riesz and Nagy [Ries 56]:

$$\det|(1+V)| = 1 + \sum_k^N V_{kk} + \frac{1}{2} \sum_{\substack{k,i \\ k \neq i}}^N V_{kk} V_{ii} - \frac{1}{2} \sum_{\substack{i,k \\ i \neq k}}^N V_{ik} V_{ki} + \ldots \, , \tag{10.90}$$

substitute (10.90) into (10.89) and calculate the lowest orders. We obtain by a straight–forward calculation for the zero order

$$\hat{Z}_\varepsilon^0 = \prod_{i,k} \exp\left( \frac{g^2}{2} Q_{ik}^2 D_{ik} \right) \, , \tag{10.91}$$

for the first order

$$\hat{Z}_\varepsilon^I = \sum_{l,h} g^2 G_{lh} D_{lh} Q_{lh} \prod_{i,k} \exp\left( \frac{g^2}{2} Q_{ik}^2 D_{ik} \right) \, , \tag{10.92}$$

and for the second order

$$\hat{Z}_\varepsilon^{II} = \left[ \frac{1}{2} \sum_{\substack{i,j,k,l \\ i \neq k}} \left( g^4 G_{ij} Q_{ij} D_{ij} G_{kl} Q_{kl} D_{kl} \right.\right. \tag{10.93}$$

$$\left.\left. - g^4 G_{il} D_{lk} Q_{lk} G_{kj} D_{ji} Q_{ji} \right) \right] \prod_{i,k} \exp \left( \frac{g^2}{2} Q_{ik}^2 D_{ik} \right) .$$

In the continuum limit we obtain by means of definitions (10.88)

$$\lim_{\varepsilon \to 0} \hat{Z}_\varepsilon^0 = \exp \left[ \frac{1}{2} g^2 \int d^4x d^4y D(x-y) Q(x,y)^2 \right] , \tag{10.94}$$

$$\lim_{\varepsilon \to 0} \hat{Z}_\varepsilon^I = g^2 \int d^4x d^4y G_0(x-y) D(x-y) Q(x,y) \times \tag{10.95}$$

$$\exp \left[ \frac{1}{2} g^2 \int d^4x d^4y D(x-y) Q(x,y)^2 \right]$$

and

$$\lim_{\varepsilon \to 0} \hat{Z}_\varepsilon^{II} = \left\{ \frac{1}{2} \left[ g^2 \int d^4x d^4y G_0(x-y) D(x-y) Q(x,y) \right]^2 \right. \tag{10.96}$$

$$\left. - \frac{1}{2} g^4 \int d^4x d^4y d^4z d^4u G_0(x-y) D(y-z) Q(y,z) G_0(z-u) D(u-x) Q(u,x) \right\} \times$$

$$\exp \left\{ \frac{1}{2} g^2 \int d^4x d^4y G_0(x-y) Q(x,y)^2 \right\} .$$

These expressions can only be interpreted if we return to the local field description (10.73). In this description the sources read

$$Q(x,y) = \sum_k b_k(X) \Gamma_k(z) . \tag{10.97}$$

Thus the $b_k(X)$ are the sources for the local fields $\varphi_k(X)$. If we substitute (10.97) into (10.94), (10.95) and (10.96) we find

$$\lim_{\varepsilon \to 0} \hat{Z}_\varepsilon^0 = \exp \left\{ -\frac{1}{2} g^2 \sum_{k,k'} d_{kk'} \int b_k(X) b_{k'}(X) d^4X \right\} \tag{10.98}$$

with $d_{kk'} := \int dz D(z) \Gamma_k(z) \Gamma_{k'}(z)$,

$$\lim_{\varepsilon \to 0} \hat{Z}_\varepsilon^I = g^2 \sum_j c_j \int b_j(X) d^4X \times \tag{10.99}$$

$$\exp \left[ \frac{1}{2} g^2 \sum_{k,k'} d_{kk'} \int b_k(X) b_{k'}(X) d^4X \right]$$

with $c_j := \int D(z)G_0(z)\Gamma_j(z)dz$, and

$$\lim_{\varepsilon\to 0} \hat{Z}^{II}_\varepsilon = \left\{\frac{1}{2}\left[g^2\sum_j c_j \int b_j(X)d^4X\right]^2 \right. \tag{10.100}$$

$$\left. -2g^4\int K_{kk'}(X,X')b_k(X)b_{k'}(X')d^4Xd^4X'\right\} \times$$

$$\exp\left[\frac{1}{2}g^2\sum_{k,k'} d_{kk'}\int b_k(X)b_{k'}(X)d^4X\right]$$

with

$$K_{kk'}(X,X') := \int G_0(x-2X+z)D(2X-2z)\Gamma_k(2X-2z)\times \tag{10.101}$$

$$G_0(z-2X'+x)D(2X'-2x)\Gamma_{k'}(2X'-2x)d^4xd^4z\ .$$

Summing up we have

$$\lim_{\varepsilon\to 0}\hat{Z}[b_k]_\varepsilon = \lim_{\varepsilon\to 0}\sum_{\alpha=0,I,II}\hat{Z}[b_k]^\alpha_\varepsilon + \dots\ , \tag{10.102}$$

and this functional contains the exact bosonic two–point functions of the theory if we define

$$\hat{Z}[0,0,0,b_k]^{\text{ren}} := \lim_{\varepsilon\to 0}\hat{Z}[b_k]_\varepsilon\ . \tag{10.103}$$

The two–point functions then are given by

$$\langle\varphi_j(X)\varphi_{j'}(X')\rangle := \frac{\delta^2}{\delta b_j(X)\delta b_{j'}(X')}\lim_{\varepsilon\to 0}\hat{Z}[b_k]_{\varepsilon|b_k=0} \tag{10.104}$$

$$= g^2 d_{jj'}\delta(X-X') + g^4 c_j c_{j'}\delta(X-X') - g^4 K_{jj'}(X,X')\ .$$

For strongly localized four–fermion interactions with $D(x)\approx\delta(x)$ we obtain

$$\langle\varphi_j(X)\varphi_{j'}(X')\rangle = g^2 d_{jj'}\delta(X-X') + g^4 c_j c_{j'}\delta(X-X') - g^4\Gamma_j(0)\Gamma_{j'}(0)G_0(X-X')^2\ . \tag{10.105}$$

This is a diverging expression due to the occurrence of products of singular distributions in $G_0(X-X')^2$.

Apart from this fact it has to be emphasized, however, that with the assumption of real $B_{ik}$ the terms (10.98), (10.99), (10.100) are exact expressions for the corresponding VEV's, as higher powers in the sources do not contribute additional terms to these lowest VEV's. The same applies to higher VEV's if they are systematically evaluated by means of the expansion (10.90).

Finally one may ask whether the expansion (10.81) gives correct results in spite of its wrong use. We thus compare (10.81) with the exact expansion (10.90). We apply (10.81) to (10.89), and restrict ourselves to the first–order term of (10.81). This gives

$$\hat{Z}[Q]_\varepsilon \overset{(9.81)}{\approx} \int \prod_{i,k} dB_{ik} \exp\left(\sum_{h,l} Q_{hl} B_{hl}\right) \times \qquad (10.106)$$

$$\exp\left(\sum_k V_{kk} - \sum_{i',k'} \frac{B_{i'k'}B_{k'i'}}{2g^2 D_{i'k'}}\right) C_\varepsilon^{-1}$$

as the first order term of the effective action. This has to be compared with the function $\hat{Z}_\varepsilon := \sum_{\alpha=0,I,II} \hat{Z}_\varepsilon^\alpha + \dots$ based on the exactly valid expansion (10.90).

Due to their definition in (10.88), and the integration over $B_{ik}$ in the range $-\infty < B_{ik} < \infty$, the matrix elements $V_{kk}$ vary in the range $-\infty < V_{kk} < \infty$. Thus obviously $\exp(\operatorname{tr} V)$ is a bad approximation of $\det(1+V)$. We demonstrate this by calculating the bosonic two–point functions resulting from (10.106).

Performing the continuum limit of (10.106) and introducing the local field description (10.73) we obtain

$$\langle \varphi_k(X)\varphi_{k'}(X')\rangle \overset{(9.81)}{=} g^2 d_{kk'}\delta(X-X') + g^4 c_k c_{k'}\delta(X-X') \qquad (10.107)$$

which has to be compared with the correct result (10.104). The comparison shows that already in the lowest order correlation function, i.e. the two–point function, the most singular and thus the most important expression $-g^4\Gamma_k(0)\Gamma_{k'}(0)G_0(X-X')^2$ has been lost. Hence the explicit calculation confirms the formal argument against the use of (10.81).

Evaluating higher order correlation functions we obtain the same result: the expansion (10.81) is not suitable for approximating the correct expansion (10.90). Thus we have to conclude that the uncritical evaluation of the fermion determinant is a further doubtful procedure in connection with the bosonization via path integral.

## 10.6 Conclusions

We summarize the results of the foregoing sections and add some further remarks with respect to composite particle dynamics and path integrals.

- i) Mathematical foundations :

  Mathematically, by the definition of the path integral several problems arise. These are mainly concerned with the necessity of analytic continuation from Minkowski space into Euclidean space, the definition of the measure, the non-existence of classical paths for field theories, etc. Rivers [Riv 88] has devoted a chapter in his book to the discussion of these difficulties, see also Cheng [Che 90] and the critical remarks of Roepstorff [Roep 91]. We do not discuss them in detail because we are exclusively interested in composite particle theory.

- ii) Field operator products and composite particles :

The use of field operator products for the description of composite particles in quantum field theory has a long history. A collection of references with short comments is given in the introductions of the papers of Stumpf [Stu 85a,b], [Stu 86a]. In relativistic quantum field theory the method of auxiliary fields in path integrals, which is closely related to the use of field operator products, was introduced by Coleman, Jackiw and Politzer [Cole 74] and Gross and Neveu [Gro 74]. Later on this method was applied in many body theory, cf. Kleinert [Klei 78]. From the beginning these techniques were critisized by various authors, see the introductions of Stumpf [Stu 85a,b], [Stu 86a].

Physically, the biggest problem is the generation of corresponding kinetic energy operators for the field operator products. With respect to path integrals Rajaraman [Raja 78] explicitly evaluated path integrals for simple models and demonstrated that the generation of kinetic energy terms for the field operator products leads to incurable inconsistencies. Along similar lines Kerler [Kerl 77], [Kerl 82] has given arguments against this path integral procedure.

- iii) Difficulties of bosonization :

  In the present chapter in addition to the arguments already given, new arguments are raised against the use of path integrals, in particular with respect to composite particles. We showed that if one wants to derive a complete quantum dynamics for the composite particles, then as a consequence the auxiliary fields have to be treated as quantum fields and field operator products in their constituent fields. Due to the interaction picture with free field representation for the subfields which is induced by the path integral representation, these composite operator fields are noncanonically quantized fields with nonorthogonal basis vectors. This is in clear contradiction to the assumption that path integrals biuniquely replace canonical quantization. Thus, if these noncanonical composite fields appear in path integrals their evaluation is not in accordance with the noncanonical quantization properties, or the path integrals must simultaneously describe fields with canonical and noncanonical quantization, which is obviously a contradiction.

  Furthermore, if one insists on evaluating the path integral for these composite fields ab initio, it was shown in Section 10.5 that the results diverge in any order of an exactly valid moment expansion of the path integral. In contrast to the result of this exact calculation, the authors cited in Section 10.3 obtain finite results for the same integral by uncritically making use of a nonconvergent series expansion of the logarithm.

- iv) Limits of applicability :

  Even if one ignores all arguments in i) — iii) against this use of path integrals in composite particle theory, one can pose the question of what can be done with this approach. From Section 10.3 it follows that this approach is closely related to bilinear field operator products. Neither can one directly treat hard core particles with more than two subfermions nor dressed particles. While dressed particles are completely unattainable, one can try to decompose $n > 2$ hard core states into parts of $n = 2$ and $n = 1$ states. However, this procedure would violate inherent antisymmetries, apart from group theoretical difficulties, and would add new drawbacks in addition to those mentioned in i) — iii).

Summarizing these results, it can be concluded that the uncritical evaluation of path integrals obscures their inherent fallacies and, if the above mentioned objections are taken seriously, path integrals are no suitable tool for quantum field theory, in particular not for composite particle field theory.

# 11 Fock Space Mappings

## 11.1 Introduction

The main fields of application of Fock space mappings have been in solid state and nuclear physics. With respect to these applications there exists a comprehensive literature which we cannot discuss all in detail. Rather as in the previous chapter we restrict ourselves to the discussion of a representative example. As such an example we choose boson mappings in nuclear physics.

Boson mappings are widely used in nuclear physics as a mathematical tool for the description of collective excitations in nuclei. They are based on the idea of combining fermions, i.e. nucleons or quasi–particles, to pairs and to simulate the physics of the multifermion system by a system of bosons, each of them carrying the quantum numbers of a fermion pair. Of course one can consider this mapping from fermions to bosons as a purely algebraic transformation without giving it the meaning of the transition to a composite particle theory. In this case the strong mapping is a formal reformulation of the Schrödinger dynamics of the system which, due to the reduced number of degrees of freedom, possibly allows a simpler solution. But this is not the point of view we are interested in. Rather we like to discuss the use of Fock space mappings for the derivation of effective theories for composite particles.

Indeed, the latter point of view is emphasized in recent publications by Meyer [Mey 92], who intends an application of Fock space mappings to quark physics in order to derive effective theories for nucleons and mesons. These papers deal with a generalization of the boson transformation to three–fermion systems where the starting point are just those considerations which for two–fermion systems we are going to discuss. Hence the interpretation of the boson transformation and its three–fermion generalization, cf. Meyer [Mey 92], as a means for the derivation of effective composite particle theories is of actual interest, and it is our intention to critically discuss this kind of interpretation of Fock space mappings.

It is a peculiarity of these boson transformations that they are referred to Fock space representations of the algebra. Although, as already mentioned in preceding chapters, from a general algebraic point of view the dynamics of systems with an infinite number of degrees of freedom cannot a priori be confined to a Fock space representation, the latter representation allows a very transparent argumentation with respect to the connection of these mappings and composite particles. In our description of the boson transformation we closely follow the paper of Janssen, Dönau, Frauendorf and Jolos [Jan 71].

## 11.2 Ideal and Physical Boson Spaces

To grasp the essence of the Fock space mapping method we suppress all complicated details of realistic physical fermion models and restrict ourselves to the discussion of an ordinary

Schrödinger equation for interacting fermions in their Fock representation. The Hamiltonian of such a system is then given by, see, e.g., Fetter and Walecka [Fett 71], Haken [Hak 73],

$$H^f = \sum_{\alpha} E_\alpha a_\alpha^+ a_\alpha + \sum_{\alpha,\beta,\gamma,\delta} V_{\alpha\beta\gamma\delta} a_\alpha^+ a_\beta^+ a_\gamma a_\delta \tag{11.1}$$

and the operators $a_\alpha^+, a_\alpha$ obey the fermionic anticommutation relations

$$\left[a_\alpha^+, a_\beta\right]_+ = \delta_{\alpha\beta}\,, \quad \left[a_\alpha, a_\beta\right]_+ = \left[a_\alpha^+, a_\beta^+\right] = 0 \tag{11.2}$$

for all $\alpha, \beta$ characterizing a single particle state. Furthermore, the Fock vacuum is defined by $a_\alpha|0\rangle = 0\ \forall\alpha$. For even numbers of fermions the basis vectors of this system are given by the set

$$|2n\rangle := a_{\alpha_1}^+ a_{\beta_1}^+ a_{\alpha_2}^+ a_{\beta_2}^+ \ldots a_{\alpha_n}^+ a_{\beta_n}^+ |0\rangle \tag{11.3}$$

for $n = 1, 2, \ldots, \infty$.

If one wants to map this fermion theory onto an equivalent boson theory one formally defines a corresponding boson Fock space by introducing boson operators $b_{\alpha\beta}, b_{\alpha\beta}^+$ with commutation relations

$$\begin{aligned} \left[b_{\alpha\beta}, b_{\gamma\delta}^+\right]_- &= \delta_{\alpha\gamma}\delta_{\beta\delta} - \delta_{\alpha\delta}\delta_{\beta\gamma}\ , \\ \left[b_{\alpha\beta}, b_{\gamma\delta}\right]_- &= \left[b_{\alpha\beta}^+, b_{\gamma\delta}^+\right]_- = 0 \end{aligned} \tag{11.4}$$

and $b_{\alpha\beta} = -b_{\beta\alpha}$. Although it is likely to assume the identification $b_{\alpha\beta} = a_\alpha a_\beta$, this step is too simple to be performed. Rather at first one considers the boson space to be completely independent of the fermion space. The corresponding set of basis vectors is then given by

$$|m) := \left(b_{\alpha_1\beta_1}^+\right)^{m_1} \left(b_{\alpha_2\beta_2}^+\right)^{m_2} \cdots \left(b_{\alpha_n\beta_n}^+\right)^{m_n} |0)\,, \tag{11.5}$$

where $|0)$ is the boson Fock vacuum with $b_{\alpha\beta}|0) = 0\ \forall\alpha,\beta$. For all integer sequences of $m_1 \ldots m_n$ this set defines the so–called ideal boson space. The independence of both those state spaces is symbolically expressed by different designation of the vacua, namely $|0\rangle$ or $|0)$ for bosons or fermions, respectively.

If one intends to use these boson state vectors for the description of the fermion pair states (11.3), the set (11.5) obviously contains nonfermionic states, as for instance $\left(b_{\alpha_1\beta_1}^+\right)^k$, $k > 1$, cannot be generated by fermion pairs, while according to the boson commutation relation $\left(b_{\alpha_1\beta_1}^+\right)^k$ is unequal zero for all $k$. Therefore, in order to obtain a correspondence between fermionic pair states and boson states one has to introduce a physical subspace which can be defined as the span of the following orthogonal basis states:

$$|2n) = N \sum_P \operatorname{sgn}(\hat{P})\hat{P}\left(b_{\alpha_1\beta_1}^+ b_{\alpha_2\beta_2}^+ \cdots b_{\alpha_n\beta_n}^+\right)|0) \tag{11.6}$$

where $N$ is a suitable normalization constant and $\hat{P}$ is the permutation operator over all subscripts $(\alpha_1\beta_1 \ldots \alpha_n\beta_n)$. It is obvious that by this definition a biunique relation between the states $|2n\rangle$ and $|2n)$ is established. In doing so all complications of the fermion statistics are transferred into the boson space, and in principle nothing is gained. Thus, to escape this conclusion a reformulation of the physical boson subspace is needed allowing a better handling of the underground fermionic statistics in boson space. By direct calculation it can be verified that the basis vectors (11.6) can be equivalently expressed by

$$|2n) = N'\langle 0| \exp\left[\frac{1}{2}\sum_{\gamma,\delta} b^+_{\gamma\delta} a_\delta a_\gamma\right] a^+_{\alpha_1} a^+_{\beta_1} \ldots a^+_{\alpha_n} a^+_{\beta_n} |0\rangle|0) . \tag{11.7}$$

The states (11.7) shall be used as the starting point for the description of the physical boson space and can be rewritten in a more convenient form. Applying the relations

$$\langle 0| \left\{ \exp\left[\frac{1}{2}\sum_{\gamma,\delta} b^+_{\gamma\delta} a_\delta a_\gamma\right]\right\} a^+_\alpha = \langle 0| \left\{ \exp\left[\frac{1}{2}\sum_{\gamma,\delta} b^+_{\gamma\delta} a_\delta a_\gamma\right]\right\} \sum_\beta b^+_{\alpha\beta} a_\beta \tag{11.8}$$

and

$$b_{\alpha\beta} \exp\left[\frac{1}{2}\sum_{\gamma,\delta} b^+_{\gamma\delta} a_\delta a_\gamma\right] |0) = a_\beta a_\alpha \exp\left[\frac{1}{2}\sum_{\gamma,\delta} b^+_{\gamma\delta} a_\delta a_\gamma\right] |0) \tag{11.9}$$

we get for (11.7)

$$|2n) = N' B^+_{\alpha_1\beta_1} \langle 0| \exp\left[\frac{1}{2}\sum_{\gamma,\delta} b^+_{\gamma\delta} a_\delta a_\gamma\right] a^+_{\alpha_2} a^+_{\beta_2} \ldots a^+_{\alpha_n} a^+_{\beta_n} |0\rangle|0) \tag{11.10}$$

with

$$B^+_{\alpha\beta} = b^+_{\alpha\beta} - \sum_{\gamma,\delta} b^+_{\alpha\gamma} b^+_{\beta\delta} b_{\gamma\delta} . \tag{11.11}$$

Successive application of equation (11.10) leads to the following expression for the set (11.7) or (11.6), respectively:

$$|2n) = N' B^+_{\alpha_1\beta_1} B^+_{\alpha_2\beta_2} \cdots B^+_{\alpha_n\beta_n} |0) \tag{11.12}$$

with

$$N' = \left[(2n-1)!!\right]^{\frac{1}{2}} . \tag{11.13}$$

With (11.12) we have obtained a formulation of the physical boson states $|2n)$ which explicitly shows the one–to–one correspondence to the fermion states $|2n\rangle$ from (11.3).

A further useful quantity is the projector $P$ onto the physical subspace which simply is

$$P := \sum_n |2n)(2n| , \tag{11.14}$$

and it can be shown that the following relations hold:

$$P = P^+ = P^2 , \tag{11.15}$$

$$\left[P, B^+_{\alpha\beta}\right]_- = 0 , \tag{11.16}$$

$$P b_{\alpha\beta} P = b_{\alpha\beta} P , \tag{11.17}$$

$$\left[P, \sum_\gamma b^+_{\alpha\gamma} b_{\beta\gamma}\right]_- = 0 . \tag{11.18}$$

Furthermore, if we define the operators

$$N \;:=\; \sum_{\alpha,\beta} b^+_{\alpha\beta} b_{\alpha\beta} , \tag{11.19}$$

$$\rho_{\alpha\beta} \;:=\; \sum_\gamma b^+_{\beta\gamma} b_{\alpha\gamma} , \tag{11.20}$$

the following relations can be derived:

$$B^+_{\alpha\beta} P = P b^+_{\alpha\beta}(1 + N) , \tag{11.21}$$

$$P b^+_{\alpha\beta} N = -P \sum_\gamma b^+_{\alpha\gamma} \rho_{\gamma\beta} =: -P(b^+\rho)_{\alpha\beta} . \tag{11.22}$$

For the proof we refer to the literature.

## 11.3 Usui Mappings

Although in the preceding section we have derived a biunique relation between the set of fermion states (11.3) and the set of physical boson states (11.6) which is simply expressed by the identical denumeration of the states, by definition both state spaces are assumed to be independent from each other. Hence, in order to establish a mapping, corresponding mapping operators have to be defined. These operators are provided by the Usui transformation, [Usui 60], and are defined in the product space of boson and fermion states:

$$U := |0\rangle\langle 0| \exp\left[\frac{1}{2}\sum_{\alpha,\beta} b^+_{\alpha\beta} a_\beta a_\alpha\right] |0)(0| \tag{11.23}$$

and

$$\hat{U} := |0)(0| \sum_{n=0}^{\infty} \frac{1}{(2n)!}\left(\sum_{\alpha,\beta} a^+_\alpha a^+_\beta b_{\alpha\beta}\right)^n |0\rangle\langle 0| \quad . \tag{11.24}$$

Applying the formulas of the preceding section we obtain the following relations, which illustrate the meaning of $U$ and $\hat{U}$:

$$U|2n\rangle|0) = [(2n-1)!!]^{\frac{1}{2}}|2n)|0\rangle , \tag{11.25}$$

$$\hat{U}|2n)|0\rangle = [(2n-1)!!]^{-\frac{1}{2}}|2n\rangle|0) , \tag{11.26}$$

$$(0|\langle 2n|\hat{U} = [(2n-1)!!]^{-\frac{1}{2}}\langle 0|(2n| , \tag{11.27}$$

$$\langle 0|(2n|U = [(2n-1)!!]^{\frac{1}{2}}(0|\langle 2n| , \tag{11.28}$$

where $|2n)|2n'\rangle := |2n) \otimes |2n'\rangle$ is used.

From these equations and the normalization properties of the fermion and boson states one obtains

$$\langle 2n|\hat{U}U|2n'\rangle = \delta_{nn'} . \tag{11.29}$$

Thus in fermion space the operator $\hat{U}U$ behaves like the unity operator. This property can be used to derive the transformation formula for fermion operators. Let $T^f$ be an operator in fermion space, then it is

$$\begin{aligned} \langle 2n|T^f|2n'\rangle &= \sum_{k,\ell} \delta_{2n,k}\langle k|T^f|\ell\rangle\delta_{\ell,2n'} \\ &= \sum_{k,\ell}\langle 2n|\hat{U}U|k\rangle\langle k|T^f|\ell\rangle\langle \ell|\hat{U}U|2n'\rangle \\ &= \langle 2n|\hat{U}UT^f\hat{U}U|2n'\rangle \quad . \end{aligned} \tag{11.30}$$

If we now apply (11.25) and (11.27) to (11.30) we obtain the relation between the matrix elements of fermion operators $T^f$ and their boson representation $T^b$:

$$\langle 2n|T^f|2n'\rangle = {}_r(2n|T^b|2n')_r \tag{11.31}$$

where

$$T^b := UT^f\hat{U} \tag{11.32}$$

and

$$\begin{aligned} |2n')_r &:= [(2n-1)!!]^{\frac{1}{2}}|2n') \\ {}_r(2n| &:= [(2n-1)!!]^{-\frac{1}{2}}(2n| . \end{aligned} \tag{11.33}$$

In particular for operators $T_c^f$ with fermion number conservation (11.31) yields

$$\langle 2n|T_c^f|2n'\rangle = (2n|T_c^b|2n') \quad . \tag{11.34}$$

In the literature, emphasis is laid on the evaluation of the operator relation (11.32). Without going into details we give only the results. One can show that

$$T^b = UT^f\hat{U} = \tilde{T}^bP \tag{11.35}$$

holds, where $\tilde{T}^b$ is a boson operator which is obtained from $T^f$ by replacing the products $a^+_\alpha a^+_\beta$, $a_\beta a_\alpha$ and $a^+_\alpha a_\beta$ by $B^+_{\alpha\beta}$, $b_{\alpha\beta}$ and $b^+_{\alpha\gamma} b_{\beta\gamma}$. If one evaluates (11.35) for $T^f \equiv H^f$ from (11.1) one obtains

$$H^b = UH^f\hat{U} = \tilde{H}^b P \tag{11.36}$$

with

$$\begin{aligned} \tilde{H}^b \; := \; & \sum_\alpha E_\alpha \sum_\gamma b^+_{\alpha\gamma} b_{\alpha\gamma} \\ & + \sum_{\alpha,\beta,\gamma,\delta} V_{\alpha\beta\gamma\delta} \Big[ B^+_{\alpha\beta} b_{\delta\gamma} + \delta_{\gamma\beta} \sum_\kappa b^+_{\delta\kappa} b_{\alpha\kappa} - \delta_{\beta\delta} \sum_\kappa b^+_{\gamma\kappa} b_{\alpha\kappa} \\ & - \sum_\kappa b^+_{\gamma\kappa} b_{\alpha\kappa} \sum_\lambda b^+_{\delta\lambda} b_{\beta\lambda} + \sum_\kappa b^+_{\delta\kappa} b_{\alpha\kappa} \sum_\lambda b^+_{\gamma\lambda} b_{\beta\lambda} \Big] . \end{aligned} \tag{11.37}$$

The Schrödinger–Fock equation

$$H^f \Psi(a^+)|0\rangle|0) = E\Psi(a^+)|0\rangle|0) \tag{11.38}$$

is then transformed into the boson representation

$$H^b \phi(b^+)|0)|0\rangle = \tilde{H}^b P\phi(b^+)|0)|0\rangle = E\phi(b^+)|0)|0\rangle \tag{11.39}$$

with

$$\phi(b^+)|0)|0\rangle = U\Psi(a^+)|0\rangle|0) \quad , \tag{11.40}$$

where the property (11.29) was used.

## 11.4 Boson Mapping and Effective Dynamics

So far, we have cited Janssen, Dönau, Frauendorf and Jolos [Jan 71], and so far the boson mapping is a formal reformulation of the fermion dynamics in terms of boson variables, and both representations are completely equivalent. As we are interested in the derivation of effective dynamics for composite particles we have to evaluate this boson mapping in terms of composite particle operators and states.

First we represent the fermion states by

$$\Psi(a^+)|0\rangle = \sum_n \psi_n|2n\rangle \tag{11.41}$$

and the boson states by

$$\phi(b^+)|0) - \sum_n \varphi_n|2n) . \tag{11.42}$$

If we substitute (11.41) and (11.42) into (11.40) and observe (11.25), we obtain

$$\varphi_n \equiv \left[(2n-1)!!\right]^{\frac{1}{2}} \psi_n \,, \tag{11.43}$$

i.e., the original wave functions for the fermions are identical with those of the bosons apart from a normalization factor. Hence a boson transformation of fermion pairs brings no progress at all. Therefore we are forced to introduce linear combinations of fermion pairs in order to simulate composite particle states.

We introduce a complete set of two–fermion wave functions $C^r_{\alpha\beta}$, $r = 1, \ldots, \infty$, with the orthogonality relation

$$\frac{1}{2} \sum_{\alpha,\beta} C^{r^*}_{\alpha\beta} C^s_{\alpha\beta} = \delta_{rs} \tag{11.44}$$

and the completeness relation

$$\sum_r C^r_{\alpha\beta} C^{r^*}_{\gamma\delta} = \delta_{\alpha\gamma}\delta_{\beta\delta} - \delta_{\beta\gamma}\delta_{\alpha\delta} \,. \tag{11.45}$$

Then, if we replace the states of the boson physical subspace (11.6) by

$$|r_1 \ldots r_n) = N C^{r_1}_{\alpha_1\beta_1} \cdots C^{r_n}_{\alpha_n\beta_n} \sum_{\hat{P}} \mathrm{sgn}(\hat{P})\hat{P}\left(b^+_{\alpha_1\beta_1} \cdots b^+_{\alpha_n\beta_n}\right) |0) \,, \tag{11.46}$$

the states $|r_1 \ldots r_n)$ describe configurations of composite particles. But this transition to composite particle states has immediate consequences for the whole formalism. Calculating the scalar product of the states (11.46) we find that in contrast to the states (11.6) or (11.3), respectively, these states are no longer orthogonal. Rather we get

$$(r'_1 \ldots r'_{n'} | r_1 \ldots r_n) = \delta_{nn'} S^{k_1 \ldots k_n}_{k'_1 \ldots k'_{n'}} \tag{11.47}$$

with

$$S^{r_1 \ldots r_n}_{r'_1 \ldots r'_{n'}} := C^{r_1}_{\{\alpha_1\beta_1} \cdots C^{r_n}_{\alpha_n\beta_n\}} \left(C^{r'_1}_{\alpha_1\beta_1}\right)^* \cdots \left(C^{r'_n}_{\alpha_n\beta_n}\right)^* \tag{11.48}$$

and $S^2 = S$, and the projector on the physical subspace now reads

$$P = \sum_{\substack{r_1 \ldots r_n \\ r'_1 \ldots r'_n}} |r_1 \ldots r_n) S^{r_1 \ldots r_n}_{r'_1 \ldots r'_{n'}} (r'_1 \ldots r'_n| \quad . \tag{11.49}$$

With respect to the definition of the Usui operators (11.23) and (11.24) no modification for composite particles is necessary. But the linear relations (11.25)–(11.28) show that the transition to composite particle states must be simultaneously performed in boson as well as in fermion space. Due to the linearity of the operation (11.46) which relates the states $|2n)$ with the states $|r_1 \ldots r_n)$ and due to the fact that an analogous relation can be established between the states $|2n\rangle$ and $|r_1 \ldots r_n\rangle$, we obtain for (11.25)–(11.28)

$$U|r_1 \ldots r_n\rangle|0) \quad = \quad \left[(2n-1)!!\right]^{\frac{1}{2}} |r_1 \ldots r_n)|0\rangle \,, \tag{11.50}$$

$$\hat{U}|r_1 \ldots r_n)|0\rangle \quad = \quad \left[(2n-1)!!\right]^{-\frac{1}{2}} |r_1 \ldots r_n\rangle|0) \,, \tag{11.51}$$

$$(0|\langle r_1 \ldots r_n|\hat{U} \quad = \quad \left[(2n-1)!!\right]^{-\frac{1}{2}} \langle 0|(r_1 \ldots r_n| \,, \tag{11.52}$$

$$\langle 0|(r_1 \ldots r_n|U \quad = \quad \left[(2n-1)!!\right]^{\frac{1}{2}} (0|\langle r_1 \ldots r_n| \,. \tag{11.53}$$

From these equations and the scalar products (11.47) which also hold for the fermion states $\langle r_1 \ldots r_n|$ one now obtains

$$\langle r_1 \ldots r_n|\hat{U}U|r'_1 \ldots r'_n\rangle = S^{r_1 \ldots r_n}_{r'_1 \ldots r'_{n'}} \tag{11.54}$$

and

$$S^{r_1 \ldots r_n}_{r'_1 \ldots r'_n}|r'_1 \ldots r'_n\rangle = |r_1 \ldots r_n\rangle \,. \tag{11.55}$$

Using these properties we can transform any fermion operator into corresponding boson operators, but we have to observe that the techniques of Section 11.3 can only be transferred to the case under consideration if we apply composite particle states like (11.46) for fermions as well as for bosons. For brevity we replace $r_1 \ldots r_n$ by $r$, etc. Then we have with $S^2 = S$:

$$\begin{aligned}
\langle r|T^f|r'\rangle &= \sum_{k,\ell} S_{rk}\langle k|T^f|\ell\rangle S_{\ell r'} \\
&= \sum_{k,k',\ell,\ell'} S_{rk'}S_{k'k}\langle k|T^f|\ell\rangle S_{\ell\ell'}S_{\ell' r'} \\
&= \langle r|\hat{U}U|k'\rangle S_{k'k}\langle k|T^f|\ell\rangle S_{\ell\ell'}\langle \ell'|\hat{U}U|r'\rangle \\
&= \langle r|\hat{U}UT^f\hat{U}U|r'\rangle \quad .
\end{aligned} \tag{11.56}$$

Therefore, observing (11.50), (11.52) and (11.35), we obtain for fermion number conserving operators

$$\langle r_1 \ldots r_n|T^f_c|r'_1 \ldots r'_n\rangle = (r_1 \ldots r_n|T^b_c|r'_1 \ldots r'_n) \quad . \tag{11.57}$$

Thus with relation (11.40) the Schrödinger–Fock equation (11.38) goes over into

$$H^b\phi(b^+)|0)|0\rangle = E\phi(b^+)|0)|0\rangle \quad . \tag{11.58}$$

Hence, due to the linearity of the relations (11.25)–(11.28) the techniques of Section 11.3 can be transferred to the case of composite particle states (11.46), resulting in the bosonic Schrödinger–Fock equation (11.58). But also in this case one has to ask for the meaning of this "bosonization".

In the present case we have to observe the expansions

$$\Psi(a^+)|0\rangle = \sum_r \psi_{r_1 \ldots r_n}|r_1 \ldots r_n\rangle \tag{11.59}$$

and

$$\phi(b^+)|0) = \sum_r \varphi_{r_1 \ldots r_n}|r_1 \ldots r_n) \,. \tag{11.60}$$

If we substitute these expansions into (11.40) and observe relations (11.50), (11.52) this yields

$$\varphi_{r_1 \ldots r_n} = \left[(2n-1)!!\right]^{\frac{1}{2}}\psi_{r_1 \ldots r_n} \tag{11.61}$$

i.e., an analogous relation to (11.43). That means that also for composite particles like (11.46) the boson transformation is only a formal reformulation of the fermionic Schrödinger equation and with respect to the derivation of a genuine composite particle dynamics this procedure brings no progress at all. The only difference compared with the original approach is now that the Schrödinger equation has to be solved in a representation by a nonorthogonal basis system.

The treatment of many–particle quantum mechanical Schrödinger equations with nonorthogonal basis systems generated by hard core states was initiated by Wheeler [Whe 37a,b] and is commonly referred to as resonating group method and (or) cluster theory. In nuclear physics it was elaborated and further developed by Wildermuth and coworkers, cf. Wildermuth and Tang [Wild 77], Schmid [Schmi 80a,b], [Schmi 81], [Schmi 83], Schmid and Spitz [Schmi 85], Faessler [Faes 84],[Faes 88]. Its group theoretical implications were studied by Kramer and coworkers, see Kramer, John and Schenzle [Kram 81]. Recently this method was extended to the treatment in astrophysics, quark physics, etc.; for a review see Langanke [Lang 92].

Without using a complicated boson transformation the treatment of many–particle Schrö"-din"-ger equations by means of nonorthogonal basis systems for the description of composite particles can be directly formulated in Fock space. Corresponding investigations were for instance performed by Koutchouk [Kou 70], who constructed Fock spaces for nonrelativistic bound states and tried to discuss the effective dynamics of these particles. Similar approaches were suggested by Girardeau [Gir 71,75,78], Gilbert [Gil 77] and Kvasnicka [Kvas 82], who studied a second quantization representation for systems containing (nonrelativistic) bound states, in particular atoms, nuclei and electrons. Rosenberg [Ros 74] developed a theory of nonrelativistic cluster approximations, and Goldberg and Puff [Gold 74] discussed the formulation of a nonrelativistic many–body theory.

# Bibliography

[Ald 80] Aldaya, V. and de Azcarraga, J.A.: J. Phys. A **13** (1980) 2545

[Alk 94] Alkofer, R., Reinhard, H. and Weigel, H.: Baryons as chiral solitons in the Nambu–Jona–Lasinio Model, preprint Univ. Tübingen, to be published

[App 75] Appelquist, F. and Carrazone, J.: Phys. Rev D **11** (1975) 2856

[Aud 92] Audretsch, J., Hehl, F.W. and Lämmerzahl, C.: Matter wave interferometry and why quantum objects are fundamental for establishing a gravitational theory, in: Relativistic gravity research, ed. by J. Ehlers and G. Schäfer, Lecture Notes in Physics, Springer Verlag, Berlin 1992, p. 368–407

[Baez 92] Baez, J.C., Segal, I.E. and Zhou, Z.: Introduction to algebraic and constructive quantum field theory, Princeton Univ. Press, Princeton 1992

[Bal 90] Balachandrian, A.P., Ercolesi, E., Morandi, G. and Srivastava, A.M.: Int. J. Mod. Phys. B **4** (1990) 2057

[Barg 48] Bargmann, V. and Wigner, E.P.: Proc. Nat. Acad. Sci. (USA) **34** (1948) 211

[Best 90] Besting, P. and Schütte, D.: Phys. Rev. D **42** (1990) 594

[Bigi 74] Bigi, I.I., Dürr, H.P. and Winter, N.J.: Nuovo Cim. **22A** (1974) 420

[Bla 54] Blatt, J.M. and Butler, S.T.: Phys. Rev. **96** (1954) 1149

[Bleu 50] Bleuler, K.: Helv. Phys. Acta **23** (1950) 567

[Bogn 74] Bognar, J.: Indefinite inner product spaces, Springer Verlag, Berlin 1974

[Bogo 75] Bogoliubov, N.N., Logunov, A.A. and Todorov, I.T.: Introduction to axiomatic quantum field theory, W.A. Benjamin Inc., Reading, Mass., 1975

[Bogo 59] Bogoliubov, N.N. and Shirkov, D.V.: Introduction to the theory of quantized fields, Interscience Publ., New York 1959

[Bopp 40] Bopp, F.: Ann. Phys. (Germ.) **38** (1940) 345

[Bor 91] Borne, Th.: Quantenfeldtheoretische Energiedarstellung und schwache Abbildung, Diplomarbeit, Universitity of Tübingen, 1991

[Bor 69] Borneas, M.: Phys. Rev. **186** (1969) 1299

[Borz 88] v. Borzeszkowski, H.H. and Treder, H.J.: The meaning of quantum gravity, Reidel Publ. Comp., Dordrecht 1988

[Brac 76] Bracci, L., Morchio, G. and Strocchi, F.: Description of symmetries in indefinite metric spaces; in: Jammer, A., Jansen, T., Boon, M. (eds): Proceed. $4^{th}$ Int. Coll. of grouptheor. Methods in Physics, Springer Verlag, Berlin 1976, p. 551

[Brat 79] Bratteli, O. and Robinson, D.W.: Operator algebras and quantum statistical mechanics I, Springer Verlag, Berlin 1979

[Brog 34] de Broglie, L.: Comt. rend. **199** (1934) 445

[Brog 54] de Broglie, L.: Théorie générale des particules à spin, Gauthier–Villars, Paris 1954, 2nd ed.

[Buch 80] Buchmüller, W. and Dietz, K.: Z. Physik C **5** (1980) 45

[Buck 77] Buckel, W.: Supraleitung, Physik Verlag, Weinheim 1977, 2nd ed.

[Cah 85] Cahill, R.T. and Roberts, C.D.: Phys. Rev. D **32** (1985) 2419

[Cah 89a] Cahill, R.T., Praschifka, J. and Burden, C.J.: Austr. J. Phys. **42** (1989) 161

[Cah 89b] Cahill, R.T.: Austr. J. Phys. **42** (1989) 171

[Cha 48] Chang, T.S.: Proc. Cambridge Phil. Soc. **44** (1948) 76

[Che 90] Cheng, H.: Lecture notes in physics **361**, (1990), 3, Springer Verlag, Heidelberg 1990

[Che 86] Cheng, Ta–Pei and Li, Ling–Fong: Gauge theory of elementary particles, Clarendon Press, Oxford 1986

[Chri 80] Christ, N.H. and Lee. T.D.: Phys. Rev. D **22** (1980) 939

[Cole 74] Coleman, S., Jackiw, R. and Politzer, H.D.: Phys. Rev. D **10** (1974) 2491

[Coll 87] Collins, J.: Renormalization, Cambridge University Press, Cambridge 1987

[Dehn 60] Dehnen, H., Hönl, H. and Westpfahl, K.: Ann. Phys. (Germany) **6** (1960) 370

[Dehn 85] Dehnen, H. and Ghaboussi, F.: Nucl. Phys. B **262** (1985) 144

[Dehn 94] Dehnen, H. and Hitzer, E.: Int. J. Theor. Phys. **33** (1994) 575

[Dick 92] Dickhoff, W.H. and Müther, H.: Rep. Progr. Phys. **11** (1992) 1947

[Dir 66] Dirac, P.A.M.: Lectures on quantum field theory; Belfer Graduate School of Science, Yeshiva Univ., New York 1966

[Dürr 77a] Dürr, H.P. and Saller, H.: Nuovo Cim. **39 A** (1977) 31

[Dürr 77b] Dürr, H.P. and Saller, H.: Nuovo Cim. **41 A** (1977) 677

[Dürr 78] Dürr, H.P. and Saller, H.: Nuovo Cim. **48 A** (1978) 505

[Dys 49a] Dyson, F.J.: Phys. Rev. **75** (1949) 486

[Dys 49b] Dyson, F.J.: Phys. Rev. **75** (1949) 1736

[Dys 51a] Dyson, F.J.: Phys. Rev. **82** (1951) 428

[Dys 51b] Dyson, F.J.: Phys. Rev. **83** (1951) 608

[Dys 51c] Dyson, F.J.: Phys. Rev. **83** (1951) 1207

[Dys 51d] Dyson, F.J.: Proc. Roy. Soc. (London) **A207** (1951) 395

[Dys 53a] Dyson, J.F.: Phys. Rev. **90** (1953) 994

[Dys 53b] Dyson, J.F.: Phys. Rev. **91** (1953) 1543

[Edd 49] Eddington, A.S.: Fundamental theory, Cambridge University Press, 1949

[Ein 15] Einstein, A.: Sitzungsberichte Preuss. Akad. Wiss. 1915, 844

[Emch 72] Emch, G.: Algebraic methods in statistical mechanics and quantum field theory, Wiley Interscience, New York 1972

[Fadd 80] Faddeev, L.D. and Slavnov, A.A.: Gauge Fields — Introduction to quantum theory, Benjamin/Cummings Publ. Co., 1980

[Faes 84] Faessler, A.: Progr. Part. and Nucl. Phys. **11** (1984) 171

[Faes 88] Faessler, A.: Progr. Part. and Nucl. Phys. **20** (1988) 151

[Fett 71] Fetter, A.L. and Walecka, J.D.: Quantum theory of many particle systems, McGraw–Hill, New York 1971

[Feyn 87] Feynman, R.P. in Hiley, B.J., Peat, F.D. (Eds): Quantum Implications, Routledge & Kegan, London, 1987

[Flei 83] Fleig, W.: Acta Phys. Austr. **55** (1983) 135

[Fon 70] Fonda, L. and Ghirardi, G.C.: Symmetry principles in quantum physics, Marcel Dekker Inc., New York 1970

[Free 51] Freese, E.: Z. Naturforsch. **8a** (1951) 776

[Frie 72] Fried, H.M.: Functional methods and models in quantum field theory, MIT Press, Cambridge (Mass.) 1972

[Frie 75] Friedrich, A., Gerling, W. and Bleuler, K.: Z. Naturforsch. **30a** (1975) 142

[Frie 76] Friedrich, A. and Gerling, W.: Z. Naturforsch. **31a** (1976) 872

[Frie 53] Friedrichs, K.O.: Mathematical aspects of the quantum theory of fields, Interscience Publ., New York 1953

[Fur 51] Furry, W.M.: Phys. Rev. **81** (1951) 115

[Gai 90] Gaigg, P., Kummer, W. and Schweda, M. (eds.): Physical and nonstandard gauges, Lecture Notes in Physics **361**, Springer Verlag, Heidelberg 1990

[Gel 43] Gelfand, I. and Naimark, M.A.: Mat. Sborn. NS **12** [54] (1943) 197

[Gil 77] Gilbert, J. D.: J. Math. Phys. **18** (1977) 791

[Gir 71] Girardeau, M. D.: Phys. Rev. Lett. **27** (1971) 1416

[Gir 75] Girardeau, M. D.: J. Math. Phys. **16** (1975) 1901

[Gir 78] Girardeau, M. D.: J. Math. Phys. **19** (1978) 2605

[Gli 87] Glimm, Y. and Jaffe, A.: Quantum Physics — A functional integral point of view, Springer Verlag, Berlin 1987 (2nd ed.)

[Gold 74] Goldberg, A. and Puff, R. D.: Phys. Rev. A **10** (1974) 323

[Gold 78] Goldstone, J. and Jackiw, R.: Phys. Lett. **74 B** (1978), 81

[Gor 59] Gorkov, L.P.: Sov. Phys. JETP **9** (1959) 1364

[Gri 91] Grimm, G., Hailer, B. and Stumpf, H.: Ann. Phys. (Germ.) **48** (1991) 327

[Gri 94] Grimm, G.: Elektro–schwache Standardtheorie als effektive Dynamik von Bindungszuständen im Rahmen einer nichtlinearen Spinortheorie, Thesis, University of Tübingen, 1994

[Gro 74] Gross, D.J. and Neveu, A.: Phys. Rev. D **10** (1974) 3235

[Gro 83] Grosser, D.: Z. Naturforsch. **38a** (1983) 1293

[Gro 87] Grosser, D.: Nuovo Cim. **97 A** (1987) 439

[Gup 77] Gupta, S.N.: Quantum electrodynamics, Gordon & Breach, New York 1977

[Gutz 90] Gutzwiller, M.C.: Chaos in classical and quantum mechanics, Springer Verlag, Berlin 1990

[Haag 92] Haag, R.: Local quantum physics, Springer Verlag, Berlin 1992

[Hail 85] Hailer, B.: Norm– und Massenberechnungen skalarer Bosonen als Bindungszustände in einheitlichen nichtlinearen Spinorfeldmodellen, Thesis, University of Tübingen, 1985

[Hak 73] Haken, H.: Quantenfeldtheorie des Festkörpers, Teubner Verlag, Stuttgart 1973

[Hall 90] Haller, K.: Lecture notes in physics **361**, (1990), Springer Verlag, Heidelberg 1990, p. 33

[Halp 67] Halpern, F.R.: J. Math. Phys. **8** (1967) 648

[Hat 94] Hatsuda, T. and Kunihiro, T.: Phys. Rep., to be published

[Hehl 70] Hehl, F.W.: Spin und Torsion in der Allgemeinen Relativitätstheorie, Habilitationsschrift, TU Clausthal, 1970

[Hehl 85] Hehl, F.W.: Found. Phys. **15** (1985) 451

[Heis 66] Heisenberg, W.: Introduction to the unified field theory of elementary particles, Interscience Publ., London 1966

[Heis 84] Heisenberg, W.: Collected works (eds.: Blum,W., Dürr, H.–P., Rechenberg, H.), Springer Verlag, Berlin 1984

[Hil 15] Hilbert, D.: Nachr. Kgl. Ges. Wiss. Göttingen 1915, 395

[Hon 90] Honegger, R. and Rieckers, A.: Publ. Research Inst. Math. Sci. **26**, no. 2, Kyoto University 1990

[Hor 86] Hornung, L.: Spin–0 und Spin–1–Bosonen als Preon–Antipreonbindungszustände in einheitlichen nichtlinearen Spinorfeldmodellen, Thesis, University of Tübingen, 1986

[Hua 92] Huang, K.: Quarks, leptons and gauge fields, World Scientific, Singapore 1992

[Itz 80] Itzykson, C. and Zuber, J.B.: Quantum field theory, McGraw–Hill Inc., New York 1980

[Jan 71] Jansen, D., Dönau, F., Frauendorf, S. and Jolos, R.V.: Nucl. Phys. A **172** (1971) 145

[Jack 80] Jackiw, R.:Rev. Mod. Physics **52** (1980), 661

[Jak 84] Jakobczyk, L.: J. Math. Phys. **25** (1984) 617

[Jak 88] Jakobczyk, L. and Strocchi, F.: Comm. Math. Phys. **119** (1988) 529

[Jel 68] Jelinek, F.: Comm. Math. Phys. **9** (1968) 169

[Jor 28] Jordan, P.: Ergebn. Ex. Naturwiss. **7** (1928) 158

[Kaus 94] Kaus, H.–J.: Gravitative Kopplung als Spin–Fusion, Thesis, University of Tübingen, 1994

[Kerl 77] Kerler, W.: Phys. Lett B **69** (1977) 355

[Kerl 82] Kerler, W.: Nucl. Phys. B **202** (1982) 437

[Ker 91] Kerschner, R. and Stumpf, H.: Ann. Phys. (Germ.) **48** (1991) 56

[Ker 94] Kerschner, R.: Effektive Dynamik zusammengesetzter Quantenfelder, Thesis, University of Tübingen, 1994

[Kib 61] Kibble, T.W.B.: J. Math. Phys. **2** (1961) 212

[Kim 91] Kim, S.K., Namgung, W., Soh, K.S. and Yee, I.H.: Phys. Rev. D **43** (1991) 2046

[Klei 78] Kleinert, H.: Fortschr. Phys. **26** (1978) 565

[Klev 92] Klevanski, S.P.: Rev. Mod. Phys. **64** (1992) 649

[Kou 70] Koutchouk, F: Nuovo Cim. **57 A** (1970) 483

[Kram 81] Kramer, P., John, G. and Schenzle, D.: Group theory and the interaction of composite nucleon systems, Vieweg, Braunschweig 1981

[Kvas 82] Kvasnicka, V.: Czech. J. Phys. B **32** (1982) 947

[Lag 64] Lagally, M. and Franz, W.: Vorlesungen über Vektorrechnung, Akad. Verlagsgesellschaft Geest & Portig, Leipzig 1964

[Land 87] Landsman, N.P. and van Weert, C.G.: Phys. Rep. (NL) **145** (1987) 141

[Lang 92] Langanke, K.: Phys. Blätter **48** (1992) 611

[Lau 86] Lauxmann, Th.: Fermionen als Dreipreonen–Bindungszustände in einheitlichen nichtlinearen Spinorfeldtheorien, Thesis, University of Tübingen, 1986

[Leh 55] Lehmann, H., Symanzik, K. and Zimmermann, W.: Nuovo Cim. **1** (1955) 205

[Leib 87] Leibbrandt, G.: Rev. Mod. Phys. **59** (1987), 1067

[Logu 93] Logunov, A.A.: Phys. Part. Nucl. **24** (1993) 545

[Luri 68] Lurié, D.: Particles and fields, Interscience Publ., New York 1968

[Lyo 83] Lyons, L.: Progress in Particle and Nuclear Physics **10** (1983) 227, Pergamon Press

[Magg 84] Magg, M.: Fortschr. Phys. **32** (1984) 353

[Mai 66] Maison, D. and Stumpf, H.: Z. Naturforsch. **21a** (1966) 1829

[Mat 60] Matsubara, T. and Blatt, J. M.: Progr. Theor. Phys. **23** (1960) 451

[Meh 73] Mehra, J. in: Mehra, J. (ed.): The Physicist's Conception of Nature, Reidel Publ. Comp., Dordrecht 1973, p. 92–178

[Mey 92] Meyer, J.: J. Math. Phys. **33** (1992) 3445, 3454, 3463

[Mic 90] Micmas, R., Ranninger, J., Robaszkiewicz, S.: Rev. Mod. Phys. **62** (1990) 113

[Mor 80] Morchio, G. and Strocchi, F.: Ann. Inst. H. Poincaré **33** (1980) 251

[Mos 90] Moschella, U. and Strocchi, F.: Lett. Math. Phys. (NL) **19** (1990) 143

[Mott 93] Mott, N.F.: Physica C **205** (1993) 191

[Nach 86] Nachtmann, O.: Phänomene und Konzepte der Elementarteilchenphysik, Vieweg, Braunschweig 1986

[Naka 90] Nakanishi, N. and Ojima, I.: Covariant operator formalism of gauge theories and quantum gravity, World Scientific, Singapore 1990

[Nish 53] Nishijima, K.: Prog. Theor. Phys. **10** (1953) 549

[Niv 72] Niven, I. and Zuckerman, H.S.: An introduction to the theory of numbers, Wiley, New York 1972

[Pau 49] Pauli, W. and Villars, F.: Rev. Mod. Phys. **21** (1949) 434

[Pfi 87] Pfister, W.: Der einzeitige Formalismus in der funktionalen Quantentheorie, Diplomarbeit, University of Tübingen, 1987

[Pfi 89] Pfister, W., Rosa, M. and Stumpf, H.: Nuovo Cim. **102A** (1989) 1449

[Pfi 90] Pfister, W.: Yang–Mills Dynamik als effektive Theorie von vektoriellen Spinor–Isospinor Bindungszuständen in einem Preonfeldmodell, Thesis, University of Tübingen, 1990

[Pfi 91] Pfister, W. and Stumpf, H.: Z. Naturforsch. **46a** (1991) 389

[Pfi 93] Pfister, W.: Nuovo Cim. **106A** (1993) 363

[Pod 40] Podolski, B.: Phys. Rev. **62** (1940) 68

[Pra 87a] Praschifka, J., Roberts, C.D. and Cahill, R.T.: Int. J. Mod. Phys. A **2** (1987) 1797

[Pra 87b] Praschifka, J., Roberts, C.D. and Cahill, R.T.: Phys. Rev. D **36** (1987) 209

[Prim 83] Primas, H.: Chemistry, quantum mechanics and reductionism, Springer Verlag, Berlin 1983

[Pry 38] Pryce, M.H.L.: Proc. Roy. Soc. London A **165** (1938) 247

[Raja 78] Rajaraman, R.: Pramana **11** (1978) 491

[Ram 65] Rampacher, H., Stumpf, H. and Wagner, F.: Fortschr. Phys. **13** (1965) 385

[Ray 48] Rayski, J.: Acta Phys. Pol. **9** (1948) 129

[Rick 65] Rickayzen, G.: Theory of superconductivity, Interscience Publ., New York 1965

[Riec 77] Rieckers, A. and Stumpf, H.: Thermodynamik, Vol. 2, Vieweg Verlag, Braunschweig 1977

[Riec 91] Rieckers, A.: Condensed Cooper pairs and macroscopic quantum phenomena, in: Gans, W. et al. (eds.): Large–scale molecular systems, Nato ASI Series in Physics, Plenum Press, New York 1991

[Riec 93] Rieckers, A.: Condensed particle structure of the inhomogeneous BCS–model and a meanfield Hubbard model, hold at Conference on Superconductivity and strongly correlated Systems, Amalfi, 1993, to appear

[Riem 97] Riemann, B.: Mathematische Werke, ed. by H. Weber, 2nd ed., Leipzig 1897

[Riem 80] Riemann, B.: Schwere Elektrizität und Magnetismus, ed. by K. Hattendorf, 2nd ed., Hannover 1880

[Ries 56] Riesz, F. and Nagy, B.S.: Vorlesungen über Funktionalanalysis, Deutscher Verlag der Wissenschaften, Berlin 1956

[Riew 72] Riewe, F. and Green, A.E.S.: J. Math. Phys. **13** (1972) 1368, 1374

[Riv 88] Rivers, R.J.: Path integral methods in quantum field theory, Cambridge University Press 1988

[Rob 87] Roberts, C.D. and Cahill, R.T.: Austr. J. Phys. **40** (1987) 499

[Rob 88] Roberts, C.D., Cahill, R.T. and Praschifka, J.: Ann. Phys. (N.Y.) **188** (1988) 20

[Rob 89] Roberts, C.D., Praschifka, J. and Cahill, R.T.: Int. J. Mod. Phys. A **7** (1989) 1681

[Rod 77] Rodrigues, P.R.: J. Math. Phys. **18** (1977) 1720

[Rod 91] Rodrigues, W.A. and Souza, Q.A.G. de, in: Olivo, J.C. d' et al. (eds.): Relativity and gravitation: classical and quantum, Proc.; World Scientific, Singapore (1991), p. 220

[Rod 84] Rodrigues, M.A. and Lorente, M.: Nuovo Cim. **83A** (1984) 249

[Roep 91] Roepstorff, G.: Pfadintegrale in der Quantenphysik, Vieweg, Wiesbaden 1991

[Rom 69] Roman, P.: Introduction to quantum field theory, Wiley, New York 1969

[Ros 40a] Rosen, N.: Phys. Rev. **57** (1940) 147

[Ros 40b] Rosen, N.: Phys. Rev. **57** (1940) 150

[Ros 74] Rosenberg, L.: Phys. Rev. D **9** (1974) 1789

[Rue 69] Ruelle, D.: Statistical mechanics, Benjamin Inc., New York 1969

[Rze 69] Rzewuski, J.: Field theory, Vol. 2, Iliffe Books, London 1969

[Sal 51] Salpeter, E.E. and Bethe, H.E.: Phys. Rev. **84** (1951) 1232

[Sall 75] Saller, H.: Nuovo Cim. **30 A** (1975) 541

[Sall 76] Saller, H.: Nuovo Cim. **34 A** (1976) 99

[Sall 89] Saller, H.: Nuovo Cim. **104 B** (1989) 291

[Sall 91] Saller, H.: Nuovo Cim. **106 B** (1991) 1319

[Sall 92] Saller, H.: Nuovo Cim. **107 B** (1992) 1355

[Sall 93] Saller, H.: Nuovo Cim. **106 A** (1993) 469

[Sand] Sand, J.: Zur Pauli–Villars–Regularisierung in der Preontheorie, Diplomarbeit 1991

[Sar 91] Sardanashvily, G. and Zakharov, O.: Rep. Math. Phys. **29** (1991) 101

[Schaf 54a] Schafroth, M.R.: Phys. Rev. **96** (1954) 1149

[Schaf 54b] Schafroth, M.R.: Phys. Rev. **96** (1954) 1442

[Schar 89] Scharf, G.: Finite quantum electrodynamics, Springer Verlag, Berlin 1989

[Schli 60a] Schlieder, S.: Z. Naturforsch. **15 a** (1960) 448

[Schli 60b] Schlieder, S.: Z. Naturforsch. **15 a** (1960) 460

[Schli 60c] Schlieder, S.: Z. Naturforsch. **15 a** (1960) 555

[Schmi 80a] Schmid, E.W.: Z. Phys. A **297** (1980) 105

[Schmi 80b] Schmid, E.W.: Phys. Rev. C **21** (1980) 691

[Schmi 81] Schmid, E.W.: Z. Phys. A **302** (1981) 311

[Schmi 83] Schmid, E.W.: Z. Phys. A **311** (1983) 67

[Schmi 85] Schmid, E.W. and Spitz, D.: Z. Phys. A **321** (1985) 581

[Schu 67] Schuler, W. and Stumpf, H.: Z. Naturforsch. **22a** (1967) 1842

[Schwe 55] Schweber, S.S., Bethe, H.A. and de Hoffmann, F.: Mesons and fields, Vol. 1, Row, Peterson and Comp., New York 1955

[Schwi 51] Schwinger, J.: Proc. Nat. Acad. Sc. **37** (1951) 452, 455

[Schwi 54a] Schwinger, J.: Phys. Rev. **93** (1954) 615

[Schwi 54b] Schwinger, J.: Phys. Rev. **94** (1954) 1362

[Sci 62] Sciama, D.W. in: Recent developments of general relativity, Pergamon, Oxford 1962, p. 415

[Seg 47] Segal, J.E.: Ann. Math. **48** (1947) 930

[Sew 86] Sewell, G.L.: Quantum theory of collective phenomena, Clarendon Press, Oxford 1986

[Sexl 87] Sexl, R.U. and Urbantke, H.K.: Gravitation und Kosmologie, B.I. Wissenschaftsverlag, Mannheim 1987

[Sil 56] Silin, W.P. and Fainberg, W.J.: Fortschr. Phys. **4** (1956) 233

[Stu 70a] Stumpf, H.: Z. Naturforsch. **25a** (1970) 575

[Stu 70b] Stumpf, H., Scheerer, K. and Märtl, H.G.: Z. Naturforsch. **25a** (1970) 1556

[Stu 72] Stumpf, H.: Z. Naturforsch. **27a** (1972) 1058

[Stu 75] Stumpf, H. and Scheerer, K.: Z. Naturforsch. **30a** (1975) 1361

[Stu 82] Stumpf, H.: Z. Naturforsch. **37a** (1982) 1295

[Stu 83a] Stumpf, H.: Quantum processes in polar semiconductors and insulators, Part 1, Vieweg Verlag, Wiesbaden 1983

[Stu 83b] Stumpf, H.: Z. Naturforsch. **38a** (1983) 1064, 1184

[Stu 85a] Stumpf, H.: Z. Naturforsch. **40a** (1985) 14, 183, 294

[Stu 85b] Stumpf, H.: Z. Naturforsch. **40a** (1985) 752

[Stu 86a] Stumpf, H.: Z. Naturforsch. **41a** (1986) 683

[Stu 86b] Stumpf, H.: Z. Naturforsch. **41a** (1986) 1399

[Stu 87] Stumpf, H.: Z. Naturforsch. **42a** (1987) 213

[Stu 88] Stumpf, H.: Z. Naturforsch. **43a** (1988) 345

[Stu 89] Stumpf, H.: Z. Naturforsch. **44a** (1989) 262

[Stu 92] Stumpf, H. and Pfister, W.: Nuovo Cim. A **105** (1992) 677

[Stu 93a] Stumpf, H., Fauser, B. and Pfister, W.: Z. Naturforsch. **48a** (1993) 765

[Stu 93b] Stumpf, H., Borne, Th., Kaus, H.–J.: Z. Naturforsch. **48a** (1993)

[Sym 54] Symanzik, K.: Z. Naturforsch. **9a** (1954) 809

[Sym 73] Symanzik, K.: Comm. Math. Phys. **34** (1973) 7

[Swi 84] Swift, A.R. and Marrero, J.L.R.: Phys. Rev. D **29** (1984) 1823

[Thir 59] Thirring, W.: Fortschr. Physik **7** (1959) 79

[Thir 67] Thirring, W. and Wehrl, A.: Comm. Math. Phys. **4** (1967) 303

[Thir 68] Thirring, W.: Comm. Math. Phys. **7** (1968) 181

[Thir 80] Thirring, W.: Lehrbuch der Mathematischen Physik Bd. 4, Springer Verlag, Wien, New York 1980

[Tre 78] Treder, H.J.: Astron. Nachr. **299** (1978) 165

[Tsai 90] Tsai, E.C.: Lecture notes in physics **361**, Springer Verlag, Heidelberg 1990, p. 41

[Usui 60] Usui, T.: Prog. Theor. Phys. **23** (1960) 787

[Uti 56] Utiyama, R.: Phys. Rev. **101** (1956) 1597

[Völ 77] Völkel, A.M.: Fields, particles and currents, Lecture Notes in Physics **66**, Springer Verlag, Heidelberg 1977

[Vogl 91] Vogl, U. and Weise, W.: Prog. Part. Nucl. Phys. **27** (1991) 195

[Wahl 84] Wahl, F., Duscher, R., Göbel, K. and Maichle, J. K.: Z. Naturforsch. **39a** (1984) 524

[Wein 80] Weinberg, S. and Witten, E.: Phys. Lett. **96 B** (1980) 59

[Weiz 85] Weizsäcker, C.F.v.: Aufbau der Physik, Hanser Verlag, München 1985

[Wet 48] de Wet, J.S.: Proc. Cambridge Phil. Soc. **44** (1948) 546

[Whe 37a] Wheeler, J.A.: Phys. Rev. **52** (1937) 1083

[Whe 37b] Wheeler, J.A.: Phys. Rev. **52** (1937) 1107

[Wigh 56] Wightman, A.S.: Phys. Rev. **101** (1956) 860

[Wign 39] Wigner, E.P.: Ann. Math. **40** (1939) 149

[Wild 50] Wildermuth, K.: Z. Naturforsch. **5a** (1950) 373

[Wild 77] Wildermuth, K. and Tang, Y.C.: A unified theory of the nucleus, Vieweg Verlag, Braunschweig 1977

[Zim 54] Zimmermann, W.: Nuovo Cim. Suppl. **11** (1954) 43

# Index

GPSR Compliance

*The European Union's (EU) General Product Safety Regulation (GPSR) is a set of rules that requires consumer products to be safe and our obligations to ensure this.*

*If you have any concerns about our products, you can contact us on ProductSafety@springernature.com*

In case Publisher is established outside the EU, the EU authorized representative is:

Springer Nature Customer Service Center GmbH
Europaplatz 3
69115 Heidelberg, Germany

**Batch number: 10577896**

Printed by Printforce, the Netherlands